Tropical Cyclone Risk and Resilience

Jennifer Collins · James Done · Yi-Jie Zhu
Editors

Tropical Cyclone Risk and Resilience

Springer

Editors
Jennifer Collins
School of Geosciences
University of South Florida
Tampa, FL, USA

Yi-Jie Zhu
Department of Geosciences
Florida Atlantic University
Boca Raton, FL, USA

James Done
Mesoscale and Microscale Meteorology Laboratory
NSF National Center for Atmospheric Research
Boulder, CO, USA

ISBN 978-3-032-28744-1 ISBN 978-3-032-28745-8 (eBook)
https://doi.org/10.1007/978-3-032-28745-8

This Springer imprint is published by the registered company Springer Nature Switzerland AG
The registered company address is: Gewerbestrasse 11, 6330 Cham, Switzerland

Foreword

Tropical cyclones are awe-inspiring forces of nature and a convergence of water, energy, motion, and power. My personal venture into the field of meteorology was inspired by these powerful storms. Though scientifically alluring, tropical cyclones are also equally infamous for massive scales of human loss of life, property damage, and other society-altering outcomes. Tropical cyclones cause multiple and cascading levels of risk through compounding impacts, differential exposure, and varying degrees of sensitivity.

Meteorological extremes and societal vulnerability are further amplified within the context of a changing climate system. Scholarly studies increasingly point to a complex web of tropical cyclone sensitivities as the atmosphere-ocean system warms. Human systems will continue to be exposed to a generation of higher-end, wetter storms traversing elevated seas. In 2025, Supertyphoon Ragasa ravaged several Asian counties, while Hurricane Melissa destroyed much of the island nation of Jamaica. Both storms represent this emerging generation of mega-tropical cyclones likely to span global seas over the next century.

While traditionally viewed as coastal or island threats, tropical cyclones are increasingly an inland risk as well. My home state of Georgia is still recovering from significant agricultural, economic, and human impacts from Hurricane Helene (2024). That storm, like Hurricane Michael (2018), was a compound disaster reaching hundreds of kilometers inland. Risk management assessment, preparedness communications, and resiliency planning have traditionally been framed from a coastal perspective, but that will need to evolve.

We know what's coming because it is already here. That's why this book, *Tropical Cyclone Risk and Resilience*, is not only timely, it's vital. During my graduate work at Florida State University, I had access to some of the earliest U.S. operational Doppler weather radar data. My research group sought to develop hurricane-tracking algorithms using the new dataset. Hurricane Andrew (1992) had just devastated South Florida. As a young scholar, I recalled having a strong sense of urgency working on that project because I always believed in the *5P's: Prior Planning Prevents Poor Performance.*

It is not a secret that the footprint of human activity and related infrastructure continues to expand around the globe, particularly in tropical cyclone-vulnerable geographies. It is very clear that tropical cyclones are feasting on ever-warming waters, yet risk planning perspectives are anchored in benchmark storms of the past. That was a key finding from a 2024 National Academies report entitled, "Compounding Disasters in Gulf Coast Communities, 2020–2021: Impacts, Findings, and Lessons Learned." I co-authored that report, and this book is the response we need for global awareness and action related to tropical cyclone risk and resiliency. We wrote, "With each successive punch, vulnerabilities increase, heightening the risk that the next event could provide a knockout blow. These are compounding disasters." Of greater concern, such compounding disasters have cross-cutting implications in policy, economics, natural ecosystems, infrastructure, and well-being.

This book applies a wholistic, multi-scale approach to tropical cyclone risk. The reader is presented with a topical mix of climate science, resilience planning, risk evaluation, economic impacts, anthropology, fisheries, infrastructure and more. It is the perfect next step forward in the ongoing series of books by this accomplished team.

Athens, Georgia, USA

Marshall Shepherd
Regents' and Georgia Athletic
Association Distinguished Professor,
University of Georgia, Director,
University of Georgia Program
in Atmospheric Sciences, Member,
National Academy of Sciences,
National Academy of Engineering,
and American Academy of Arts
and Sciences, 2013 President,
American Meteorological Society

Preface

This book provides a source reference for researchers and risk managers for topics at the interface between tropical cyclones, risk and resilience. The book opens with physical and statistical modeling of the compound tropical cyclone hazard and examines management approaches for the compound hazard across diverse global regions. The book then takes us through economic impacts of the compound hazard and resilience of the built and natural environments. The book then expands our view of resilience by exploring the power of systems thinking, the value of communication and preparation, and the importance of community narratives for resilience. The book closes with a human-focused assessment of protective actions.

Critical questions of the physical science of the tropical cyclone hazard are laid out in the first two chapters. Chapter "Relationship Between Tropical Cyclones and Upper Ocean Heat Contents in the Global Ocean" connects modes of global tropical cyclone behavior to modes of variability in the tropical cyclone energy source of ocean heat. Chapter "Contemporary Methods for Probabilistic Coastal Compound Flood Hazard Analysis" then invites us to consider tropical cyclone flood and its compound drivers from surge, riverine flooding and inland deluges. Compound flood hazard analysis methods are outlined, and future improvements are identified. We next explore planning for the compound hazard in Chapters "One Storm, Many Hazards: Multi-Hazard Planning for Wind, Surge, and Flood Risks in the Gulf Coast" and "Tropical Cyclone Risk in the Bay of Bengal: Advancing Multi-Hazard Planning and Resilience Strategies", spanning the U.S. Gulf Coast and the Bay of Bengal. These chapters make a compelling case for the need to advance beyond traditional risk assessments that address hazards in isolation towards an integrated multi-hazard assessment and planning framework.

Chapters "Hurricane Impacts Compounded by Sea Level Rise: An Illustrative Case Study of Economic Impacts and Workforce Resilience Analysis"–"Nature's Barrier: Strategic Wetland Restoration for Tropical Cyclone Flood Risk Mitigation" examine the economic impacts of the compound U.S. hurricane and sea-level rise along the low-lying Louisiana Coast and use historical case studies to show the role of public policy in driving new engineering design ideas to reduce impacts. The role of natural environments in storing flood water and attenuating wave energy and height is then examined. Challenges for wetland restoration are discussed, and the chapter highlights the role of cost-benefit analyses in motivating action.

The second half of the book encourages broad resilience thinking that includes systems analysis, the role of communication, and specifically the role of human experiences, community narratives, and informal adaptation. These chapters highlight the complexity of the resilience challenge and the multiple and interacting resilience drivers interwoven through our built and natural environments and through our communities. The value of local experience and community knowledge is highlighted. The integration of local knowledge with artificial intelligence demonstrates how artificial intelligence can be thought of as civic infrastructure that can be place-based and culturally attuned. Local knowledge is shown to contribute to driving local adaptive capacity and informing national resilience strategies. Social connections and trusted relationships are also shown to bring tangible resilience benefits.

In closing, this book explores several advances in the science, policy, and overall resilience building that can reduce risks and sustain thriving communities in the face of today's tropical cyclone threats.

Tampa, USA Jennifer Collins
Boulder, USA James Done
Boca Raton, USA Yi-Jie Zhu

Acknowledgements

The editors would like to thank the expert reviewers for their time and careful review of the chapters. The following experts in their fields reviewed one of the chapters (most anonymously at the time of review): Dionne V. Frank, Joost Santos, Natalie Kruse Daniels, Naresh Krishna Vissa, J'Anna-Mare Lue, Indah Susilowati, Chris Massey, Madison C. Yawn, Yanfeng Wu, Fariha Islam Mou, Patrick Market, Angie B. Lindsey, Swagata Payra, and Ajit Tyagi. In addition, we thank additional reviewers who remain anonymous. We also thank Anthony Rivera for his careful review of each chapter. We would like to acknowledge the following people who co-organized the 2025 Symposium on Tropical Cyclone Risk in a Changing Climate in Clearwater, FL: Jane Baldwin, Jennifer Collins, James Done, Neva Duncan Tabb, David Henderson, Anthony Johnson, Phil Klotzbach, Chia-Ying Lee, Christos Mitas, Jo Muller, Dail Rowe, Suz Tolwinski-Ward, Charles Paxton, Gabriel Vecchi, Angelika Werner, Andrew Robson, Yijie Zhu. This symposium was the inspiration for this book, and we likewise acknowledge our symposium sponsors: FM, Moody's, MS Amlin, Reask, RenaissanceRe, Gallagher, the American Association of Geographers, University of South Florida (USF), USF College of Marine Science, Florida FloodHub, University of Miami School of Communication, Florida Atlantic University Department of Geosciences, and the City of St. Pete/Clearwater (visitspc). Finally, the editors deeply appreciate the productive collaboration with the professionals at Springer.

THE AMERICAN ASSOCIATION OF GEOGRAPHERS • 1904 • AAG

UNIVERSITY of SOUTH FLORIDA
College of Marine Science

FLORIDA
FloodHub
FOR APPLIED RESEARCH AND INNOVATION

FLORIDA ATLANTIC UNIVERSITY
Department of Geosciences
Charles E. Schmidt
College of Science

UNIVERSITY OF MIAMI
SCHOOL of
COMMUNICATION

ST.PETE
CLEARWATER
FLORIDA

Contents

Relationship Between Tropical Cyclones and Upper Ocean Heat Contents in the Global Ocean 1
Akiyoshi Wada

Contemporary Methods for Probabilistic Coastal Compound Flood Hazard Analysis 39
Aaron J. Dewar, Nathan B. Geldner, and David R. Johnson

One Storm, Many Hazards: Multi-Hazard Planning for Wind, Surge, and Flood Risks in the Gulf Coast 73
Robert V. Rohli, Nazla Bushra, and Rubayet Bin Mostafiz

Tropical Cyclone Risk in the Bay of Bengal: Advancing Multi-Hazard Planning and Resilience Strategies 111
Tanvir Hossain and Rubayet Bin Mostafiz

Hurricane Impacts Compounded by Sea Level Rise: An Illustrative Case Study of Economic Impacts and Workforce Resilience Analysis 139
Juita-Elena Yusuf, Yin-Hsuen Chen, Robert M. McNab, and George M. McLeod

Fortifying the Gulf Coast: Adapting Infrastructure and Housing to Intensifying Storms 165
Rubayet Bin Mostafiz, Nazla Bushra, Robert V. Rohli, and Fahmida Mukarroma

Nature's Barrier: Strategic Wetland Restoration for Tropical Cyclone Flood Risk Mitigation 199
Rose E. Noggle and Rubayet Bin Mostafiz

Tropical Cyclones in Systems Thinking: Testing Perspectives in Resilience Connections 229
Justin Nguyen and Kesley J. Richardson

Hurricane Season in the Caribbean: The Case of Preparing for the Known Unknown 249
Nicola Williams

Hurricane-Induced Agricultural Disruptions in the Gulf Region: A Framework for Risk Mitigation and Resilience 269
Mohammad A. Shovon and Rubayet Bin Mostafiz

Advancing Cyclone Resilience Through Community-Centered and Glocally-Informed AI Systems 299
Poulomi Chakravarty, Sai Gattupalli, Urjani Chakravarty, Gulab Chand, and Betty Bitengo Atandi

Cyclone Risk and Livelihood Recovery in Coastal South India: Narratives of Small-Scale Fishers of Andhra Pradesh 337
Bala Raju Nikku, Dalibandhu Pukkalla, Srinivasu Kodi, and Navya Vikraman Nair

Changing Climate and Gendered Dynamics in Tropical Cyclone Evacuation Behaviors in Bangladesh: Who Will Leave Before the Storm? 361
Musabber Ali Chisty

Relationship Between Tropical Cyclones and Upper Ocean Heat Contents in the Global Ocean

Akiyoshi Wada

Abstract The Intergovernmental Panel on Climate Change (IPCC) Sixth Assessment Report indicated that the global proportion of Saffir-Simpson category 3–5 tropical cyclones (TCs) has increased over the past four decades. Warming in the upper ocean is considered one of the factors. However, since the number, lifetime maximum intensity, phase transition, and track of TCs vary depending on atmospheric internal factors, it is uncertain whether the increase in upper ocean heat content (UOHC) will have a similar effect on worldwide TCs. This chapter aims to present the representative modes of atmospheric geopotential heights and upper ocean heat contents mainly in the tropical and subtropical oceans. In addition, the representative modes in the inner core of TCs were obtained from the empirical orthogonal analysis with infrared brightness temperatures. The similarities and differences in the relationship among the representative modes, TC central pressure, the lowest temporally average brightness temperature in the inner core from the genesis of a TC to first reaching minimum central pressure, and the spatiotemporally average upper-ocean heat content were examined. Each representative mode for geopotential heights and UOHC was related to climate indices based on sea surface temperature variability, but the relationship varied depending on the region. They were unrelated to the representative modes of infrared brightness temperature in the inner core of TCs. However, the axisymmetric pattern in the inner core was correlated with UOHC beneath a TC and infrared brightness temperature in the inner core of TCs, although their contribution rates varied depending on the region. Therefore, the TC intensity is not determined by the combination of climate indices but is directly influenced by the upper ocean warming beneath a TC.

Keywords Tropical cyclone · Upper ocean heat content · Ocean warming

---

A. Wada (✉)
Department of Typhoon and Severe Weather Research, Meteorological Research Institute, Tsukuba, Ibaraki, Japan
e-mail: awada@mri-jma.go.jp

J. Collins et al. (eds.), *Tropical Cyclone Risk and Resilience*,
https://doi.org/10.1007/978-3-032-28745-8_1

1 Introduction

Sea levels and sea surface temperatures (SSTs) are rising due to anthropogenic global warming (Merchant et al., 2025; Terhaar et al., 2025; World Economic Forum, 2025). As a result of anthropogenic global warming, the number of tropical cyclones (TCs) will decrease globally in the future (Chand et al., 2022) due to a significant downturn in western North Pacific TC activity (Klotzbach et al., 2022). However, recent increases in SSTs have been linked to an increase in the proportion of intense TCs (Camargo et al., 2023; Hong & Wu, 2021; Kossin et al., 2020; Wang & Toumi, 2021) and thereby to making TCs fiercer (Choi et al., 2024). Global damage from TCs has also increased significantly, largely due to population growth and the increased value of coastal assets (Klotzbach et al., 2022). To understand the factors resulting in the aggravation of natural disasters caused by increases in SSTs and thereby fiercer TCs, it is necessary to comprehend the role of internal variability in the upper ocean on interannual to decadal time scales, as well as the impact of anthropogenic factors on TC activity and maximum TC intensity.

Regarding the role of the ocean in TCs, TCs usually develop over the ocean by utilizing heat and moisture transported from the ocean to the atmosphere as energy (Emanuel, 1986). The increases in SSTs or high values of SSTs are generally considered to serve as one of the positive factors to TCs (Ginis, 2005; Gray, 1968; Shay et al., 2000). In contrast, as a negative factor, the more intense a TC becomes, the stronger the surface wind stress is, resulting in a large decrease in SSTs beneath and after the passage of a TC (Ginis, 2005; Price, 1981; Shay et al., 1998; Wada, 2002). However, in recent years, the extent of decreases in SSTs and diapycnal mixing associated with TCs has been decreasing (Zhang et al., 2022). The recent upper ocean warming may be caused by decreases in diapycnal mixing.

Previous studies have reported that TC activity was influenced by internal variability in the ocean on interannual to decadal scales as well as on the seasonal time scale (Hong & Wu, 2021). This is usually explained by regional SST variability such as El Niño-Southern Oscillation (ENSO), the Atlantic meridional mode (AMM), Atlantic multidecadal oscillation (AMO), Pacific decadal oscillation (PDO), the Pacific meridional mode (PMM), the Indian Ocean basin (IOB) mode, and the Indian Ocean Dipole (IOD) (e.g., Camargo et al., 2010; Li et al., 2022; Zhao et al., 2025). In addition, Wada and Chan (2021) have reported the influence of the ENSO Modoki index (EMI) (Ashok et al., 2007) and dipole mode index (DMI) (Saji et al., 1999) on Typhoons FAXAI and HAGIBIS in 2019 in the western North Pacific Ocean (WP). The relationship between TCs and these climate indices has been studied in each basin, particularly in the Easten Pacific Ocean (EP) (Caron et al., 2015; Jin et al., 2014), in the North Atlantic Ocean (NA) (Wang et al., 2017), in the North Indian Ocean (NI) (Albert & Bhaskaran, 2020; Nirmal et al., 2023; Singh & Roxy, 2022; Vissa et al., 2013), in the south Indian Ocean (SI) (Malan et al., 2013; Mawren & Reason, 2017), and in WP (Li & Zhou, 2012; Liu & Chan, 2020; Wada & Chan, 2008, 2021; Wada & Yanase, 2024). Chan and Liu (2022) demonstrated the relation of the recent decrease in the difference in the annual number of TCs between WP and

NA to AMO and the interdecadal Pacific Oscillation (Li et al., 2015). Although the following indices on intraseasonal variations are not based on regional SSTs, the relation of Boreal Summer Intraseasonal Oscillation (BSISO) (Kikuchi, 2021) and the Madden Julian Oscillation (MJO) (Madden & Julian, 1972) to tropical cyclone heat potential (TCHP) (Leipper & Volgenau, 1972) and TC activity has been investigated. Weng and Hsu (2017) have reported that mechanical factors, such as convection and lower-level atmospheric convergence, exert a greater influence on TC activity in WP than TCHP does. Chacko and Balaguru (2025) showed that the IOD provides more favorable TCHP conditions for TC activity in NI than the MJO.

In the climatology of TCs, TC best track datasets provide us with historical data, which has been used to study the relationship between TCs and the ocean. To date, TC best tracks have been analyzed by several meteorological agencies and so on, but these analyses are based on their own procedures that include subjective processes depending on forecasters' skill. Since the TC best track data archived in one agency includes errors attributed to the differences between several agencies (Kruk et al., 2010; Schreck III et al., 2014), differences in the analysis results on the relationship between TCs and the ocean may fundamentally stem from the errors. In fact, Wada and Yanase (2024) compared the relationship between TCs and multiple oceanic metrics among the four TC best track datasets in the western North Pacific and discussed the similarities and differences among them. However, there may be another similarity/difference in the relationship between TCs and the ocean when comparing the relationship between TCs and the ocean in other regions and all tropical and subtropical oceans.

In this chapter, as a global collection of TCs, the International Best Track Archive for Climate Stewardship (IBTrACS) dataset compiled by meteorological agencies worldwide (Knapp et al., 2010; Kurk et al., 2010) was used to investigate the relationship between TCs and the ocean. Also, a unified global satellite data set (Knapp et al., 2011) was used to extract the representative modes on a cloud pattern in the inner core of a TC and obtain the lowest value of spatiotemporally average infrared brightness temperature as a metric of the intensity of TCs, such as their central pressure (minimum sea-level pressure). By using the Japan Coastal Ocean Predictability Experiment (JCOPE)-Forecast Global Ocean (JCOPE-FGO) reanalysis data (Kido et al., 2022) from January 1993 to August 2024, the upper ocean heat content was calculated mainly in the tropical and subtropical oceans, and then the relationship between TCs and their upper ocean heat content was statistically examined. A discussion will be provided on the similarities and differences in the relationship in each region and in all tropical and subtropical oceans.

The rest of this paper consists of the following sections. Section 2 described data and analysis methods. Section 3 showed the results on the representative modes of geopotential heights and TCHP, as well as the representative modes of infrared brightness temperature collocated at the TC center, which were regarded as representative cloud patterns. In addition, the results of the linear regression analysis were presented to show the relationship between the climate indices, minimum central pressures, the lowest value of spatiotemporal average infrared brightness temperature, and representative modes for geopotential heights, Upper Ocean heat contents, and the inner-core cloud patterns. Section 4 discussed the validity of the analysis method in applying the concept of 'spatiotemporal average' from the genesis of a TC to first reaching a peak intensity. Section 4 also discussed the uncertainty of datasets used in this chapter, particularly on the TC best track and satellite infrared brightness temperature. Section 5 is devoted to the concluding remarks.

2 Data and Methods

2.1 Data

This chapter used the World Meteorological Organization (WMO) data archived in IBTrACS from January 1993 to August 2024. The WMO data aggregates the best track data for TCs issued by agencies responsible for monitoring TCs as one of the regional specialized meteorological centers, including the Central Pacific Hurricane Center Honolulu and National Hurricane Center Miami in EP and NA, India Meteorological Department New Delhi in NI, Météo-France La Reunion in SI, Bureau of Meteorology, Fiji Meteorological Service Nadi, MetService of New Zealand Wellington in the South Pacific Ocean (SP) and Japan Meteorological Agency in WP. Most of the centers are in charge of the regional specialized meteorological centers. The 6-hourly latitude, longitude, and central pressure archived in IBTrACS were used for analyses. Among them, the 6-h central pressure (minimum sea-level pressure) was used as a measure of TC intensity. Also, the maximum sustained wind speed was used to determine the occurrence time when the TC first reached 35 knots.

The SST dataset used in this study is the Merged Satellite and In-situ Data Global Daily Sea Surface Temperatures (MGDSST) dataset (Kurihara et al., 2006). This is a daily SST product with a spatial resolution of 0.25°. The MGDSST is analyzed based on SSTs derived from satellite's infrared sensors (National Oceanic and Atmospheric Administration (NOAA)/Advanced Very High Resolution Radiometer (AVHRR), Meteorological Operational satellite (MetOp)/AVHRR, and National Polar-orbiting Operational Environmental Satellite System Preparatory Project (NPP)/Visible Infrared Imaging Radiometer Suite (VIIRS)), microwave sensors (AQUA/Advanced Microwave Scanning Radiometer (AMSR)-E, Coriolis/WINDSAT, and Global Change Observation Mission (GCOM)-W/AMSR-2), and in-situ SST (from buoys and ships) (Japan Meteorological Agency, 2014). AQUA is a National Aeronautics and Space Administration (NASA) scientific research satellite in orbit around the Earth, and Coriolis/WINDSAT is a polarimetric microwave radiometer developed by the U.S. Navy and NASA.

To calculate the daily TCHP, the daily JCOPE-FGO reanalysis dataset (Kido et al., 2022) is used. The horizontal resolution is 0.1°, which covers the global ocean. The vertical layer consists of 44 layers and employs a sigma coordinate system. The 5-min Gridded Global Relief Data (ETOPO5) (Sloss, 1988) bathymetry dataset is used as the bottom topography after undergoing spatial smoothing.

Gridded Satellite (GridSat) data used in this chapter are gridded on a 0.07° latitude equal-angle grid, approximately 8 km (Knapp et al., 2011). GridSat Data obtained from different satellites is merged by selecting the nadir-most observations for each grid point. The Brightness Temperature with a unit of Kelvin (BTK) in the climate data record-quality InfRared WINdow (IRWIN) channel was used. It is important to note that the ocean reanalysis data is on a daily time scale, whereas the TC data set is on a 6-h time scale, and the GridSat data set is originally on a 3-h time scale. This means that ocean reanalysis data cannot capture rapid decreases in water temperature that occur on time scales shorter than a day, as noted in Wada and Yanase (2024). In this chapter, the 6-h GridSat data (at 00, 06, 12, 18 UTC) are used.

Geopotential heights at the three isobaric levels of 850-, 500-, and 300-hPa were used for conducting empirical orthogonal function (EOF) analysis. These data were obtained from the Japanese Reanalysis for Three Quarters of a Century (JRA-3Q) (Kosaka et al., 2024) with latitudinal/longitudinal resolution of 1.25°. The temporal resolution was 6 h.

Oceanic monitoring indices such as the difference in SST from the reference averaged over 5°N – 5°S and 150°W – 90°W (NINO3) and the PDO index were obtained from the Japan Meteorological Agency (JMA) webpage (Japan Meteorological Agency,). The EMI and DMI were downloaded from the Application Laboratory (APL) virtual earth webpage (Japan Agency for Marine-Earth Science and Technology, 2020). The AMM and PMM SST indices (Chiang & Vimont, 2004) were downloaded from the Physical Sciences Laboratory website (Physical Sciences Laboratory,). NINO3 is used as an index of ENSO. For the monthly climate indices, daily data for the middle day in each month were matched to the monthly index.

2.2 Analysis Method

The regions for TCs addressed in this chapter are six basins, including EP, NA, NI, SI, SP, and WP, introduced in Sect. 1. The six abbreviations are the same as those used in the IBTrACS dataset. The upper panel of Fig. 1 shows a global map of the six regions with GridSat IRWIN BTK at 00 UTC on October 8 in 2019, and the lower panel of Fig. 1 shows daily TCHP calculated from JCOPE-FGO data on October 8 in 2019, at JRA-3Q 500-hPa geopotential heights at 00 UTC on October 8 in 2019.

For each TC, a lifetime maximum intensity was determined during the lifecycle of a TC: from the time when the maximum sustained wind speed first reached 35 knots to the time when the central pressure first reached its minimum value during the lifecycle. Using the position at which the central pressure first reached its minimum value as the TC center, a 5° by 5° area collocated at the TC center was extracted at the interval of 6 h from the GridSat IRWIN BTK dataset. Figure 2 is an example of a 5° by 5° area in the case of Typhoon Hagibis (2019). Since the resolution of GridSat was finer than 0.1 degrees, the lowest value within a 0.1° grid cell was applied to make gridded data (Fig. 2). The gridded data were also used to make an EOF analysis for all TCs from January 1993 to August 2024.

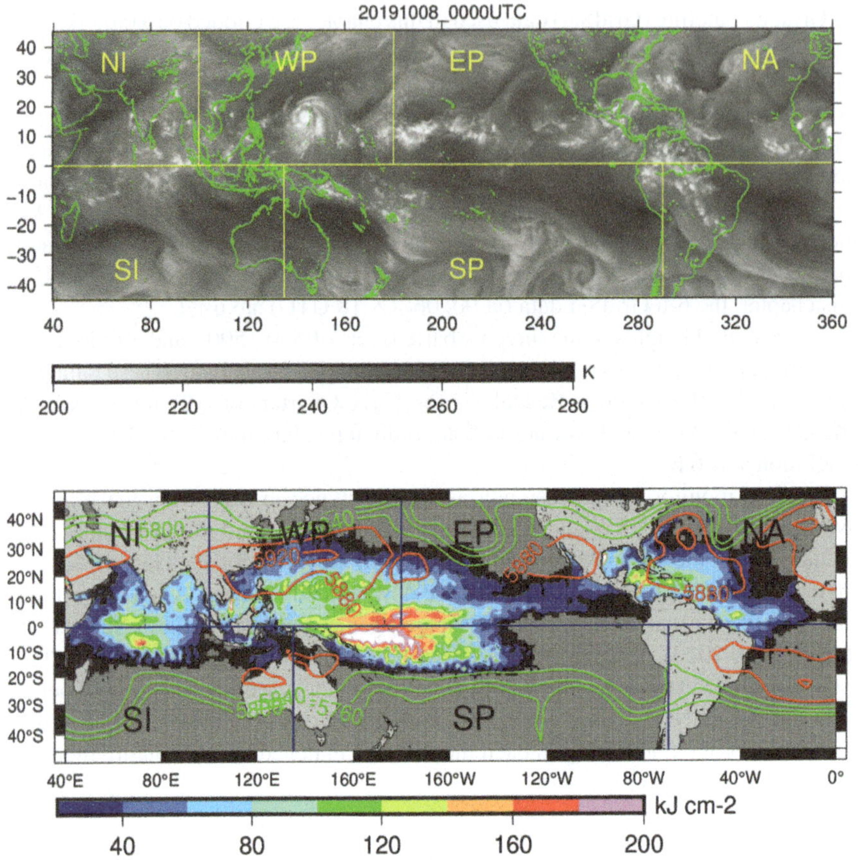

Fig. 1 The upper panel shows a global map of GridSat IRWIN BTK at 00 UTC on October 8 in 2019. The lower panel shows a global map of TCHP (kJ cm^{-2}, shades) at the 500-hPa altitude. The two letters (yellow in the upper panel and black in the lower panel) in the yellow (upper panel) and blue (lower panel) boxes indicate the region specified in the IBTrACS. The contour interval is 40 m. The thick red contour indicates the altitude equal to or higher than 5880 m, whereas the thin pink contours indicate the altitude lower than 5880 m

2.3 Tropical Cyclone Heat Potential

TCHP (Q_{TCHP}) is a kind of upper ocean heat content (UOHC) specified for TC activity and is calculated in each grid using Eq. (1) with the daily JCOPE-FGO global oceanic reanalysis data (see Sect. 2.1).

$$Q_{TCHP} = \sum_{i=1}^{i=iD26} \rho_i C_p (T_i - 26) \cdot Z_i, \tag{1}$$

Fig. 2 An example of a 5° by 5° area. Horizontal distribution of IRWIN BTK at 00 UTC on October 8 in 2019, collocated at the center of Typhoon Hagibis (2019)

where i is the number of vertical layers, ρ is the water density, C_p is the specific heat capacity of liquid water at constant pressure, T_i is the water temperature at the ith layer, and ΔZ_i is the layer thickness of the ith layer. TCHP is used as a measure of UOHC from the surface to the 26 °C isotherm depth (Leipper & Volgenau, 1972). When the 26 °C isotherm depth falls within the range of ΔZ_i, the 26 °C isotherm depth was calculated by linear interpolation of the upper and lower depths.

In previous studies (Moon et al., 2022; Wada & Usui, 2007; Wada et al., 2012; Wada & Yanase, 2024), an accumulated TCHP was used for their research. In this chapter, the temporal average from the time when the maximum sustained wind speed first reached 35 knots to the time when the central pressure reached its minimum value for the first time in the lifecycle of a TC was used (Wada & Yanase, 2024). Here, $Q_{TCHP}(t)$ is a grid value of TCHP in the 5° by 5° area collocated at the TC center at a time t. The equation is

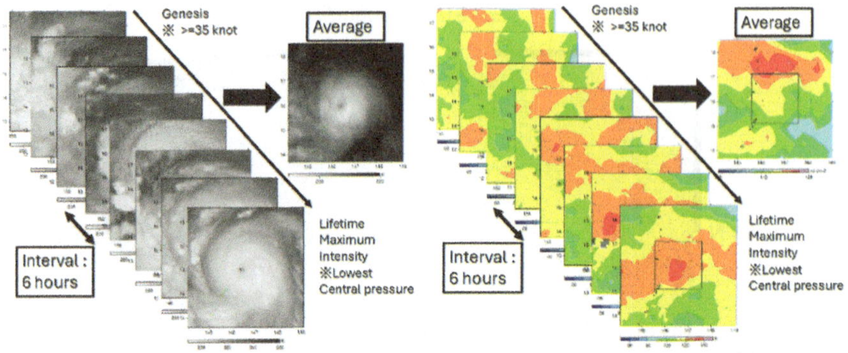

Fig. 3 Schematic diagram for calculating the temporal average of (left panel) IRWIN BTK and (right panel) the temporal average TCHP in a 5° by 5° area collocated at the center of a TC accumulated from the time when the maximum sustained wind speed first reached 35 knots to the time when the central pressure first reached its minimum value. A square in the right panel indicates the spatiotemporal average area for SST and TCHP, which is enclosed by a 2° by 2° square collocated at the TC center

$$\overline{Q}_{TCHP} = \frac{1}{T}\sum_{t=0}^{T} Q_{TCHP}(t), \tag{2}$$

where the overbar denotes the temporal average from the initial time to the time T. Although the oceanic reanalysis was a daily product, the interval of summation was 6 h since the accumulated TCHP in the 5° by 5° area collocated at the TC center changes due to a movement of a TC (Fig. 3). The accumulated SST was calculated in a similar way. Hereafter, the accumulated value is called 'average' or 'temporal average'. The procedure was also applied for creating the average IRWIN BTK. The temporal average SST, TCHP, and IRWIN BTK corresponded to the peak TC intensity during its lifetime, which was determined by the central pressure archived in IBTrACS.

To make statistical analyses for investigating the relation to TC intensity, the values of the average SST and TCHP were spatially averaged over the 2° by 2° area collocated at the TC center at a time, while the value of the average IRWIN was spatially averaged over the 5° by 5° area collocated at the TC center at a time t (Fig. 3). In other analyses, the lowest value of the temporal average IRWIN BTK in the 5° by 5° area at a peak TC intensity was used for the timeseries analysis. The lowest IRWIN BTK indicates the depth of convection in areas where deep convection persisted and is closely related to the central pressure of a TC. These data and the TC central pressure were used to make the regression function analysis. Regarding the 6-h geopotential heights, the relatively coarse horizontal resolution of 1.25 degrees makes it more suitable for analyzing synoptic variations rather than for TC intensity. Therefore, we do not calculate the accumulated value like the average SST, TCHP, and IRWIN BTK. However, the application of accumulation for 6-h geopotential heights may be worth considering when the horizontal resolution of atmospheric reanalysis data improves in the future.

2.4 Empirical Orthogonal Function (EOF) Analysis

EOF analyses were performed for daily TCHP and 6-h geopotential heights at the isobaric surfaces of 850, 500, and 300 hPa from January 1993 to August 2024. In addition, an EOF analysis was made for the temporal average IRWIN BTK in the 5° by 5° area for all TCs worldwide by changing a Fortran code used by Wada and Chan (2021) and Wada et al. (2022). To save computational resources, the EOF analysis was performed on TCHP and the three geopotential heights using a 5-day interval. In addition, the EOF analysis for the temporal average IRWIN BTK was performed only for TCs with the lifetime maximum intensity.

Six days (1, 6, 11, 16, 21, 26) at 00 UTC were selected in a month to perform an EOF analysis for daily TCHP and the three 6-h geopotential heights from January 1993 to August 2024. The number of data for making the EOF analysis for TCHP and the three geopotential heights was 2817. As for the EOF analysis for the temporal average IRWIN BTK, the number of TCs with the lifetime peak intensity was 2614 from January 1993 to August 2024: 584 in EP, 514 in NA, 263 in NI, 446 in SI, 261 in SP, and 746 in WP.

Using the climate indices introduced in Sect. 2.1 and normalized amplitudes obtained from the EOF analyses, the correlation coefficients between monthly climate indices and TCHP at 11th in a month and between monthly climate indices and the three geopotential heights at 00 UTC of 11th in a month were calculated respectively after applying a six-month moving average to the timeseries for normalized amplitudes of the representative three EOF modes.

3 Results

3.1 Trend and Variability of SST, TCHP, and IRWIN BTK

In this section, the annual changes in daily SSTs from 1993 to 2023 were overviewed in each region (Fig. 4). Due to the large seasonal variation in a year, the 95% confidence intervals for the mean SST value in a year were indicated as vertical bars. The sign of the slope of the linear regression function indicates the linear trend relative to the mean SST value in 1993: a positive sign means an increase, whereas a negative sign means a decrease. The slopes in all six regions showed a positive trend, with correlation coefficients ranging from 0.06 (SP) to 0.18 (NI), which were significant at the 95% significance level. However, the regions of NA, NI, and SP may have shown a positive trend due to extraordinarily high values in 2023, which may be attributable to extraordinary regional oceanic variability rather than a linear trend. Moreover, the mean SST value in EP was abnormally high in 2015, and that in WP was relatively high from 1998 to 2001. The abnormally high SST in WP was due to the 1997–98 El Niño (McPhaden, 1999). The above results suggest that the mean SST value increased in recent years in all six regions while exhibiting extraordinary regional oceanic variability, and that the relationship between the oceanic variability and the linear increase is different among the six regions.

Next, the annual changes in daily TCHPs from 1993 to 2023 were shown in each region (Fig. 5). Climatologically, TCHP was the highest in SP, and TCHP in WP was the second in the tropical and subtropical oceans. This is different from SST in that SST was the highest in NI. This is due to the influence of the northern limit of latitude in each average area. Particularly in NI, the northernmost latitude of the analysis domain is relatively low, so increases in daily SSTs may be more readily apparent. Unlike the annual changes in daily SSTs, the correlation coefficient for the linear regression ranged from 0.15 (EP) to 0.46 (WP), indicating a higher correlation coefficient than that in daily SSTs. In addition, the rate of increase in daily TCHPs varied significantly across regions, ranging from 1.5 kJ cm^{-2} (EP) to 4.6 kJ cm^{-2} (WP) over a 10-year period. In all six regions, since the standard deviations were 19.9 kJ cm^{-2} in EP, 15.3 kJ cm^{-2} in NA, 29.3 kJ cm^{-2} in NI, 15.4 kJ cm^{-2} in SI, 21.0 kJ cm^{-2}in SP, and 23.7 kJ cm^{-2} in WP, the changes in daily TCHPs on interannual to decadal timescales were large against the linear trend, which was similar to the result on daily SSTs. It should be noted that in EP, the slope of the linear function became nearly flat due to the extraordinarily high TCHP value in 2015. In fact, when the data in 2015 were extracted, the correlation coefficient was 0.12, the slope was 0.12, and the standard deviation was 19.3 kJ cm^{-2} in EP. The factors contributing to the formation of abnormally high SST and TCHP in EP remain a subject for future research.

Using 6-h GridSat IRWIN BTKs with a horizontal resolution of approximately 8 km, we conducted the time-series analyses, which were similar to those for SST and TCHP except for the amount of data included in a year (Fig. 6). As a prerequisite for interpreting the result, the value of the IRWIN BTK itself is considered to show a similar trend to daily SSTs in clear-sky conditions since it shows the effect of radiation from the sea surface. Therefore, the difference from Fig. 4 is considered due to the existence of clouds covering a region. In fact, the correlation was low (0.04) in WP, while it was the highest (0.16) in NA among the six regions. In NI, except for the 1990s, the linear slope decreased, and the variability of 6-h IRWIN BTKs became clear. The influence of atmospheric variability against the linear trend was also clear in EP and WP. When daily SSTs increased and the cloudy areas decreased, it can be simply assumed that 6-h IRWIN BTKs might increase. However, for regions where the linear trends were not clear, changes in cloud properties, such as the nature of convection, may affect the linear trends of annual mean 6-h IRWIN BTKs against an increase in daily SSTs. The details are beyond the scope of this chapter and will be a challenging theme in the future.

The analysis results for 6-h BTKs across six ocean basins include cloud variability spanning a wide range of spatial scales from synoptic-scale phenomena like

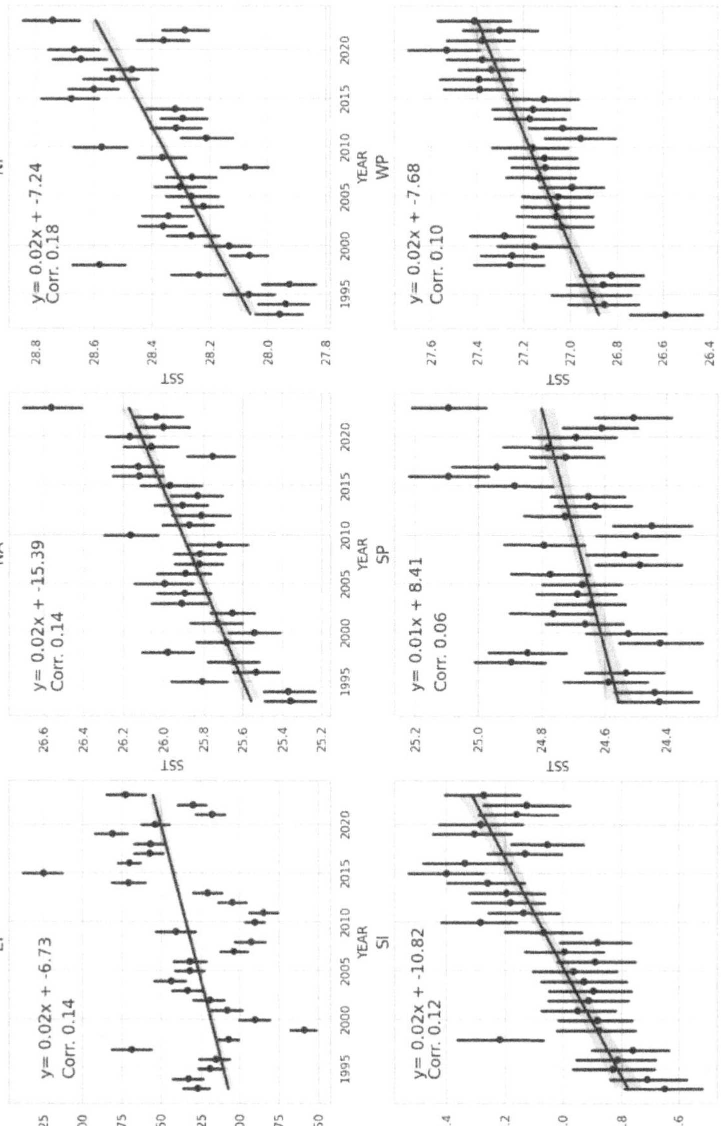

Fig. 4 Annual changes in daily SSTs in six regions. Circles indicate the mean value in each year. Vertical bars indicate the significance range at the 95% confident level. A linear regression function and correlation coefficient are shown in each panel

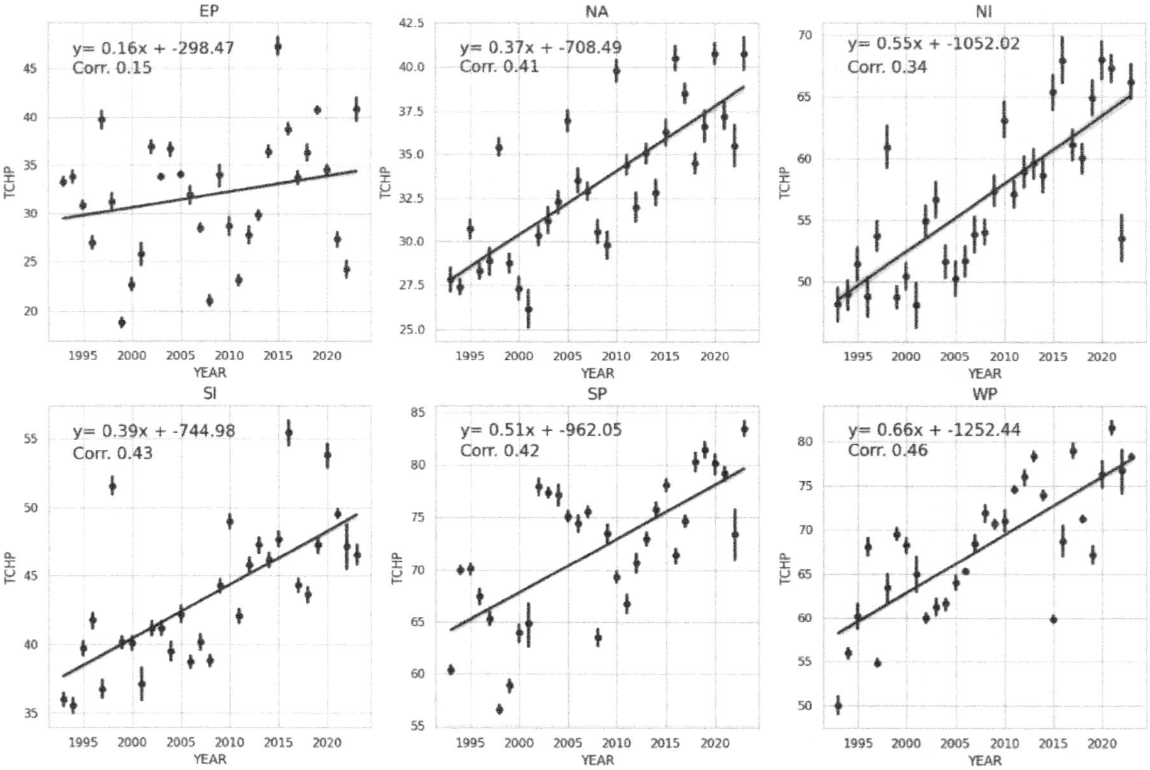

Fig. 5 Annual changes in daily TCHPs in six regions. Circles indicate the mean value in each year. Vertical bars indicate the significance range at the 95% confidence level. A linear regression function and correlation coefficient are shown in each panel

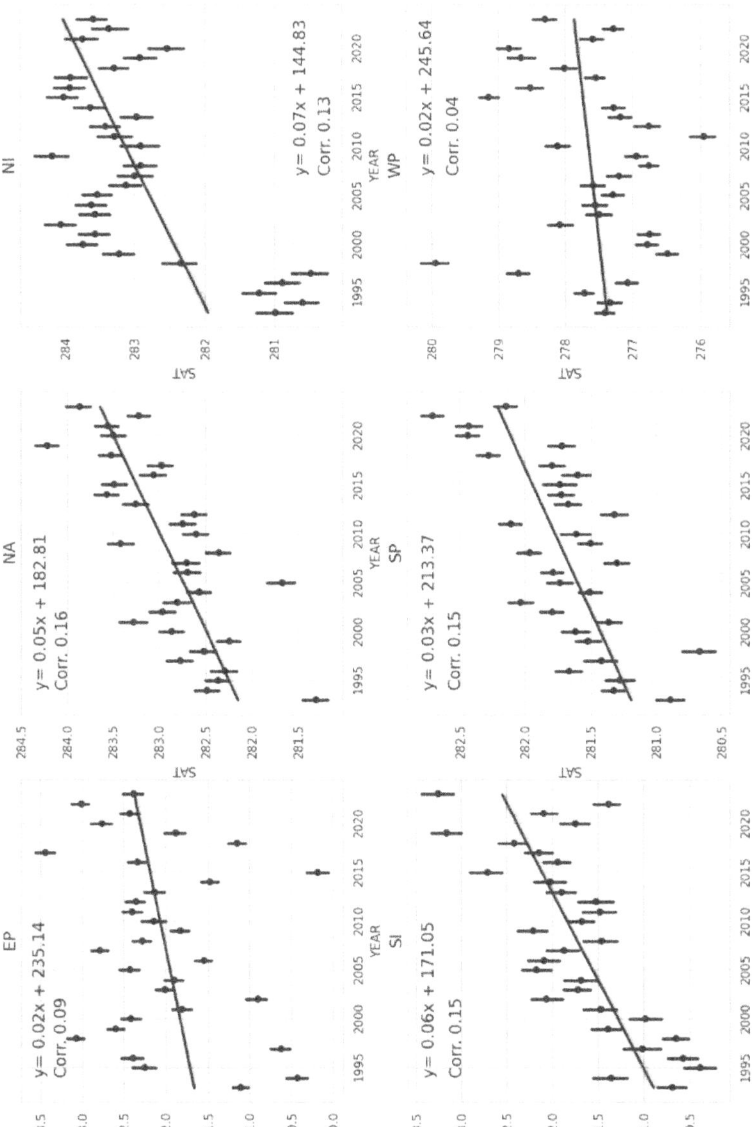

Fig. 6 Annual changes in 6-h GridSat IRWIN BTKs in six regions. Circles indicate the mean value in each year. Vertical bars indicate the significance range at the 95% confidence level. A linear regression function and correlation coefficient are shown in each panel

monsoons and the intertropical convergence zone to cumulus convection. In other words, the analysis results are not limited to TCs. However, for EP and WP, the increase in 6-h BTKs was moderate against the increase in daily TCHPs. In contrast, in NA and SI, 6-h BTKs appear to increase alongside the increase in daily TCHPs.

3.2 Spatial Representative Modes by EOF Analysis

The previous subsection showed variability and linear trends of daily SSTs, TCHPs, and 6-h IRWIN BTKs for six regions (EP, NA, NI, SI, SP, WP) on a year-by-year timescale. This subsection focuses on representative variational modes in the global ocean derived from the EOF analysis from 1 January 1993 to 26 August 2024. Regarding variations in SST, however, the climate indices were used instead because these were calculated based on historical SST data for the target area. It should be noted that the climate indices introduced in Sect. 2.1 were not always based on MGDSST.

Figure 7 shows the spatial patterns of the first to third EOF modes (EOF1, EOF2, and EOF3, respectively) obtained for daily TCHPs with a horizontal grid space of 0.1°, which covered the zonal area around 35°S-35°N. Table 1 shows the correlation coefficients between the normalized amplitudes of the first three EOF modes for daily TCHPs and the six climate indices: EOF1 showed relatively high correlations with EMI, PDO, and AMO. EOF2 has high correlations with NINO3 and PMM. EOF1 and EOF2 were clearly distinguished by highly correlated climate indices. EOF3 has correlation coefficients exceeding 0.10 with climate indices except DMI and AMO. Regarding DMI, the correlations with the first three EOF modes were relatively low. It should be noted that these SST indices were monthly, so that the 11th of each month was used to calculate the correlation coefficient with normalized amplitudes derived from the EOF analysis (see subsect. 2.4). Unlike the results of Wada and Chan (2021), no significant correlation was obtained between EOF1 and NINO3. This is presumed to be due to differences in the computational domains. EOF1 explains approximately 37.9%, EOF2 approximately 26.8%, and EOF3 approximately 9.8% of total variance. The sum of the three leading EOF modes explains approximately 74.5% of total variance.

Figures 8, 9 and 10 show the results of the EOF analysis for the 6-h geopotential heights at the isobaric surfaces of 850, 500, and 300 hPa. It should be noted that, unlike TCHP, where the value is zero when SST is lower than 26 °C, the results of the EOF analysis for the 6-h geopotential heights include the atmospheric variability not only in the tropical and subtropical oceans but also in the mid-latitude. The correlation coefficients between the normalized amplitudes and the climate indices are shown in Tables 2, 3 and 4 for each geopotential height: EOF1 showed a higher correlation of the climate indices with the normalized amplitude of the 850-hPa geopotential height (Fig. 8), while EOF3 showed a higher correlation with the normalized amplitude of the 300-hPa geopotential height (Fig. 10). The results at 500-hPa geopotential

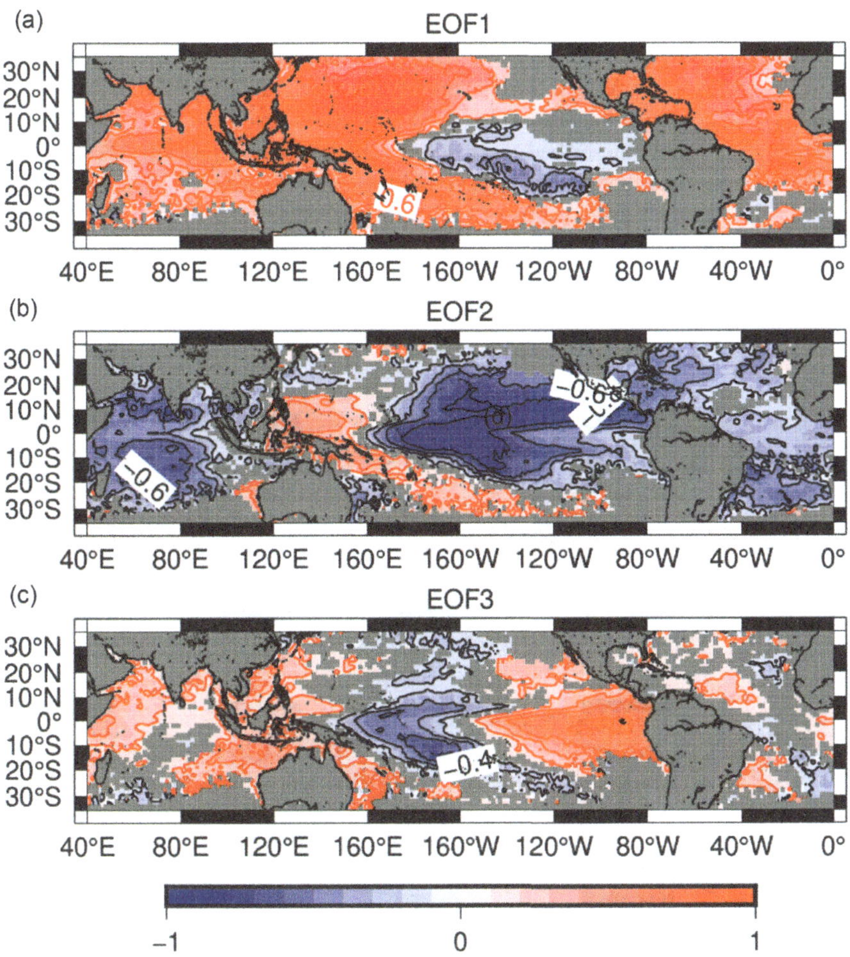

Fig. 7 The horizontal map of the correlation coefficients of **a** the first (EOF1), **b** the second (EOF2), **c** the third (EOF3) modes for daily TCHPs

Table 1 Correlation coefficients between the normalized amplitudes of the three EOF modes (EOF1, EOF2, EOF3) for the original TCHP and climate indices introduced in Sect. 2.1. No value means the correlation was not significant at the 99.9% confidence level based on a *t*-test

	NINO3	EMI	PDO	DMI	PMM	AMM
EOF1	–	− 0.20	− 0.39	0.10	− 0.10	0.25
EOF2	0.23	0.06	− 0.05	0.02	− 0.30	− 0.03
EOF3	− 0.19	− 0.32	− 0.21	0.07	− 0.18	− 0.02

heights (Fig. 9) were intermediate between the two results (850- and 300-hPa geopotential heights). This feature was particularly pronounced for NINO3, EMI, and PDO. However, PMM and AMM showed relatively low correlations with the normalized amplitudes of geopotential heights at the three isobaric levels.

Thus far, we have described the results of EOF analysis for daily TCHPs and 6-h geopotential heights in the tropical and subtropical oceans. At the end of this subsection, we present the results of EOF analysis for the 5° by 5° average IRWIN BTK collocated at the TC center at the time when the TC first reached its minimum central pressure (left panel of Fig. 3). Unlike previous EOF analyses, this analysis aims to obtain representative modes for cloud patterns within the inner core of a

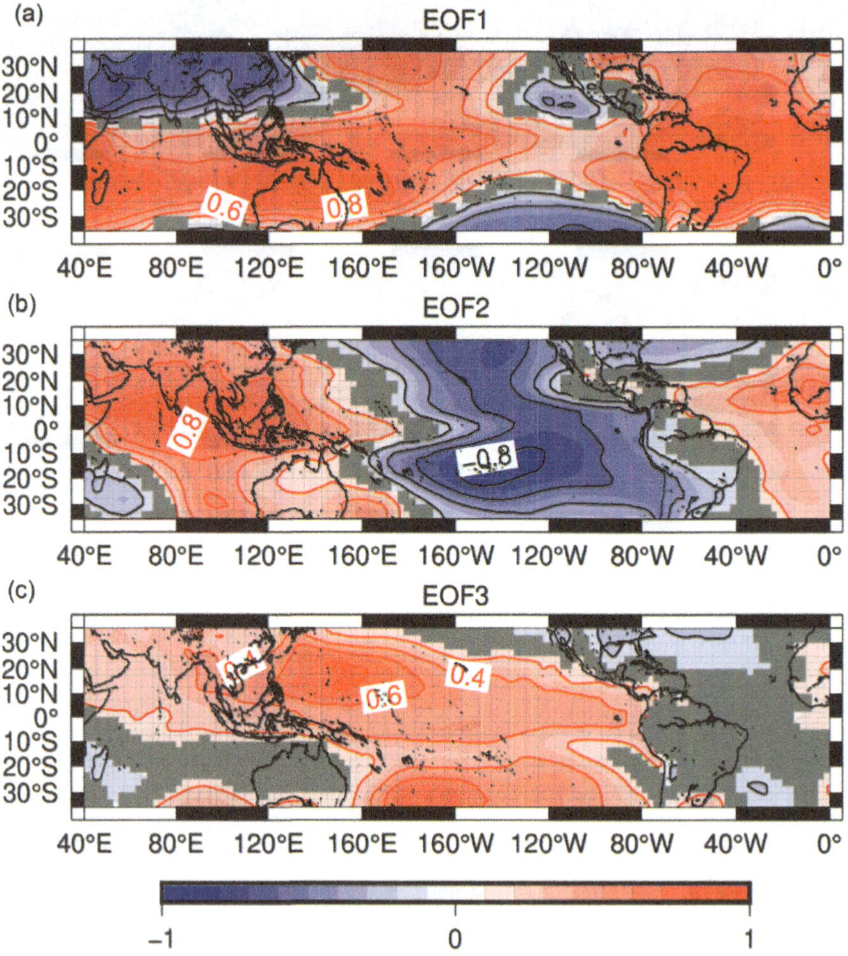

Fig. 8 The horizontal map of the correlation coefficient of **a** the first, **b** the second, and **c** the third EOF modes for 850-hPa geopotential heights

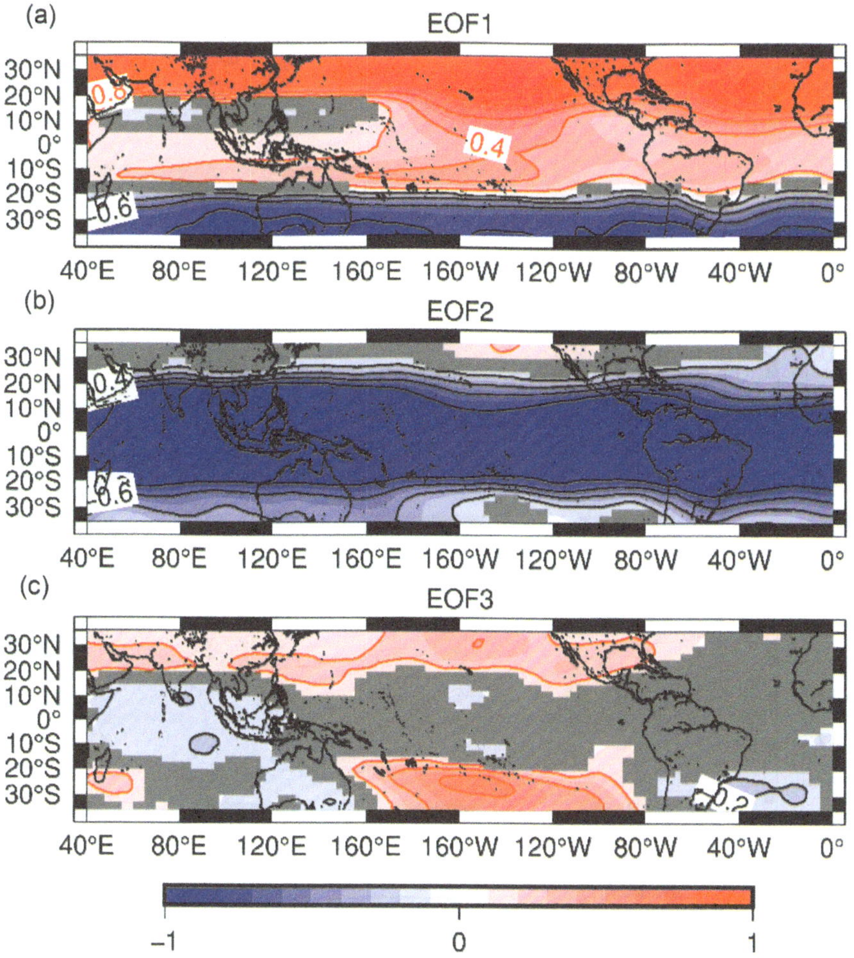

Fig. 9 The horizontal map of the correlation coefficient of **a** the first (EOF1), **b** the second (EOF2), and **c** the third (EOF3) modes for 500-hPa geopotential heights

TC, which explains the variance derived from the average IRWIN BTK distribution at peak TC intensity for all TCs used in the analysis. The spatial representative EOF pattern specified in the TC inner core was quite different from the EOF mode shown in Figs. 7, 8, 9 and 10 in that the EOF modes for the average IRWIN BTK did not contain interannual to decadal variability in the atmosphere and the ocean. Figure 11 shows the spatial representative patterns of the first to third EOF modes for all worldwide TCs from January 1993 to August 2024. The spatial pattern of EOF1 indicated an axisymmetric structure (Fig. 11a). In contrast, the second and third EOF modes exhibited a wavenumber-1 non-axisymmetric structure (Fig. 11b, c). These patterns are consistent with the result of Wada et al. (2022) only when their simulated TC underwent a late intensification phase: An axisymmetric pattern

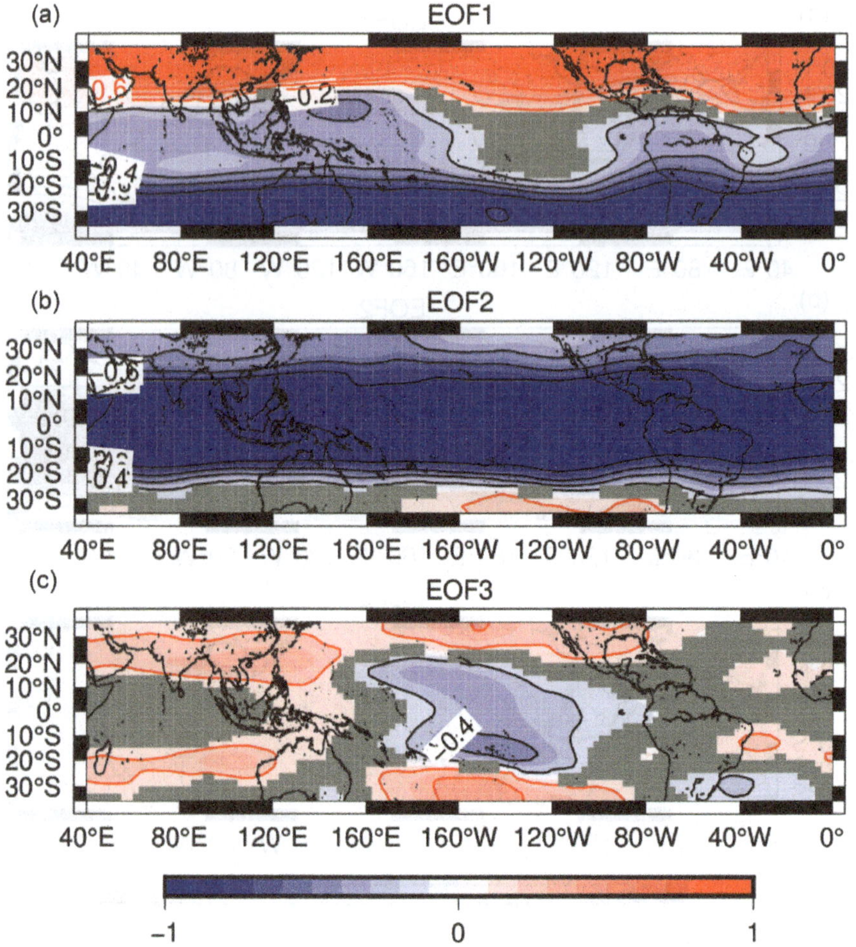

Fig. 10 The horizontal map of the correlation coefficient of **a** the first (EOF1), **b** the second (EOF2), and **c** the third (EOF3) modes for 300-hPa geopotential heights

Table 2 Correlation coefficients between the three EOF modes (EOF1, EOF2, EOF3) for 850-hPa geopotential heights and climate indices introduced in Sect. 2.1. No value means the correlation was not significant at the 99.9% significance level based on a *t*-test

	NINO3	EMI	PDO	DMI	PMM	AMM
EOF1	0.42	0.20	0.24	0.16	–	– 0.20
EOF2	0.73	0.50	0.53	0.31	0.13	– 0.10
EOF3	0.09	– 0.29	– 0.14	0.02	– 0.16	0.22

Table 3 Correlation coefficients between the three EOF modes (EOF1, EOF2, EOF3) for 500-hPa geopotential heights and climate indices introduced in Sect. 2.1

	NINO3	EMI	PDO	DMI	PMM	AMM
EOF1	0.14	0.06	0.03	0.03	0.02	− 0.02
EOF2	− 0.50	− 0.21	− 0.19	− 0.21	− 0.05	− 0.04
EOF3	− 0.62	− 0.58	− 0.55	− 0.28	− 0.11	0.21

Table 4 Correlation coefficients between the three EOF modes (EOF1, EOF2, EOF3) for 300-hPa geopotential heights and climate indices introduced in Sect. 2.1

	NINO3	EMI	PDO	DMI	PMM	AMM
EOF1	− 0.07	− 0.03	− 0.04	− 0.06	− 0.01	− 0.04
EOF2	− 0.49	− 0.21	− 0.16	− 0.21	− 0.05	− 0.07
EOF3	− 0.65	− 0.59	− 0.62	− 0.25	− 0.10	0.23

and two wave-number 1 patterns were extracted from the EOF analysis of Wada et al. (2022). However, it should be noted that in Wada et al. (2022), ensemble simulation results with a horizontal resolution of 1 km in the intensification phase of numerically simulated Typhoon Hagibis (2019) at a certain integration time were used in conjunction with Himawari-8 IRWIN BTK (Bessho et al., 2016). In other words, the result of EOF analysis for the average IRWIN BTK shown in Fig. 11 demonstrated that similar patterns were extracted irrespective of a horizontal resolution when a TC first reaches a minimum central pressure during the lifecycle. Table 5 shows each contribution rate to the total variation of the first to third EOF modes. In addition to the results of EOF analysis overall the tropical and subtropical oceans, the results calculated for each of the six regions were also presented. The sum of the contribution rates from the first to third EOF modes was approximately 70%, ranging from 66.9% (EP) to 72.2% (WP). The results show the similarity of TC inner core structures among the six regions in that the axisymmetric spatial pattern was the most dominant mode, and the second was the wavenumber-1 pattern.

From a global view, the contribution rates in all tropical and subtropical oceans were 45.5%, 14.2%, and 12.8% for the first to third EOF modes, respectively, totaling 72.5%. Although the summations of the three EOF contribution rates are similar across the six ocean regions, the contribution rate in each mode differs by region. Looking at the contribution rates for each region in detail, the contribution rate of EOF1 in EP and NI was relatively low, even though EOF1 was significantly dominant in the two regions. This indicates a low rate of TCs, mainly characterized by showing an axisymmetric pattern to total TCs in EP and NI. In NI, the rate of TCs with wavenumber 1 patterns (both EOF2 and EOF3) was relatively high, while in EP, the rate was also relatively low. In SI, the contribution rate of EOF1 was relatively high, while the rates of EOF2 and EOF3 were the lowest among the six regions. Therefore, the spatial patterns of the average IRWIN BTK collocated at the TC center exhibited their own characteristics in each region, and do not necessarily have to be

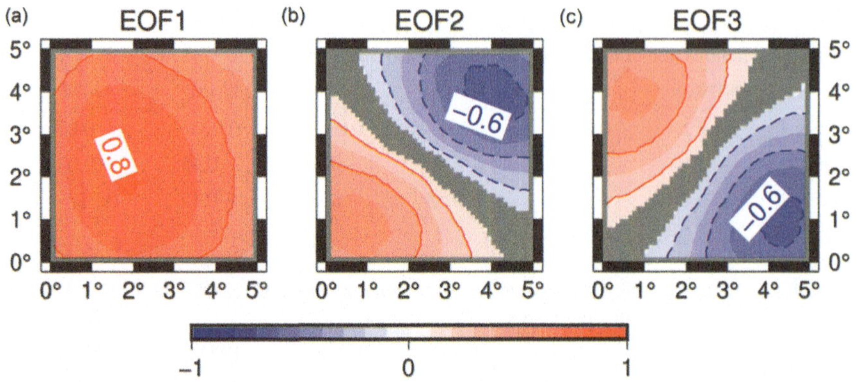

Fig. 11 The horizontal map of the correlation coefficient of **a** the first (EOF1), **b** the second (EOF2), and **c** the third (EOF3) modes for the average IRWIN BTK. The horizontal/vertical axes indicate longitude/latitude. The center of each panel corresponds to the TC center

Table 5 Ratios of the variance in the first (EOF1), second (EOF2), and third (EOF3) modes and the sum of the ratios to total variance

	EP (%)	NA (%)	NI (%)	SI (%)	SP (%)	WP (%)
EOF1	38.5	44.4	39.4	44.2	45.1	44.2
EOF2	16.6	14.1	16.3	14.8	14.4	16.3
EOF3	11.8	13.0	15.2	10.8	11.1	11.7
SUM	66.9	71.5	70.9	69.8	70.6	72.2

homogeneous in all the tropical and subtropical oceans. Factors contributing to the own characteristics of the inner-core cloud pattern across the six regions include the TC size, TC potential intensity, differences in vertical shear and static stability in the atmospheric environment, and SST and TCHP in the oceanic environments (Moon et al., 2022; Wada et al., 2012), as well as differences in operational procedures by forecasters across the agencies (Kruk et al., 2010; Schreck III et al., 2014). Therefore, clarifying the factors will be a challenging subject. It should be noted that the analysis was confined only to the time at first reaching the minimum central pressure of a TC during the life cycle, which is different from the two intensification phases of Wada et al. (2022). In addition, the current analysis could not capture cloud variations in the TC inner core on spatiotemporal scales that cannot be represented at the horizontal resolution of 0.1° and at the interval of 6 h.

3.3 Relation of the EOF Modes for IRWIN BTK to TC Intensity, SST, and TCHP

In the previous subsection, we identified the representative EOF modes for daily TCHPs and 6-h geopotential heights at the isobaric surfaces of 850, 500, and 300 hPa. In addition, we showed that the spatial variability of daily TCHPs and 6-h geopotential heights had a unique relationship with the climate indices in each region. Each EOF mode derived from the daily TCHP and 6-h geopotential heights was not one-to-one with a specified climate index, but rather, the mode contained its own interannual to decadal variability, which was combined with multiple indices introduced in subsect. 2.1. We also identified the representative EOF modes for average IRWIN BTKs collocated at the TC center and exhibited their own characteristics in each region. Next, we examined whether the spatial variability of daily TCHPs and 6-h geopotential heights was related to the spatial patterns of the temporal average IRWIN BTK collocated at the TC center. Furthermore, whether interannual to decadal variations in the atmosphere and ocean did affect the cloud patterns in the inner core of a TC will be examined. The objective of this subsection is to clarify whether the unique spatial patterns of clouds in each of the ocean basins at the TC peak intensity are related to interannual to decadal variations in the atmosphere and ocean.

Results showed that the first to three EOF modes of the temporal average IRWIN BTK had little or no significant correlation with the first to three modes of daily TCHPs (not shown). However, the first EOF mode of the three 6-h geopotential heights showed a correlation with the first to three EOF modes of the temporal average IRWIN BTK, but the value is relatively low (not shown). This suggests that the spatial patterns of the average IRWIN BTK in the inner core of a TC are unrelated to interannual to decadal variability in the atmosphere and ocean in all the tropical and subtropical oceans. For example, ENSO is unrelated to the axisymmetric pattern of the temporal average IRWIN BTK shown in Fig. 11a. Wada and Chan (2008) have reported that in WP, there exist TCHP modes that show a significant correlation with accumulated cyclone energy (Bell et al., 2000) and power dissipation index (Emanuel, 2005), which seems to contradict the above-mentioned results, even though we do not know the relation between TC intensity and the spatial patterns of the average IRWIN BTK in the inner core of a TC.

Next, we reexamined the relationship between a lifetime minimum central pressure of TC, average BTK, and the normalized amplitude of the average IRWIN BTK. We focus on the extent to which the three EOF patterns at the peak TC intensity are related to a lifetime minimum central pressure of TC and the lowest IRWIN BTK. In Fig. 12, the central pressure shows a significant correlation with the normalized amplitudes of EOF1, while correlations with those of EOF2 and EOF3 are not significant. The result reveals that TC axisymmetry is associated with the central pressure. The lowest IRWIN BTK shows a higher correlation with the normalized amplitude of EOF1 than the central pressure. That is, TC axisymmetry and the lowest IRWIN BTK exhibit a stronger correlation than TC axisymmetry and the central pressure. In addition, we examine the extent to which the three EOF patterns at the peak TC intensity are related to the average SST and TCHP (Fig. 13). Although there is a significant correlation between average TCHP and the normalized amplitude of EOF1 for the average IRWIN BTK, correlations of average SST and TCHP with those of EOF2 and 3 were low in addition to the correlation of average SST with the normalized amplitude of EOF1 (Fig. 13). Therefore, stronger axisymmetry is related to lower central pressure, smaller values of the lowest IRWIN BTK, and higher average TCHP. It should be emphasized that the lowest IRWIN BTK is a better indicator than TC central pressure for explaining whether a TC possesses an axisymmetric structure.

Then, we compared the first to third EOF modes for the temporal average IRWIN BTK with central pressures of TCs (Table 6) and the lowest IRWIN BTK (Table 7). In Tables 6 and 7, the normalized amplitude of EOF1-3 for the average IRWIN BTK was used for calculating the correlation with the central pressure and the lowest IRWIN BTK. The normalized amplitudes of EOF1 showed relatively high positive correlations with the central pressures and the lowest IRWIN BTK in all the tropical and subtropical oceans, and also in each of the six regions. In contrast, the normalized amplitudes of EOF2 and EOF3 showed lower correlations than the first EOF mode did. The differences in the correlation coefficients (Tables 6 and 7) suggest that the depth of convection itself in the inner core of a TC, as described in subsect. 2.3, did not completely or tightly correspond to the central pressure. If the errors were caused by the differences in TC intensity analysis methods and forecasters' skill among meteorological agencies, the differences in correlation coefficients between the first to third EOF modes and the central pressures or the lowest IRWIN BTK across regions would demonstrate one aspect of errors in TC intensity analysis.

3.4 Relation of Central Pressure and the Lowest IRWIN BTK with SST and TCHP

The previous subsections showed the significant relation of the EOF1 for the temporal average IRWIN BTK to central pressures of TCs, the lowest IRWIN BTK, the average TCHP in all the tropical and subtropical oceans and in each of the six regions. In this subsection, we examined the correlations between TC intensity (central pressures and the lowest average BTK) and the ocean (the average SST and TCHP). To the author's knowledge, this is the first attempt to investigate the relationship between TC intensity and average SST and TCHP using the same high-resolution (0.1 degrees) global dataset for each ocean region over 30 years. Figure 14 shows a scatter plot of average SSTs and central pressures obtained from the IBTrACS dataset. The confidence interval in the linear regression equation was set to 95%. In NI, higher average SSTs were associated with lower central pressures, whereas in the other five regions, lower average SSTs were associated with lower central pressures. Particularly in NA and WP, TCs with relatively high minimum central pressures are frequently found in regions with high average SST. In other regions, likely in mid-latitude areas, TCs

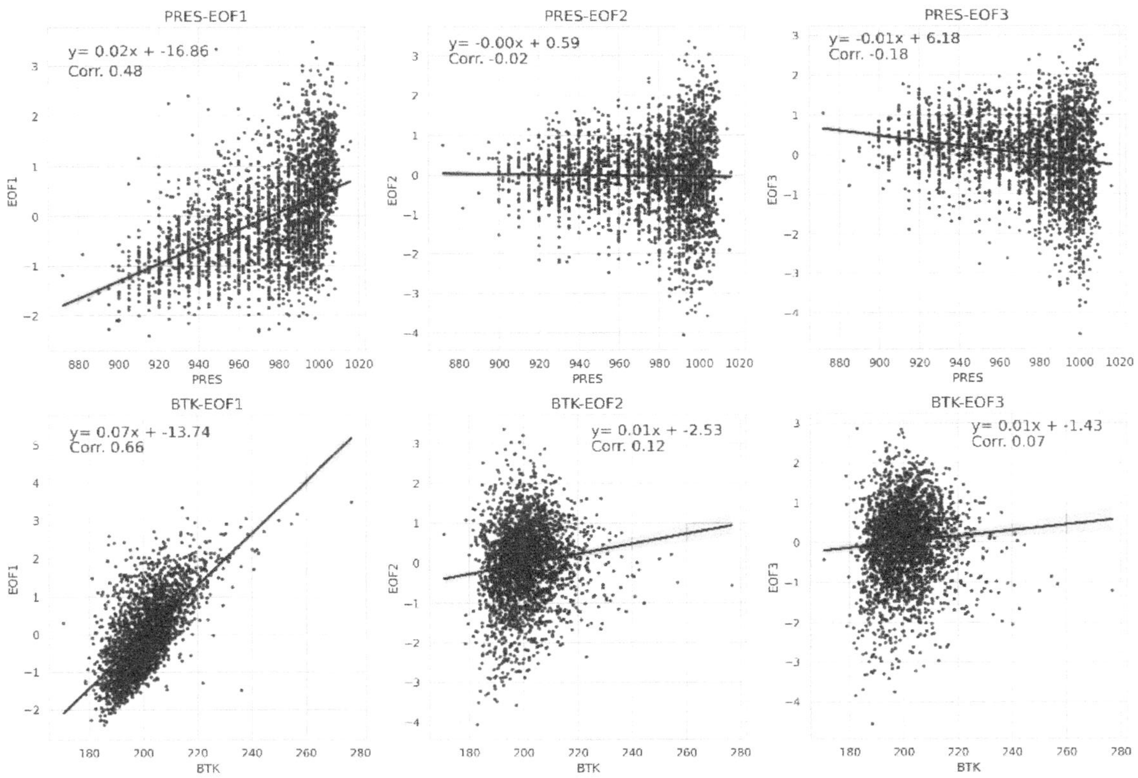

Fig. 12 Scatter diagram with a linear regression function between the EOF1-3 normalized amplitudes of the temporal average IRWIN BTK and (upper panel) central pressure and (lower panel) the lowest IRWIN BTK in the 5° by 5° area collocated at the TC center. A linear regression function and correlation coefficient are shown in each panel

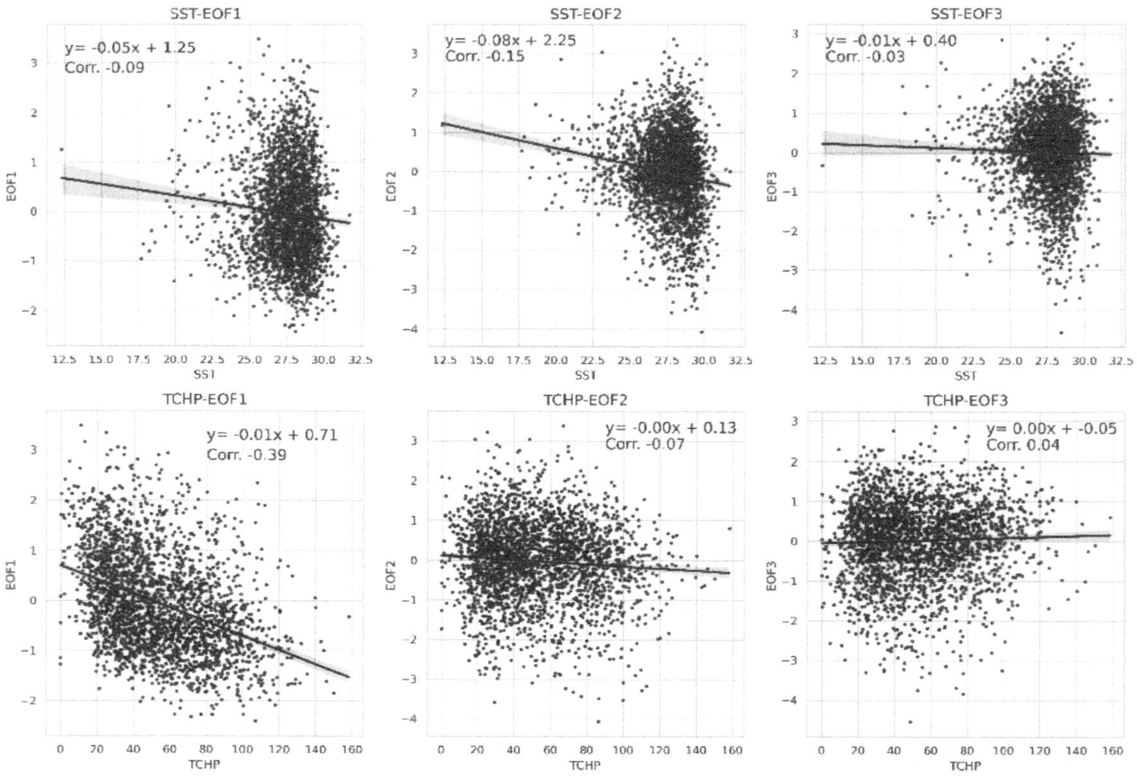

Fig. 13 Scatter diagram with a linear regression function between the EOF1-3 normalized amplitudes of the temporal average IRWIN BTK and (upper panel) the average SST and (lower panel) the spatiotemporal average TCHP in the 2° by 2° area collocated at the TC center. A linear regression function and correlation coefficient are shown in each panel

Table 6 Correlation coefficients between the first three EOF modes for the temporal average IRWIN BTK and central pressures. Bold fonts indicate that the correlation is significant at the 99% confidence level based on the *t*-test

	GLOBAL	EP	NA	NI	SI	SP	WP
EOF1	**0.48**	**0.44**	**0.58**	**0.30**	**0.41**	**0.35**	**0.47**
EOF2	− 0.02	**− 0.17**	0.07	0.00	0.04	0.08	− 0.01
EOF3	**− 0.18**	**− 0.14**	− 0.01	**− 0.28**	**− 0.34**	− 0.13	**− 0.22**

Table 7 Correlation coefficients between the first three EOF modes for the temporal average IRWIN BTK and the lowest IRWIN BTK over the 5° by 5° domain. Bold fonts indicate that the correlation is significant at the 99% confidence level based on the *t*-test

	GLOBAL	EP	NA	NI	SI	SP	WP
EOF1	**0.66**	**0.61**	**0.65**	**0.39**	**0.69**	**0.74**	**0.63**
EOF2	**0.12**	**0.17**	**− 0.14**	0.05	0.07	0.00	**0.15**
EOF3	**0.07**	**0.10**	**− 0.17**	**0.22**	− 0.05	− 0.02	**0.16**

with relatively low minimum central pressures exist in regions with low average SST. Consequently, the sign of slopes except in NI was not consistent with the thermodynamic theory (Emanuel, 1986, 1988) and positive feedback mechanisms in a TC-ocean interaction framework (Ginis, 2005). However, since the correlation was not statistically significant, the results suggest that it is not appropriate to analyze the relationship between the average SSTs and central pressures by using a linear regression. The results of this study also supported the fact that the relationship between SST and maximum lifetime potential intensity differed between EP, NA, and WP shown in Fig. 2 in Wada and Usui (2007).

Figure 15 shows the relationship between average TCHPs and central pressures in each of six regions. Although there was a tendency for lower central pressure values with higher average TCHP values, the rate of change varied significantly depending on the region, ranging from 0.4 hPa (SI) to 3.4 hPa (WP) per 10 kJ cm^{-2}. Unlike the relationship between average SSTs and central pressures, the extraordinarily high value of the average TCHP in EP exceeding 140 kJ cm^{-2} was found when the central pressure was approximately 1000 hPa. Although the average TCHP exceeded 100 kJ cm^{-2} and the central pressure around 1000 hPa was observed in other regions, the average TCHP directly beneath a TC never exceeded 80 kJ cm^{-2} in EP except for this case. In particular, in EP, where the number of TCs is relatively low compared to that in other regions, it should be noted that only one unusual case regarded as an outlier can have a significant impact on the interpretation of statistical significance.

The scatter diagram and a linear regression function with a 95% confidence level in Fig. 16 showed the relationship between average SSTs and the lowest IRWIN BTKs. Unlike the result in Fig. 14, which showed the relationship between average SSTs and central pressures, the lowest IRWIN BTK tended to be low as the average SST tended to be high in all regions. In other words, the WMO standard central pressure archived

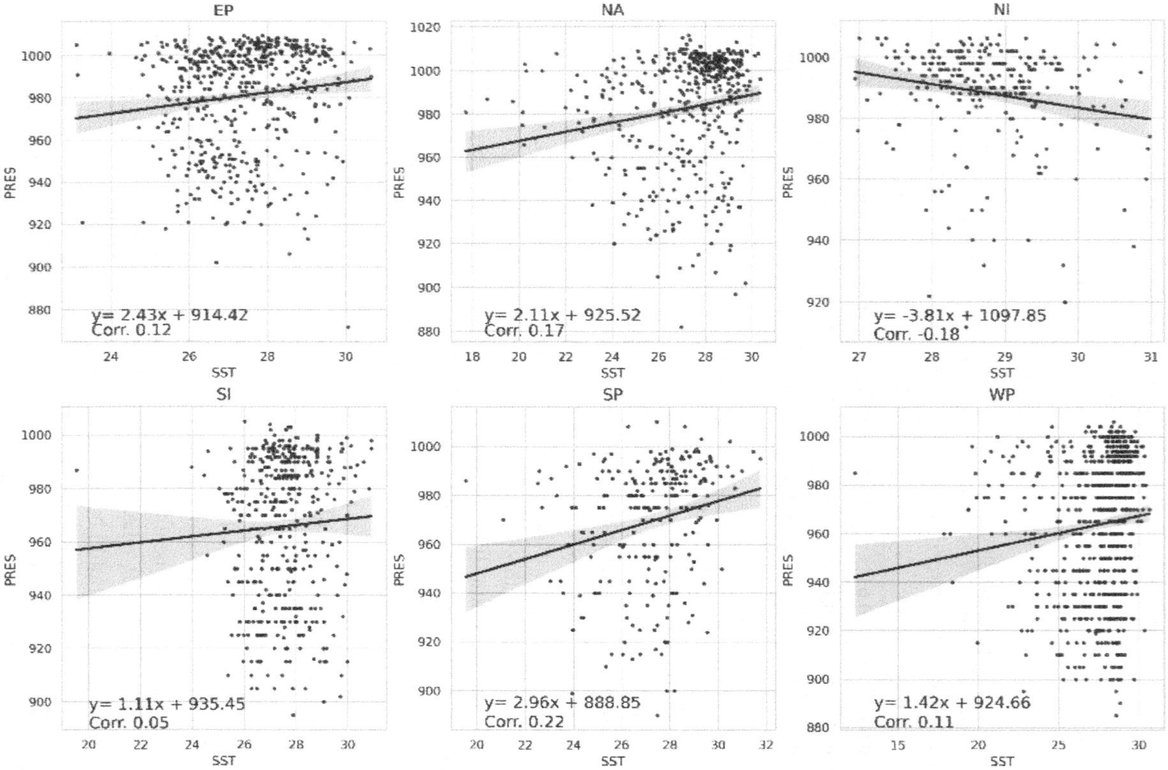

Fig. 14 Scatter diagram on the relationship between central pressures and average SSTs with a linear regression in each of six regions. Gray shades indicate the significance of the linear regression at the 95% confidence level. A linear regression function and correlation coefficient are shown in each panel

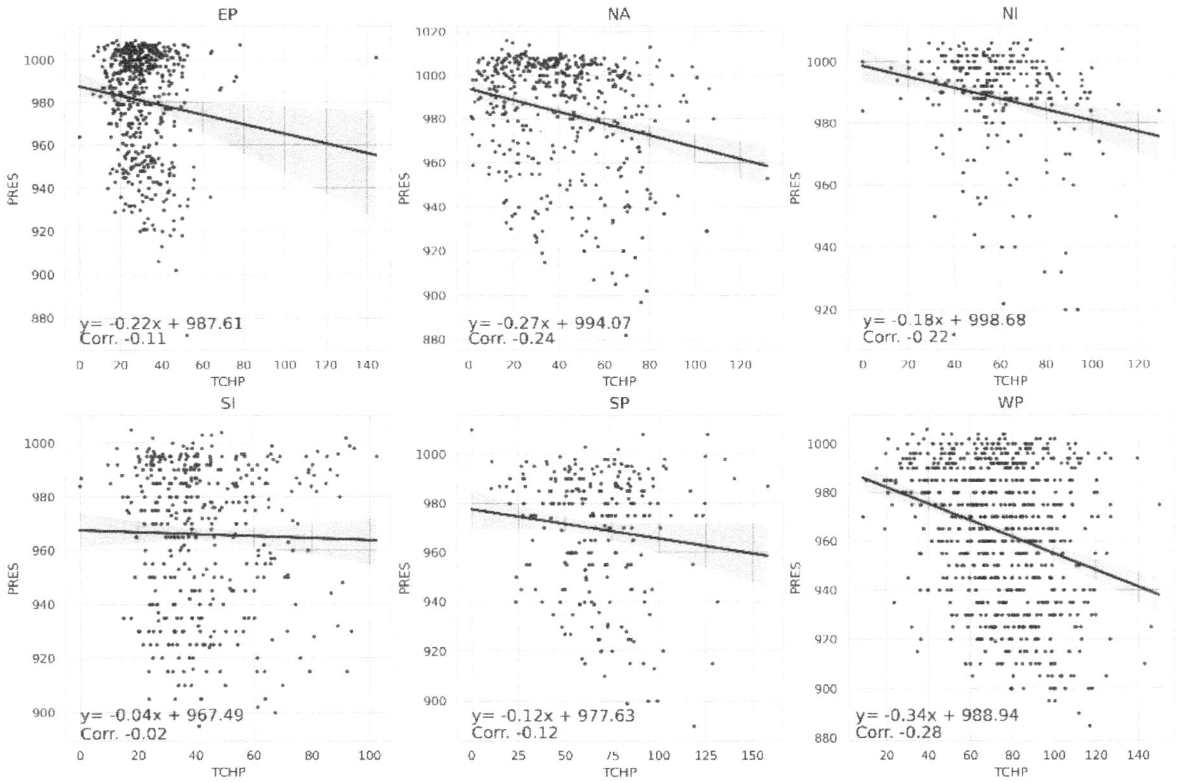

Fig. 15 Scatter diagram on the relationship between central pressures and average TCHPs with a linear regression in each of six regions. Gray shades indicate the significance of the linear regression at the 95% confidence level. A linear regression function and correlation coefficient are shown in each panel

in IBTrACS was poorly correlated with the lowest IRWIN BTK. The relationship of the lowest IRWIN BTK with the average SST was consistent with the well-known facts that thermodynamics control TC intensity (Emanuel, 1986, 1988) and that the positive trend was connected to a warming world (Kossin et al., 2020).

The relationship between average TCHPs and the lowest IRWIN BTK at the location of the TC center is shown in Fig. 17 to compare the results with those shown in Fig. 15. Unlike the results in Fig. 15, and similar to those in Fig. 16, the average TCHP increased as the lowest IRWIN BTK decreased, with relatively high correlations. In cases where the average TCHP in EP exceeded 140 kJ cm^{-2}, the lowest IRWIN BTK was lower than 200 K even if the value did not fall within the 95% confidence interval of the linear regression. However, when compared to the relationship between average TCHPs and central pressures, the results on the relationship between average TCHPs and the lowest IRWIN BTKs were more reasonable in explaining the thermodynamic role of the warm upper ocean in the development of convection and thus TC intensity.

4 Discussion

In this chapter instead of using a daily SST and TCHP at a specific time when the central pressure of a TC reached the lowest value for the first time, the spatiotemporal average value over the 2° by 2° area collocated at the TC center was used for analysis, which was calculated by dividing the accumulated value from the time when the maximum sustained speed first reached 35 knots to the time when the central pressure first reached its minimum value (Fig. 3). The spatiotemporal average SST and TCHP were found to be effective metrics for demonstrating the linear correlation between central pressures and the lowest IRWIN BTK. This is because the UOHC just beneath a TC does not directly reflect the intensity, but rather, a TC can reach its peak intensity through a series of dynamic and thermodynamic processes called 'second circulation', such as frictional convergence near the surface and in the atmospheric boundary layer, the ascent of air masses at around the eyewall, the release of condensational heating as diabatic heating, and outflow in the upper troposphere mainly in the intensification phase (Wada et al., 2018). During the enhancement of secondary circulation, on the contrary, sea surface cooling induced by a TC occurs, resulting in decreases in TCHP just beneath a TC (Ginis, 2005; Price, 1981; Wada, 2002). The average TCHP is thought to yield good correlations because the effects of rapid decreases in SSTs accumulate during the intensification phase and then are continuously reflected in a cloud pattern within the inner core of a TC through continuous secondary circulation and thus TC intensity. In fact, the correlation between accumulated TCHP and TC intensity has been reported in EP, NI, and WP (Moon et al., 2022; Vissa et al., 2013; Wada & Usui, 2007; Wada et al., 2012). The difference from the previous studies is that this chapter addresses the relationship between TCs and the ocean by using a global TCHP dataset in addition to global TC and satellite datasets. Although Jangir et al (2020) made intercomparisons among various

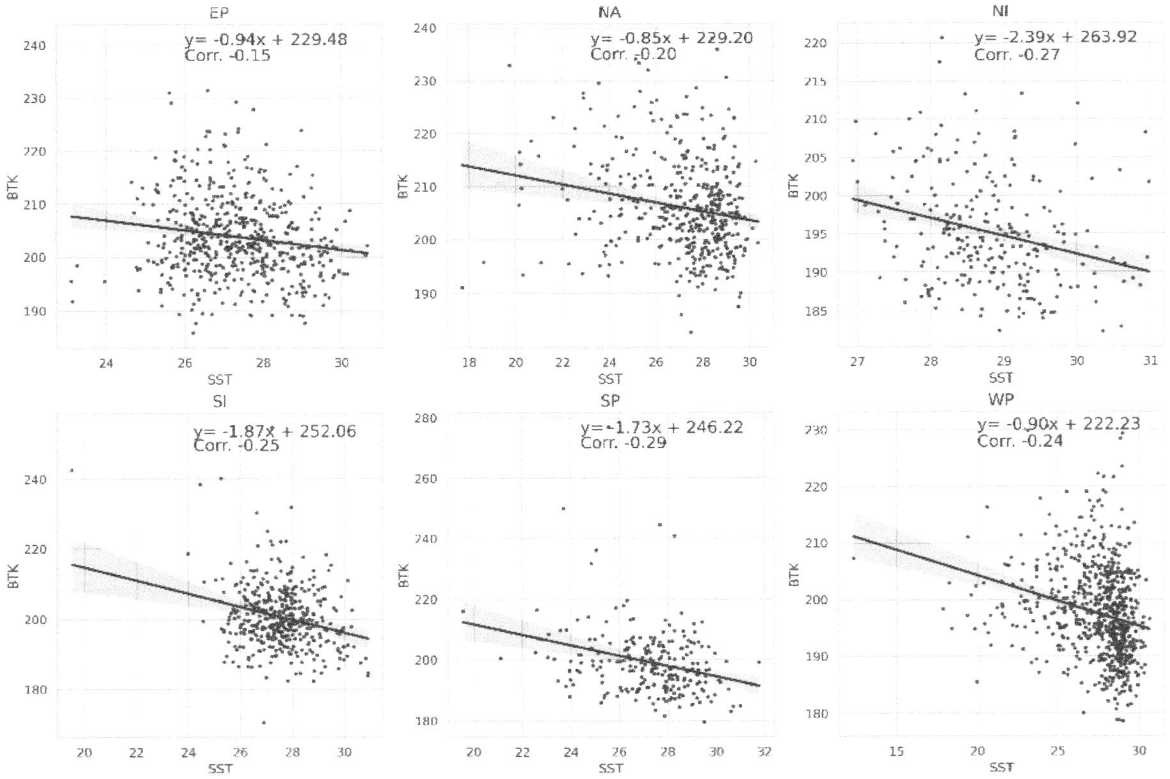

Fig. 16 Scatter diagram on the relationship between the lowest IRWIN BTKs and average SSTs with a linear regression in each of six regions. Gray shades indicate the significance of the linear regression at the 95% confidence level. A linear regression function and correlation coefficient are shown in each panel

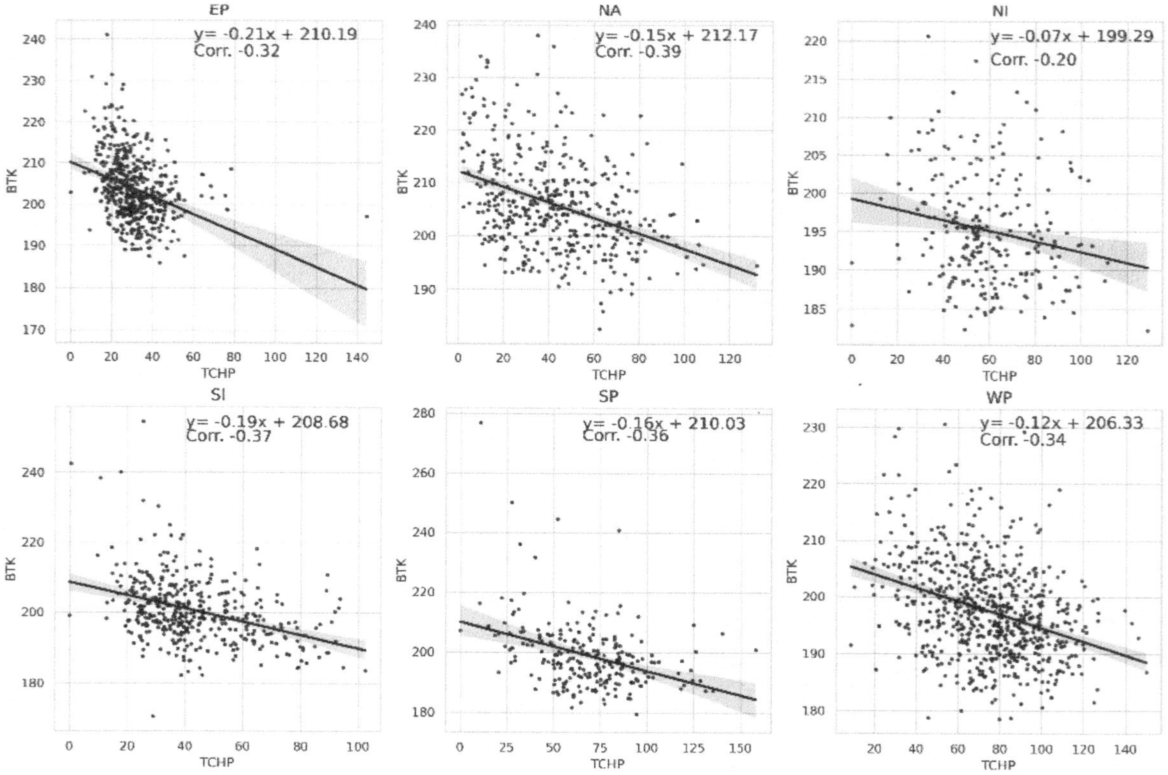

Fig. 17 Scatter diagram on the relationship between the lowest IRWIN BTKs and average TCHPs with a linear regression in each of six regions. Gray shades indicate the significance of the linear regression at the 95% confidence level. A linear regression function and correlation coefficient are shown in each panel

TCHP products in NI, there is no other global TCHP derived from advanced global ocean reanalysis spanning over 30 years with a horizontal resolution of 01 degree. By constructing the global TCHP dataset, it becomes possible to comprehensively analyze the relationship with TC intensity across six regions.

Moreover, a new challenge is the application of the concept of the accumulated SST and TCHP (spatiotemporal average SST and TCHP) to GridSat IRWIN BTK in a 5° by 5° domain collocated at the TC center. One purpose of using the dataset is to determine the lowest IRWIN BTK, and the other is to investigate the relation to central pressures, the average SSTs, and TCHPs. Furthermore, an EOF analysis was conducted for the temporal average IRWIN BTK to extract the represented spatial modes around the TC center. In fact, the difference between the lowest and highest gridded IRWIN BTK became small in a 5° by 5° area when the temporal average IRWIN BTK was applied to show the horizontal pattern at around the center of a TC (Fig. 3). In other words, the continuous occurrence of convection in a specified area in the inner core of a TC was easily detected without being influenced by the intermitted sporadic convection at a specified time during the intensification. In an operational aspect, satellite images around the TC center prior to the analysis time are usually referenced in TC intensity analysis using the Dvorak method (Dvorak, 1984), although each meteorological agency has its own rules. However, in operational analysis, the most recent observation tends to be prioritized over the lifetime from the genesis to the peak of a TC. Rather, the most recent TC analysis tends to impose constraints on the TC analysis at the current analysis time. This differs from the concept of average TCHPs, which is uniformly averaged in a certain area during a period from the genesis to the peak of a TC.

Based on the analysis results of this study, it is speculated that central pressures archived in IBTrACS are not necessarily influenced by the lowest IRWIN BTK in a 5° by 5° area collocated at the TC center. The other processes, such as adiabatic heating at the TC center (e.g., Wada, 2012), may also be influenced in determining the value of central pressure in operation. However, as mentioned in the previous section, spatially unresolved small TCs could not be captured in the representative spatial EOF modes. Otherwise, the spatial patterns of small-sized TCs may have been calculated as a high-order mode with a relatively low contribution to variations, which is not significant. This issue will also affect the result on the relationship between TC intensity and the ocean if the number of small TCs increases, particularly under anthropogenic global warming.

This chapter revealed that the relationship among central pressures, the lowest IRWIN BTK, average SSTs, and TCHPs shows similarities to some extent across the six regions. However, the differences were evident. The differences included those reported previously between EP and WP (Moon et al., 2022; Wada et al., 2012). In certain regions, future ocean warming may lead to increases in TCs with enhanced axisymmetric structures caused by the sustenance of enhanced convection in their inner core and thus may increase TC intensity. As for the central pressures, however, the differences may be attributed to unique procedures of each agency and the forecasters' skills, as pointed out by Kossin et al., 2020, and then may be reflected in the discrepancies in the relation of average SSTs and TCHPs between central pressures and the lowest IRWIN BTKs. Additionally, regarding the 6-h IRWIN BTK dataset, GridSat is a merged dataset from different satellites, so it may not be spatiotemporally homogeneous. This is likely to remain an unavoidable challenge in future studies investigating the relationship between ocean warming and TC intensity.

5 Concluding Remarks

This chapter addressed the relation of TC intensity to the ocean by using a global TCHP dataset, MGDSST, GridSat IRWIN BTK, JRA-3Q geopotential heights, and IBTrACS TC best track data. To investigate the relationship between TC intensity and the ocean in all the tropical and subtropical oceans, WMO TC data recorded as minimum sea-level pressures were used as central pressures of TCs. In addition, spatiotemporal average SSTs and TCHPs over a 2° by 2° area collocated at the TC center and the lowest value of the temporal average IRWIN BTKs in a 5° by 5° area collocated at the TC center were calculated from the time when the maximum sustained speed of a TC first reached 35 knots to the time when the central pressure first reached its minimum value. The application of the average values follows the accumulated SST and TCHP reported in previous studies (Moon et al., 2022; Wada & Usui, 2007; Wada et al., 2012; Wada and Yanase, 2024). Furthermore, EOF analyses were conducted for daily TCHP, 6-h geopotential heights at the 300-, 500-, and 850-hPa isobar levels to investigate the representative modes of the atmosphere and ocean and then to show their relationship with representative climate indices (NINO3, PDO, DMI, EMI, AMM, and PMM) derived from regional SST data in the target region. An EOF analysis was made to extract the spatial representative modes for temporal average IRWIN BTKs in the 5° by 5° area to investigate the relation to central pressures, the lowest IRWIN BTKs, average SSTs, and TCHPs. From these results, we explored the differences between central pressures and the lowest IRWIN BTKs as metrics of TC intensity, as well as the differences among the six regions (EP, NA, NI, SI, SP, WP).

The outline of this chapter is as follows:

- Climatologically, SST was the highest in NI, but this result was largely influenced by the northern limit of latitude in the average area. TCHP was the highest in SP, and TCHP in WP was the second in all the tropical and subtropical oceans. The time series of the original SST and TCHP in the tropical and subtropical oceans and each of the six regions indicated recent increases in the upper ocean except in EP, but the rate of the increase differed by region: the increasing rate of daily SSTs was the highest in NI, whereas that of the original TCHP was the highest in WP.
- Most of the climate indices based on regional SST data were correlated with the first to third EOF representative modes in both the atmosphere and the ocean in all the tropical and subtropical oceans, but the correlation coefficients differed by region.

- Global-scale atmospheric and oceanic variability derived from the first to third EOF modes in the atmosphere and the ocean has little effect on the first to third EOF modes of the temporal average IRWIN BTK and the lowest IRWIN BTK in a 5° by 5° area collocated at the TC center. This suggests that global atmospheric and oceanic variability has little effect on the structure of the inner core of a TC.
- Regarding the result of linear regression analysis, the slope of the regression function for daily TCHPs showed a positive value, which was irrespective of regions and differed from the slope for daily SSTs. This result provided a reasonable explanation for the thermodynamic theory: the upper ocean warming leads to deepening the center pressure through intensification of convection in the inner core of a TC.
- Regarding the representative cloud patterns derived from the EOF analysis for the temporal average IRWIN BTK in the 5° by 5° area, the axisymmetric structure in the first EOF mode was significantly related to strong TC represented by the lowest IRWIN BTK and low central pressure.

Since the TC best track central pressure data were not homogeneous worldwide due to the differences in the analysis method and forecasters' skill among meteorological agencies, it might be preferable to use the 6-h/temporal average IRWIN BTK dataset instead of TC central pressure to study the climatology of TCs in all the tropical and subtropical oceans. However, since it is currently impossible to monitor the entire globe with a single satellite at once, it is unavoidable to make a composite dataset from multiple satellites, particularly one that contains an issue with the inhomogeneous data, particularly with the boundary between two satellites. Also, differences in the quality of historical data between satellites themselves may become a serious issue for TC analysis and climate studies.

The inner core structure and intensity of a TC had no relation to the representative atmospheric and oceanic interannual to decadal variability, while the upper ocean warming represented by an increase in the spatiotemporal average TCHP will possibly enhance the axisymmetrization of a TC, resulting in strengthening a TC. However, it cannot be denied that there are differences in TC structure and intensity caused by regional atmospheric and oceanic environmental conditions on daily to seasonal timescales, which were not taken into account in this chapter. Due to the spatial and temporal limitations, the result of this chapter did not capture the characteristics of small TCs. How to address such small TC cases in TC climate studies will be a future challenge to explore the role of ocean warming on TCs.

Acknowledgements The author is grateful for editors and two anonymous reviewers for their instructive comments and is also grateful for Dr. Wataru Yanase, Ms. Nao Takamura and Ms. Haruka Sasaki for their helpful discussions. Generic Mapping Tools software (http://gmt.soest.hawai.edu/) was used to draw figures. The research introduced in this chapter was funded by Grants-in-Aid for Scientific Research (JP22K03725) from the Japan Society for the Promotion of Science.

References

Albert, J., & Bhaskaran, P. K. (2020). Ocean heat content and its role in tropical cyclogenesis for the Bay of Bengal basin. *Climate Dynamics, 55*, 3343–3362. https://doi.org/10.1007/s00382-020-05450-9

Ashok, K., Behara, S. K., Rao, S. A., Weng, H., & Yamagata, T. (2007). El Niño Modoki and its possible teleconnection. *Journal of Geophysical Research, 112*, C11007.

Bell, G. D., Halpert, M. S., Schnell, R. C., Higgins, R. W., Lawrimore, J., Kousky, V. E., Tinker, R., Thiaw, W., Chelliah, M., & Artusa, A. (2000). Climate assessment for 1999. *Bulletin of the American Meteorological Society, 81*, S1–S50. https://doi.org/10.1175/1520-0477(1999)080%3c1040:CAF%3e2.0.CO;2

Bessho, K., Date, K., Hayashi, M., Ikeda, A., Imai, T., Inoue, H., Kumagai, Y., Miyakawa, T., Murata, H., Ohno, T., Okuyama, A., Oyama, R., Sasaki, Y., Shimizu, Y., Shimoji, K., Sumida, Y., Suzuki, M., Taniguchi, H., Tsuchiyama, H., … Yoshida, R. (2016). An introduction to Himawari–8/9—Japan's new–generation geostationary meteorological satellites. *Journal Meteorological Society Japan, 94*, 151–183.

Camargo, S. J., Sobel, A. H., Barnston, A. G., & Klotzbach, P. J. (2010). The influence of natural climate variability, and seasonal forecasts of tropical cyclone activity. In J. C. L. Chan & J. D. Kepert (Eds.), *Global perspectives on tropical cyclones, from science to mitigation* (2nd ed., pp. 325–362). World Scientific.

Camargo, S. J., H. Murakami, N. Bloemendaal, S. S. Chand, M. S. Deshpande, C. Dominguez-Sarmiento, J. J. Gonzalez-Aleman, T. R. Knutson, I. I. Lin, I. J. Moon, C. M. Patricola, K. A. Reed, M. J. Roberts, E. Scoccimarro, C. Y. (Francis) Tam, E. J. Wallace, L. Wu, Y. Yamada, W. Zhang, and H. Zhao, 2023. An update on the influence of natural climate variability and anthropogenic climate change on tropical cyclones. *Trop. Cyclone Res. Rev.*, 12. 216–239, https://doi.org/10.1016/j.tcrr.2023.10.001

Caron, L., Boudreault, M., & Camargo, S. J. (2015). On the variability and predictability of Eastern Pacific tropical cyclone activity. *Journal of Climate, 28*, 9678–9696. https://doi.org/10.1175/JCLI-D-15-0377.1

Chacko, N., & Balaguru, K. (2025). Impact of Indian Ocean Dipole on the Arabian Sea tropical cyclone activity. *Climate Dynamics, 63*, 197. https://doi.org/10.1007/s00382-025-07676-x

Chan, J. C. L., & Liu, K. S. (2022). Recent decrease in the difference in tropical cyclone occurrence between the Atlantic and the western North Pacific. *Advances in Atmospheric Sciences, 39*, 1387–1397. https://doi.org/10.1007/s00376-022-1309-x

Chand, S. S., Walsh, K. J. E., Camargo, S. J., Kossin, J. P., Tory, K. J., Wehner, M. F., Chan, J. C. L., Klotzbach, P. J., Dowdy, A. J., Bell, S. S., Ramsay, H. A., & Murakami, H. (2022). Declining tropical cyclone frequency under global warming. *Nature Clinical Practice Endocrinology and Metabolism, 12*, 655–661. https://doi.org/10.1038/s41558-022-01388-4

Chiang, J. C. H., & Vimont, D. J. (2004). Analogous meridional modes of atmosphere-ocean variability in the tropical Atlantic and tropical Pacific. *Journal of Climate, 17*, 4143–4158. https://doi.org/10.1175/JCLI4953.1

Choi, H. Y., Park, M. S., Kim, H. S., & Lee, S. (2024). Marine heatwave events strengthen the intensity of tropical cyclones. *Communications Earth Environment, 5*(69), 1–8. https://doi.org/10.1038/s43247-024-01239-4

Dvorak, V. F. (1984). Tropical cyclone intensity analysis using satellite data. *NOAA Technical Reports, 11*, 45.

Elsner, J. B. (2020). Continued increases in the intensity of strong tropical cyclones. *Bulletin American Meteorological Society, 101*, E1301–E1303. https://doi.org/10.1175/BAMS-D-19-0338.1

Elsner, J. B., Kossin, J. P., & Jagger, T. H. (2008). The increasing intensity of the strongest tropical cyclones. *Nature, 455*, 92–95. https://doi.org/10.1038/nature07234

Emanuel, K. A. (1986). An air-sea interaction theory for tropical cyclones. Part I: Steady-state maintenance. *Journal of Atmospheric Science, 43*, 585–605. https://doi.org/10.1175/1520-0469(1986)043%3c0585:AASITF%3e2.0.CO;2

Emanuel, K. A. (1988). The maximum intensity of hurricanes. *Journal of Atmospheric Science, 45*, 1143–1155. https://doi.org/10.1175/1520-0469(1988)045%3c1143:TMIOH%3e2.0.CO;2

Emanuel, K. A. (2005). Increasing destructiveness of tropical cyclones over the past 30 years. *Nature, 436*, 686–688. https://doi.org/10.1038/nature03906

Ginis, I. (2005). Tropical cyclone-ocean interactions. In W. Perrie (Ed.), *Atmosphere-ocean interactions* (pp. 83–114). Advances in Fluid Mechanics Series, No. 33. WIT Press.

Gray, M. (1968). Global overview of the origin of tropical disturbances and storms. *Monthly Weather Review, 96*, 669–700. https://doi.org/10.1175/1520-0493(1968)096%3c0669:GVOTOO%3e2.0.CO;2

Hong, J., & Wu, Q. (2021). Modulation of global sea surface temperature on tropical cyclone rapid intensification frequency. *Environmental Research Communications, 3*, Article 041001. https://doi.org/10.1088/2515-7620/abf39b

IPCC. (2021). Climate change 2021, *The physical science basis. Contribution of working group I to the sixth assessment report of the Intergovernmental panel on climate change* [V. Masson-Delmotte, P. Zhai, A. Pirani, S. L. Connors, C. Péan, S. Berger, N. Caud, Y. Chen, L. Goldfarb, M. I. Gomis, M. Huang, K. Leitzell, E. Lonnoy, J. B. R. Matthews, T. K. Maycock, T. Waterfield, O. Yelekçi, R. Yu, B. Zhou (Eds.),]. Cambridge University Press, Cambridge, In press, https://doi.org/10.1017/9781009157896

Jangir, B., Swain, D., Ghose, S. K., Goyal, R., & Bhaskar, T. V. S. U. (2020). 2020, Inter-comparison of model, satellite and in situ tropical cyclone heat potential in the North Indian Ocean. *Natural Hazards, 102*, 557–574. https://doi.org/10.1007/s11069-019-03756-4

Japan Agency for Marine-Earth Science and Technology. (2020). APL-virtualearth. (Available online at: http://www.jamstec.go.jp/virtualearth/general/en/index.html, Accessed 5 Nov 2025)

Japan Meteorological Agency. (2014). NEAR-GOOS regional real time data base. (Available online at: data.jma.go.jp/goos/data/pub/JMA-product/mgd_sst_glb_D/Readme_mgd_sst_glb_D, Accessed 5 Nov 2025)

Japan Meteorological Agency. (2020a). Download El Niño Monitoring Indices. (Available online at: http://ds.data.jma.go.jp/tcc/tcc/products/elnino/index/sstindex/base_period_8110/Nino_3/anomaly, Accessed 5 Nov 2025)

Japan Meteorological Agency. (2020b). Pacific Decadal Oscillation (PDO) index. (Available online at: http://ds.data.jma.go.jp/tcc/tcc/products/elnino/decadal/annpdo.txt, Accessed 5 Nov 2025)

Jin, F. F., Boucharel, J., & Lin, I. I. (2014). Eastern Pacific tropical cyclones intensified by El Niño delivery of subsurface ocean heat. *Nature, 516*, 82–85. https://doi.org/10.1038/nature13958

Kido, S., Nonaka, M., & Miyazawa, Y. (2022). JCOPE-FGO: An eddy-resolving quasi-global ocean reanalysis product. *Ocean Dynamics, 72*, 599–619. https://doi.org/10.1007/s10236-022-01521-z

Kikuchi, K. (2021). The boreal summer intraseasonal oscillation (BSISO): A review. *Journal Meteorological Society Japan, 99*, 933–972. https://doi.org/10.2151/jmsj.2021-045

Klotzbach, P. J., Wood, K. M., Schreck III, C. J., Bowen, S. G., Patricola, C. M., & Bell, M. M. (2022). Trends in global tropical cyclone activity: 1990–2021. *Geophysical Research Letters, 49*, e2021GL095774. https://doi.org/10.1029/2021GL095774

Knapp, K. R., Kruk, M. C., Levinson, D. H., Diamond, H. J., & Neumann, C. J. (2010). The international best track archive for climate stewardship (IBTrACS): Unifying tropical cyclone best track data. *Bulletin American Meteorological Society, 91*, 363–376. https://doi.org/10.1175/2009BAMS2755.1

Knapp, K. R., Ansari, S., Bain, C. L., Bourassa, M. A., Dickinson, M. J., Funk, C., Helms, C. N., Hennon, C. C., Holmes, C. D., Huffman, G. J., Kossin, J. P., Lee, H.-T., Loew, A., & Magnusdottir, G. (2011). Globally gridded satellite (GridSat) observations for climate studies. *Bulletin American Meteorological Society, 92*, 893–907. https://doi.org/10.1175/2011BAMS3039.1

Knapp, K. R., Knaff, J. A., Sampson, C. R., Riggio, G. M., & Schnapp, A. D. (2013). A pressure-based analysis of the historical western North Pacific tropical cyclone intensity record. *Monday Weather Review, 141*, 2611–2631. https://doi.org/10.1175/MWR-D-12-00323.1

Kosaka, Y., Kobayashi, S., Harada, Y., Kobayashi, C., Naoe, H., Yoshimoto, K., Harada, M., Goto, N., Chiba, J., Miyaoka, K., Sekiguchi, R., Deushi, M., Kamahori, H., Nakaegawa, T., Tanaka, T. Y., Tokuhiro, T., Sato, Y., Matsushita, Y., & Onogi, K. (2024). The JRA-3Q reanalysis. *Journal Meteorological Society Japan, 102*, 49–109. https://doi.org/10.2151/jmsj.2024-004

Kossin, J. P., Knapp, K. R., Olander, T. L., & Velden, C. S. (2020). Global increase in major tropical cyclone exceedance probability over the past four decades. *Proceedings of the National Academy of Sciences, 117*, 11975–11980. https://doi.org/10.1073/pnas.1920849117

Kruk, M. C., Knapp, K. R., & Levinson, D. H. (2010). A technique for merging global tropical cyclone best track data. *Journal Atmospheric Oceanic Technology, 27*, 680–692. https://doi.org/10.1175/2009JTECHA1267.1

Kurihara, Y., Sakurai, T., & Kuragano, T. (2006). Global daily sea surface temperature analysis using data from satellite microwave radiometer, satellite infrared radiometer and in-situ observations. *Weather Service Bulletin, 73*, S1–S18. (in Japanese).

Leipper, D., & Volgenau, D. (1972). Upper ocean heat content of the Gulf of Mexico. *Journal of Physical Oceanography, 2*, 218–224. https://doi.org/10.1175/1520-0485(1972)002%3c0218:HHPOTG%3e2.0.CO;2

Li, R. C. Y., & Zhou, W. (2012). Changes in western Pacific tropical cyclones associated with the El Nino-Southern oscillation cycle. *Journal of Climate, 25*, 5864–5878. https://doi.org/10.1175/JCLI-D-11-00430.1

Li, W., Li, L., & Deng, Y. (2015). Impact of the interdecadal Pacific oscillation on tropical cyclone activity in the North Atlantic and Eastern North Pacific. *Science and Reports, 5*, 12358. https://doi.org/10.1038/srep12358

Li, S., Mei, W., & Xie, S. (2022). Effects of tropical sea surface temperature variability on northern hemisphere tropical cyclone genesis. *Journal of Climate, 35*, 4719–4739. https://doi.org/10.1175/JCLI-D-21-0084.1

Liu, K. S., & Chan, J. C. L. (2020). Recent increase in extreme intensity of tropical cyclone making landfall in South China. *Climate Dynamics, 55*, 1059–1074. https://doi.org/10.1007/s00382-020-05311-5

Liu, K. S., & Chan, J. C. L. (2022). Growing threat of rapidly-intensifying tropical cyclones in East Asia. *Advances in Atmospheric Sciences, 39*, 222–234. https://doi.org/10.1007/s00376-021-1126-7

Madden, R. A., & Julian, P. R. (1972). Description of global-scale circulation cells in the tropical with a 40–50 day period. *Journal of Atmospheric Science, 29*, 1109–1123. https://doi.org/10.1175/1520-0469(1972)029%3c1109:DOGSCC%3e2.0.CO;2

Malan, N., Reason, C. J. C., & Loveday, B. R. (2013). Variability in tropical cyclone heat potential over the Southwest Indian Ocean. *Journal Geophysical Research Oceans, 118*, 6734–6746. https://doi.org/10.1002/2013JC008958

Mawren, D., & Reason, C. J. C. (2017). Variability of upper-ocean characteristics and tropical cyclones in the South West Indian Ocean. *Journal Geophysical Research Oceans, 122*, 2012–2028. https://doi.org/10.1002/2016JC012028

McPhaden, M. J. (1999). Genesis and evolution of the 1997–98 El Niño. *Science, 283*, 950–954. https://doi.org/10.1126/science.283.5404.950

Merchant, C. J. R. P. Allan, & Rmbury, Q. (2025). Quantifying the acceleration of multidecadal global sea surface warming driven by Earth's energy imbalance. *Environmental Research Letters, 20*, 024037. https://doi.org/10.1088/1748-9326/adaa8a

Moon, I., Knutson, T. R., Kim, H., Babanin, A. V., & Jeong, J. (2022). Why do eastern North Pacific hurricanes intensify more and faster than their western-counterpart typhoons with less ocean energy? *Bulletin American Meteorological Society, 103*, E2604–E2627. https://doi.org/10.1175/BAMS-D-21-0131.1

Nirmal, A. C. S., Abhilash, S., Martin, M., Sankar, S., Mohapatra, M., & Sahai, A. K. (2023). Changes in the thermodynamical profiles of the subsurface ocean and atmosphere induce cyclones to congregate over the Eastern Arabian Sea. *Science and Reports, 13*, 15776. https://doi.org/10.1038/s41598-023-42642-9

Physical Sciences Laboratory, 2020: Monthly Climate Timeseries: Pacific Meridional Mode (PMM) SST Index. (2020). (Available online at: https://psl.noaa.gov/data/timeseries/month/DS/PMM/pmm.data, Accessed 12 May 2025)

Physical Sciences Laboratory. (2020). Monthly climate time series: Atlantic meridional mode (AMM) SST index. (Available online at: https://psl.noaa.gov/data/timeseries/month/DS/AMM/amm.data, Accessed 12 May 2025)

Price, J. F. (1981). Upper ocean response to a hurricane. *Journal of Physical Oceanography, 11*, 153–175. https://doi.org/10.1175/1520-0485(1981)011%3c0153:UORTAH%3e2.0.CO;2

Saji, N. H., Goswami, B. N., Vinayachandran, P. N., & Yamagata, T. (1999). A dipole mode in the tropical Indian Ocean. *Nature, 401*, 360–363.

Schreck, C. J., III., Knapp, K. R., & Kossin, J. P. (2014). The impact of best track discrepancies on global tropical cyclone climatologies using IBTrACS. *Monthly Weather Review, 142*, 3881–3899. https://doi.org/10.1175/MWR-D-14-00021.1

Shay, L. K., Mariano, A. J., Jacob, S. D., & Ryan, E. D. (1998). Mean and near-inertial ocean current response to hurricane Gilbert. *Journal of Physical Oceanography, 28*, 858–889. https://doi.org/10.1175/1520-0485(1998)028%3c0858:MANIOC%3e2.0.CO;2

Shay, L. K., Goni, G. J., & Black, P. G. (2000). Effect of a warm ocean ring on hurricane Opal. *Monthly Weather Review, 128*, 1366–1383. https://doi.org/10.1175/1520-0493(2000)128%3c1366:EOAWOF%3e2.0.CO;2

Singh, V. K., & Roxy, K. M. (2022). A review of ocean-atmosphere interactions during tropical cyclones in the north Indian Ocean. *Earth-Science Reviews, 226*, Article 103967. https://doi.org/10.1016/j.earscirev.2022.103967

Sloss, P.W. (1988). ETOPO5 global surface relief. Distributed by the Pacific Islands Ocean Observing System (PacIOOS). (Available online at: http://pacioos.org/metadata/etopo5.html. Accessed 4 Nov 2025)

Terhaar, J., Burger, F. A., Vogt, L., Frölicher, T. L., & Stocker, T. F. (2025). Record sea surface temperature jump in 2023–2024 unlikely but not unexpected. *Nature, 639*, 942–946. https://doi.org/10.1038/s41586-025-08674-z

Vissa, N. K., Satyanarayana, A. N. V., & Prasad Kumar, B. (2013). Intensity of tropical cyclones during pre- and post-monsoon seasons in relation to accumulated tropical cyclone heat potential over Bay of Bengal. *Natural Hazards, 68*, 351–371. https://doi.org/10.1007/s11069-013-0625-y

Wada, A. (2002). The processes of SST cooling by typhoon passage and case study of Typhoon Rex with a mixed layer ocean model. *Papers Meteorology Geophysics, 52*, 31–66. https://doi.org/10.2467/mripapers.52.31

Wada, A., & Chan, J. C. L. (2008). Relationship between typhoon activity and Upper Ocean heat content. *Geophysical Research Letters, 35*, L17603. https://doi.org/10.1029/2008GL035129

Wada, A., & Chan, J. C. L. (2021). Increasing TCHP in the western North Pacific and its influence on the intensity of FAXAI and HAGIBIS in 2019. *SOLA, 17A*, 29–32. https://doi.org/10.2151/sola.17A-005

Wada, A., & Usui, N. (2007). Importance of tropical cyclone heat potential for tropical cyclone intensity and intensification in the western North Pacific. *Journal of Oceanography, 63*, 427–447. https://doi.org/10.1007/s10872-007-0039-0

Wada, A., Usui, N., & Sato, K. (2012). Relationship of maximum tropical cyclone intensity to sea surface temperature and tropical cyclone heat potential in the North Pacific Ocean. *Journal of Geophysical Research, 117*, D11118. https://doi.org/10.1029/2012JD017583

Wada, A., Kanada, S., & Yamada, H. (2018). Effect of air–sea environmental conditions and interfacial processes on extremely intense typhoon Haiyan (2013). *Journal of Geophysical Research, 123*, 10379–10405. https://doi.org/10.1029/2017JD028139

Wada, A., Hayashi, M., & Yanase, W. (2022). Application of empirical orthogonal function analysis to 1 km ensemble simulations and Himawari–8 observation in the intensification phase of Typhoon Hagibis (2019). *Atmosphere, 13*, 1559. https://doi.org/10.3390/atmos13101559

Wada, A., & Yanase, W. (2024). In J. Collins, J. Done, Y.-J. Zhu, P. Wilson (Eds.), Is the increasing trend in the environmental upper-ocean heat content in the Northwest Pacific strengthening tropical cyclone intensity? Hurricane risk in a changing climate (pp. 97–124). Springer Nature. https://doi.org/10.1007/978-3-031-63186-3.

Wada, A. (2012). In I. Yucel (Ed.), *Numerical study on the effect of the ocean on tropical-cyclone intensity and structural change.* Atmospheric Model Applications, In Tech (pp. 43–68). https://doi.org/10.5772/34418

Wang, S., & Toumi, R. (2021). Recent migration of tropical cyclones toward coasts. *Science, 371*, 514–517. https://doi.org/10.1126/science.abb9038

Wang, C., Wang, X., Weisberg, R. H., & Black, M. L. (2017). Variability of tropical cyclone rapid intensification in the North Atlantic and its relationship with climate variations. *Climate Dynamics, 49*, 3627–3645. https://doi.org/10.1007/s00382-017-3537-9

Weng, C. H., & Hsu, H. H. (2017). Intraseasonal oscillation enhancing C5 typhoon occurrence over the tropical western North Pacific. *Geophysical Research Letters, 44*, 3339–3345. https://doi.org/10.1002/2017GL072743

World Economic Forum. (2025). Global Risks Report 2025. World Economic Forum, link. https://www.weforum.org/publications/global-risks-report-2025/

Zhang, X., Xu, F., Zhang, J., & Lin, Y. (2022). Decrease of annually accumulated tropical cyclone-induced sea surface cooling and diapycnal mixing in recent decades. *Geophysical Research Letters, 49*, e2022GL099290. https://doi.org/10.1029/2022GL099290

Zhao, H., Duan, X., Raga, G. B., & Klotzbach, P. J. (2018). Changes in characteristics of rapidly intensifying western North Pacific tropical cyclones related to climate regime shifts. *Journal of Climate, 31*, 8163–8179. https://doi.org/10.1175/JCLI-D-18-0029.1

Zhao, Y., Song, F., Sun, D., Dong, L., & Capotondi, A. (2025). Different roles of Indian Ocean Basin and Dipole modes in tropical Pacific climate variability. *Npj Climate Atmospheric Science, 8*, 163. https://doi.org/10.1038/s41612-025-01065-1

Contemporary Methods for Probabilistic Coastal Compound Flood Hazard Analysis

Aaron J. Dewar, Nathan B. Geldner, and David R. Johnson

Abstract Standard methods, models and input datasets for probabilistic flood hazard analysis vary widely depending on whether the flooding stems from storm surge, rainfall, riverine overflows, or other mechanisms. This has historically posed a challenge to estimating compound flood hazard, where flooding derives from two or more source mechanisms simultaneously or sequentially. Consequently, a wealth of statistical approaches and multi-model frameworks have evolved for analyses of hazard from tropical cyclones in coastal zones and transition zones where hazard shifts from being primarily surge-driven to rainfall-driven. In this chapter, we review the state of the art in compound flood hazard analysis. We outline the basic components of hazard modeling frameworks, describing typical input data, simulation models, and statistical relationships used to estimate hazard associated with multiple flood outcomes (e.g., inundation depths and duration, surge velocity, wave heights). We then also evaluate methods for statistically aggregating observed and/or simulated data to estimate multivariate probability distributions for those outcomes. Finally, the chapter concludes by outlining opportunities for future improvements to current compound flood hazard modeling approaches.

Keywords Compound hazard · Probabilistic risk analysis · Flood modeling frameworks

A. J. Dewar · D. R. Johnson (✉)
Edwardson School of Industrial Engineering, Purdue University, West Lafayette, IN, USA
e-mail: davidjohnson@purdue.edu

N. B. Geldner
Seattle, WA, USA

D. R. Johnson
Department of Political Science, Purdue University, West Lafayette, IN, USA

J. Collins et al. (eds.), *Tropical Cyclone Risk and Resilience*,
https://doi.org/10.1007/978-3-032-28745-8_2

1 Introduction

The floodplain is a representation of flood risk depicting the extent or water surface elevation with a specified annual probability. Traditionally, estimates of the frequency of flooding from all sources (e.g., precipitation, riverine overflows, storm surge, etc.) have been made by fitting observational data sets to extreme value distributions. In cases where water surface elevations are not directly observed, other variables (e.g., rainfall or river stage) are combined with physics-based models to translate such proxy measures to flooding outcomes. Probabilistic studies of flood hazard rely on modeling due to the low spatial coverage and reliability of high-water mark data, as well as the need to estimate tail risks outside of the historic record (i.e., at the extreme tail of the underlying probability distribution). Model-based studies have traditionally focused on representing the predominant source(s) of flooding in an area independently (i.e., pluvial/fluvial inland, storm surge in coastal zones) without connected considerations of the interdependencies between different modes. In the United States (US), for example, floodplains in the coastal zone published by the Federal Emergency Management Agency (FEMA) typically show the risk of coastal storm surge-based flooding or the risk of riverine flooding separately, but not combined.

Treating flood risks separately in the coastal transition zone ignores the interactions between fluvial, pluvial and coastal flood drivers. Coastal transition zones are the areas between land and ocean where, for example, waves and storm surge converge with river discharge and heavy rainfall, i.e., where flood hazard transitions from being primarily surge- and wave-driven to primarily pluvial and/or fluvial. The interactions between these drivers can lead to greater flood risks than the sum of their individual impacts.

However, the disparities between floodplain maps based on traditional analyses and actual observations have become greater, indicating the importance of accounting for compound events where multiple sources of flooding co-occur (Feng et al., 2023; Ganguli et al., 2020; Gori et al., 2020; Hendry et al., 2019). Climate change can also produce nonstationarity in the distribution of flood hazards, invalidating or introducing bias in estimates from studies that assume a stationary historical record that also carries forward. Both of these factors can lead to substantive inaccuracies in risk analyses. A study of the insurance claims in the Clear Creek watershed showed that over half of the insurance claims in an 11-year period fell outside the floodplain boundaries, demonstrating the deficiencies of floodplain maps for coastal Texas (Brody et al., 2013). The disparity between the 100-year FEMA floodplain map (i.e., area with at least a 1-in-100 chance of flooding in a given year) and flood loss outside the floodplain map can exceed fifty percent in flood-prone areas, and approximately twenty-five percent of insured losses occur outside of specified floodplains in the US (Blessing et al., 2017). A US Department of Homeland Security study found that 58% of FEMA flood maps were out of date and potentially inaccurate at the time of their analysis, despite regulations specifying that maps should be evaluated every five years to determine if they should be updated (US Department of Homeland Security, 2017); illustrating this point, some communities affected by Hurricane Harvey in 2017 were living under flood maps effective from the 1980s (US Department of Homeland Security, 2023). Finally, a study of the uncertainty in flood exposure associated with randomness in the year flood maps are updated found that the irregular frequency of updates could lead to many thousands of individuals in Louisiana alone being misclassified as being inside or outside of the best-estimate 100-year floodplain (Meyer & Johnson, 2019) which is used to inform flood insurance regulations; thousands of homeowners may be unintentionally required to carry flood insurance or be given a false sense of security, depending on the type of misclassification.

Random Variables Versus Random Fields

Probabilistic compound flood hazard models represent uncertain variables pertaining to the outcomes of discrete flood events (e.g., peak inundation) as random variables or as random fields. Deeper discussion of the properties of random variables and random fields is outside of the scope of this chapter, and can be found in several recently released textbooks (Christakos, 2017; Hristopulos, 2020). For the purposes of this chapter, a random variable is to a scalar quantity as a random field is to a scalar-valued surface: the peak flood depths of a discrete storm event mapped over some geographic region can be represented as a 2D random field, and the time series of depths across the entire region can be represented as a 3D random field. For every set of 2D coordinates in the region, the peak flood depth of a discrete storm event is a random variable, and the time series of flood depths is a random field. At any fixed, 2D coordinate, the time series is a random scalar function.

Probabilistic compound hazard models represent uncertain quantities pertaining to outcomes of discrete flood events (e.g., peak inundation) as random variables or random fields. These can encompass parameters of a tropical/extratropical cyclone (either over time or at landfall) or convective storm (e.g., atmospheric instability and wind shear), parameters defining the antecedent conditions within the study domain (e.g., soil moisture, riverine baseflow rates), or parameters defining the domain's boundary conditions at the time of an event (e.g., mean sea level, astronomical tide, transient riverine flow rates). They can more generally include arbitrary functions of input parameters which characterize physically relevant processes, such as riverine discharge estimates from a computationally inexpensive model (e.g., a basin-level rainfall-runoff or flow routing model). The frequency of these discrete events can be modelled as a constant scalar frequency parameter. It can also be modelled as a periodic function representing intra-annual or predictable and periodic changes in frequency, i.e., hurricane frequency is only non-zero in and around hurricane season, may vary over the course of hurricane season, and is affected by El Niño and La Niña. It can also be modelled as an arbitrary function encoding both periodic variability and climate non-stationarity. The choice of how to encode these quantities as random variables, functions, or fields is necessarily circumscribed by data availability and format. Fundamentally, a probabilistic flood hazard modeling framework involves assessing the relative likelihoods of storm events that span a space of plausible storm parameters, estimating the outcome(s) of interest associated with those storm events, and statistically aggregating them to estimate the probability distribution(s) of the outcomes of interest.

In the past, flood hazard analyses typically did not explicitly address joint hazards. They often addressed the less technically demanding problem of estimating annual exceedance probability (AEP) distributions of single hazard-drivers. For example, a univariate model of rainfall hazard can be characterized in isolation from other flood mechanisms by estimating the AEP distribution of millimeters of rainfall over a 24-h period; a univariate model of fluvial hazard can be expressed as the AEP distribution of streamflow rates or river stage. Practitioners should exercise caution before assuming univariate methods are appropriate even in single-driver contexts. For example, a univariate model of pluvial hazard focused on 24-h rainfall peaks provides no guarantee of capturing flashy dynamics driven by 3-h peaks and is therefore liable to underestimate peak flood depths in areas dominated by flash-flooding dynamics. More generally, univariate hazard analyses by necessity are suitable only for estimating quantities that are multiple steps away from inundation, loading or other outcomes more relevant to engineering design or policy analyses. In reality, these outcomes are typically driven by multiple hazards.

In this chapter, we review the state of the art in compound flood hazard analysis. Section 2 outlines the basic components of hazard modeling frameworks, describing typical input data and simulation models; Sect. 3 details the statistical relationships used to estimate hazard associated with multiple flood outcomes (e.g., inundation depths, surge velocity, wave heights). Section 4 describes methods for statistically aggregating observed and/or simulated data to estimate multivariate probability distributions for those outcomes. Finally, the chapter concludes in Sect. 5 by presenting opportunities for future improvements to current compound flood hazard modeling approaches.

2 The Anatomy of a Probabilistic Compound Flood Hazard Model

Purely empirical analysis is unable to effectively estimate compound flood depths associated with return periods on the order of 100 years or more, especially when considering flooding associated with relatively rare events like TCs. There are more possible storms with distinct combinations of features with the potential to lead to catastrophic flooding than have ever been recorded in any one location due to finite periods of record. For this reason, parametric and semi-parametric models are used to characterize the likelihood of each possible combination of features.

These statistical models are typically continuous, and, therefore, characterize the likelihood of an uncountably infinite number of storms. However, analytical solutions capable of mapping the entire space of possible storms to their corresponding outcomes via fundamental physics are unavailable. Analysts are then left with two options, both of which require a physical simulation model of the system. They can (a) generate discrete sets of storm events for simulation and assign each of them a corresponding probability mass, with the intention of approximating the true, continuous distribution of outcomes as closely as possible, or (b) generate discrete sets of storm events for simulation and use them to train a metamodel efficient enough to permit numerical integration of estimated flood depth outcomes over the continuous distribution of storms. The anatomy of a probabilistic model of flood risk can therefore be broken down as follows (although other perspectives exist):

- A physics-based simulation, typically a finite element hydraulic model, which maps forcings to flood outcomes such as flood depths (expressed as scalar peaks or time series) or other possibly salient features such as velocity, salinity, or sediment flux.
- A scheme for using the simulation, which only operates on a discrete set of flood events, to represent the flood response of the (usually) uncountably infinite set of storm events whose likelihoods are characterized by the probabilistic model of storm events and their associated forcings.
- A probabilistic model of storm events and their associated forcings which can be used as inputs to hydraulic simulation models, which either explicitly or implicitly characterizes storms as continuous random vectors or fields of arbitrary dimension.

This chapter addresses these three components in order, from the relatively "concrete" physics-based simulations to the more abstract probabilistic models of storm events and associated forcings and boundary conditions.

2.1 Finite-Element Hydrodynamic Simulation of Discrete Flood Events

Finite-element hydrodynamic simulations can be 1D, 1.5D, or 2D, although 2D simulations are most appropriate for compound flood risk analysis. A 1D hydraulic simulation partitions a river channel into discrete segments and numerically solves a simplified 1D set of Navier–Stokes equations over those segments, without accounting for dynamics outside of the river channel. A 1.5D hydraulic simulation is effectively a 1D simulation along the river channel which has been coupled to additional model components to represent the behavior of water spilling out of the river channel; it does not generally support sheet flow outside of river systems, however. A 2D hydraulic model carves the model domain into polygonal cells (together referred to as a mesh), and assumes some level of homogeneity in flood depth, velocity, etc., within each cell. A simplified set of Navier–Stokes equations are applied to model the fluxes of water across the faces of the polygonal cells. 2D hydrodynamic simulation packages used in probabilistic flood risk analysis include the ADvanced CIRCulation (ADCIRC); the Sea, Lake and Overland Surge from Hurricanes (SLOSH); the 2D Hydrologic Engineering Center's River Analysis System (HEC-RAS); the Super-Fast INundation of CoastS (SFINCS); the Adaptive Hydraulics (AdH); and the D-Flow Flexible Mesh models. Note that strictly speaking, ADCIRC and SLOSH do not simulate wave behavior but are often coupled (Dietrich et al., 2012; Massey et al., 2011) with wave models such as the Simulating Waves Near-Shore (SWAN) model (Booij et al., 1999) or Steady State Spectral Wave (STWAVE) model (Massey et al., 2011). Finite element hydraulic models can also be steady or unsteady, in terms of whether flow depths and velocity in channels are constant or variable with time, but unsteady solutions are the ones most relevant to the modeling of discrete compound flood events (Jelesnianski, 1992; Luettich Jr et al., 1992; Brunner, 2020; Sebastian et al., 2021). While other modeling packages exist, much of the discussion to follow discusses considerations common across many if not all salient modeling packages; the aforementioned modeling packages will be used as illustrative examples where appropriate to highlight variability in features and assumptions across commonly used software.

Finite-element hydraulic simulations are typically constructed using Digital Elevation Model (DEM) data derived from surveys, lidar, or satellite imagery. Additional information pertaining to land use, land cover, preliminary estimates of surface roughness, vegetation, soil type, etc. are all used to determine the salient hydraulic parameters associated with each cell. These are frequently used as "tuning" or calibration parameters during model calibration.

The process of constructing finite-element hydraulic models for flood risk varies dramatically between different modeling packages. HEC-RAS and SLOSH are principally graphical user interface (GUI)-based and offer limited support for script-driven workflows. ADCIRC and SFINCS are principally tools for script-driven analysis and offer limited support for GUI-driven workflows. The benefits of script-driven analysis include greater reproducibility and reduced labor overhead. GUI-driven model development can easily suffer from reproducibility issues due to a large number of manual adjustments made to model elements and greater requirement for user inputs. The chief benefit of GUI-driven model development is related to the ease of adjusting model elements, e.g., adjustments to the model mesh which are often necessary due to the performance limitations of automated mesh generation algorithms. These are most easily implemented in a click-and-drag fashion in a GUI. Further work remains to be done to develop modeling packages which offer robust support for both development modalities.

After the model is created as a function of landscape data, the model is typically calibrated. This means that certain landscape attributes associated with the cells of the model mesh, generally those which are estimated with low confidence from landscape data such as Manning's roughness coefficient (Medeiros et al., 2012) are adjusted such that simulated flood outcomes more closely resemble measured flood outcomes across a set of training events for which measured input and output values (e.g., high-water marks) are known. Out-of-sample performance can then be estimated in a process referred to as model validation, typically conducted by calibrating the model on only a subset of available events and/or observed outputs and retaining the remaining out-of-sample events for comparison. Alternatively, analysts can use resampling methods such as k-fold cross validation, in which different versions of the model are trained on different, overlapping subsets of the training events in order to estimate out-of-sample validity, and the final model is calibrated to all available calibration storms.

It is worth noting that model calibration and validation, like model construction, varies significantly between GUI-driven and script-driven workflows. HEC-RAS, which principally supports GUI-driven workflows, does not currently support automatic calibration over more than one event (Brunner, 2020). Calibration over multiple storms must therefore involve laborious and difficult-to-reproduce manual tuning, or 3rd party software specifically designed to add functionality not intended by the developers (Abouelsaad et al., 2024; Dysarz, 2018; Philippus et al., 2021). By contrast, SFINCS, which principally supports script-driven workflows, does not natively support any sort of calibration or validation methods but can be driven in Python via the HydroMT package (Eilander et al., 2023). Automatic calibration can be handled via optimization methods supported by commonly used open-source Python packages such as SciPy (Virtanen et al., 2021). Cross-validation can be handled by commonly used open-source Python packages such as Scikit-Learn (Pedregosa et al., 2011).

2.2 *Physical Input Data for Storm Hazard Models*

Another critical criterion in selecting a finite element simulation modeling package is the range of input data and physical dynamics supported by that package. ADCIRC and SLOSH take spatiotemporal wind and pressure fields as a primary input and are mostly used for storm surge modeling (often coupled to a wave model). They have not traditionally supported rain-on-grid (i.e., direct addition of variable rainfall over time at the grid scale) or soil infiltration but do support upstream and downstream boundary conditions and resolve the impacts of wind-driven flow in greater detail than other models; recent modifications to ADCIRC allow the incorporation of precipitation (Szpilka et al., 2022). They are therefore currently unsuited for primary use in compound flood hazard modeling where pluvial effects are salient; they can however be used to resolve downstream boundary conditions to be fed into other models such as SFINCS or 2D HEC-RAS. By contrast, SFINCS and 2D HEC-RAS support rain-on-grid and soil infiltration as well as upstream and downstream boundary conditions. However, recent releases of HEC-RAS do not permit users to apply nonzero velocity (equivalently, momentum) to water introduced at boundary conditions, nor to include other physics like wave radiation stress gradients (important in coastal zones); they may therefore systematically over- or underestimate the impact of surge effects applied at the downstream boundary.

In many cases the choice of simulation follows directly from the desired input data; a purely fluvial model driven by flow frequency analysis of an upstream outlet would be best served by a simulation that resolves channel dynamics and control structures in high detail, while a purely pluvial model would be best served by a model that resolves rainfall-runoff dynamics and overland sheet flow in high detail. For compound flood risk applications, the authors broadly recommend the use of a simulation designed for coastal hazard such as ADCIRC or SLOSH to resolve any salient coastal forcings at the downstream boundary of the region of interest, and hydraulic simulations of the region of interest supporting rain-on-grid, wind forcing, upstream and downstream boundary conditions which may represent coastal and riverine inputs, soil infiltration, etc., such as HEC-RAS and SFINCS. Other modeling packages, such as AdH, allow for coupling in coastal and riverine flood transition zones. Models also vary in their computational requirements; this is partly due to the application-driven choice of 1D, 1.5D, or 2D configuration, but other factors may outweigh native differences in model performance. These include the necessary mesh resolution; the number of events, landscapes, and uncertainties in an experimental design to support planning; and the size of the model domain.

A non-exhaustive summary of common input parameters to models used in probabilistic compound flood hazard analyses is found in Tables 1, 2 and 3. Note that not all models use all parameters, and some can be derived directly from others. Other important inputs include the characteristics of flood control structures and other built infrastructure, as well as assumptions about their performance under extreme loads and the logic used for operating them (e.g., scheduled flow releases, regulations on floodgate closures).

Coastal models are a unique case that bears further scrutiny. They are less often driven by empirical frequency analysis and are more often driven by a multivariate probability distribution of several parameters characterizing discrete TC events such as central pressure, radius of maximum winds, and forward velocity as well as track information. The parameterization of each TC must then be mapped to a spatiotemporally explicit wind and pressure field before it can be simulated via finite-element hydraulic simulation.

2.3 Wind Models

The wind model used in most, if not all, of the largest coastal flood risk modeling exercises in the United States is the Planetary Boundary Layer (PBL) model owned by Ocean weather Incorporated (OWI) (Cialone et al., 2015; Fischbach et al., 2017; Nadal-Caraballo et al., 2022; Johnson et al., 2023; Bartlett et al., 2023). It is described as principally an analytical model of wind fields as a function of several free parameters (Cardone & Cox, 2009), but its precise behavior is hard to verify due to the model's closed-source, proprietary status. The most recent available validation exercise found that the PBL model, applied as a hindcast of Hurricane Katrina without

Table 1 Commonly modeled meteorological parameters for tropical cyclones

Characteristic	Input variables	Notation	Units	Temporally variable	Description
Wind	Radius of maximum winds, Average wind speed, Surface Wind speed, etc[1]	r_{max} (radius); ff (speed)	km, nm; kts, km/hr	✓	Distance from the center of a tropical cyclone (TC) to the point of maximum windspeed; velocity of winds at various locations or aggregated in various ways
Intensity	Minimum central pressure	c_p	mbar, hPa	✓	Minimum atmospheric pressure at the center of a TC
Intensity	Maximum Wind speed	v_w	km/hr, kt	✓	Maximum sustained windspeed of a TC
Intensity	Central pressure deficit	ΔP	mbar	✓	Difference between ambient pressure and central pressure
Intensity	Rainfall intensity		mm/hr	✓	Amount of rainfall per unit of time
Position	Latitude	y	Degrees	✓	Angular distance north or south of the equator
Position	Longitude	x	Degrees	✓	Angular distance east or west of the prime meridian
Velocity	Forward velocity	v_f	kts, km/hr	✓	Translational speed of the storm in the direction of its heading
Direction	Heading	θ	Degrees	✓	Direction of storm movement
Temperature	Temperature	T	°F, C	✓	Ambient air or water temperature

augmentation from measured wind data, underestimated peak surge at the one available gage station by just over 30% (Cardone & Cox, 2009). Substantial research has been conducted in the area of TC wind field modeling in the time since OWI's proprietary model was developed (Kepert, 2023; Olfateh et al., 2017; Ruiz-Salcines

[1] Other commonly used components include quantities such as peak gust wind speed, highest instantaneous wind speed, u- v- wind component, instantaneous 10 wind gust surface northward (v- component), surface eastward (u- component), 10 m wind speed, and maximum wind speed.

Table 2 Commonly modeled landscape parameters

Characteristic	Input variables	Notation	Units	Temporally variable	Description
Topography	Elevation		m	Static	Height above a reference point (e.g., mean sea level, NAVD88)
Topography	Soil slope	–	Degrees	Static	Inclination of soil surface
Bottom roughness	Manning's n	*n*	–	Static	Dimensionless coefficient representing bottom roughness in open channel flow
Surface roughness	Free surface roughness	*z*0	–	Static	Surface roughness affecting water flow
Land use	Land cover	–	Categorical	Static	
Surface canopy	Surface canopy coefficient		–	✓	Reduction in wind stress on water surfaces produced by vegetation
Soil	Soil type	–	Categorical	Static	Soil type classification
Soil	Hydraulic conductivity	–		Static	Capability of soil to absorb water
Soil moisture	Degree of saturation	–	–	✓	Proportion of soil saturated with water
Drainage	Drainage density			Static	Density of drainage network

Table 3 Commonly modeled boundary conditions

Input variables	Units	Temporally variable	Description
Mean sea level	m	✓	Average height of the sea surface over a specific period
Sea level rise	m	✓	Increase in average sea level over a time period
Astronomical tide	m	✓	Natural periodic rising and falling of the sea level due to astronomical influences
Streamflow	cfs	✓	Flow of water in a natural or artificial channel
Radiation stress	–	✓	Surface wave radiation stress gradients (i.e., momentum fluxes) from wave models

et al., 2019; Yan & Zhang, 2022; Young, 2017), but it remains in common use for consistency with prior or neighboring studies.

Another influential, physics-based wind field model used in hurricane and flood risk analysis is the model put forward by Vickery (Vickery et al., 2000). This wind field model uses the full solution to the equations of motion for translating hurricanes and solves for the PBL using a finite-difference scheme. It accounts for sea surface roughness and air-sea temperature differences by altering the peakedness of the wind field; a Fourier fitting approach can then be used to characterize wind fields with a few parameters.

2.4 Semiparametric Rainfall Generation

Several of the modeling frameworks described previously model large-scale behaviors of TCs and map them to explicit wind fields, but do not account for variability in rainfall. Accounting for rainfall can be difficult, because it is difficult to predict and highly stochastic. One frequently used empirical model is the Rainfall Climatology and Persistence (R-CLIPER) Model, which requires only the position of the storm and maximum wind speed to model rainfall, but which fails to account for vertical wind shear when producing precipitation fields (Marks & DeMaria, 2003; Tuleya et al., 2007). The Parametric Hurricane Rainfall Model (PHRaM) improves upon the R-CLIPER model by including the effects of topography and wind shear, thus removing the assumption of azimuthal symmetry (Lonfat et al., 2007).

The Interagency Performance Evaluation Taskforce (IPET) Rainfall Analysis employed a parametric model to produce a symmetric rainfall field (Interagency Performance Evaluation Taskforce, 2009). The symmetrical component of mean rainfall, is represented as:

$$m_I(r) = \begin{cases} 1.14 + 0.12\Delta P, & r \leq R_{max} \\ (1.14 + 0.12\Delta P)e^{-0.3\left(\frac{r-R_{max}}{R_{max}}\right)}, & r > R_{max} \end{cases}$$

where ΔP is the central pressure deficit, r is the distance from the hurricane center and R_{max} is the radius of maximum wind speed. The rainfall field, $m_I(r)$, is multiplied by a factor of 1.5 if it is to the right of the direction of motion and remains unaltered otherwise. This multiplication provides a crude method of incorporating the effect of wind shear on rainfall quantities.

A probabilistic approach to precipitation forecasting requires the quantification of uncertainty, which is an essential part of decision-making for emergency preparedness. Building on the IPET model, Villarini et al. (2022) developed a probabilistic rainfall generator to generate large ensembles of rainfall for landfalling TCs for the state of Louisiana. The relationship between reference data (i.e., Stage IV, a gridded rainfall reference product based on rain gauge measurements and weather radar data) and IPET predicted rainfall for 12 landfalling TCs was represented by a multiplicative error model composed of a deterministic component, to correct for rain-dependent bias, and a stochastic component for residual fluctuations in rainfall. The marginal distribution of multiplicative rainfall error at each pixel was modelled as a mixture of four Gaussians with an empirically derived linear measure of rank-correlation between pixels enforced via meta-Gaussian copula. It should be noted however, that these error fields are generated for pixel-wise rainfall totals rather than rates of rainfall over time. As a consequence, any time series data generated by this model reflects a pixel-wise scalar multiple of the IPET model, and, therefore, has the potential to underestimate short-duration peaks driven by the visually observable rotating radial bands of precipitation observed in real TCs. The Louisiana Watershed Initiative utilized this rainfall ensemble and downstream boundary conditions from ADCIRC as inputs to over 300,000 HEC-RAS simulations for a pilot study estimating compound flood hazard in the state's Amite River Basin (Bartlett et al., 2023).

Other contemporary models include a Bayesian forecasting system (BFS) developed for precipitation forecasting with relative air humidity used as a meteorological indicator (Reggiani & Weerts, 2008). This system uses a binary model to determine what areas will be subject to rainfall and a continuous model to approximate total rainfall in areas where rainfall is predicted to occur. The BFS performs better in predicting the occurrence of rainfall than in predicting rainfall depths. As another example, the developers of the Hurricane Intensity and Track Simulator (HITS) algorithm extended their model with stochastic precipitation fields (Nakamura et al., 2024). This model performed well in predicting precipitation peaks, but in some cases underpredicted non-peak rainfall. Still another rainfall model is the Tropical Cyclone Rainfall Model (TCRM), a physics-based model of hurricane precipitation that uses atmospheric and topographic parameter inputs to produce results that agree with Weather Research and Forecasting (WRF) rainfall projections (Lu et al., 2018). Further, TCRM can be coupled with hydrological models to produce accurate estimates of flood peaks.

In transition zones subject to compound flooding, rainfall models must be appropriately convolved with models of other drivers such as streamflow and storm surge. The spatiotemporal dependence of rainfall and other hazards must be rigorously evaluated in order to ensure the accuracy and validity of models of compound flooding (Keef et al., 2009; Kim et al., 2023).

2.5 Rainfall-Runoff Relationships

Integration of rainfall into hydraulic simulations requires a characterization of rainfall-runoff dynamics, or the respective degrees to which rainfall is absorbed by the soil and to which it is not absorbed and "runs off" to flow downhill. Runoff occurs both when soil is impermeable or fully saturated, and when the rate of rainfall exceeds the maximum rate at which the soil can absorb water. A variety of rainfall-runoff model types are employed for water resource management, ranging from purely conceptual models, to those which explicitly model physical processes either deterministically or stochastically, to those based entirely on empirical relationships without consideration for the underlying dynamics of physical processes. Many models occupy intermediate positions between these three perspectives (Jehanzaib et al., 2022). Runoff can directly lead to flooding or compound with other drivers.

Probabilistic and spatiotemporally explicit rainfall models are therefore necessary to adequately characterize the probability distribution of flood outcomes (Bartlett et al., 2025b).

3 Connecting Discrete Event Simulations to Continuous Probability Models

As noted previously, a finite-element hydraulic simulation can be used in two primary ways to convert a probabilistic model of compound flood forcings to a probability distribution of flood outcomes.

1. The distribution can be discretized for simulation: a countable set of storms is generated for simulation, usually through an optimal or adaptive sampling scheme, and probability masses are assigned to them in a manner to best approximate the continuous distribution.
2. A response surface can be used: a countable set of storms is simulated, the results are used to train a highly efficient metamodel that interpolates outcomes between the simulated storms, and this response surface is used to numerically integrate flood outcomes over the continuous distribution.

In this section, we discuss various approaches for identifying a sample of events to analyze, distinguishing between schemes where events stand alone as a discrete set of simulations or data and schemes where sampled events inform development of a response surface. Finally, we also describe the current state of research on response surfaces in the context of estimating compound flood hazard from TCs.

3.1 Sample Identification

The first step in sampling events for hazard modeling is the selection of thresholds to distinguish what constitutes a hazardous event. For single-hazard analysis, only one threshold is required (e.g., 1-h rainfall greater than x), but compound extreme event analysis contains multiple concurrent or consecutive hazards that interact (Sadegh et al., 2018). As such, compound flood hazard analysis requires consideration of multiple hazard thresholds and the relationships between the hazards and extreme events. The selection of thresholds (or more generally, inclusion and exclusion criteria) impacts every subsequent step in building the flood risk analysis framework. This is complicated by interactions between hazards, wherein two hazards that are both below their respective thresholds might occur simultaneously and produce the same or more impact than a qualifying single-hazard event would have produced. Once criteria have been established to identify qualifying events, analysts then must also choose what events to simulate and the modeling platform to best simulate the event outcomes.

Levee design in the 1960s was based on estimated system performance evaluated against a single, relatively extreme Standard Project Hurricane (U.S. Army Corps of Engineers, 1965). The notion of a reference Standard Project Flood, or "design storm(s)," is still used for flood planning in some cases but is vulnerable to inaccurate estimation of the chosen design load's return period or the multiplicity of events that could generate a design load but otherwise still have consequential differences. As computers have become more powerful and methodologies have become more sophisticated, alternative approaches have been employed to evaluate an ever-larger number of storms. It should be noted, though, that high-fidelity, state-of-the-art models have also leveraged computing advances and become increasingly expensive; this introduces a tradeoff between the number of simulations to run and the accuracy of models under a fixed computational budget. Several methods exist to provide guidance on this aspect of flood risk assessment. Due to the relatively short period of historic storm records to draw upon in a given geographic region, each method also addresses the challenges of limited data. These methods include (Resio et al., 2017b):

Joint Probability Method (JPM)—a method that uses a set of TC parameters, including central pressure, radius of maximum winds, storm forward velocity, and the angle of the storm track relative to the coast, together with a multivariate probability distribution to describe the relative likelihood of possible storms (Resio et al., 2009). Feasible parameter combinations are selected and the corresponding storm surge is then simulated. This provides useful guidance to generate a suite of synthetic TCs that span the range of plausible storm characteristics, but a drawback of this method is that JPM requires the simulation of thousands of synthetic storms (Yang et al., 2019).

Historical Storm Method (HSM)—a method that parameterizes historical data to obtain an empirical cumulative distribution function or fit to a parametric distribution for storm surge elevation (or whatever outcome variable is to be analyzed). This approach is analytically simple, but it is severely limited by the set of historic events, likely to be much smaller than needed to empirically approximate a smooth function in coastal settings, much less to cover the range of potential flood outcomes. Meyer and Johnson (2019) found that tens of thousands of people in coastal Louisiana may be misclassified as inside or outside of the 100-year floodplain due to stochasticity in the historical record and the year in which an analysis is done. Another drawback is that parameterized distributions may not always provide a good fit to historic outcomes (i.e., a realistic approximation of the underlying distribution).

Empirical Simulation Technique—a method similar to HSM, but which incorporates bootstrapping (sampling with replacement). It does not assume any parametric relationships but instead "uses the joint probability relationships inherent in the database" (Scheffner et al., 1996). It utilizes bootstrapping and nearest neighbor interpolation to obtain larger datasets than HSM, making it more appealing from a statistical standpoint; however, this assumes that current events will be similar in size and frequency to the original dataset (i.e., nonstationarity in TC parameters) and that the historic record is broadly representative of the underlying set of plausible events. Identical storms that occur in a bootstrap sample may or may not need to be simulated multiple times, depending on whether the risk study involves running events in time series under changing or state-dependent conditions (e.g., sea level rise, prior storms altering landscape morphology).

The Stochastic-Deterministic Track Method (SDTM)—a physically-based method that randomly seeds TC formations rather than adhering to predefined tracks. Empirical relationships determine parameter values and storm tracks for each simulation (Emanuel et al., 2006). Since SDTM includes physics relationships within it, it was intended to be better at simulating storms than statistical approaches or the use of idealized synthetic events. However, this SDTM requires long periods of time to run, or large numbers of events or simulated years, potentially limiting its practical usage if the simulations are computationally expensive.

Joint Probability Method with Optimal Sampling (JPM-OS)—an improved version of JPM that prescribes a "small" subset of synthetic storm events to approximate the information within the full parameter space of plausible events, using the input parameters of central pressure, radius of maximum windspeed, forward velocity, location of landfall, and landfall heading (Resio, 2007). The smaller suite of events to simulate can be identified using techniques such as Bayesian quadrature (JPM-OS-BQ) or developing a set of events to train a response surface to approximate outcomes over the full space (JPM-OS-RS) (Niedoroda et al., 2012; Resio et al., 2009). When compared with a baseline given by the SLOSH model, both approaches were shown to closely match the reference model (Niedoroda et al., 2012). The Probabilistic Coastal Hazard Analysis (PCHA) framework has employed JPM-OS with Gaussian process metamodels to predict storm hazards from storm parameters while maintaining high fidelity (Nadal-Caraballo et al., 2020a). Thus, the models have provided suitable estimates of the joint probability of surge inundation. The Coastal Louisiana Risk Assessment model has modified JPM-OS to better propagate parametric uncertainties as described in more detail in Sect. 4.4. To date, however, the optimal sampling procedure is based on the set of TC parameters at landfall (or closest bypassing point to landfall), assuming a single landfall and ignoring subsequent time periods.

A summary of the basic JPM-OS-RS procedure is shown in Fig. 1. Past events from the historical record are fit to a joint probability function using maximum likelihood methods and structural assumptions about the family of distributions appropriate for each TC parameter. A response surface trained on a corpus of hydrodynamic simulations is used to predict surge, wave, and potentially other (e.g., rainfall) characteristics for a suite of synthetic storms. The flood outcomes associated with those synthetic events are combined with estimates of their relative likelihoods from the joint probability function, and the underlying frequency of storm events, to generate an AEP distribution for the outcomes of interest (Fig. 1).

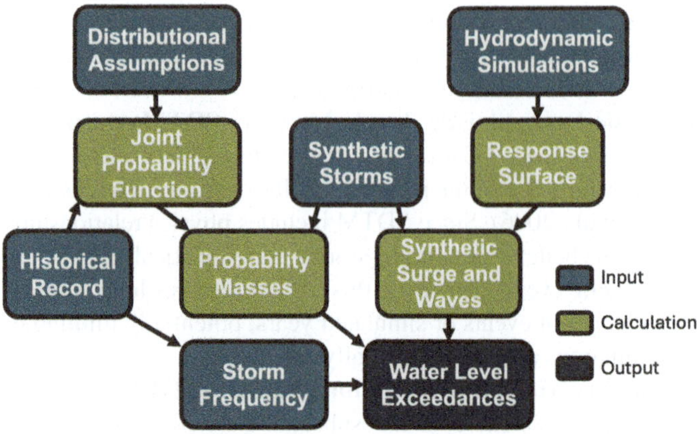

Fig. 1 Conceptual workflow of the JPM-OS-RS procedure

3.2 *Discretization and Optimal Sampling*

Generation of discrete storm events is typically achieved through some combination of altered historic storms (e.g., with shifted track or under different antecedent conditions), synthetic storms (i.e., TCs that follow an idealized smooth track and evolution of parameters over time), storms produced by a physical process-based model, or direct Monte Carlo sampling of the continuous probabilistic model of storm events. These approaches are sometimes accompanied by "optimal sampling" methods. Optimal sampling for discretization selects storm events to simulate in order to minimize the difference between the discrete and continuous distributions. Optimal sampling for response surfaces selects storms to simulate in order to maximize the accuracy of the resulting response surface.

There has been relatively little research in optimal sampling for discretization, with the canonical method being via Bayesian quadrature. This approach assumes a covariance structure between modeled drivers and storm-parameter-conditioned flood depth exceedance probability, guided by expert opinion. This covariance structure is generally assumed to be constant across depth thresholds and storm parameter space. It relies on that assumption to identify an "optimal" set of flood events for simulation with associated probability weights. This assumed covariance structure is trivially incorrect—for a given depth threshold, there exist broad regions of TC parameter space in which the true covariance is locally zero because the parameter-conditioned exceedance probability is effectively zero or one (Toro et al., 2010).

Clustering-driven optimal sampling is a recently developed alternative. It requires clustering on a set of Monte Carlo draws of TC events expressed as vectors of hydraulic model inputs and derivative values thereof expected to have a smooth functional relationship with flood severity, accompanied by appropriate standardization, *k*-means clustering, and assignment of total cluster probability masses to the cluster-nearest centroids (i.e., medoids). This method is guaranteed to converge to a discretization that produces integrated L2 error in the random field of flood depths within a finite neighborhood of the global optimum with radius dependent on smoothness conditions (Bartlett et al., 2023).

The JPM-OS-BQ approach has shown several benefits over the original JPM approach. With a storm ensemble of 149, the Bayesian quadrature approach was shown to closely match the JPM approach with an ensemble of 4108 storms for the Mississippi Coastal Analysis Project (Federal Emergency Management Agency, 2008). This method employs Bayesian quadrature to discretize the joint probability of distribution of TC parameters which leads to a smaller set of synthetic TCs that are still representative of the larger set. In practice, the JPM-OS-BQ approach obtained a near-zero (− 0.006 m) average error when predicting the 100-year AEP still water level (Gonzalez et al., 2019).

3.3 Response Surfaces and Optimal Sampling

The literature on the use of response surfaces or metamodels to numerically integrate estimates of flood response over continuous probabilistic models of storm events is rich and varied. A response surface can be as simple as a multivariate ordinary least squares regression with a few regression coefficients and an intercept, or it can be as complex as a deep neural network leveraging advanced computer vision techniques.

One promising form of surrogate modeling is Gaussian Process Regression (GPR), also referred to as kriging (Fischbach et al., 2017; Kyprioti et al., 2021b; Zhang et al., 2018). Kriging can approximate computationally expensive functions while taking spatial correlation and uncertainty quantification into account. The uncertainty is captured by the distribution over the functions, and spatial correlations are captured by the covariance term (i.e., spatial variogram). Kriging metamodels in conjunction with principal component analysis (PCA) have successfully approximated hurricane responses (Jia & Taflanidis, 2013; Jia et al., 2016). Since TC outcomes consist of high-dimensional output across multiple locations, PCA was considered to account for spatial and temporal correlation of the output. Separate metamodels for each latent output from PCA were built, which permitted greater accuracy than would a single metamodel. However, over large geospatial grids wherein hundreds of thousands of grid points need hazard curve estimations, *k*-means clustering might be incorporated with PCA and kriging by leveraging geospatial distributions and correlations (Kyprioti et al., 2021a). This is done by aggregating similar nodes through *k*-means clustering and then interpolating via kriging, with PCA used to reduce dimensionality of the input.

A comparison of artificial neural networks (ANN) (Chen et al., 2012), GPRs–typically kriging—and support vector regression (SVR) showed interesting properties of each method (Al Kajbaf & Bensi, 2020). All three models showed increases in errors such as RMSE when predicting surge height for greater values of surge, though ANNs and GPR models were not as strongly impacted as SVR models. Errors also increased when many surge values close to zero relative to local elevation (i.e., within the local variability in elevation from a DEM relative to the elevation assigned to a given mesh cell) were included, highlighting a need for approaches to dry-node correction (Kyprioti et al., 2021c; Nadal-Caraballo et al., 2020a; Shisler & Johnson, 2020). Two truncation methods were employed to remove the effect of the many near-zero surge values. Keeping the small surge values in the training datasets but truncating small surge values for the test dataset improved results for ANN and GPR. However, the SVR models saw improved accuracy when the dataset was truncated before testing. GPR was shown to be the most robust to changes in training set size, as errors were consistently smaller for GPR than for ANN or SVR when the training set size was reduced.

Many other studies have focused on the use of surrogate models for forecasting future water surface elevations over the course of a storm (De Oliveira et al., 2009; Kim et al., 2019; Lee, 2006, 2009; Rajasekaran et al., 2008; Sztobryn, 2003). Recent works have incorporated sea level rise into predictions on a static landscape (Kyprioti et al., 2021b), but landscape morphology also plays a significant role in modeling flood inundation and flood risk accurately (Bates et al., 2010). Gharehtoragh and Johnson (2024) addresses this by incorporating temporally variable landscape features (e.g., topo/bathy elevation, roughness) and boundary conditions (e.g., mean sea level) to train surrogate models that can infer changes to storm surge under future environmental change; we note, however, that using estimates of future landscape morphology change introduces additional uncertainties into the overall modeled system which may not be easily quantified to incorporate into estimates of flood hazard.

When constructing metamodels for storm surge estimation, the parameter values at one specific time point are often used. However, two storms can have similar values at one point and then deviate from each other quite drastically. One method to address this issue is to train an ANN on a network of spatial points across a period a time with different input parameter values. One such model was tested on two historical hurricanes, Katrina and Gustav, and was shown to be accurate for points near landfall (Kim et al., 2015). More recently, another method incorporated time series information through the use of convolutional neural networks (CNN) to predict peak storm surge at point locations from time series data of storm parameters (Lee et al., 2021). Before building the neural networks, the data output (peak storm surges at spatial points) was grouped together via k-means clustering, and then PCA is applied to each cluster to reduce dimensionality before a CNN is trained. The CNN takes spatial and temporal dependencies into account through 3D convolutional kernels. Both of these methods successfully utilize the full time series of data to obtain greater accuracy for storm surge estimation.

The effectiveness of a machine-learning driven surrogate model rests in the ability of the underlying machine learning technique to reliably yet implicitly infer real physical relationships which are explicitly resolved in physically driven simulations. This can be tricky. Consider a metamodel of flood outcomes in a large physical area which includes a small stream and a large stream. The simulation model "knows" that the physics are the same between the large and small rivers, but the metamodel doesn't necessarily. It would improve our metamodel performance if we could "teach" it that the physics are in fact the same between the large and small river, so that patterns inferred for the large river can be used to understand the small river and vice versa. It is hard to "teach" that to a metamodel; the magnitudes of the inputs (e.g., the water entering the stream) and outputs (e.g., the nominal quantity of discharge) are on completely different levels of scale, and patterns inferred by machine learning methods are typically local.

This can be addressed by training metamodels on data which expresses not nominal quantities but rather dimensionless parameters expressed as physically salient ratios. Returning to the example of the large and small river, this allows us to effectively "factor out" the size of the river, allowing the underlying machine learning model to better understand cross-scale dynamics. As an example, the base-flow index in hydrologic modeling, which is typically defined as the ratio of total river flow to baseflow, allows us to compare the relative intensity of flows between streams of different sizes. A dimensionless dryness index of an area of land can be derived as the ratio of potential evapotranspiration (e.g., in millimeters) to precipitation (in the same units) per unit area, which allows us to compare the relative aridity of areas with profoundly different meteorologic behavior. This allows machine learning techniques to better infer patterns in the data which are consistent across scales without explicit use of physics-constrained loss functions. Dimensionless parameters have been utilized in flood risk modeling for coastal flooding (Resio, 2007), pluvial flooding (Bartlett et al., 2025b; Sivapalan et al., 1990), and riverine dynamics (Porporato, 2022; Sivapalan et al., 1990). The use of dimensionless parameters represents a subset of a broader area of study in machine learning referred to as "feature engineering." Rather than using direct observational data as inputs to a surrogate model, the observations are first transformed into some "feature," a quantity that is more physically informative to the desired outputs. For example, to construct a metamodel of a simulation that maps rain-on-grid to compound flood outcomes as mediated by interacting pluvial and fluvial mechanisms, using rainfall as a predictor seems obvious. It is perhaps less obvious to use a computationally faster dendritic hydrologic model to create a first-order estimate of river flows throughout the system and use those as an additional set of inputs. In that way, the metamodel may only have to infer local interactions, rather than making more complex inferences about the complex relationships between upstream rainfall and downstream discharge.

4 Methods for Estimating Multivariate Probability Distributions

Previous sections have described approaches for how storm events in a probabilistic risk analysis are characterized and how the specific events to simulate (or estimate, in the case of using a response surface) are selected. This section details methods for aggregating those events to estimate probability distributions for the flood outcomes of interest.

In constructing a probabilistic joint flood hazard model, variable system parameters that can influence outcomes of interest are treated as random variables. Such variables encompass parameters of a tropical or extratropical cyclone (either over time or at landfall), parameters defining the antecedent conditions of the study domain (e.g., soil moisture, riverine flow rates), or parameters defining the domain's boundary conditions at the time of an event (e.g., mean sea level, astronomical tide). Estimating the relative likelihoods or probabilities of different extreme events occurring requires a statistical framework to estimate covariance structures between all of these informative parameters. The three main methods of modeling these relationships, outlined in this section, are copulas, physically-driven Monte Carlo simulations, and the joint probability method with Bayesian factorization.

4.1 Copula Methods

Copulas are a widely used method for modeling dependence between factors in multivariate distribution estimation. These probability models were first introduced by Sklar (1959) as a means for modeling joint probability distributions. The copula is a multivariate cumulative distribution function (CDF) C for which the marginal probability density function (PDF) of each variable is uniform on the interval [0, 1]. Sklar's Theorem shows that any multivariate joint probability distribution can be expressed in terms of its variables' marginal distributions, transformed by a copula that accounts for the dependencies between the variables. For F as the joint distribution of random variables $x_1, \ldots, x_k$, there exists copula C such that

$$F(x_1, x_2, \ldots, x_k) = C(u_1, u_2, \ldots, u_k)$$

where u_i is the cumulative probability $F_i(X_i \leq x_i)$ and $\mathrm{F_i}$ is the marginal CDF of x_i. The probability distribution function for the above CDF is thus

$$f(x_1, \ldots, x_k) = c(u_1, \ldots, u_k)\left[\prod_{i=1}^{k} f_i(x_i)\right]$$

where f is the PDF of the joint probability distribution, c is the copula density, and f_i is the marginal density.

Sklar's Theorem assumes that variables observed as time series are stationary and continuous, so a joint CDF estimated via copula may possess an inaccurate dependence structure. In practice, copula methods for estimating joint distributions over variables such as TC parameters can suffer from a greater issue, which is that analysts do not derive the unique copula which satisfies the equality given by Sklar. Instead, common practice is to evaluate the goodness of fit of a variety of commonly used and tractably fit copulas, then utilize whichever provides the best match to the observed record. Like any distribution fitting exercise, such approaches require parametric or resampling methods to rigorously characterize the standard error of resulting quantile estimators, a step that is commonly omitted either in execution or reporting.

While copulas are commonly used due to the ease of application, only requiring marginal distributions and empirically estimated copula parameters, there are drawbacks to their use. Copulas are a simplified means of modeling dependence that do not require strong parametric assumptions. As such, copulas can often only capture one or two forms of dependence. For example, Gaussian copulas can account for rank correlation, while Gumbel copulas can account for tail dependence. Commonly used copula families do not account well for nonlinear covariance structures.

Another limitation of copulas is that they are often limited to bivariate distributions. This is potentially problematic in cases where more than two scalar quantities are necessary to uniquely identify a storm; a bivariate model of precipitation and storm surge can characterize only peak precipitation and peak surge depth, while neglecting residual variability in the relative timing of the two forcings as well as the spatiotemporal structure of precipitation and runup and drawdown rates of storm surge. As mentioned previously, there are methods of circumventing the bivariate limitations without using more cumbersome multivariate distributions. However, the copulas still require certain dependence assumptions that may or may not hold.

Copulas have a range of families that can be chosen to represent dependence between variables with distributions such as Gaussian, Clayton, Gumbel, and others. For scenarios with only two factors, x_1 and x_2, a bivariate copula can be used. Situations with more than two significant flood drivers–typical in compound flood risk analysis–require the use of a narrower set of copulas supporting more than two random variables, including meta-Gaussian and vine copulas. Linear-circular copulas have also been employed to model the joint dependence between linear variables and circular variables, such as between central pressure deficit and storm heading (Liu et al., 2024).

A comparison of different copula constructions found that both vine copula construction and the Heffernan and Tawn (2004) approach outperformed higher-dimensional copulas in explaining the dependence structure of ocean-side water level, rainfall, and groundwater level (Jane et al., 2020). Vine copulas are hierarchical structures of bivariate copulas that are used to model higher-dimensional dependence, i.e., a multivariate copula that is constructed with bivariate copulas as building blocks (Czado & Nagler, 2022). Within these structures, each node contains a pair of variables and the edges represent bivariate copulas.

Different forms of copulas, such as C- and D-vine copulas, have been investigated with different probabilistic ensembles of forcing factors of storm surge to estimate dependence relationships (Zhang & Wang, 2021). Due to the many possible vine structures, and in order to make robust forecasts, Bayesian model averaging was used to generate probabilities. These ensembles were shown to be robust when compared with principal component regression and random forests as benchmarks.

A promising approach for constructing multi-hazard designs using copulas can be found in Sadegh et al. (2018). The framework first requires selecting marginal distributions for each forcing factor using the Bayesian Information Criterion (BIC). Subsequently, the Multivariate Copula Analysis Toolbox (MVCAT) is employed to fit bivariate copulas to each pairwise combination of forcing factors (Sadegh et al., 2017). For each pair, one of 26 possible bivariate copulas is fitted and then optimized via a hybrid-evolution Markov Chain Monte Carlo algorithm. Once the model is obtained, it is typical to use the most likely scenario of a critical joint return period by selecting the combination of hazards with the greatest density (Salvadori et al., 2014). However, Sadegh et al. (2018) utilized the more accurate approach of drawing a weighted random sample of compound events with the copula probability density function as the weight.

The notion of a joint return period in a copula model is often misunderstood and warrants elaboration. Generally speaking, planning and design exercises require a characterization of flood depths associated with a variety of return periods, e.g., the 100-year flood depths. Broadly, the 100-year flood depths are taken to mean the flood depths which we expect to see exceeded with a 1% annual probability. In univariate modeling contexts, the 100-year flood depths are uniquely specified by the 100-year forcing, whether that forcing be streamflow, precipitation, or storm surge. These forcings are often referred to as corresponding to the "100-year storm." This is highly cost-effective in contexts where it is appropriate, because it allows analysis to identify a single realization of an event producing the 100-year flood depth (or other outcome(s)) with only one finite-element hydraulic simulation run if the domain of interest is sufficiently localized.

The correspondence between forcing return periods and flood depth return periods breaks down in the multivariate, compound case. There are multiple ways to characterize the joint return period of a storm's forcings (e.g., surge, river flow, rainfall), but among the most popular is the bivariate-AND exceedance formulation (Amorim et al., 2025; Bender et al., 2016; Jane et al., 2020, 2022; Kim et al., 2023; Maduwantha et al., 2025). A storm has a bivariate-AND exceedance probability of 1%, for example, if there is a 1% chance in a given year that a storm will be more severe with respect to *both* forcings of interest. This is to say that in a bivariate model of river flow and storm surge, the 100-year storm under the bivariate-AND formulation has a 1% chance in a given year of seeing *both* its storm surge and river flow exceeded. Under the bivariate-AND formulation of joint exceedance, the 100-year storm is not unique. Rather there exists a curve (or isoquant) over the two-dimensional support of the model in which all points on the curve reflect distinct combinations of the two drivers, all of which have the same bivariate-AND exceedance probability. The problematic nature of the bivariate-AND exceedance formulation can be inferred directly from the definitional properties of the bivariate-AND isoquant and a corresponding isoquant of flood depth. The 1% AEP flood depth isoquant definitionally has its flood depth exceeded with a 1% annual probability, and therefore the area outside of the isoquant must integrate to a 1% annual probability. The 1% AEP bivariate-AND isoquant is defined such that every point on the isoquant has an integrated annual probability of 1% in the rectangular region in which both of its forcings are exceeded. A necessary consequence of these definitions is that if the bivariate model were the ground truth without any residual variance attributable to unmodelled parameters, a design storm associated with a fixed bivariate-AND exceedance probability deterministically produces less flooding than the flood severity associated with the same exceedance probability. In other words, a 100-year storm generated under a bivariate-AND exceedance formulation will always produce less severe flooding than the actual 100-year flood severity, except in cases where a bivariate model is inadequate to capture the true variability in flood dynamics and the generated storm happens to be more intense than average for its bivariate characterization.

Bivariate or otherwise multivariate distributions do not need to be characterized using problematic joint exceedance definitions like the bivariate-AND formulation described above. They can instead be densely sampled from and simulated via finite-element hydraulic simulation, resulting in an estimated distribution of flood outcomes, from which depths corresponding to select return periods can be derived, and the events which generate them enumerated. While joint probability methods can use copulas to model dependence between event parameters, other approaches apply them to event outcomes that influence inundation. For example, the D-Flow Flexible Mesh model has been employed to obtain storm surge predictions based on storm parameters and used copulas to model the dependence between peak water level and accumulated rainfall (Xu et al., 2022). This allows analysis of scenarios where either of the flood drivers exceeds a predefined hazard threshold or when both exceed predefined hazard thresholds. This is particularly relevant for compound flooding scenarios, where multiple contributing factors, such as runoff and high sea levels, result in flooding.

The Probabilistic Coastal Hazards Analysis (PCHA) framework was developed for the Coastal Hazards System (CHS) for determining TC hazard (Nadal-Caraballo et al., 2020a). The framework incorporates a meta-Gaussian copula with correlation matrix constructed from the correlations of all possible pairs of forcing parameters of synthetic events. The PCHA has a high-resolution set of cyclone simulations from which a high-resolution response surface is estimated through the use of a Gaussian process metamodel (Nadal-Caraballo et al., 2020b). The input parameters for the storm simulations are fed into the planetary boundary layer (PBL) that are then used in high-fidelity surge and wave models (Nadal-Caraballo et al., 2022). The PCHA statistical framework is model- and geography-agnostic and can be designed to include or exclude sources of compound flooding as desired, but depending on the spatiotemporal scale of a study and the need for rigorous uncertainty analysis, it can require substantial computational resources to provide high-fidelity training data for the meta-models.

4.2 Bayesian Factorization

An alternative approach that addresses some of the deficiencies of copulas is Bayesian factorization. Bayesian factorization fuses Bayesian probability with matrix factorization for the estimation of dependencies between variables represented as a joint distribution. Bayesian factorization decomposes a joint distribution into the marginal distribution of one parameter, multiplied by the distribution of another parameter conditioned on the first, multiplied by the distribution of yet another parameter conditioned on the previous two, and so on until all variables are represented. A naïve Bayesian factorization grows exponentially in its data requirements as more parameters are added but can be simplified by leveraging theoretically well-established or empirically validated conditional independence relationships. An illustrative example is the joint probability distribution of TC parameters, $\Lambda\left(c_p, r, v_f, \theta_l, x\right)$, employed by the Coastal Louisiana Risk Assessment (CLARA) model in Louisiana's 2023 Coastal Master Plan (Johnson et al., 2023). This joint distribution was identified empirically by Resio (2007) and modified by the CLARA team:

$$\Lambda\left(c_p, r, v_f, \theta_l, x\right) = \Lambda_1 \cdot \Lambda_2 \cdot \Lambda_3 \cdot \Lambda_4 \cdot \Lambda_5$$

$$\Lambda_1 = f\left(c_p|x\right) = \frac{\delta}{\delta c_p}\left\{\exp\left\{-\exp\left[-\frac{c_p - (a_0(x) + a_1(x)t)}{a_2(x)}\right]\right\}\right\}$$

$$\Lambda_2 = f\left(r|c_p\right) = \frac{1}{\sigma\left(c_p\right)\sqrt{2\pi}} e^{\frac{\left(\ln r - \overline{r}\left(c_p\right)\right)^2}{2\sigma^2\left(c_p\right)}}$$

$$\Lambda_3 = f\left(v_f|\theta_l\right) = \frac{1}{\sigma\sqrt{2\pi}} e^{\frac{\left(v_f - \overline{v_f}(\theta_l)\right)^2}{2\sigma^2}}$$

$$\Lambda_4 = f\left(\theta_l|x\right) = \frac{1}{\sigma(x)\sqrt{2\pi}} e^{\frac{\left(\theta_l - \overline{\theta_l}(x)\right)^2}{2\sigma^2(x)}}$$

$$\Lambda_5 = f(x) = \Phi(x)$$

The relationships among five key storm parameters—central pressure c_p, radius of maximum wind speed r, forward velocity v_f, landfall heading θ_l, and location of landfall x—are characterized structurally by a joint distribution composed of marginal distributions that are empirically, approximately independent when conditioned on other parameters (Λ_1 through Λ_5, respectively). For example, the equation $\Lambda_1 = f\left(c_p|x, r, v_f, \theta_l\right) = f\left(c_p|x\right)$ indicates that the distribution of central pressure is dependent on the landfall location x but not on the heading, maximum wind speed or forward velocity at landfall. Maximum likelihood estimation is then applied to the historic record to fit the various distribution parameters (a_i, $\overline{r}$, etc.), using a moving window to identify included events or kernel smoothing to apply weights when conditioning on, for example, a specific value of x.

The PCHA system used by the US Army Corps of Engineers is similar (Nadal-Caraballo et al., 2020a, 2020b). It instead considers storm frequencies at a range of points along the coast, and in an area around each point it fits marginal distributions to central pressure, radius of maximum wind speed, heading direction, and forward velocity. The correlation structure between them is handled via a pressure-stratified meta-Gaussian copula, which is simple to implement but only resolves rank-linear correlation structures. It therefore may be less suitable for use in areas where data to support more flexibly structured Bayesian factorization models with conditional independence are available (Nadal-Caraballo et al., 2020a).

4.3 Semiparametric and Mixed Methods

The dependence structure among hydroclimatic extremes can be analyzed via a hybrid approach between empirical analysis and peak-over-threshold distribution fitting (Sharma & Mujumdar, 2022). Generalized Pareto distributions are then employed to fit values above the threshold, while empirical distributions are used to model values below the threshold. Then the distributions are transformed in a way that captures dependence between multiple existent hazards.

Other approaches to characterizing the likelihood of different kinds of storm events do not address the probability distribution of the storms directly, but rather permit some source of randomness in a more detailed parametric or semi-parametric physical model. The Hybrid WRF Cyclone Model (HWCM) is an example of a model which addresses this challenge by creating simulated, physically explicit TCs based on user-specified features (Bruyère et al., 2019). The HWCM uses physically driven models to spin up a certain number of TCs and then sets them into a real-world domain where it can develop and make landfall. This allows the generation of detailed storm events that can serve as the basis of discretization for an arbitrary probabilistic model of a small number of salient parameters. The HWCM has also been used to evaluate the sensitivity of TC dynamics to sea surface temperature, moisture, and other climate parameters.

Efforts to simulate TCs for risk assessment that maintain memory over time have resulted in the HITS algorithm, a nonparametric stochastic model (Nakamura et al., 2015). Unlike Markovian methods which assume that the frequencies, locations, or characteristics of individual storms are independent of one another, the HITS algorithm is not spatially or temporally memoryless (Nakamura et al., 2015). The HITS algorithm consists of ten steps wherein the number of storms for a season, the storm birth location, and the storm track lifetime are simulated. Once the storm has an initial starting point, it can jump to neighboring tracks by drawing from a conditional kernel density function. Thus, each segment of the simulated storm is part of a historical track, with the distance between neighbors utilized being (1) distance from track to current position, (2) vector difference in direction of movement relative to the current track, and (3) the age relative to the current track. These distance measures were used to cluster storms by behavior and birth location (Nakamura et al.,

2009). The HITS algorithm was later expanded and improved to become the Cluster-Based Climate-Conditioned Hurricane Intensity and Track Simulator, or C^3-HITS algorithm (Nakamura et al., 2021). The C^3-HITS algorithm utilizes the original HITS algorithm within itself but precedes the HITS algorithm with two other steps. The first step is to cluster the data with a k-means clustering algorithm (Nakamura et al., 2009). The second is to build a Poisson process model to simulate the number of storms forming in each cluster, conditioned on early season climate covariates such as sea surface temperature (SST) anomaly indices based on the TC main development region and the Niño-3.4 index. The inclusion of these early season climate covariates for conditioning the storm-track generation on SST indices produced more accurate simulated storms and reduced RMSE compared to the original HITS algorithm when examining both the number of basin formations and the number of landfalls.

4.4 Addressing Uncertainties

Amongst the shortcomings of some contemporary statistical hazard models, an inability to account for natural structure is one that is in need of deeper study (Resio et al., 2017a, 2017b). For example, omnidirectional sampling can fail to account for relationships between storm heading and intensity. Another limitation of many present models of flood hazard estimates is the failure to incorporate potentially inundated regions when calculating AEP values. Epistemic uncertainty of predicted inundation can reveal areas that are "nearly inundated" according to the predictive model. For example, if flood levels are predicted to be below but close to ground level for a region, the uncertainty from the predictive model might be great enough for there to be a statistically significant probability that the region would be inundated.

Uncertainty can take many forms and, when accounted for, often leads to significantly different estimates of inundation risk. Two salient forms of uncertainty are parametric, which is uncertainty in model parameter values, and structural uncertainties, which are the uncertainties related to the physical representation of systems under consideration. A comparison of different storm surge models applied to the classical dike heightening problem (Oddo et al., 2020) showed that while introducing parametric uncertainties lead to small changes in the estimated results, structural uncertainties around scenario assumptions such as sea level rise lead to much greater disparities in cost estimates. For TC storm surge hazard, uncertainty can cascade to impacts on AEP estimates, with previous analyses showing that neglecting uncertainty when estimating AEPs can lead to under-predicted storm surge levels by as much as twenty percent (Resio et al., 2013). On the other hand, overestimating uncertainty can also lead to inaccurate AEP estimates, as in the case of applying a uniform relative uncertainty to all modeled surge values (Nadal-Caraballo et al., 2022).

The Tropical Cyclone Forecasting Framework (TC-FF) was developed to address uncertainty in storm characteristics such as wind speed, intensity, and storm track (Nederhoff et al., 2024). This framework generates storms through Monte Carlo sampling of errors for along-track, cross-track, and intensity components of storms.

The sampled errors are added back to the best forecasted track to obtain an ensemble of potential realizations.

The CLARA model utilizes a combination of analytic uncertainty propagation, Monte Carlo sampling, and bootstrapping to generate AEP estimates including confidence bounds around each return period (Johnson et al., 2023). The model accounts for a range of uncertainties including randomness in the observed record of historic TC events, noise in DEM elevations, standard errors in ADCIRC + SWAN surge and wave outputs estimated from the calibration and validation process, physical variability in surge and wave overtopping rates, and the possibility of breaches or catastrophic failure of levees, floodwalls, and gates.

In regard to uncertainty in forcing data such as precipitation, Bayesian Model Averaging (BMA) has seen success in improving accuracy in flood modeling (Gomez et al., 2024). Multiple forcing datasets with differing spatial or temporal resolutions can be integrated with BMA. The resultant data can then be input to hydrodynamic models to generate more accurate flood maps.

As uncertainty analysis is quite a broad topic, we refer readers to Gonzalez et al. (2019) for a more comprehensive discussion of uncertainty analysis in the context of JPM-based methods for flood hazard analysis.

5 Opportunities for Improvements to Probabilistic Compound Flood Hazard Analysis

Bivariate analyses of joint probabilities of flooding from any pairwise combination of flood drivers, including storm surge, rainfall, and river discharge, have become more common as policy makers recognize the complexity of interactions between multiple flood drivers. Rigorous probabilistic analyses of all three of these drivers are still less common, however (often due to constraints on data availability). The use of copulas to estimate covariance between independently modeled or estimated factors is convenient, given existing software libraries for fitting computationally tractable copula families, but it limits the representation of non-linear, complex dependency structures and nonstationarity in underlying hazards. Modeling and computational capabilities have reached a point where more physically-driven analyses are relatively accessible. As such, it is recommended to pursue studies that extend joint probability methods, historically used to estimate surge hazard, to account for variability in other flooding mechanisms such as rainfall and extreme stream flows. These approaches can also be easily adapted to incorporate stochastic uncertainty in antecedent conditions (e.g., soil saturation, sea level and tides) that vary in importance by local geography (Bartlett et al., 2025a). This also presents an opportunity for advancements and improvements in the physical modeling systems themselves, such as better coupling coastal and inland processes that are being forced by the same meteorological drivers.

Existing extensions of JPM-OS have generally focused on flooding from TCs, with some studies incorporating hazard from non-tropical frontal events. The impact of convective storms has been implicitly addressed by empirical analysis of antecedent conditions to which convective storms may have contributed, but more attention could be given to explicitly model interdependencies between such storms and TCs from the opposite direction, i.e., when flooding from convective systems is impacted by recent cyclonic events.

References

Abouelsaad, O., Hassan, A., Omar, M., & Hinkelmann, R. (2024). Identifying manning roughness coefficient using automatic calibration method and simulation of pollution incidents in the Nile River, Egypt. *Journal Hydrology Regional Studies, 55*, Article 101908.

Al Kajbaf, A., & Bensi, M. (2020). Application of surrogate models in estimation of storm surge: A comparative assessment. *Applied Soft Computing, 91*, Article 106184. https://doi.org/10.1016/j.asoc.2020.106184

Amorim, R., Villarini, G., Kim, H., Jane, R. A., & Wahl, T. (2025). A practitioner's approach to process-driven modeling of compound rainfall and storm surge extremes for coastal Texas. *Journal of Hydrologic Engineering, 30*, 04025025. https://doi.org/10.1061/JHYEFF.HEENG-6482

Bartlett, M., Cobell, Z., Diaz, O., Geldner, N., Grimley, L., Johnson, D., Kim, H., McCorquodale, A., McManus, M., Misra, S., Partida, L., Roberts, H., Saharia, A., Villarini, G., Wang, Y., Zou, S., Yuill, B., & Young, N. (2023). Compound flood transition zone pilot study for the Amite River Basin. Water Institute of the Gulf.

Bartlett, M., Geldner, N., Cobell, Z., Partida, L., Diaz, O., Johnson, D., Kim, H., McMann, B., Villarini, G., Misra, S., Roberts, H., Narayanaswamy, M. (2025a). Extending the Joint Probability Method to Compound Flooding: Statistical Delineation of Transition Zones and Design Event Selection. arXiv preprint. https://doi.org/10.48550/arXiv.2511.03871

Bartlett, M. S., Van Blitterswyk, J., Farella, M., Li, J., Smith, C., Parolari, A. J., Krishnamoorthy, L., & Mrad, A. (2025b). Physically based dimensionless features for pluvial flood mapping with machine learning. *Water Resources Research, 61*, e2024WR039086. https://doi.org/10.1029/2024WR039086

Bates, P. D., Horritt, M. S., & Fewtrell, T. J. (2010). A simple inertial formulation of the shallow water equations for efficient two-dimensional flood inundation modelling. *Journal of Hydrology, 387*, 33–45. https://doi.org/10.1016/j.jhydrol.2010.03.027

Bender, J., Wahl, T., Müller, A., & Jensen, J. (2016). A multivariate design framework for river confluences. *Hydrological Sciences Journal, 61*, 471–482. https://doi.org/10.1080/02626667.2015.1052816

Blessing, R., Sebastian, A., & Brody, S. D. (2017). Flood risk delineation in the united states: how much loss are we capturing? *Nature Hazards Review, 18*, 04017002. https://doi.org/10.1061/(ASCE)NH.1527-6996.0000242

Booij, N., Ris, R. C., & Holthuijsen, L. H. (1999). A third-generation wave model for coastal regions: 1. Model description and validation. *Journal Geophysics Research Oceans, 104*, 7649–7666. https://doi.org/10.1029/98JC02622

Brody, S. D., Blessing, R., Sebastian, A., & Bedient, P. (2013). Delineating the reality of flood risk and loss in Southeast Texas. *Nature Hazards Review, 14*, 89–97. https://doi.org/10.1061/(ASCE)NH.1527-6996.0000091

Brunner, G. W. (2020). HEC-RAS, River analysis system hydraulic reference manual. US Army Corps of Engineers Hydrologic Engineering Center (HEC).

Bruyère, C. L., Done, J. M., Jaye, A. B., Holland, G. J., Buckley, B., Henderson, D. J., Leplastrier, M., & Chan, P. (2019). Physically-based landfalling tropical cyclone scenarios in support of risk assessment. *Weather and Climate Extremes, 26*, 100229.

Cardone, V. J., & Cox, A. T. (2009). Tropical cyclone wind field forcing for surge models: Critical issues and sensitivities. *Natural Hazards, 51*, 29–47. https://doi.org/10.1007/s11069-009-9369-0

Chen, W.-B., Liu, W.-C., & Hsu, M.-H. (2012). Predicting typhoon-induced storm surge tide with a two-dimensional hydrodynamic model and artificial neural network model. *Natural Hazards and Earth Systems Sciences, 12*, 3799–3809. https://doi.org/10.5194/nhess-12-3799-2012

Christakos, G. (2017). *Spatiotemporal random fields: Theory and applications*. Elsevier.

Cialone, M., Massey, C., Anderson, M., Grzegorzewski, A., Jensen, R., Cialone, A., Mark, D., Pevey, D., Gunkel, B., Mcalpin, T., Nadal Caraballo, N. C., Melby, J., & Ratcliff, J. (2015). North Atlantic coast comprehensive study (NACCS)—Coastal storm model simulations: Waves and water levels.

Czado, C., & Nagler, T. (2022). Vine copula based modeling. *Annual Review Statistics Its Applications, 9*, 453–477. https://doi.org/10.1146/annurev-statistics-040220-101153

De Oliveira, M. M. F., Ebecken, N. F. F., De Oliveira, J. L. F., & De Azevedo, S. I. (2009). Neural network model to predict a storm surge. *Journal of Applied Meteorology and Climatology, 48*, 143–155. https://doi.org/10.1175/2008JAMC1907.1

Dietrich, J. C., Dawson, C. N., Proft, J. M., Howard, M. T., Wells, G., Fleming, J. G., Luettich, R. A., Westerink, J. J., Cobell, Z., Vitse, M., Lander, H., Blanton, B. O., Szpilka, C. M., & Atkinson, J. H. (2012). Real-time forecasting and visualization of hurricane waves and storm surge using SWAN+ADCIRC and FigureGen. In *Computational Challenges in the Geosciences* (pp. 49–70). New York, NY: Springer New York.

Dysarz, T. (2018). Application of python scripting techniques for control and automation of HEC-RAS simulations. *Water, 10*, 1382.

Eilander, D., Boisgontier, H., Bouaziz, L. J. E., Buitink, J., Couasnon, A., Dalmijn, B., Hegnauer, M., De Jong, T., Loos, S., Marth, I., & Van Verseveld, W. (2023). HydroMT: Automated and reproducible model building and analysis. *Journal Open Source Software, 8*, 4897. https://doi.org/10.21105/joss.04897

Emanuel, K., Ravela, S., Vivant, E., & Risi, C. (2006). A statistical deterministic approach to hurricane risk assessment. *Bulletin of the American Meteorological Society, 87*, 299–314. https://doi.org/10.1175/BAMS-87-3-299

Federal Emergency Management Agency. (2008). Mississippi coastal analysis project. Final Report: HMTAP Task Order 18. U.S. Department of Homeland Security, Washington, DC.

Feng, D., Tan, Z., Xu, D., & Leung, L. R. (2023). Understanding the compound flood risk along the coast of the contiguous United States. *Hydrology and Earth System Sciences, 27*, 3911–3934. https://doi.org/10.5194/hess-27-3911-2023

Fischbach, J. R., Johnson, D. R., Kuhn, K., Pollard, M., Stelzner, C., Costello, R., & Cobell, Z. (2017). Coastal master plan modeling: Attachment C3–25—Storm surge and risk assessment. Louisiana Coastal Protection and Restoration Authority, Baton Rouge, Louisiana.

Ganguli, P., Paprotny, D., Hasan, M., Güntner, A., & Merz, B. (2020). Projected changes in compound flood hazard from riverine and coastal floods in Northwestern Europe. *Earths Future, 8*, e2020EF001752. https://doi.org/10.1029/2020EF001752

Gharehtoragh, M. A., & Johnson, D. R. (2024). Using surrogate modeling to predict storm surge on evolving landscapes under climate change. *Npj Nat Hazards, 1*, 33. https://doi.org/10.1038/s44304-024-00032-9

Gomez, F. J., Jafarzadegan, K., Moftakhari, H., & Moradkhani, H. (2024). Probabilistic flood inundation mapping through copula Bayesian multi-modeling of precipitation products. *Natural Hazards and Earth Systems Sciences, 24*, 2647–2665. https://doi.org/10.5194/nhess-24-2647-2024

Gonzalez, V. M., Nadal-Caraballo, N. C., Melby, J. A., & Cialone, M. A. (2019). Quantification of uncertainty in probabilistic storm surge models: Literature review. Coastal and Hydraulics Laboratory (U.S.)

Gori, A., Lin, N., & Xi, D. (2020). Tropical cyclone compound flood hazard assessment: From investigating drivers to quantifying extreme water levels. *Earths Future, 8*, e2020EF001660. https://doi.org/10.1029/2020EF001660

Heffernan, J. E., & Tawn, J. A. (2004). A conditional approach for multivariate extreme values (with discussion). *Journal of the Royal Statistical Society, Series B: Statistical Methodology, 66*, 497–546. https://doi.org/10.1111/j.1467-9868.2004.02050.x

Hendry, A., Haigh, I. D., Nicholls, R. J., Winter, H., Neal, R., Wahl, T., Joly-Laugel, A., & Darby, S. E. (2019). Assessing the characteristics and drivers of compound flooding events around the UK coast. *Hydrology and Earth System Sciences, 23*, 3117–3139. https://doi.org/10.5194/hess-23-3117-2019

Hristopulos, D. T. (2020). *Random fields for spatial data modeling: A primer for scientists and engineers*. Springer.

Interagency Performance Evaluation Task Force. (2009). Performance Evaluation of the New Orleans and Southeast Louisiana Hurricane Protection System, Vol. VIII–Engineering and Operational Risk and Reliability Analysis, Final Report of the Interagency Performance Evaluation Task Force. US Army Corps of Engineers, Washington, DC.

Jane, R., Cadavid, L., Obeysekera, J., & Wahl, T. (2020). Multivariate statistical modelling of the drivers of compound flood events in south Florida. *Natural Hazards and Earth Systems Sciences, 20*, 2681–2699. https://doi.org/10.5194/nhess-20-2681-2020

Jane, R., Wahl, T., Santos, V. M., Misra, S. K., & White, K. D. (2022). Assessing the potential for compound storm surge and extreme river discharge events at the catchment scale with statistical models: sensitivity analysis and recommendations for best practice. *Journal of Hydrologic Engineering, 27*, 04022001. https://doi.org/10.1061/(ASCE)HE.1943-5584.0002154

Jehanzaib, M., Ajmal, M., Achite, M., & Kim, T. W. (2022). Comprehensive review: Advancements in rainfall-runoff modelling for flood mitigation. *Climate, 10*(10), 147.

Jelesnianski, C. P. (1992). SLOSH: Sea, lake, and overland surges from hurricanes. US Department of Commerce, National Oceanic and Atmospheric Administration.

Jia, G., & Taflanidis, A. A. (2013). Kriging metamodeling for approximation of high-dimensional wave and surge responses in real-time storm/hurricane risk assessment. *Computer Methods in Applied Mechanics and Engineering, 261–262*, 24–38. https://doi.org/10.1016/j.cma.2013.03.012

Jia, G., Taflanidis, A. A., Nadal-Caraballo, N. C., Melby, J. A., Kennedy, A. B., & Smith, J. M. (2016). Surrogate modeling for peak or time-dependent storm surge prediction over an extended coastal region using an existing database of synthetic storms. *Natural Hazards, 81*, 909–938. https://doi.org/10.1007/s11069-015-2111-1

Johnson, D. R., Fischbach, J. R., Geldner, N. B., Wilson, M. T., Story, C., & Wang, J. (2023). 2023 Coastal Master Plan: CLARA Model Summary, Attachment E4. Louisiana Coastal Protection and Restoration Authority, Baton Rouge, Louisiana.

Keef, C., Svensson, C., & Tawn, J. A. (2009). Spatial dependence in extreme river flows and precipitation for Great Britain. *Journal of Hydrology, 378*, 240–252. https://doi.org/10.1016/j.jhydrol.2009.09.026

Kepert, J. D. (2023). A parametric model of tropical cyclone surface winds for sea and land. *Weather and Forecasting, 38*, 1739–1757.

Kim, S.-W., Melby, J. A., Nadal-Caraballo, N. C., & Ratcliff, J. (2015). A time-dependent surrogate model for storm surge prediction based on an artificial neural network using high-fidelity synthetic hurricane modeling. *Natural Hazards, 76*, 565–585. https://doi.org/10.1007/s11069-014-1508-6

Kim, S., Pan, S., & Mase, H. (2019). Artificial neural network-based storm surge forecast model: Practical application to Sakai Minato, Japan. *Applications Ocean Research, 91*, Article 101871. https://doi.org/10.1016/j.apor.2019.101871

Kim, H., Villarini, G., Jane, R., Wahl, T., Misra, S., & Michalek, A. (2023). On the generation of high-resolution probabilistic design events capturing the joint occurrence of rainfall and storm surge in coastal basins. *International Journal of Climatology, 43*, 761–771. https://doi.org/10.1002/joc.7825

Kyprioti, A. P., Taflanidis, A. A., Nadal-Caraballo, N. C., & Campbell, M. (2021a). Storm hazard analysis over extended geospatial grids utilizing surrogate models. *Coastal Engineering, 168*, Article 103855. https://doi.org/10.1016/j.coastaleng.2021.103855

Kyprioti, A. P., Taflanidis, A. A., Nadal-Caraballo, N. C., & Campbell, M. O. (2021b). Incorporation of sea level rise in storm surge surrogate modeling. *Natural Hazards, 105*, 531–563. https://doi.org/10.1007/s11069-020-04322-z

Kyprioti, A. P., Taflanidis, A. A., Plumlee, M., Asher, T. G., Spiller, E., Luettich, R. A., Blanton, B., Kijewski-Correa, T. L., Kennedy, A., & Schmied, L. (2021c). Improvements in storm surge surrogate modeling for synthetic storm parameterization, node condition classification and implementation to small size databases. *Natural Hazards, 109*, 1349–1386. https://doi.org/10.1007/s11069-021-04881-9

Lee, T.-L. (2006). Neural network prediction of a storm surge. *Ocean Engineering, 33*, 483–494. https://doi.org/10.1016/j.oceaneng.2005.04.012

Lee, T.-L. (2009). Predictions of typhoon storm surge in Taiwan using artificial neural networks. *Advances in Engineering Software, 40*, 1200–1206. https://doi.org/10.1016/j.advengsoft.2007.06.005

Lee, J.-W., Irish, J. L., Bensi, M. T., & Marcy, D. C. (2021). Rapid prediction of peak storm surge from tropical cyclone track time series using machine learning. *Coastal Engineering, 170*, Article 104024. https://doi.org/10.1016/j.coastaleng.2021.104024

Liu, Z., Carr, M. L., Nadal-Caraballo, N. C., Aucoin, L. A., Yawn, M. C., & Bensi, M. T. (2024). Comparative analysis of joint distribution models for tropical cyclone atmospheric parameters in probabilistic coastal hazard analysis. *Stochastic Environmental Research and Risk Assessment, 38*(5), 1741–1767.

Lonfat, M., Rogers, R., Marchok, T., & Marks Jr, F. D. (2007). A parametric model for predicting hurricane rainfall. *Monthly Weather Review, 135*(9), 3086–3097.

Lu, P., Lin, N., Emanuel, K., Chavas, D., & Smith, J. (2018). Assessing hurricane rainfall mechanisms using a physics-based model: Hurricanes Isabel (2003) and Irene (2011). *Journal of the Atmospheric Sciences, 75*(7), 2337–2358.

Luettich Jr, R. A., Westerink, J. J., & Scheffner, N. W. (1992). ADCIRC: An advanced three-dimensional circulation model for shelves, coasts, and estuaries. Report 1. Theory and Methodology of ADCIRC-2DDI and ADCIRC-3DL.

Maduwantha, P., Wahl, T., Santamaria-Aguilar, S., Jane, R., Dangendorf, S., Kim, H., & Villarini, G. (2025). Generating boundary conditions for compound flood modeling in a probabilistic framework. *Egusphere, 2025*, 1–29.

Massey, T. C., Anderson, M. E., Smith, J. M., Gomez, J., & Jones, R. (2011). STWAVE: Steady-state spectral wave model user's manual for STWAVE, Version 6.0 (No. ERDCCHLSR111).

Marks, F. D., & DeMaria, M. (2003). *Development of a tropical cyclone rainfall climatology and persistence (RCLIPER) model*. National Oceanic and Atmospheric Administration. Miami, Florida. https://www.nhc.noaa.gov/jht/final_rep/R-CLIPER_final.pdf

Medeiros, S. C., Hagen, S. C., & Weishampel, J. F. (2012). Comparison of floodplain surface roughness parameters derived from land cover data and field measurements. *Journal of Hydrology, 452*, 139–149.

Meyer, M. R., & Johnson, D. R. (2019). Variability of best-estimate flood depth return periods in coastal Louisiana. *Journal Marine Science Engineering, 7*, 145.

Murphy, W. M., & Geelan, C. W. (1965). Development of hurricane flood protection for Texas City, Texas. U.S. Army Corps of Engineers: Galveston, Texas.

Nadal-Caraballo, N. C., Campbell, M. O., Gonzalez, V. M., Torres, M. J., Melby, J. A., & Taflanidis, A. A. (2020). Coastal hazards system: A probabilistic coastal hazard analysis framework. *Journal of Coastal Research, 95*, 1211. https://doi.org/10.2112/SI95-235.1

Nadal-Caraballo, N. C., Yawn, M. C., Aucoin, L. A., Carr, M. L., Melby, J. A., Ramos-Santiago, E., Gonzalez, V. M., Taflanidis, A. A., Kyprioti, A. A., Cobell, Z., & Cox, A. T. (2022) Coastal Hazards System-Louisiana (CHS-LA). Engineer Research and Development Center (U.S.).

Nakamura, J., Lall, U., Kushnir, Y., & Camargo, S. J. (2009). Classifying North Atlantic tropical cyclone tracks by mass moments. *Journal of Climate, 22*, 5481–5494. https://doi.org/10.1175/2009JCLI2828.1

Nakamura, J., Lall, U., Kushnir, Y., & Rajagopalan, B. (2015). HITS: Hurricane intensity and track simulator with North Atlantic Ocean applications for risk assessment. *Journal of Applied Meteorology and Climatology, 54*, 1620–1636. https://doi.org/10.1175/JAMC-D-14-0141.1

Nakamura, J., Lall, U., Kushnir, Y., Harr, P. A., & McCreery, K. (2021). Early season hurricane risk assessment: Climate-conditioned HITS simulation of North Atlantic tropical storm tracks. *Journal of Applied Meteorology and Climatology, 60*, 559–575. https://doi.org/10.1175/JAMC-D-20-0237.1

Nakamura, J., Lall, U., Kushnir, Y., & Harr, P. A. (2024). A saturated stochastic simulator: Synthetic US Gulf coast tropical cyclone precipitation fields. *Natural Hazards, 120*, 1295–1318. https://doi.org/10.1007/s11069-023-06245-x

Nederhoff, K., van Ormondt, M., Veeramony, J., van Dongeren, A., Antolínez, J. A. Á., Leijnse, T., & Roelvink, D. (2024). Accounting for uncertainties in forecasting tropical-cyclone-induced compound flooding. *Geoscientific Model Development, 17*, 1789–1811. https://doi.org/10.5194/gmd-17-1789-2024

Nelsen, R. B. (2005). Copulas and quasi-copulas: An introduction to their properties and applications. In *Logical, algebraic, analytic and probabilistic aspects of triangular norms* (pp. 391–413). Elsevier.

Niedoroda, A. W., Resio, D.T., Toro, G., Divoky, D., & Reed, C. (2012) Efficient strategies for the joint probability evaluation of storm surge hazards (pp. 242–255). https://doi.org/10.1061/40968(312)22

Oddo, P. C., Lee, B. S., Garner, G. G., Srikrishnan, V., Reed, P. M., Forest, C. E., & Keller, K. (2020). Deep uncertainties in sea-level rise and storm surge projections: Implications for coastal flood risk management. *Risk Analysis, 40*, 153–168. https://doi.org/10.1111/risa.12888

Olfateh, M., Callaghan, D. P., Nielsen, P., & Baldock, T. E. (2017). Tropical cyclone wind field asymmetry—Development and evaluation of a new parametric model. *Journal Geophysical Research Oceans, 122*, 458–469. https://doi.org/10.1002/2016JC012237

Pedregosa, F., Varoquaux, G., Gramfort, A., Michel, V., Thirion, B., Grisel, O., Blondel, M., Prettenhofer, P., Weiss, R., & Dubourg, V. (2011). Scikit-learn: Machine learning in Python. *Journal of Machine Learning Research, 12*, 2825–2830.

Philippus, D., Wolfand, J. M., Abdi, R., & Hogue, T. S. (2021). Raspy-Cal: A genetic algorithm-based automatic calibration tool for HEC-RAS hydraulic models. *Water, 13*, 3061.

Porporato, A. (2022). Hydrology without dimensions. *Hydrology and Earth System Sciences, 26*, 355–374.

Rajasekaran, S., Gayathri, S., & Lee, T.-L. (2008). Support vector regression methodology for storm surge predictions. *Ocean Engineering, 35*, 1578–1587. https://doi.org/10.1016/j.oceaneng.2008.08.004

Reggiani, P., & Weerts, A. H. (2008). Probabilistic quantitative precipitation forecast for flood prediction: An application. *Journal of Hydrometeorology, 9*, 76–95. https://doi.org/10.1175/2007JHM858.1

Resio, D. T., Irish, J., & Cialone, M. (2009). A surge response function approach to coastal hazard assessment—part 1: Basic concepts. *Natural Hazards, 51*, 163–182. https://doi.org/10.1007/s11069-009-9379-y

Resio, D. T., Irish, J. L., Westerink, J. J., & Powell, N. J. (2013). The effect of uncertainty on estimates of hurricane surge hazards. *Natural hazards, 66*(3), 1443–1459.

Resio, D. T., Asher, T. G., & Irish, J. L. (2017a). The effects of natural structure on estimated tropical cyclone surge extremes. *Natural Hazards, 88*, 1–29. https://doi.org/10.1007/s11069-017-2935-y

Resio, D. T., Asher, T. G., & Irish, J. L. (2017b). The effects of natural structure on estimated tropical cyclone surge extremes. *Natural Hazards, 88*, 1609–1637. https://doi.org/10.1007/s11069-017-2935-y

Resio, D. T. (2007). White paper on estimating hurricane inundation probabilities. In *Performance evaluation of the New Orleans and Southeast Louisiana hurricane protection system. Appendix 8: Hazard analysis*. U.S. Army Corps of Engineers

Ruiz-Salcines, P., Salles, P., Robles-Díaz, L., Díaz-Hernández, G., Torres-Freyermuth, A., & Appendini, C. M. (2019). On the use of parametric wind models for wind wave modeling under tropical cyclones. *Water, 11*, 2044.

Sadegh, M., Ragno, E., & AghaKouchak, A. (2017). Multivariate copula analysis toolbox (MvCAT): Describing dependence and underlying uncertainty using a Bayesian framework. *Water Resources Research, 53*, 5166–5183. https://doi.org/10.1002/2016WR020242

Sadegh, M., Moftakhari, H., Gupta, H. V., Ragno, E., Mazdiyasni, O., Sanders, B., Matthew, R., & AghaKouchak, A. (2018). Multihazard scenarios for analysis of compound extreme events. *Geophysical Research Letters, 45*, 5470–5480. https://doi.org/10.1029/2018GL077317

Salvadori, G., Tomasicchio, G. R., & D'Alessandro, F. (2014). Practical guidelines for multivariate analysis and design in coastal and off-shore engineering. *Coastal Engineering, 88*, 1–14. https://doi.org/10.1016/j.coastaleng.2014.01.011

Scheffner, N. W., Borgman, L. E., & Mark, D. J. (1996). Empirical simulation technique based storm surge frequency analyses. *Journal of Waterway, Port, Coastal and Ocean Engineering. American Society of Civil Engineers, 122*, 93–101. https://doi.org/10.1061/(ASCE)0733-950X(1996)122:2(93)

Sebastian, A., Bader, D. J., Nederhoff, C. M., Leijnse, T. W. B., Bricker, J. D., & Aarninkhof, S. G. J. (2021). Hindcast of pluvial, fluvial, and coastal flood damage in Houston, Texas during Hurricane Harvey (2017) using SFINCS. *Natural Hazards, 109*, 2343–2362. https://doi.org/10.1007/s11069-021-04922-3

Sharma, S., & Mujumdar, P. P. (2022). Modeling concurrent hydroclimatic extremes with parametric multivariate extreme value models. *Water Resources Research, 58*(2), e2021WR031519.

Shisler, M. P., & Johnson, D. R. (2020). Comparison of methods for imputing non-wetting storm surge to improve hazard characterization. *Water, 12*, 1420. https://doi.org/10.3390/w12051420

Sivapalan, M., Wood, E. F., & Beven, K. J. (1990). On hydrologic similarity: 3. A dimensionless flood frequency model using a generalized geomorphologic unit hydrograph and partial area runoff generation. *Water Resources Research, 26*, 43–58. https://doi.org/10.1029/WR026i001p00043

Sklar, M. (1959). Fonctions de répartition à n dimensions et leurs marges. In Annales de l'ISUP (Vol. 8, No. 3, pp. 229–231).

Szpilka, C., Dresback, K., Kolar, R., Moghimi, S., & Myers, E. (2022). *Modifications to the ADCIRC hydrodynamic model to incorporate precipitation and inland hydrology*. National Oceanic and Atmospheric Administration, Silver Spring, Maryland

Sztobryn, M. (2003). Forecast of storm surge by means of artificial neural network. *Journal of Sea Research, 49*, 317–322. https://doi.org/10.1016/S1385-1101(03)00024-8

Toro, G. R., Niedoroda, A. W., Reed, C. W., & Divoky, D. (2010). Quadrature-based approach for the efficient evaluation of surge hazard. *Ocean Engineering, 37*, 114–124.

Tuleya, R. E., DeMaria, M., & Kuligowski, R. J. (2007). Evaluation of GFDL and simple statistical model rainfall forecasts for US landfalling tropical storms. *Weather and Forecasting, 22*(1), 56–70.

US Department of Homeland Security. (2017). FEMA Needs to Improve Management of Its Flood Mapping Programs.

US Department of Homeland Security. (2023). FEMA flood map service center. https://msc.fema.gov/portal/home. Accessed 1 Dec 2023

Vickery, P. J., Skerlj, P. F., Steckley, A. C., & Twisdale, L. A. (2000). Hurricane wind field model for use in hurricane simulations. *Journal of Structural Engineering, 126*(10), 1203–1221.

Villarini, G., Zhang, W., Miller, P., Johnson, D. R., Grimley, L. E., & Roberts, H. J. (2022). Probabilistic rainfall generator for tropical cyclones affecting Louisiana. *International Journal of Climatology, 42*, 1789–1802. https://doi.org/10.1002/joc.7335

Virtanen, P., Gommers, R., Burovski, E., Oliphant, T. E., Weckesser, W., Cournapeau, D., Peterson, P., Reddy, T., Haberland, M., & Wilson, J. (2021). scipy/scipy: SciPy 1.6. 0. Zenodo

Xu, H., Tian, Z., Sun, L., Ye, Q., Ragno, E., Bricker, J., Mao, G., Tan, J., Wang, J., Ke, Q., Wang, S., & Toumi, R. (2022). Compound flood impact of water level and rainfall during tropical cyclone periods in a coastal city: The case of Shanghai. *Natural Hazards and Earth Systems Sciences, 22*, 2347–2358. https://doi.org/10.5194/nhess-22-2347-2022

Yan, D., & Zhang, T. (2022). Research progress on tropical cyclone parametric wind field models and their application. *Regional Studies Marine Science, 51*, Article 102207.

Yang, K., Paramygin, V., & Sheng, Y. P. (2019). An objective and efficient method for estimating probabilistic coastal inundation hazards. *Natural Hazards, 99*, 1105–1130. https://doi.org/10.1007/s11069-019-03807-w

Young, I. R. (2017). A review of parametric descriptions of tropical cyclone wind-wave generation. *Atmosphere, 8*, 194.

Zhang, J., Taflanidis, A. A., Nadal-Caraballo, N. C., Melby, J. A., & Diop, F. (2018). Advances in surrogate modeling for storm surge prediction: Storm selection and addressing characteristics related to climate change. *Natural Hazards, 94*, 1225–1253.

Zhang, B., & Wang, S. (2021). Probabilistic characterization of extreme storm surges induced by tropical cyclones. *Journal Geophysical Research Atmospheres, 126*, e2020JD033557. https://doi.org/10.1029/2020JD033557

One Storm, Many Hazards: Multi-Hazard Planning for Wind, Surge, and Flood Risks in the Gulf Coast

Robert V. Rohli, Nazla Bushra, and Rubayet Bin Mostafiz

Abstract A landfalling tropical cyclone presents multiple perils, including extreme winds, storm surge, torrential rain, and tornadoes. Traditionally, however, risk assessments and emergency plans address these dangers in silos. For example, engineering standards not necessarily with the same rigor are applied to the wind vs. flood hazards, and to surge vs. river flood evacuation zones, with little consideration for the unique setting in which the threat occurs. Multi-hazard planning is needed to manage the tropical cyclone risk, especially along the multi-threat landscape of the U.S. Gulf Coast. The chapter argues for integrated planning that accounts for compound scenarios (e.g., a strong storm with heavy rainfall coinciding with high tide). Cutting-edge modeling tools are highlighted, with explanations of how planners and infrastructure managers use such tools to identify overlapping hazard zones and fortify weak points (e.g., assessing how wind-driven debris could compromise levees during a surge event). Policy and coordination are key themes; agencies can coordinate flood control, wind resistance standards, and evacuation planning into one comprehensive strategy. By planning for concurrent hazards, municipalities can better prepare for worst-case scenarios, minimizing cascading failures when the next storm strikes. This chapter thus bridges meteorology, engineering, and emergency management to promote holistic hurricane preparedness.

Keywords Gulf coast · Louisiana · Multi-hazard · Compound flooding · Storm surge · Wind · Risk communication · Resilience

R. V. Rohli (✉) · N. Bushra
Louisiana State University, Baton Rouge, LA 70803, USA
e-mail: rohli@lsu.edu

R. B. Mostafiz
Louisiana State University Agriculture Center, Baton Rouge, LA, USA

J. Collins et al. (eds.), *Tropical Cyclone Risk and Resilience*,
https://doi.org/10.1007/978-3-032-28745-8_3

1 Introduction

Hurricanes are not isolated meteorological events; rather, they include an ensemble of hazards, each with devastating potential, and even more dangerous with interaction. When a tropical cyclone makes landfall, it unleashes violent winds, storm surge, riverine and flash flooding, and tornado outbreaks. Historically, however, risk assessments, mitigation strategies, and emergency management and response efforts have treated these threats as distinct phenomena (Cegan et al., 2022), perhaps even managed by separate tools, models, and agencies. This siloed approach is increasingly inadequate in the face of compound and cascading risks, both of which consider that multiple hazards occurring together or in sequence, respectively, can generate impacts that are far greater than the sum of their parts. For example, during Hurricane Harvey (2017), catastrophic rainfall totals exceeding 1,000 mm in parts of Texas coincided with storm surge along the coast, producing simultaneous inland and coastal flooding (van Oldenborgh et al., 2017). In Louisiana, even moderate hurricanes can generate this "rain-on-surge" dynamic, with surge blocking drainage channels and rainfall accumulating behind levees, causing backwater flooding in areas thought to be protected.

Over the past two decades, science and practice have shifted toward systems thinking and hazard interaction theory, emphasizing the interconnected nature of extreme events (Gill & Malamud, 2016; Kappes et al., 2012). Compound and multi-hazard science now highlights dependencies among drivers and impacts that extend beyond single-hazard design (Zscheischler et al., 2018, 2020). For instance, regional analyses show that storm surge coinciding with extreme rainfall is increasingly consequential for U.S. coasts (Wahl et al., 2015), while sea-level rise amplifies fluvial flooding in backwater zones typical of coastal Louisiana, including in nuisance flooding (Moftakhari et al., 2017).

Nowhere is this multi-hazard reality more acute than the U.S. Gulf Coast, where low-lying geography, dense development, and exposure to Atlantic-borne cyclones create an extreme convergence of risks (Anderson et al., 2021). Yet much of the built environment, from levees to housing stock, was designed under assumptions of single-hazard exposure. Recent storms such as Katrina (2005), Harvey (2017), and Ida (2021) revealed how underestimating compound scenarios produces cascading failures across infrastructure and society. Attribution studies underscore this new normal, with human influence increasing the likelihood and magnitude of extreme hurricane rainfall, such as during Harvey, thereby intensifying rain-surge compound risk (van Oldenborgh et al., 2017).

Advances in coupled hydrodynamic modeling have improved understanding of these dynamics. Tools such as the Advanced CIRCulation (ADCIRC) + SWAN system capture interactions between waves, surge, and circulation (Dietrich et al., 2011a) and are validated for both design and real-time forecasting in Louisiana (Bunya et al., 2010; Dietrich et al., 2011b). Flood-zone transition analyses in low-gradient watersheds highlight nonlinear inundation responses when surge blocks drainage, a critical process in southern Louisiana (Bilskie & Hagen, 2018). Similarly, modeling platforms like Hazus-MH, Bayesian networks, and emerging machine learning approaches allow planners to evaluate wind, surge, and rainfall in an integrated way to avoid underestimating risks. Science alone, however, is insufficient. Risk communication research demonstrates how people interpret hurricane products and storm-surge graphics, with message framing significantly shaping protective action (Bostrom et al., 2016; Morss et al., 2018).

It is also important to consider the disproportionate burden on lower-income and marginalized communities, often widening wealth gaps (Howell & Elliott, 2019; Wing et al., 2022). Catastrophic storms such as Katrina reveal the disparities in New Orleans, Louisiana, particularly well, with disproportionate burden by race (e.g., Durant & Sultan, 2008; Sastry & Gregory, 2013), gender (e.g., Sastry & Gregory, 2013), and age (e.g., Maltz, 2019; Sastry & Gregory, 2013).

Similar patterns have been observed for post-Harvey Houston, with those with low incomes (e.g., Bodenreider et al., 2019) and disabilities (e.g., Chakraborty et al., 2019), and minority (e.g., Collins et al., 2019; Lieberknecht et al., 2021) and elderly (e.g., Bodenreider et al., 2019) populations. Aging infrastructure often associated with industry or sewage plants, all too often located in marginalized communities, is often a problem during severe flooding associated with Hurricanes, including Laura in 2021 (Cegan et al., 2022).

The 2023 Louisiana Coastal Master Plan (Coastal Protection and Restoration Authority (CPRA) 2023) reflects this integrated, multi-hazard approach at scale, combining engineered protection, nature-based solutions, and social considerations under scenarios of sea-level rise and subsidence. Wetland and reef studies further highlight the value of ecological infrastructure, demonstrating the potential of marsh and barrier-island restoration to attenuate storm surge and waves while providing co-benefits such as biodiversity and carbon storage (Barbier et al., 2011; Narayan et al., 2017). Such natural and nature-based solutions designed to restore coastal marsh through levee diversions are particularly helpful in both southeastern Louisiana and the Chenier Plain of southwestern Louisiana (Norman et al., 2022), especially when effectiveness of such features is assessed (Meselhe et al., 2020).

Collectively, these ideas point toward a paradigm shift in hurricane risk management. Effective resilience lies not in stronger responses to isolated hazards but in integrated planning for the full spectrum of threats that a single storm may unleash. This requires bridging meteorology, engineering, social science, and policy to design strategies that are scientifically robust and operationally feasible. By embracing multi-hazard planning, Gulf Coast and other communities can move from fragmented defense to coordinated resilience.

2 Understanding Hurricane Hazards

The multiple interacting hazards including extreme winds, storm surge, heavy rainfall leading to riverine and pluvial flooding, and, in many instances, tornadoes and infrastructure failure, that make hurricanes so complex and dangerous are important to consider first separately. Recognizing the distinct mechanisms, impacts, and interaction effects of these hazards is critical to the development of integrated, multi-hazard planning frameworks. This section provides an overview of the primary hazards associated with hurricanes and their compounding behavior in real-world scenarios.

2.1 Wind Hazards

Extreme wind is the most visibly dramatic hazard associated with hurricanes and often the first to be experienced, well before landfall. Sustained winds and gusts can exceed 100 mph in major storms, capable of toppling trees, tearing off roofs, and turning debris into deadly projectiles. Wind loading is a primary consideration in civil and structural engineering standards, particularly in hurricane-prone regions. However, damage assessments and resilience planning have traditionally focused on wind in isolation, often underestimating the impact when it coincides with other hazards.

Wind-induced damage can compromise critical infrastructure systems. For example, strong winds frequently down power lines and cellular towers, disabling early warning systems and hampering coordination during a disaster (Tupper & Fearnley, 2023; Unnikrishnan & Barbato, 2016). Buildings and bridges designed only for wind loads without accounting for concurrent effects of water inundation may suffer structural failure when storm surge or flooding follows.

Wei et al. (2024) demonstrated that building codes often fail to account for how wind speeds interact with topographical features and urban geometry, such as channeling effects between high-rise structures. In urban zones, wind can exacerbate fire risk during a storm event by damaging electrical systems and hindering firefighting capabilities. Furthermore, Lin (2010) emphasized that wind hazard mapping often lacks granularity in rapidly urbanizing coastal areas, leaving new developments especially vulnerable.

2.2 Storm Surge

Storm surge is the abnormal rise of sea level accompanying a hurricane, primarily driven by strong winds on the right forward side of the storm (in the Northern Hemisphere) pushing seawater toward the shore. Unlike gradual tidal fluctuations, storm surges can inundate low-lying coastal areas within minutes, causing extensive structural damage and drowning risks. Surge levels are particularly deadly when they coincide with high tides or are amplified by the shape and bathymetry of the coastline, such as funneling effects in bays and estuaries. Sahoo and Bhaskaran (2018) modeled storm surge interactions with seabed bathymetry and inland topography, affecting not only surge height but also inland penetration. Their research stresses the importance of dynamic surge modeling in risk maps and highlights that many current surge simulations underestimate compound flood risks when coupled with rainfall.

Storm surge poses a different set of engineering and emergency management challenges than wind. For example, levee and seawall systems must withstand immense hydrostatic pressure, wave action, and potential scouring at their base. In some cases, surge overtopping or infrastructure failure results in catastrophic inundation, as was the case during Hurricane Katrina (Pine et al., 2005). Moreover, storm surge frequently disables coastal transport systems and therefore the supply chain. Ports, rail networks, and coastal highways are particularly vulnerable, compromising disaster response logistics. When surge models are developed in isolation, they often fail to consider the timing of overlap with rainfall events or the loss of drainage due to backwater effects.

2.3 *Inland and Pluvial Flooding*

While storm surge tends to dominate the media coverage of hurricanes, rainfall-induced flooding often causes more widespread and longer-lasting damage, especially inland. Torrential rains can cause both riverine (i.e., capacity exceedance of rivers and streams) and pluvial (i.e., surface water accumulation through local rainfall, overwhelming drainage systems) flooding.

Ming et al. (2022) showed that hurricanes are increasingly delivering more rainfall per event in recent years, which increases atmospheric water vapor capacity. Urban expansion has also led to more impervious surfaces, reducing infiltration and increasing surface runoff. The combination creates a high likelihood of suburban and urban flooding, even miles inland from the storm's eye.

Thakur and Mohanty (2023) noted that rainfall severity and distribution are heavily influenced by storm track speed. Slow-moving hurricanes, such as Harvey in 2017, can stall over an area for days, depositing up to 50 inches of rain and overwhelming flood defenses built for much lower thresholds. In such cases, the failure of small dams, culverts, or storm water networks becomes a significant compounding hazard. Furthermore, rainfall can exacerbate storm surge by increasing the volume of water in estuaries and delaying the drainage of rivers into the ocean. This feedback loop significantly increases flood depth and duration, especially in coastal plains where elevation gradients are minimal.

2.4 Compound and Cascading Hazards

The most dangerous and least understood dimension of hurricane risk lies in the interaction between hazards. These compound events are more than the sum of their parts and often result in nonlinear impacts. Compound hazards can be simultaneous (e.g., storm surge and pluvial flooding), sequential (e.g., wind damage followed by rain infiltration into weakened structures), or cascading (e.g., power loss leading to pump failure, triggering urban inundation).

Gill and Malamud (2016) outlined a taxonomy of multi-hazard interactions, noting that planning and mitigation efforts rarely account for synergistic effects. For example, emergency evacuation planning based solely on storm surge zones might inadvertently route evacuees through areas at high risk of flash flooding due to rainfall. Lee et al. (2024) used machine learning to simulate compound risk scenarios in Gulf Coast communities and found that damage levels increase significantly when two or more hazards co-occur. Risk communication and modeling systems must incorporate these interactions to provide realistic assessments.

Cascading hazards are particularly insidious in urban infrastructure systems. The interdependence of power, potable water, transport, and communication systems means that a failure in one system can propagate quickly through others, amplifying societal disruption and danger. Despite the growing recognition of these interactions, hazard-specific planning frameworks still dominate most local emergency management strategies. The challenges lie in data integration, coordination across agencies, and the technical complexity of modeling multiple systems simultaneously.

3 Problems with Traditional Siloed Risk Planning

The treatment of each hazard in isolation that is typical of conventional disaster planning and management approaches is rooted in institutional structures, disciplinary specialization, and the historical development of emergency management systems. This section explores in greater detail the structural, institutional, and methodological shortcomings of single-hazard frameworks and discusses how they fall short in the context of real-world disaster scenarios.

3.1 Sectoral Fragmentation of Planning Institutions

One of the primary issues with traditional risk management lies in the institutional separation of hazard domains. Agencies responsible for hurricane response, flood management, infrastructure resilience, and emergency communications often operate independently, guided by distinct mandates, funding streams, and regulatory standards. During a July 2025 interview, a local non-profit administrator lamented to a member of the author team the inefficiencies of working with governmental disaster relief agencies because the government officer is only concerned with mitigating or assessing the value relative to the hazard under their purview.

From an administrative perspective, wind risk in the U.S. is largely addressed through building codes regulated by state or municipal governments, often informed by organizations like the American Society of Civil Engineers (ASCE). Flood risk, by contrast, is managed through federal programs such as (U.S.) Federal Emergency Management Agency's (FEMA's) National Flood Insurance Program (NFIP), which uses 100-year floodplain maps and prioritizes riverine and coastal flooding. Storm surge modeling may be handled by the National Hurricane Center (NHC) or the U.S. Army Corps of Engineers, while local storm water drainage systems fall under the jurisdiction of municipal departments of public works.

This compartmentalization creates gaps and overlaps that hinder effective risk reduction. As Julià and Ferreira (2021) noted, sectoral fragmentation leads to inconsistent risk assessments, conflicting priorities, and poorly coordinated responses. For example, a municipality might approve a development based on minimal wind hazard risk, unaware that the same area is prone to pluvial flooding when drainage systems fail due to surge backflow. Without interagency coordination, comprehensive resilience becomes impossible.

3.2 Limitations of Single-Hazard Modeling Frameworks

Traditional hazard assessment tools such as wind maps, floodplain delineations, or surge simulations are designed to address one and only one threat. These models are built on simplifying assumptions and historical data, often excluding the compounding or interacting nature of hazards. For instance, wind damage models may assume that structural failure occurs under wind load alone, without accounting for weakening due to water infiltration or inundation. Similarly, storm surge models often do not incorporate rainfall inputs or river discharge rates, failing to represent combined flood depths accurately.

Moreover, risk assessments based on separate hazard return periods, such as the 1-in-100-year flood or the 1-in-50-year windstorm, assume statistical independence, which does not reflect observed compound or cascading events. Wahl et al. (2015) and Zscheischler et al. (2018) demonstrated that compound flooding events occur more frequently than previously estimated, especially under conditions of tropical cyclones where heavy rainfall and surge coincide. These modeling limitations result in policy blind spots. Development decisions, insurance pricing, and infrastructure investments based on under representative risk models can amplify exposure and vulnerability. As a result, communities may believe they are prepared for certain hazards when, in fact, their systems are only resilient to those considered in isolation, which exacerbates the insurance crisis that U.S. society is currently facing.

3.3 *Inefficient Emergency Response Coordination*

Siloed planning extends into emergency operations. In many jurisdictions, emergency response is divided into functional units of fire, police, health, transportation, each with its own protocols, resource allocations, and lines of authority. During a multi-hazard event like a hurricane, this disjointed structure impedes communication, decision-making, and logistical coordination. For instance, flooding may block critical evacuation routes designated by emergency managers focused on wind zones, or power outages from wind damage may disable flood warning systems downstream. When agencies fail to share common operating pictures or real-time data, they may issue conflicting guidance, increasing public confusion and response delays. Siloed emergency coordination led to inefficient resource deployment in the aftermath of Hurricanes Harvey, Irma, and Maria (National Academies of Science, Engineering, and Medicine (NASEM) 2020). Lack of interoperability among systems used by different agencies, such as separate GIS platforms, communications networks, and mapping software, resulting in overlapping evacuation orders and duplicated relief efforts, are often problematic.

Further, silos often hinder adaptive response. In a dynamic hurricane event, response plans must evolve as conditions change. A rigid, department-specific plan may not permit flexible resource reallocation or cross-functional problem-solving. This problem is compounded in urban areas, where interdependent systems, such as power for water pumps or transportation for hospital access, demand coordinated, multi-sector responses.

3.4 *Over-Reliance on Historical Hazard Paradigms*

Another significant limitation of traditional planning approaches is the heavy reliance on historical precedent. Floodplain maps, wind zone maps, and storm surge models are frequently based on past events, which no longer represent future risks in a changing climate. This retrospective bias embeds systemic vulnerabilities into current planning practices. Climate change is altering hazard profiles. Hurricanes are becoming wetter, slower, and more intense, increasing the likelihood of compound flooding (Mousavi et al., 2011). Sea-level rise amplifies storm surge impacts, while urban expansion into previously unpopulated areas exposes new assets to extreme wind and flood threats. Yet many regulatory frameworks and emergency planning tools remain anchored in outdated datasets and hazard assumptions.

This over-reliance on the past limits anticipatory planning. For example, FEMA flood maps are updated slowly and do not always include pluvial flood zones, even as these become more prevalent. Similarly, storm surge barriers and levees may be built to outdated design storms that fail to reflect future compound risks. Without forward-looking, dynamic planning methods, traditional strategies will increasingly leave communities exposed to the elevated risk.

3.5 *Implications of Siloed Planning*

Finally, single-hazard approaches often fail to account for how compound risks disproportionately affect vulnerable populations. For example, affordable housing developments might meet wind resistance standards but be in areas without adequate flood drainage or surge protection. The failure to integrate multi-hazard considerations into zoning and permitting decisions often places low-income or minority populations at greater risk of compound events. Likewise, recovery programs are often tailored to single hazards. Insurance programs like NFIP may only cover flood damage, excluding wind-related losses. FEMA aid categories may not capture losses from cascading effects, such as mold from water intrusion after wind damage. These programmatic silos can delay recovery and deepen socioeconomic disparities.

4 Joint-Hazard Scenario Tools for Parish Planning

Joint-hazard scenario tools help local officials recognize, visualize, and stress-test operations under the overlapping conditions associated with multiple simultaneous hazards in progress. Figure 1 shows a conceptual schematic of compound flooding (Moftakhari et al., 2017); rainfall, storm surge, and background sea-level rise interact to threaten levee crests and pumping operations. For example, even if rainfall alone is within pump capacity, a concurrent surge can raise water levels in canals, blocking outflow and creating "backwater flooding." Similarly, sea-level rise reduces the freeboard margin of levees, meaning conditions that were once manageable may now result in overtopping.

Rainfall runoff accumulates behind levees while storm surge and sea-level rise elevate external water levels, reducing freeboard and overtopping safety margins. Strong winds both intensify surge and threaten pumps and power supply, creating cascading risks where drainage capacity fails just as rainfall peaks. Backwater flooding (rain-on-surge) occurs when elevated surge blocks river and canal outflow, inundating areas otherwise considered safe. Pumps attempt to evacuate interior water but remain vulnerable to wind-driven power outages. This schematic illustrates how concurrent hazards—wind, surge, rainfall, and sea-level rise—interact with levee and pump operations to generate nonlinear impacts.

To translate such interactions into actionable planning, Table 1 provides an illustrative *joint-hazard scenario matrix*. This tool allows emergency managers and engineers to test their system's performance under different combinations of wind, surge, and rainfall. Each row represents a distinct hazard configuration, paired with operational or planning priorities.

Scenario A (Moderate) might involve a slow-moving Category 1 hurricane. While such a storm may not threaten levee overtopping, it could still deliver as much as 150 mm of rainfall over a 48-h period. This stresses internal drainage systems and allows officials to test whether existing pumps can clear water from canals quickly

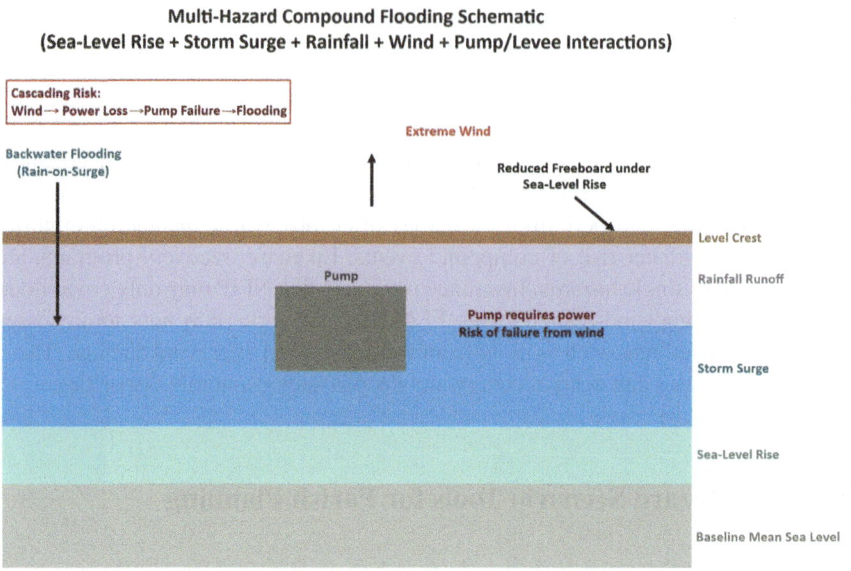

Fig. 1 Multi-hazard compound-flooding schematic (sea-level rise + storm surge + rainfall + wind + pump/levee interaction), developed by the authors for illustrative purposes. Informed by multi-hazard planning concepts, the matrix highlights how overlapping hazards can stress-test parish systems and guide operational decision-making

enough, or whether debris accumulation might clog culverts and trigger localized neighborhood flooding.

Scenario B (Severe) represents a more intense Category 3 storm producing a 2.5 m surge in combination with 200 mm of rain. In this case, municipal decision-makers would need to consider whether navigation gates should be closed to prevent surge from pushing inland through waterways, and whether contraflow operations should be activated on major highways to facilitate evacuation. This scenario is particularly valuable for stress-testing both hydrodynamic and traffic systems together, highlighting where conflicts or bottlenecks may arise.

Scenario C (Extreme) illustrates the risks posed by a Category 4 storm, bringing a 3–4 m surge along with 300–400 mm of rainfall. This type of event challenges the integrity of levee systems, which may be subject to erosion, seepage, or even failure under prolonged stress. It also tests whether assets such as sandbags, portable pumps, and backup generators can be pre-positioned quickly enough to reinforce vulnerable sites. Moreover, mutual aid agreements between neighboring parishes and state agencies become essential in this scenario, as no single jurisdiction may have sufficient capacity to respond independently. Scenario C therefore emphasizes proactive logistics and resource mobilization to maintain critical services during extreme but increasingly plausible events.

At the upper end, Scenario D (Catastrophic) depicts a Category 5 storm with surge heights exceeding 4 m and rainfall totals surpassing 400 mm. This situation

Table 1 Illustrative joint-hazard scenario matrix for a hypothetical parish, integrating wind, storm surge, and rainfall conditions with corresponding operational and planning priorities

Scenario	Wind (Saffir-Simpson)	Surge at lake/pass	24–48 h Rainfall	Operational/planning focus
A—Moderate	Cat 1	1–2 m	100–150 mm	Check pump capacity under sustained rainfall; clear canal debris; public information campaigns to reduce complacency
B—Severe	Cat 3	2–3 m	200–250 mm	Close navigation canals to prevent surge intrusion; stage sheltered operations for first responders; implement traffic contraflow to support evacuation
C—Extreme	Cat 4	3–4 m	300–400 mm	Reinforce or armor vulnerable levee sections; pre-position emergency assets (sandbags, generators); activate mutual-aid agreements with neighboring parishes
D—Catastrophic	Cat 5	> 4 m	> 400 mm	Initiate full evacuation for surge + rainshed; secure hospital continuity plans; ensure long-term sheltering capacity
E—Rain-on-surge	Cat 2–3	2–3 m	350–450 mm	Prioritize staging of portable pumps; utilize upstream storage basins; emphasize public messaging that heavy rainfall on surge can flood "safe" areas

pushes protective and emergency systems to their limits. Under such conditions, leaders must ask whether hospital backup generators are positioned above projected flood levels, whether nursing homes have pre-arranged evacuation plans, and whether shelter capacity is adequate to sustain a large, displaced population. This scenario exposes the "breaking points" in continuity and life-safety plans.

Scenario E (Rain-on-surge) is particularly relevant for Louisiana, where surge frequently blocks rivers and canals, impeding drainage. Heavy rainfall on top of surge can flood areas that are otherwise considered safe, extending flood risks well inland. This scenario highlights the importance of temporary storage basins, staged pumping strategies, and above all, effective public communication. Residents must understand why flooding may occur even if their area is outside traditional surge or flood maps.

Cycling through these scenarios in tabletop exercises or computer-based simulations allows parishes to identify where critical infrastructure such as pumps, levees, hospitals, and shelters, fails under combined stresses. It also provides an opportunity to test coordination protocols, particularly in situations where protective actions overlap or conflict, such as when contraflow traffic plans coincide with shelter operations or canal closures interfere with shipping.

These exercises further reveal which neighborhoods or populations face the highest residual risks under compound flooding scenarios, such as rain-on-surge. Integrating these findings into planning aligns local preparedness efforts with state-scale frameworks like the *Louisiana Coastal Master Plan* (CPRA 2023), which promotes structural, nonstructural, and nature-based approaches under projected sea-level rise and subsidence. In sum, joint-hazard scenario tools transform abstract multi-hazard science into practical, parish-scale decision support. They enable leaders to prepare for events that rarely follow neat, single-hazard boundaries and to build resilience that reflects the true complexity of Gulf Coast hurricane risk.

5 Multi-Hazard Risk Frameworks

To overcome the limitations of siloed planning, disaster risk researchers and practitioners have increasingly advocated for multi-hazard risk frameworks. These approaches seek to understand, assess, and manage the cumulative and interacting risks posed by co-occurring or sequential hazards. In the context of hurricanes, multi-hazard frameworks offer a more realistic and effective basis for resilience planning. This section explores the conceptual foundations of multi-hazard frameworks, outlines various methodological approaches to modeling and assessment, and reviews emerging tools and best practices that enable integrated risk management across hazards and sectors.

5.1 Conceptual Foundations of Multi-Hazard Thinking

The systems perspective of multi-hazard risk frameworks shifts the focus away from treating floods, hurricanes, or heatwaves as discrete events and instead frames them as interconnected processes within human–environment systems. The United Nations Sendai Framework for Disaster Risk Reduction (Pearson & Pelling, 2015) emphasizes the need for integrated, multi-hazard approaches across all government levels, with policymakers, engineers, and communities anticipating overlapping and cascading hazards rather than planning for them in isolation.

Louisiana-specific hydrodynamic research reinforces this reality. Bilskie and Hagen (2018) showed that nonlinear inundation responses emerge in low-gradient watersheds when surge blocks river drainage, producing flooding well inland of expected surge zones. Similarly, as described earlier, coupled wave-surge-circulation models such as SWAN + ADCIRC have been validated for Louisiana (Dietrich et al., 2011b) and have been used extensively for both design standards and real-time forecasting (Suh et al., 2015). These models demonstrate how surge, tide, wind, and rainfall interact dynamically across the state's estuaries and floodplains, underscoring the inadequacy of single-hazard planning. Infrastructure interdependencies in Louisiana also exacerbate risk; Hurricane Katrina (2005) not only generated an unprecedented storm surge that overtopped and breached levees in New Orleans but also knocked out pump stations and electrical supply lines (IPET, 2006). As a result, the city's capacity to remove floodwaters collapsed just when it was most needed, prolonging inundation and amplifying social and economic losses. Similarly, Hurricane Ida (2021) demonstrated how sustained winds of over 150 mph toppled nearly the entire transmission network serving southeast Louisiana. This outage left hundreds of thousands without power for weeks, halting water pumping operations and limiting the functionality of hospitals and cooling centers during extreme heat (Entergy, 2021). In both cases, the interaction between physical hazards and infrastructure fragility transformed the disaster into a cascading crisis.

Sequential hazards present another foundation for consideration of compounding and cascading risk in the Gulf region. Hurricanes Harvey, Irma, and Maria struck in rapid succession in 2017, exhausting federal and local response capacities. Puerto Rico's devastation from Maria was magnified by the island's fragile infrastructure and by resources already stretched thin following Irma (Pasch et al., 2023). While not as tightly clustered, Louisiana has also experienced back-to-back disasters that exacerbated vulnerability, such as the nearby landfalls of Hurricanes Laura and Delta within six weeks in 2020 in southwestern Louisiana. Communities already reeling from Laura's Category 4 destruction faced renewed flooding, extended displacement, and slower recovery under Delta (Cangialosi & Latto, 2021).

These dynamics highlight that resilience planning cannot focus solely on "worst-case" scenarios within individual hazard categories, such as the maximum surge or the highest wind speed. Instead, planning must account for plausible multi-hazard scenarios that reflect real-world complexity, such as cases of overlap of surge, wind, and local rainfall damage, cases in which a sequence of storms affects the same area in a single hurricane season, or cases of wet antecedent conditions prior to an extreme event. Such planning can avoid or minimize the possibility of overwhelming housing recovery and shelter systems.

By incorporating these compound and sequential risks into planning, multi-hazard frameworks provide a more realistic view of disaster risk as systemic, interconnected, and unevenly distributed. They reveal how cascading failures of levees, pumps, or power grids disproportionately affect the communities with fewer resources to mitigate, evacuate, rebuild, or adapt. Improved predictive accuracy from such approaches also yields the benefit of more comprehensive design of resilience strategies. Adopting multi-hazard frameworks means moving from fragmented defenses toward coordinated resilience strategies that acknowledge the true complexity of the risks posed by a changing climate.

5.2 Typologies of Multi-Hazard Relationships

A crucial step in designing effective multi-hazard frameworks is understanding the relationships among hazards, since the way they interact determines the nature and scale of disaster impacts. Kappes et al. (2012) outlined a descriptive matrix of multi-hazard interactions, based on earlier work by many, leading to the identification of five multi-hazard interaction categories. The first category involves multiple events that have little or no direct influence on one another. For example, an earthquake in California and a hurricane in Louisiana may occur in the same season but remain largely unrelated in their causes and impacts. Such independent hazards can also occur in the same region over time, such as a flood and tornado in consecutive years, requiring communities to prepare for multiple threats without assuming direct connections between them.

The second category is triggered events, in which one disaster initiates another, such as hurricane-induced flooding and landslides, or earthquake-induced avalanches or landslides that complicate response efforts. A related category is that of amplifying effects, in which one event intensifies the severity or likelihood of another, such as hurricane storm surge amplifying pluvial flooding, elevating stream water levels, hindering drainage, and exacerbating the upstream impacts. Similarly, the loss of electricity after a summer storm could exacerbate the effect of subsequent high temperatures on human comfort and safety.

The fourth category is compounding events, in which two or more occurrences (and any amplifying impacts) occur simultaneously or in close succession, increasing the cumulative risk. For example, in the 2017 Atlantic hurricane season, Hurricanes Harvey, Irma, and Maria struck in rapid succession, exhausting resources and leaving communities less capable of absorbing subsequent shocks. In Louisiana, Hurricanes Laura and Delta in 2020 similarly illustrate compounding effects, as the second storm struck before full recovery from the first was possible.

Finally, cascading effects occur when one event sets off a chain of secondary impacts *across systems*. Katrina (2005) remains a stark case: storm surge led to levee breaches, which caused citywide inundation, which in turn disabled pumping stations and power infrastructure. These failures then cascaded into public health crises, mass displacement, and long-term socioeconomic disruption. Cascading effects remind planners that disasters unfold as evolving processes simultaneously at different spatial and temporal scales, rather than isolated shocks, with initial damage often triggering a series of secondary crises.

Anticipating these possibilities allows planners and decision-makers to be more effective in designing targeted mitigation strategies and prioritizing investments where they will be most effective in interrupting the domino-like chain of impacts. For example, recognizing amplifying relationships can guide the reinforcement of drainage systems ahead of surge events, while acknowledging cascading risks may emphasize protecting critical infrastructure nodes, such as hospitals or power substations, that can prevent small failures from becoming systemic crises. In this way, typologies of multi-hazard interactions are not just conceptual tools, but practical guides for building resilience in hazard-prone regions like the Gulf Coast.

5.3 Multi-Hazard Risk Modeling Approaches

Advances in geospatial modeling, remote sensing, and computational simulation have transformed how researchers and practitioners analyze compound risks. These innovations are critical for translating the conceptual logic of multi-hazard frameworks into operational planning and decision-making (AghaKouchak et al., 2020; Zscheischler et al., 2018). Several approaches have become particularly influential in this regard, especially in Louisiana's coastal parishes.

One widely used method is probabilistic risk assessment (PRA), which uses statistical models to estimate the likelihood of different hazard combinations and their consequences (Paté-Cornell, 1996). Instead of considering rainfall, storm surge, or wind in isolation, PRA builds joint probability distributions that capture how these variables may co-occur. For example, Wahl et al. (2015) demonstrated how joint distributions of rainfall and storm surge can quantify compound flood risk in U.S. coastal cities. This approach has direct relevance for Louisiana, where parish planners increasingly need to evaluate not just whether levees can withstand a given surge height, but how moderate surge combined with heavy rainfall could overwhelm interior drainage systems. PRA outputs provide risk curves that reveal the probability of these compound extremes, guiding investments in pump stations, storage basins, and levee elevations.

A second approach is scenario-based modeling and planning (Schweizer & Kriegler, 2012), which creates representative multi-hazard narratives to test resilience under realistic but challenging conditions. In Louisiana, scenario-based modeling has been central to CPRA's Coastal Master Plan (), which uses a suite of storm and sea-level rise scenarios to evaluate the performance of structural measures (e.g., levees, surge barriers) and nonstructural strategies (e.g., elevating homes, wetland restoration, other forms of nature-based solutions). Local parishes also employ scenario exercises to test evacuation logistics under stress, such as simulating a slow-moving Category 4 hurricane coinciding with heavy rainfall. These scenarios account for both physical impacts and social variables, including household mobility, income, and access to vehicles. For example, Jefferson Parish authorities have tested explicitly whether vulnerable populations, such as nursing home residents, can be evacuated under simultaneous surge and rainfall-driven flooding (CPRA, 2017).

Another innovative tool is agent-based modeling (ABM). Unlike purely physical simulations, ABMs depict the actions and interactions of individual agents, such as households, emergency responders, or infrastructure operators (Crooks & Heppenstall, 2012). These models capture how human decisions and system dependencies influence hazard outcomes (Murray-Tuite & Wolshon, 2013). In a Louisiana context, ABMs can simulate hurricane evacuation under different communication strategies, showing how misinformation or delayed warnings can produce traffic congestion on evacuation routes such as Interstate Highways 10 and 12. Other ABMs explore cascading failures, such as how a power outage in Plaquemines Parish at the "toe" of Louisiana's "boot" shape could disable pump stations, exacerbating flooding and delaying recovery. By linking human decision-making with physical hazard models, ABMs provide insights into the social dimensions of compound risk (Tonn & Guikema, 2018).

A fourth approach is integrated hazard indexing, which combines multiple hazard layers into composite measures of exposure. Using geographic information systems (GIS), hazard indices integrate wind speed zones, floodplain maps, and surge extents to identify multi-hazard hotspots (Cutter et al., 2003). FEMA's HAZUS-MH platform is already being applied in Louisiana to estimate potential economic losses from hurricane wind and flood hazards (FEMA, 2020a). For example, planners in Orleans Parish have used HAZUS-MH outputs to identify neighborhoods where high wind exposure coincides with high flood vulnerability, prioritizing them for resilience investments such as roof retrofits and elevation programs. Similarly, integrated hazard mapping has been used in the Lower Ninth Ward to show how storm surge zones overlap with high social vulnerability, guiding both infrastructure improvements and community outreach (Cutter & Emrich, 2006).

While these modeling approaches differ in scope, complexity, and data requirements, they share a commitment to integrating multiple hazards into unified risk perspectives. Yet significant challenges remain. Harmonizing disparate datasets, such as ADCIRC + SWAN hydrodynamic outputs, rainfall-runoff models, and demographic surveys, requires careful calibration across temporal and spatial scales (Dietrich et al., 2011a). Furthermore, each method carries uncertainties that must be clearly communicated to decision-makers. For example, ADCIRC + SWAN storm surge forecasts for Louisiana provide high-resolution hydrodynamics but must be interpreted alongside rainfall uncertainties and future sea-level rise projections (Bilskie et al., 2016). If uncertainty is poorly conveyed, results may be misinterpreted as deterministic predictions rather than probabilistic guidance with inherent error bars.

5.4 Institutional and Governance Dimensions

The effectiveness of multi-hazard frameworks depends as much on institutional design and governance capacity as on technical models or scientific tools. Translating analytical insights into real-world resilience requires coordinated governance structures that can bridge administrative boundaries and political regimes, synchronize policies, and integrate diverse actors. In practice, this means that institutions must not only adapt existing mandates but also realign governance systems to reflect the complexity of multi-hazard risk.

Two types of integration are particularly critical. Horizontal integration refers to coordination across agencies operating at the same level of government. For example, at the parish level in Louisiana, public works departments responsible for drainage must coordinate closely with emergency management agencies, public health officials, and housing authorities. Without such alignment, responses may be fragmented, such as pumps being operated without considering evacuation routes or health services being unprepared for cascading outages. Vertical integration, on the other hand, involves coordination across different levels of governance, from federal to local. Federal flood insurance and mitigation programs (e.g., FEMA's, 2020b) NFIP and BRIC grants) are most effective when they work in tandem with state coastal restoration efforts and parish-level emergency planning. Without vertical alignment, investments risk duplication or misallocation, such as when federal resources target levee improvements while local governments emphasize housing retrofits without cross-communication.

Achieving this kind of integration requires shared data standards, interoperable platforms, and joint training exercises (Colmer et al., 2025). For example, the Netherlands' Delta Programme (Bloemen et al., 2019) demonstrates how climate adaptation, spatial planning, and water management can be brought together under one institutional umbrella. The program uses integrated models to evaluate long-term risks, while aligning investments across ministries, provinces, and municipalities. Similar efforts are emerging on the Gulf Coast, where Louisiana's Coastal Master Plan offers a statewide framework to integrate federal hydrodynamic modeling, state restoration projects, and parish mitigation strategies in a common planning horizon.

Beyond government, public–private partnerships play a crucial role in multi-hazard resilience. Critical infrastructure systems such as electricity, water, and communications are often managed by private utilities, yet they are highly vulnerable to hazards like storm surge, wind, and cascading blackouts. Partnerships between public agencies and utility providers can ensure coordinated investments in grid hardening or backup power for pumps and hospitals. The insurance sector also shapes resilience by influencing building codes and incentivizing risk reduction. Developers and community-based organizations, meanwhile, are essential for ensuring that resilience strategies are fair. For example, collaborations between local governments, developers, and nonprofits in New Orleans have supported the elevation and retrofitting of affordable housing units in neighborhoods with overlapping wind and flood risks.

Finally, adaptive governance provides a guiding principle for managing uncertainty in a changing hazard regime. Unlike static regulatory approaches, adaptive governance emphasizes flexibility, learning, and iterative adjustment. This means that multi-hazard plans cannot remain fixed but must be regularly updated as new data, technologies, and lived experiences emerge. For instance, pump capacity assumptions in Louisiana parishes must be revisited considering observed increases in extreme rainfall, while evacuation planning must adapt to changing demographics and mobility patterns. By institutionalizing flexibility through after-action reviews, periodic plan updates, and cross-sector learning exercises, adaptive governance ensures that multi-hazard planning remains dynamic rather than reactive.

5.5 Emerging Tools and Best Practices

As multi-hazard frameworks become more widely adopted, a growing set of tools and frameworks is helping practitioners operationalize this complex planning approach. These platforms integrate hazard data, vulnerability indicators, and planning scenarios into decision-making processes that can be applied across scales, from local parishes to national governments.

One influential platform is The Coastal Resilience Tool, developed by The Nature Conservancy. This tool integrates data on storm surge, shoreline erosion, sea-level rise, and ecological assets such as wetlands and oyster reefs. By mapping both hazards and natural defenses, it allows planners to visualize where ecological restoration projects, such as marsh creation or barrier island reinforcement, can provide flood protection while enhancing biodiversity. In Louisiana, where coastal wetlands are disappearing at alarming rates, such tools highlight the dual role of ecological infrastructure as both a buffer against hurricanes and a foundation for long-term coastal resilience.

FEMA's (2020c) Risk Rating 2.0 (RR2.0) represents another important consideration in multi-hazard integration. Unlike the traditional flood insurance rate maps that focused narrowly on riverine or coastal flooding, RR2.0 uses actuarial models that account for multiple flood sources including pluvial flooding from intense rainfall, as well as property-specific conditions such as distance to water, elevation, and rebuilding costs. NOAA's Digital Coast offers a complementary suite of data, visualization, and training resources for multi-hazard planning. Its tools enable users to model storm surge impacts, assess land-use changes, and evaluate coastal vulnerability. For example, planners in Plaquemines Parish can overlay land subsidence data, projected surge extents, and critical infrastructure maps to identify neighborhoods where compounded risks are highest. Because Digital Coast includes training modules, it also helps build local capacity to interpret and use data effectively, bridging the gap between technical analysis and governance practice.

At the international scale, the Global Risk Assessment Framework (GRAF) developed by United Nations Office for Disaster Risk Reduction (UNDRR) provides structured guidance for integrating multi-hazard and systemic risk perspectives into national and regional risk assessments. GRAF emphasizes the importance of understanding systemic vulnerabilities, such as supply chain interdependencies or health system fragilities, that compound hazard impacts. Its recommendations echo in Louisiana's ongoing efforts to incorporate health, energy, and housing systems into statewide resilience planning, recognizing that disasters are not just natural phenomena but socio-technical crises.

Alongside these tools, a set of best practices emerges from both the academic literature and applied field experience. First, multi-hazard vulnerability assessments that combine physical hazard exposure with social and infrastructure indicators are increasingly recognized as foundational. For instance, overlaying flood hazard maps with census data and social vulnerability indices reveals how low-income neighborhoods face disproportionately high risks for prioritization of mitigation resources.

Engaging directly with communities to shape response strategies such as through ensuring that evacuation planning accounts for households without vehicles or that flood mitigation subsidies are accessible to renters and low-income homeowners, are also important strategies. Second, resilient infrastructure design is a cornerstone of multi-hazard planning. Redundancy and flexibility are critical features. Hardening of energy substations, backup power systems, promotion of the use of Insurance Institute for Business and Home Safety's FORTIFIED® construction standards to promote wind resilience to residential structures are some examples. Third, nature-based solutions have gained traction as interventions capable of mitigating multiple hazards simultaneously while providing ecological and social co-benefits. Restored wetlands can attenuate storm surge, reduce wave energy, and store floodwaters from heavy rainfall, while also supporting fisheries and carbon sequestration. Urban examples, such as green roofs, permeable pavements, and bioswales, such as in Baton Rouge and New Orleans, demonstrate how nature-based approaches can reduce both heat stress and urban flooding. Together, these tools and best practices underscore that operationalizing multi-hazard frameworks requires governance approaches that are socially attuned, ecologically grounded, and institutionally coordinated. In tandem, they provide a comprehensive playbook for managing the complex risks facing coastal regions such as Louisiana.

6 Integrated Modeling of Hurricane Risk

As described earlier, effectively planning for multi-hazard hurricane events requires models capable of capturing the full complexity of interacting hazard drivers such as wind, storm surge, and precipitation, as well as their cascading impacts on both built and natural systems. Integrated modeling frameworks have emerged as central tools in advancing this capacity. Unlike traditional single-hazard approaches that consider each process in isolation, integrated models simulate the joint behavior of multiple hazards and their combined consequences. These frameworks often couple physically based models with socio-economic exposure data and dynamic vulnerability metrics, thereby moving beyond purely physical forecasts to provide actionable insights for risk reduction (Wing et al., 2022).

6.1 Evolution of Hurricane Risk Modeling

Historically, hurricane risk assessment was fragmented across disciplinary boundaries. Meteorologists focused primarily on storm tracks and wind fields, hydrologists analyzed rainfall-runoff processes, oceanographers developed storm surge models, and emergency managers relied heavily on static floodplain maps. While each of these specialized approaches generated detailed insights within its respective domain, critical interdependencies, such as how wind-driven surge interacts with rainfall-induced flooding, and their uncertainties, are often neglected (Hall & Solomatine, 2008).

The inadequacy of such siloed approaches (IPCC, 2022) was brought into sharp focus by catastrophic events such as Hurricane Katrina in 2005, Superstorm Sandy in 2012, and Hurricane Harvey in 2017. Each revealed that focusing on one hazard in isolation grossly underestimated the scale of disaster impacts. For example, Harvey demonstrated how prolonged rainfall could overwhelm drainage systems while coinciding with surge-related backflow in Houston's bayous, creating a devastating compound flood event (van Oldenborgh et al., 2017). Advances in computational power, real-time data collection from satellites and radar, and the growing availability of open-source hazard and risk datasets further accelerated the shift toward integrated modeling (Baró et al., 2017).

6.2 Core Components of Integrated Hurricane Models

Modern integrated hurricane risk models consist of several interlinked modules that represent both the hazard chain and the exposure landscape. At their foundation lies the meteorological forcing module, which ingests key hurricane parameters such as storm track, central pressure, wind intensity, radius of maximum winds, and forward speed. Models such as the Hurricane Weather Research and Forecasting (HWRF) system, the Global Forecast System (GFS), and the ECMWF Ensemble Prediction System (EPS) are widely used to generate surface wind fields, precipitation rates, and pressure gradients (Tallapragada et al., 2014; Magnusson et al., 2019).

Hydrodynamic surge models form the next critical component. Tools simulate the ocean's response to hurricane forcing such as ADCIRC (Bunya et al., 2010), the FEMA-National Weather Service (NWS)-USACE-collaborative Sea, Lake, and Overland Surges from Hurricanes (SLOSH) model (Jelesnianski, 1992), and more complex platforms like SCHISM (Zhang et al., 2016) and Delft3D (Deltares Systems, 2018). These models incorporate bathymetry, tides, and wave interactions, and in advanced configurations can account for dynamic coastal changes such as erosion of barrier islands or wetland loss. For example, ADCIRC was used in post-Katrina studies to evaluate how levee breaches and surge dynamics combined to inundate New Orleans.

Equally important are hydrological rainfall-runoff models that simulate inland flooding from extreme precipitation. Tools like the Hydrologic Engineering Center's HEC-HMS, the Storm Water Management Model (SWMM), and distributed systems such as GSSHA or LISFLOOD can represent watershed responses, drainage systems, and soil saturation (Arnold et al., 2015). In coastal regions, these are increasingly coupled with surge models to capture compound flood dynamics.

Outputs from surge and runoff simulations are then passed to hydraulic inundation models, which project flood extents and depths across landscapes. Models such as HEC-RAS 2D, FLO-2D, and CityCAT integrate digital elevation models to simulate how water flows across urban and rural terrain, sometimes including sediment and debris transport (Neal et al., 2012). The ability to capture compound flooding is especially critical.

To connect physical hazards with societal consequences, integrated models also require exposure and vulnerability modules. These incorporate geospatial datasets on population distribution, infrastructure networks, buildings, and critical facilities such as hospitals and emergency shelters. Vulnerability is quantified through fragility curves, damage functions, and social vulnerability indices such as the CDC's Social Vulnerability Index or the Baseline Resilience Indicators for Communities (Cutter et al., 2010). By linking hazard intensity with exposure, models can generate meaningful estimates of damage and risk distribution across communities.

The final stage of integrated modeling lies in loss estimation and impact assessment. Tools such as HAZUS-MH, RISKScape, and CAPRA aggregate hazard and exposure data to estimate direct structural damages, indirect economic losses, service disruptions, and human consequences (Scawthorn et al., 2006). These outputs are often probabilistic, offering planners multiple scenarios of potential impacts. For example, New York City's Coastal Storm Modeling System (CSMS), developed after Sandy, integrates surge, rainfall, and wind projections to estimate infrastructure vulnerabilities and guide investments in coastal defenses (Rosenzweig et al., 2019).

6.3 Challenges in Integrated Hurricane Modeling

Despite their sophistication, integrated hurricane models face several technical and institutional challenges. High-resolution data on terrain, land use, and exposure remain unevenly available, particularly in low-resource regions, yet model accuracy is highly sensitive to such inputs. Uncertainty propagates across the modeling chain, from meteorological forecasts through surge and runoff models, making it difficult to communicate reliable outcomes to policymakers (Beven, 2016; Hall & Solomatine, 2008). Computational demands also remain high, with coupled simulations requiring significant processing time, which often limits their operational use during live storm events. Interoperability is another hurdle, as models developed in different disciplines use varying assumptions, formats, and resolutions (Gregersen et al., 2007). Frameworks like the Open Modeling Interface (OpenMI) aim to address this, but widespread adoption remains limited. Finally, inaccessibility of sophisticated models to local governments and community planners due to technical or financial barriers underscores the need for simplified platforms, web-based interfaces, and targeted training.

6.4 Future Directions in Integrated Modeling

Looking ahead, integrated hurricane modeling is poised to evolve in response to intensifying climate extremes and growing recognition of systemic risk. Machine learning techniques are being incorporated to accelerate model calibration, quantify uncertainty, and build surrogate models capable of near real-time assessment (Mosavi et al., 2018). Cloud-based platforms, such as Google Earth Engine and Microsoft's Planetary Computer, are enabling large-scale, multi-hazard simulations that bypass the computational limitations of local servers. Dynamic exposure modeling, which integrates temporally variable data such as evacuation patterns or telecommunication flows, is increasingly allowing risk assessments that account for shifting population distributions during an event (Krausmann & Cruz, 2013). Moreover, socio-technical system modeling is gaining prominence, with efforts to simulate cascading failures such as power grid blackouts triggered by flood-induced substation failures (Pant et al. 2018). Finally, participatory approaches that involve local communities in co-producing models are being recognized as critical for building trust, fairness, and buy-in, thereby ensuring effectiveness, into decision-making (Mechler & Bouwer, 2015).

7 Planning Tools, Policy Frameworks, and Implementation Gaps

The development of robust planning tools and supportive policy frameworks is critical to operationalizing multi-hazard approaches in hurricane risk management. While integrated models and risk assessments provide the technical foundation, their practical utility depends on institutional capacity, governance arrangements, and the ability to translate complex risk information into actionable decisions and long-term investments (Hall & Solomatine, 2008). This section reviews the landscape of multi-hazard planning tools, examines policy instruments that shape resilience efforts, and highlights persistent gaps that continue to hinder the operationalization of holistic disaster risk reduction.

7.1 Multi-Hazard Planning Tools and Platforms

Several analytical and decision-support platforms have been developed to help planners, emergency managers, and policymakers assess and mitigate compound hurricane risks. Among the most widely used in the U.S. is HAZUS-MH, developed by FEMA (2020a), which estimates losses from hurricanes, floods, and earthquakes by integrating hazard data with population, building stock, and infrastructure datasets (Scawthorn et al., 2006). While standardized and supported by FEMA training, HAZUS-MH lacks dynamic modeling capacity for compound events such as rain-on-surge flooding, limiting its application for integrated risk assessments.

The NOAA Digital Coast platform described earlier provides visualization tools such as the Sea-Level Rise Viewer, enabling users to map future inundation under different climate scenarios (NOAA, 2022). Its GIS-enabled datasets are user-friendly and widely adopted by coastal planners. However, the platform does not fully capture interactions between coastal surge and inland hydrological processes, underscoring the limitations of siloed hazard modeling. Similarly, FEMA's Benefit–Cost Analysis Toolkit plays a central role in evaluating projects for federal mitigation funding. Although critical to decision-making, it often underestimates the benefits of multi-hazard projects due to narrowly framed loss estimation methods that treat hazards in isolation (Mechler & Bouwer, 2015).

Emerging tools developed by research institutions also expand multi-hazard planning capacity. The NIST Community Resilience Planning Guide and Resiliency Mapper enable communities to evaluate vulnerabilities across infrastructure and social systems, providing a framework for recovery planning (NIST 2015). At the international scale, platforms such as ThinkHazard!, developed by the World Bank's Global Facility for Disaster Reduction and Recovery, help development projects identify hazard exposure globally (World Bank, 2021). Other initiatives like RISKScape in New Zealand and InaSAFE in Indonesia demonstrate the growing democratization of open-source platforms that combine hazard and exposure datasets to generate locally tailored impact scenarios. Despite these advances, most tools are not systematically integrated into decision-making due to institutional silos, lack of interoperability, and limited customization for local contexts.

7.2 Policy and Governance Frameworks for Multi-Hazard Planning

Policy frameworks shape the enabling environment within which these tools are applied. At the global level, the "all-hazard approaches advocated by the United Nations Sendai Framework for Disaster Risk Reduction (2015–2030) described earlier is one example. The Paris Agreement reinforces these goals by encouraging risk-informed adaptation, while the sustainable development goals (SDGs; Kelman, 2017) embed disaster resilience within targets on sustainable cities, infrastructure, and climate action (Rosenzweig et al., 2019). These frameworks have spurred collaboration and funding but often lack enforcement mechanisms, limiting their translation into national policy and practice.

At the national and subnational scales, several U.S. policies have made progress toward multi-hazard integration. The Disaster Recovery Reform Act (2018) authorized greater investments in pre-disaster mitigation and introduced the BRIC program, which supports projects that address multiple hazards (FEMA, 2020b). NFIP remains central to floodplain management, though it has historically treated riverine and coastal flooding separately (Wing et al., 2022). Many states, including Louisiana, Florida, and Texas, have adopted resilience master plans that consider wind, surge, and flood scenarios. However, these plans often fall short of embedding multi-hazard assessments into land-use regulation and building code enforcement (Cutter et al., 2010).

Building codes illustrate another critical governance challenge. Standards such as the International Building Code (IBC) and ASCE 7 include provisions for wind and flood resilience. Yet adoption varies widely across jurisdictions, floodplain maps are often outdated, and zoning regulations rarely incorporate multi-hazard overlays or cumulative risk analysis (Aerts et al., 2014. Updating codes to reflect future conditions, including sea-level rise and changing hurricane climatology, is a pressing necessity.

8 Case Studies of Multi-Hazard Hurricane Planning in Practice

While theory and models provide the conceptual and technical backbone of risk reduction, the realities of political structures, institutional capacity, socioeconomic vulnerabilities, and funding constraints shape how multi-hazard frameworks are operationalized. Examining real-world experiences not only reveals best practices but also exposes persistent gaps and contradictions that continue to limit resilience outcomes. The following examples illustrate variations in scale, governance, and approaches while highlighting lessons for global application.

8.1 The Louisiana Watershed Initiative (United States)

Louisiana is an important case study given its repeated exposure to hurricanes, with Katrina (2005), Gustav (2008), Isaac (2012), and Laura (2020) each underscoring the interlinked challenges of storm surge, heavy rainfall, and riverine flooding (Burby, 2018; Kates et al., 2006). Recognizing the limitations of fragmented, parish-level flood management, the state launched the Louisiana Watershed Initiative (LWI) in 2018 (Louisiana Office of Community Development (OCD) 2019). This marked a paradigm shift toward watershed-based planning, moving beyond political boundaries to align governance with hydrologic systems.

The LWI strategy centers on cross-jurisdictional coordination, integration of structural and non-structural measures, and data-driven prioritization of investments. Regional watershed coalitions were created to manage eight hydrologically defined regions, each employing common risk modeling frameworks that evaluate interactions between multiple flood drivers such as rainfall, surge overtopping, and backwater effects (Louisiana OCD 2019). Technological innovations such as 2D hydraulic models have allowed more accurate simulations of surge-rain interactions, while the BayouSmart online platform enhances stakeholder engagement by providing accessible watershed data for collaborative planning.

However, institutional challenges remain. Aligning land use regulations across parishes has proven politically contentious, particularly where rural parishes face resource constraints compared to their urban counterparts. This case illustrates that technical advances must be paired with governance innovations that bridge institutional silos.

8.2 *New York City Post-Sandy Resilience Planning*

The devastation of Hurricane Sandy in 2012 exposed the vulnerabilities of one of the world's most densely populated urban regions to compound coastal hazards. More than 43 lives were lost, and the city suffered economic damage exceeding $19 billion, largely due to storm surge, infrastructure failures, and cascading power outages (Rosenzweig & Solecki, 2014). In response, New York City (NYC) launched a series of resilience strategies that explicitly addressed multiple hazards, including sea-level rise, storm surge, high winds, extreme precipitation, and even secondary stressors such as heat waves and prolonged power disruptions (Fainstein, 2018). The strategies included stringent building code reforms, requiring freeboard elevation and enhanced wind resistance, alongside floodproofing of critical infrastructure like hospitals and transit hubs. Academic collaborations, particularly with Columbia University and regional climate scientists, yielded high-resolution flood models that accounted for both tidal and rainfall-driven flooding (Aerts et al., 2014). At the community level, the Neighborhood Recovery Program engaged local leaders to co-develop neighborhood-specific rebuilding plans, ensuring that cultural and social contexts were respected.

Yet, implementation challenges remain. Large-scale structural defenses such as the East Side Coastal Resiliency Project have faced prolonged delays, while limited integration of green infrastructure with hard defenses reflects ongoing tensions between engineering and ecological approaches. Moreover, gentrification pressures in redeveloped coastal neighborhoods risk displacing low-income residents, raising questions about whether all voices are heard in resilience planning (Anguelovski et al., 2016).

8.3 *British Virgin Islands (BVI) After Hurricanes Irma and Maria*

The 2017 Atlantic hurricane season devastated the Caribbean, with the British Virgin Islands (BVI) experiencing a one-two punch from Hurricanes Irma and Maria. Irma's

Category 5 winds destroyed housing and power systems, while Maria's heavy rains triggered flash floods and landslides. This sequence revealed the compounded vulnerability of small island developing states (SIDS), where fragile economies, tourism dependency, and limited institutional capacity exacerbate disaster risks (Kelman, 2018). The BVI government's Recovery and Development Plan (2018–2028) emphasized multi-hazard resilience, anchored in hurricane-resistant construction enforced through the Caribbean Uniform Building Code, integrated watershed management to reduce flash flooding, and diversification of energy systems to minimize dependence on centralized grids (BVI Government, 2018). International agencies such as United Nations Development Program (UNDP; Jacobsson, 2019) and Caribbean Disaster Emergency Management Agency supported GIS-based hazard mapping and resilience planning, while public–private partnerships rebuilt tourism assets with resilience features like elevated structures and decentralized energy systems.

Despite these innovations, local implementation was hindered by financial constraints, reliance on external aid, and limited technical expertise to operationalize advanced planning tools. However, BVI's recovery demonstrated that post-disaster windows of opportunity can catalyze systemic transformation when long-term resilience is explicitly integrated into rebuilding.

8.4 Jakarta, Indonesia: Urban Resilience to Flood-Typhoon Complex Events

Although located outside the Atlantic hurricane belt, Jakarta faces its own multi-hazard challenges from seasonal cyclones, heavy rainfall, rapid urbanization, and subsidence-induced flooding. Jakarta's resilience strategy includes large-scale structural defenses such as the Giant Sea Wall Project, intended to buffer against coastal surge, alongside urban drainage upgrades to handle intense rainfall. Regulations on groundwater extraction aim to address land subsidence, while community-based programs build local disaster preparedness. The city also relies on advanced technological platforms such as JakSAFE, a multi-hazard risk tool that integrates hydrologic models with socioeconomic and infrastructure data to guide planning decisions.

However, persistent barriers such as political fragmentation between municipal and provincial authorities, conflicting priorities between development and conservation, and difficulties relocating informal settlements from high-risk zones have complicated implementation (Texier, 2008). Jakarta's experience highlights the importance of aligning technical capacity with institutional reforms and socioeconomic considerations.

8.5 Coastal Bangladesh Cyclone Preparedness Programme

Perhaps the most serious case in the world, yet one with great recent evidence for mitigation, is in the low-lying impoverished nation of Bangladesh. Vulnerability in a multi-hazard context takes the form of cyclones superimposed on storm surge, pluvial flooding, and tidal waters (Islam et al., 2021). Shortly after cyclones Bhola (the deadliest tropical cyclone on record) and another unnamed storm, in 1970 and 1991, respectively, that claimed approximately 300,000 and 138,000 lives, respectively, the country's establishment of the Cyclone Preparedness Programme (CPP) including an early warning system, in partnership with domestic and international relief agencies and heavily reliant on volunteers (Habib et al., 2012), has saved countless lives. Recent work that includes elaborate decision-making tools in a big geospatial data context (e.g., Aruf et al. 2023) and analytical mapping (Islam et al., 2017) and visualization techniques (Easha et al., 2025) that encompass both natural and anthropogenic dimensions further enhance the protection of life and property.

8.6 Lessons from Comparative Analysis

Taken together, these case studies underscore several cross-cutting lessons for multi-hazard planning. First, hazards rarely occur in isolation, and governance must therefore be organized at appropriate scales, whether through watershed coalitions in Louisiana, citywide frameworks in Jakarta, or the CPP in Bangladesh. Second, the integration of compound-event models, such as NYC's surge-rainfall simulations or Jakarta's real-time flood monitoring, demonstrates the necessity of dynamic and data-driven tools. Third, institutional coordination emerges as equally vital as technical innovation; the BVI, NYC, and Bangladesh initiatives reveal that interagency collaboration can determine the success of recovery. Finally, participatory approaches that involve communities, such as NYC's Neighborhood Recovery Program or Jakarta's local resilience initiatives, ensure plans remain socially legitimate and reduce the risk of unfair outcomes.

Yet, implementation challenges remain consistent across contexts. Funding shortfalls, regulatory misalignments, political resistance, and capacity deficits continue to slow the translation of multi-hazard concepts into resilient outcomes. Addressing these barriers requires a commitment to collaboration, interoperable data systems, and long-term investment in both technical and institutional capacities.

9 Concluding Ideas for Advancing Multi-Hazard Hurricane Planning for a More Resilient Future

As urban populations expand into increasingly hazard-prone areas, the need for multi-hazard hurricane planning has become urgent. Traditional siloed approaches, such as treating wind, flooding, or storm surge as discrete threats, are proving insufficient in the face of compound, cascading, and chronic risks (IPCC, 2022). Hurricanes increasingly trigger interlinked hazards, such as storm surge combining with extreme rainfall or cascading failures of power, health, and transportation systems. Considering these challenges, advancing multi-hazard hurricane planning requires forward-looking strategies that are scientifically rigorous, fair, and institutionally robust.

9.1 Adopt Risk-Based, Not Hazard-Based, Frameworks

Conventional hazard management frameworks are often built around single-event design standards, such as floodplain maps or wind load codes. Yet risk is not determined solely by hazards, but also by exposure, vulnerability, and adaptive capacity (Cardona et al., 2012). Moving forward, risk-based frameworks should integrate modeling that captures compound hazards such as rainfall-induced flooding coinciding with storm surge. For example, the LWI (Louisiana OCD 2019) has already begun experimenting with two-dimensional hydraulic models to simulate surge-rain interactions, a departure from traditional, single-hazard analyses. Updating building codes to combine flood and wind loads in structural design is another step toward ensuring that infrastructure is resilient under concurrent hazard conditions (Davis & Ryan, 2020). This shift also necessitates the adoption of probabilistic methods that explore a range of plausible scenarios rather than relying on deterministic, "design-event" standards, such as for tropical cyclone-induced rainfall in Louisiana (Villarini et al., 2022).

9.2 Institutionalize Cross-Sectoral and Cross-Scale Governance

Institutional fragmentation remains a persistent barrier to effective multi-hazard planning. As case studies from New York City, Jakarta, and the BVI demonstrate, coordination across both sectors (housing, energy, transportation, and health) and scales (local, regional, and national) is critical for effective resilience (Kelman, 2017; Rosenzweig & Solecki, 2014). Institutionalizing resilience requires multi-agency task forces with mandates that extend beyond emergency management, embedding hazard mitigation into planning, housing, and capital investment systems. The creation of watershed-based coalitions in Louisiana and regional resilience authorities in parts of the Caribbean exemplify governance models that transcend traditional political boundaries to manage risks at the scale of hydrologic and ecological systems.

9.3 Enhance Data Integration and Modeling Capacity

The foundation of multi-hazard planning lies in a shared understanding of risk, but this is often undermined by fragmented data systems, inconsistent formats, and unequal access to technical expertise. Efforts such as NOAA's Digital Coast and FEMA's RR2.0 highlight the value of integrating meteorological, hydrological, and socio-economic data into accessible platforms (NOAA, 2021; FEMA, 2020c). Emerging technologies, including "digital twins" of cities that allow planners to simulate compound hazard impacts in real time, offer additional opportunities for more dynamic risk assessment (Batty, 2018). Ensuring open access to such datasets also democratizes planning by enabling researchers, local governments, and communities to participate meaningfully in decision-making.

9.4 Integrate Climate Change Projections into All Phases of Planning

Historical baselines no longer provide reliable guidance for future risk. Rising sea levels, intensifying hurricanes, and changing precipitation patterns necessitate climate-informed planning at every stage of development (IPCC, 2022). For example, New York City's post-Sandy building codes incorporate sea-level rise projections into storm surge planning, while Jakarta's master plan integrates land subsidence forecasts into urban drainage design. Infrastructure must now be designed not only for current risks but for a range of plausible futures. Adaptive pathways approaches, which allow for phased investments and modifications as conditions evolve, represent an important paradigm shift in long-term resilience planning.

9.5 Center Social Vulnerability Considerations in Risk Reduction

Many disasters are not "natural," but instead are socially constructed, as the impacts of hazards are mediated by patterns of vulnerability and justice (Blaikie et al., 2004). Low-income households, elderly populations, people with disabilities, and historically marginalized groups face disproportionate risks during hurricanes, both in immediate impacts and long-term recovery (Cutter et al., 2003). Prioritizing resources based on Social Vulnerability Index analyses, supporting relocation assistance and retrofit grants for at-risk households, and avoiding maladaptive outcomes such as green gentrification, which displaces vulnerable residents in the name of resilience (Shokry et al., 2022), must be at the core of multi-hazard planning, both as a moral imperative and as a matter of practical necessity for effective governance.

9.6 Foster a Culture of Preparedness and Risk Literacy

Multi-hazard planning ultimately relies on public engagement and trust. Risk communication has too often been fragmented, overly technical, or narrowly focused on immediate hazards. To foster long-term preparedness, communities require sustained education and culturally grounded engagement (Paton, 2019). Examples include participatory mapping initiatives in the Caribbean, which have helped residents visualize compound risks, or local "trusted grapevines" that counter misinformation during crises (Houston et al., 2015). Educational curricula, early-warning systems that communicate compound risks in accessible terms, and partnerships with community organizations are essential to building durable risk literacy.

9.7 Link Pre-Disaster Planning to Post-Disaster Recovery

Periods following major hurricanes often provide rare political and financial opportunities for reform. Yet without pre-disaster plans, these opportunities are often wasted on reactive rebuilding that often repeats the mistakes of the past (Olshansky & Chang, 2009). Developing just, shelf-ready recovery frameworks that integrate multi-hazard assessments and community participation ensures that reconstruction aligns with long-term resilience goals. The BVI's post-Irma recovery efforts, which incorporated resilience principles into rebuilding tourism infrastructure, exemplify the potential for transformative recovery when pre-disaster frameworks are in place (Jacobsson, 2019).

9.8 Promote Innovation Through Policy and Funding Alignment

Finally, advancing multi-hazard planning requires coherence across policy frameworks and funding streams. Too often, mitigation, recovery, and adaptation programs operate in isolation, leading to inefficiencies and missed opportunities. Aligning disaster recovery funding with long-term resilience goals, incorporating performance-based planning metrics, and reducing regulatory barriers to nature-based or adaptive solutions are all critical steps (Mechler & Hochrainer-Stigler, 2014). Partnerships between researchers and practitioners also play a vital role in translating cutting-edge science into actionable policy.

Taken together, these recommendations underscore the need to reconceptualize hurricane planning as a dynamic, inclusive, and anticipatory process that includes a convergence of interrelated threats that magnify each other's impacts. By embedding risk-based analysis, innovation, and governance reform into multi-hazard frameworks, communities can move toward aligning scientific knowledge, engineering practice, land-use planning, community engagement, and climate adaptation strategies into a cohesive and anticipatory system. At the same time, there is a critical need for innovation in leadership, and buy-in from the public, as society adapts in real time to incipient disasters. Although emergency managers may have scenarios to guide their messages, it is inevitable that unanticipated events dictate a need for adapting. Moreover, the scenarios must be customized and continually refined to the evolving needs of the particular study area. Such a system would support a resilient future that reduces disaster losses while enhancing social and ecological well-being.

Ultimately, multi-hazard planning is about more than managing storms. It is about rethinking our relationship with risk, nature, and each other, and it involves creating cities and communities that are not only safe, but sustainable, inclusive, and thriving, even in the face of the fiercest storms. The tools are available, the knowledge is advancing, and the need is urgent. The time to act is now.

Acknowledgements The authors greatly appreciate the careful review and helpful suggestions of reviewers Gregory Carbone and Paul Hudak.

References

Aerts, J. C. J. H., Botzen, W. J. W., Emanuel, K., Lin, N., de Moel, H., & Michel-Kerjan, E. O. (2014). Evaluating flood resilience strategies for coastal megacities. *Science, 344*(6183), 473–475. https://doi.org/10.1126/science.1248222

AghaKouchak, A., Chiang, F., Huning, L. S., Love, C. A., Mallakpour, I., Mazdiyasni, O., Moftakhari, H., Papalexiou, S. M., Ragno, E., & Sadegh, M. (2020). Climate extremes and compound hazards in a warming world. *Annual Review of Earth and Planetary Sciences, 48*(1), 519–548. https://doi.org/10.1146/annurev-earth-071719-055228

Anderson, C., Renaud, F., Hagenlocher, M., & Day, J. (2021). Assessing multi-hazard vulnerability and dynamic coastal flood risk in the Mississippi Delta: The global delta risk index as a social-ecological systems approach. *Water, 13*(4), 577. https://doi.org/10.3390/w13040577

Anguelovski, I., Shi, L., Chu, E., Gallagher, D., Goh, K., Lamb, Z., Reeve, K., & Teicher, H. (2016). Equity impacts of urban land use planning for climate adaptation: Critical perspectives from the Global North and South. *Journal of Planning Education and Research, 36*(3), 333–348. https://doi.org/10.1177/0739456x16645166

Arif, M. S. I., Mahdi, I., Rafi, M. A., Khan, S. J., & Rahman, M. M. (2023). Cyclone exposure mapping in coastal Bangladesh: A multi-criteria decision analysis. *Heliyon, 9*(10), Article e21259. https://doi.org/10.1016/j.heliyon.2023.e21259

Arnold, J. G., Moriasi, D. N., Gwilliam, J., Srinivasan, R., Kannan, N., & Jha, M. (2015). SWAT: Model use, calibration, and validation. *Transactions of the ASABE, 55*(4), 1491–1508. https://doi.org/10.13031/2013.42256

Barbier, E. B., Hacker, S. D., Kennedy, C., Koch, E. W., Stier, A. C., & Silliman, B. R. (2011). The value of estuarine and coastal ecosystem services. *Ecological Monographs, 81*(2), 169–193. https://doi.org/10.1890/10-1510.1

Baró, R., Palacios-Peña, L., Baklanov, A., Balzarini, A., Brunner, D., Forkel, R., Hirtl, M., Honzak, L., Pérez, J. L., Pirovano, G., San José, R., Schröder, W., Werhahn, J., Wolke, R., Žabkar, R., & Jiménez-Guerrero, P. (2017). Regional effects of atmospheric aerosols on temperature: An evaluation of an ensemble of online coupled models. *Atmospheric Chemistry and Physics, 17*(15), 9677–9696. https://doi.org/10.5194/acp-17-9677-2017

Batty, M. (2018). Digital twins. *Environment and Planning b: Urban Analytics and City Science, 45*(5), 817–820. https://doi.org/10.1177/2399808318796416

Beven, K. (2016). Facets of uncertainty: Epistemic uncertainty, non-stationarity, likelihood, hypothesis testing, and communication. *Hydrological Sciences Journal, 61*(9), 1652–1665. https://doi.org/10.1080/02626667.2015.1031761

Bilskie, M. V., & Hagen, S. C. (2018). Defining flood zone transitions in low-gradient coastal regions. *Geophysical Research Letters, 45*(6), 2761–2770. https://doi.org/10.1002/2018gl077524

Bilskie, M. V., Hagen, S. C., Medeiros, S. C., & Cox, A. T. (2016). A comprehensive evaluation of hurricane surge and tide hydrodynamic modeling in the northern Gulf of Mexico. *Journal of Geophysical Research: Oceans, 121*(5), 3568–3592. https://doi.org/10.1002/2015EF000347

Blaikie, P., Cannon, T., Davis, I., & Wisner, B. (2004). *At risk: Natural hazards, people's vulnerability and disasters* (2nd ed.). Routledge. https://doi.org/10.4324/9780203714775

Bloemen, P. J., Hammer, F., van der Vlist, M. J., Grinwis, P., & van Alphen, J. (2019). DMDU into practice: Adaptive delta management in the Netherlands. In V. A. W. J. Marchau, W. E. Walker, P. J. T. M. Blowmen, & S. W. Popper (Eds.), *Decision making under deep uncertainty: From theory to practice* (pp. 321–351). Springer International Publishing. https://link.springer.com/book/https://doi.org/10.1007/978-3-030-05252-2

Bodenreider, C., Wright, L., Barr, O., Xu, K., & Wilson, S. (2019). Assessment of social, economic, and geographic vulnerability pre-and post-Hurricane Harvey in Houston, Texas. *Environmental Justice, 12*(4), 182–193. https://doi.org/10.1089/env.2019.0001

Bostrom, A., Morss, R. E., Lazo, J. K., Demuth, J. L., Lazrus, H., & Hudson, R. (2016). A mental models study of hurricane forecast and warning production, communication, and decision-making. *Weather, Climate, and Society, 8*(2), 111–129. https://doi.org/10.1175/wcas-d-15-0033.1

Bunya, S., Dietrich, J. C., Westerink, J. J., Ebersole, B. A., Smith, J. M., Atkinson, J. H., Jensen, R., Resio, D. T., Luettich, R. A., Dawson, C., Cardone, V. J., Cox, A. T., Powell, M. D., Westerink, H. J., & Roberts, H. J. (2010). A high-resolution coupled riverine flow, tide, wind, wind wave, and storm surge model for southern Louisiana and Mississippi. Part I: Model development and validation. *Monthly Weather Review, 138*(2), 345–377. https://doi.org/10.1175/2009mwr2906.1

Burby, R. J. (2018). *Hurricane Katrina and the paradoxes of government disaster policy: Bringing about wise governmental decisions for hazardous areas*. Springer. https://www.jstor.org/stable/25097787

BVI Government. (2018). *Recovery to development plan 2018–2028*. British Virgin Islands Government. https://bvi.gov.vg/sites/default/files/resources/recovery_to_development_plan_document.pdf

Cangialosi, J. P., & Latto, A. S. (2021). Tropical cyclone report: Hurricane delta (October 4–10, 2020). National Hurricane Center. https://www.nhc.noaa.gov/data/tcr/AL262020_Delta.pdf

Cardona, O. D., van Aalst, M. K., Birkmann, J., Fordham, M., McGregor, G., Perez, R., Pulwarty, R. S., Schipper, E. L. F., & Sinh, B. T. (2012). Determinants of risk: Exposure and vulnerability. In F. C. B. V. Barros, T. F. Stocker, D. Qin, D. J. Dokken, K. L. Ebi, M. D. Mastrandrea, K. J. Mach, G.-K. Plattner, S. K. Allen, M. Tignor, & P. M. Midgley (Eds.) *Managing the risks of extreme events and disasters to advance climate change adaptation (SREX)* (pp. 65–108). Cambridge University Press.

Cegan, J. C., Golan, M. S., Joyner, M. D., & Linkov, I. (2022). The importance of compounding threats to hurricane evacuation modeling. *NPJ Urban Sustainability, 2*(1), 2. https://doi.org/10.1038/s42949-021-00045-7

Chakraborty, J., Grineski, S. E., & Collins, T. W. (2019). Hurricane Harvey and people with disabilities: Disproportionate exposure to flooding in Houston, Texas. *Social Science and Medicine, 226*, 176–181. https://doi.org/10.1016/j.socscimed.2019.02.039

Coastal Protection and Restoration Authority (CPRA). (2017). *Louisiana's comprehensive master plan for a sustainable coast*. State of Louisiana. https://coastal.la.gov/our-plan/2017-coastal-master-plan/

Coastal Protection and Restoration Authority (CPRA). (2023). *Louisiana's comprehensive master plan for a sustainable coast*. State of Louisiana. https://coastal.la.gov/our-plan/2023-coastal-master-plan/

Collins, T. W., Grineski, S. E., Chakraborty, J., & Flores, A. B. (2019). Environmental injustice and Hurricane Harvey: A household-level study of socially disparate flood exposures in Greater Houston, Texas, USA. *Environmental Research, 179*, Article 108772. https://doi.org/10.1016/j.envres.2019.108772

Colmer, J. (2025). We need to predict the people disasters will hit, not just the places. *Nature, 642*(8069), 842. https://doi.org/10.1038/d41586-025-01968-2

Crooks, A. T., & Heppenstall, A. J. (2012). Introduction to agent-based modelling. In A. J. Heppenstall, A. T. Crooks, L. M. See, & M. Batty M (Eds.), *Agent-based models of geographical systems* (pp. 85–105). Springer. https://link.springer.com/book/https://doi.org/10.1007/978-90-481-8927-4

Cutter, S. L., & Emrich, C. T. (2006). Moral hazard, social catastrophe: The changing face of vulnerability along the hurricane coasts. *The ANNALS of the American Academy of Political and Social Science, 604*(1), 102–112. https://doi.org/10.1177/0002716205285515

Cutter, S. L., Boruff, B. J., & Shirley, W. L. (2003). Social vulnerability to environmental hazards. *Social Science Quarterly, 84*(2), 242–261. https://doi.org/10.1111/1540-6237.8402002

Cutter, S. L., Burton, C. G., & Emrich, C. T. (2010). Disaster resilience indicators for benchmarking baseline conditions. *Journal of Homeland Security and Emergency Management, 7*(1). https://doi.org/10.2202/1547-7355.1732

Davis, C., & Ryan, J. T. (2020). Building codes: the foundation for resilient communities. In R. Colker (Ed.), *Optimizing community infrastructure* (pp. 211–220), Elsevier. https://www.sciencedirect.com/science/article/abs/pii/B9780128162408000124

Deltares Systems. (2018). Delft3D-FLOW User Manual. https://usermanual.wiki/Pdf/Delft3DFLOWUserManual.885467064/help

Dietrich, J. C., Westerink, J. J., Kennedy, A. B., Smith, J. M., Jensen, R. E., Zijlema, M., Holthuijsen, L. H., Dawson, C., Luettich, R. A., Powell, M. D., Cardone, V. J., Cox, A. T., Stone, G. W., Pourtaheri, H., Hope, M. E., Tanaka, S., Westerink, L. G., Westerink, H. J., & Cobell, Z. (2011b). Hurricane Gustav (2008) waves and storm surge: Hindcast, synoptic analysis, and validation in southern Louisiana. *Monthly Weather Review, 139*(8), 2488–2522. https://doi.org/10.1175/2011mwr3611.1

Dietrich, J., Zijlema, M., Westerink, J., Holthuijsen, L., Dawson, C., Luettich, R., Jensen, R., Smith, J., Stelling, G., & Stone, G. (2011a). Modeling hurricane waves and storm surge using integrally-coupled, scalable computations. *Coastal Engineering, 58*(1), 45–65. https://doi.org/10.1016/j.coastaleng.2010.08.001

Durant, T. J., & Sultan, D. (2008). The impact of Hurricane Katrina on the race and class divide in America. In M. Marable, & K. Clarke (Eds.), *Seeking higher ground: The Hurricane Katrina crisis, race, and public policy reader* (pp. 191–201). Palgrave Macmillan. https://doi.org/10.1057/9780230610095_13

Easha, R. I., Mitu, M., & Haque, H. A. (2025). Visualization of multi-hazard interactions in Bangladesh through analyzing natural, anthropogenic, and technological hazards. *Natural Hazards, in Press.* https://doi.org/10.1007/s11069-025-07686-2

Entergy. (2021). Hurricane Ida restoration progress. Entergy Corporation.

Fainstein, S. S. (2018). Resilience and justice: Planning for New York City. *Urban Geography, 39*(8), 1268–1275. https://doi.org/10.1080/02723638.2018.1448571

FEMA. (2020a). HAZUS-MH: Multi-hazard loss estimation methodology. Federal Emergency Management Agency. https://www.fema.gov/sites/default/files/2020-09/fema_hazus_flood-model_user-manual_2.1.pdf

FEMA. (2020b). Building resilient infrastructure and communities. Federal Emergency Management Agency. https://www.fema.gov/grants/mitigation/learn/building-resilient-infrastructure-communities

FEMA. (2020c). Risk Rating 2.0: Equity in action. Federal Emergency Management Agency. https://www.fema.gov/flood-insurance/risk-rating

Gall, M., & Friedland, C. J. (2020). If mitigation saves $6 per Every $1 spent, then why are we not investing more? A Louisiana perspective on a national issue. *Natural Hazards Review, 21*(1), 04019013. https://doi.org/10.1061/(ASCE)NH.1527-6996.0000342

Gill, J. C., & Malamud, B. D. (2016). Hazard interactions and interaction networks (cascades) within multi-hazard methodologies. *Earth System Dynamics, 7*(3), 659–679. https://doi.org/10.5194/esd-7-659-2016

Gregersen, J. B., Gijsbers, P. J. A., & Westen, S. J. P. (2007). OpenMI: Open modelling interface. *Journal of Hydroinformatics, 9*(3), 175–191. https://doi.org/10.2166/hydro.2007.023

Habib, A., Shahidullah, M., & Ahmed, D. (2012). The Bangladesh cyclone preparedness program. A vital component of the nation's multi-hazard early warning system. In M. Golnaraghi (Ed.), *Institutional partnerships in multi-hazard early warning systems.* Springer. https://doi.org/10.1007/978-3-642-25373-7_3

Hall, J., & Solomatine, D. (2008). A framework for uncertainty analysis in flood risk management decisions. *International Journal of River Basin Management, 6*(2), 85–98. https://doi.org/10.1080/15715124.2008.9635339

Houston, J. B., Hawthorne, J., Perreault, M. F., Park, E. H., Goldstein Hode, M., Halliwell, M. R., Turner McGowen, S. E., Davis, R., Vaid, S., McElderry, J. A., & Griffith, S. A. (2015). Social media and disasters: A functional framework for social media use in disaster planning, response, and research. *Disasters, 39*(1), 1–22. https://doi.org/10.1111/disa.12092

Howell, J., & Elliott, J. R. (2019). Damages done: The longitudinal impacts of natural hazards on wealth inequality in the United States. *Social Problems, 66*(3), 448–467. https://doi.org/10.1093/socpro/spy016

IPCC. (2022). Climate change 2022: Impacts, adaptation and vulnerability. *Contribution of working group II to the sixth assessment report of the intergovernmental panel on climate change.* Cambridge University Press.

IPET. (2006). Performance evaluation of the New Orleans and Southeast Louisiana hurricane protection system. Interagency Performance Evaluation Task Force, U.S. Army Corps of Engineers. https://usace.contentdm.oclc.org/digital/collection/p266001coll1/id/2844/. Last Accessed 19 Augs 2025.

Islam, M. M., Mostafiz, M., Begum, P., Talukder, A., & Ahamed, S. (2021). Vulnerability to disaster in a multi-hazard coastal environment in Bangladesh. In M. Babel, A. Haarstrick, L. Ribbe, V. R. Shinde, & N. Dichtl (Eds.), *Water security in Asia: Opportunities and challenges in the context of climate change* (pp. 675–686). Springer International Publishing. https://doi.org/10.1007/978-3-319-54612-4_50

Islam, M. M., Rahman, M., Li, X., & Islam, N. (2017). Development of multi-hazards map for Bangladesh using GIS technique. *IUBAT Review, 1*(2), 77–85.

Jacobsson, J. (2019). United Nations Development Programme (UNDP) and UN Conference on Trade and Development (UNCTaD). In *Research handbook on the European union and international organizations*. Edward Elgar Publishing. https://doi.org/10.4337/9781786438935.00015

Jelesnianski, C. P. (1992). SLOSH: Sea, lake, and overland surges from hurricanes (Vol. 48). US Department of Commerce, National Oceanic and Atmospheric Administration, National Weather Service. https://www.google.com/books/edition/SLOSH_Sea_Lake_and_Overland_Surges_from/Wdg8mQfzkVcC?hl=en&gbpv=0

Julià, P. B., & Ferreira, T. M. (2021). From single- to multi-hazard vulnerability and risk in historic urban areas: A literature review. *Natural Hazards, 108*(1), 93–128. https://doi.org/10.1007/s11069-021-04734-5

Kappes, M. S., Keiler, M., von Elverfeldt, K., & Glade, T. (2012). Challenges of analyzing multi-hazard risk: A review. *Natural Hazards, 64*(2), 1925–1958. https://doi.org/10.1007/s11069-012-0294-2

Kates, R. W., Colten, C. E., Laska, S., & Leatherman, S. P. (2006). Reconstruction of New Orleans after Hurricane Katrina: A research perspective. *Proceedings of the National Academy of Sciences, 103*(40), 14653–14660. https://doi.org/10.1073/pnas.0605726103

Kelman, I. (2017). Linking disaster risk reduction, climate change, and the sustainable development goals. *Disaster Prevention and Management: An International Journal, 26*(3), 254–258. https://doi.org/10.1108/dpm-02-2017-0043

Kelman, I. (2018). Islandness within climate change narratives of Small Island developing states (sids). *Island Studies Journal, 13*(1), 149–166. https://doi.org/10.24043/isj.52

Krausmann, E., & Cruz, A. M. (2013). Impact of the 11 March 2011, great east Japan earthquake and tsunami on the chemical industry. *Natural Hazards, 67*(2), 811–828. https://doi.org/10.1007/s11069-013-0607-0

Lee, R., White, C. J., Adnan, M. S. G., Douglas, J., Mahecha, M. D., O'Loughlin, F. E., Patelli, E., Ramos, A. M., Roberts, M. J., Martius, O., Tubaldi, E., van den Hurk, B., Ward, P. J., & Zscheischler, J. (2024). Reclassifying historical disasters: From single to multi-hazards. *Science of the Total Environment, 912*, Article 169120. https://doi.org/10.1016/j.scitotenv.2023.169120

Lieberknecht, K., Zoll, D., Jiao, J., & Castles, K. (2021). Hurricane Harvey: Equal opportunity storm or disparate disaster? *Local Environment, 26*(2), 216–238. https://doi.org/10.1080/13549839.2021.1886063

Lin, N. (2010). *Multi-hazard risk analysis related to hurricanes*. Doctoral Dissertation, Princeton University.

Louisiana Office of Community Development (OCD). (2019). *Louisiana Watershed Initiative: Creating a more resilient Louisiana*. https://watershed.la.gov/

Magnusson, L., Bidlot, J.-R., Bonavita, M., Brown, A. R., Browne, P. A., De Chiara G., et al. (2019). ECMWF activities for improved hurricane forecasts. *Bulletin of the American Meteorological Society, 100*(3), 445–458. https://doi.org/10.1175/BAMS-D-18-0044.1

Maltz, M. (2019). Caught in the eye of the storm: The disproportionate impact of natural disasters on the elderly population in the United States. *Elder LJ, 27*, 157.

Mechler, R., & Bouwer, L. M. (2015). Understanding trends and projections of disaster losses and climate change: Is vulnerability the missing link? *Climatic Change, 133*, 23–35. https://doi.org/10.1007/s10584-014-1141-0

Mechler, R., & Hochrainer-Stigler, S. (2014). Revisiting arrow-lind: Managing sovereign disaster risk. *Journal of Natural Resources Policy Research, 6*(1), 93–100. https://doi.org/10.1080/19390459.2013.873186

Meselhe, E., Wang, Y., White, E., Jung, H., Baustian, M. M., Hemmerling, S., Barra, M., & Bienn, H. (2020). Knowledge-based predictive tools to assess effectiveness of natural and nature-based solutions for coastal restoration and protection planning. *Journal of Hydraulic Engineering, 146*(2), 05019007. https://doi.org/10.1061/(ASCE)HY.1943-7900.0001659

Ming, X., Liang, Q., Dawson, R., Xia, X., & Hou, J. (2022). A quantitative multi-hazard risk assessment framework for compound flooding considering hazard inter-dependencies and interactions. *Journal of Hydrology, 607*, Article 127477. https://doi.org/10.1016/j.jhydrol.2022.127477

Moftakhari, H. R., AghaKouchak, A., Sanders, B. F., & Matthew, R. A. (2017). Cumulative hazard: The case of nuisance flooding. *Earth's Future, 5*(2), 214–223. https://doi.org/10.1002/2016EF000494

Morss, R. E., Demuth, J. L., Bostrom, A., Lazo, J. K., & Lazrus, H. (2018). Flash flood risks and warning decisions: A mental models study of forecasters, public officials, and media broadcasters in Boulder, Colorado. *Risk Analysis, 35*(11), 2009–2028. https://doi.org/10.1111/risa.12403

Mosavi, A., Ozturk, P., & Chau, K. (2018). Flood prediction using machine learning models: Literature review. *Water, 10*(11), 1536. https://doi.org/10.3390/w10111536

Mousavi, M. E., Irish, J. L., Frey, A. E., Olivera, F., & Edge, B. L. (2011). Global warming and hurricanes: The potential impact of hurricane intensification and sea level rise on coastal flooding. *Climatic Change, 104*, 575–597. https://doi.org/10.1007/s10584-009-9790-0

Murray-Tuite, P., & Wolshon, B. (2013). Evacuation transportation modeling: An overview of research, development, and practice. *Transportation Research Part C: Emerging Technologies, 27*, 25–45. https://doi.org/10.1016/j.trc.2012.11.005

Narayan, S., Beck, M. W., Reguero, B. G., Losada, I. J., van Wesenbeeck, B., Pontee, N., Sanchirico, J. N., Ingram, J. C., Lange, G.-M., & Burks-Copes, K. A. (2017). The effectiveness, costs and coastal protection benefits of natural and nature-based defences. *PLoS ONE, 11*(5), Article e0154735. https://doi.org/10.1371/journal.pone.0154735

National Academies of Sciences, Engineering, and Medicine. (2020). Strengthening Post-Hurricane Supply Chain Resilience: Observations from Hurricanes Harvey, Irma, and Maria. The National Academies Press. https://doi.org/10.17226/25490

National Institute of Standards and Technology (NIST). (2015) *Community resilience planning guide for buildings and infrastructure systems.* U.S. Department of Commerce.

Neal, J. C., Schumann, G.J.-P., Bates, P. D., Buytaert, W., Matgen, P., & Pappenberger, F. (2012). Flood inundation model validation using crowd-sourced data. *Hydrology and Earth System Sciences, 16*, 4357–4377.

NOAA. (2021). *Digital coast: Tools and data for coastal management.* National Oceanic and Atmospheric Administration. https://coast.noaa.gov/digitalcoast

NOAA. (2022). *Digital coast: Sea level rise viewer.* National Oceanic and Atmospheric Administration. https://coast.noaa.gov/slr

Norman, C., Hunter, R. G., Day, J. W., Clark, H. C., Sanner, C., Kemp, G. P., Fucile-Sanchez, E. B., & Hodder, J. H. (2022). A review of issues related to formation, deterioration and restoration of the Chenier Plain, Mississippi River Delta, LA-Combining nature based and engineered approaches. *Nature-Based Solutions, 2*, Article 100037. https://doi.org/10.1016/j.nbsj.2022.100037

Olshansky, R. B., & Chang, S. (2009). Planning for disaster recovery: Emerging research needs and challenges. *Progress in Planning, 72*(4), 200–209. https://doi.org/10.1016/j.progress.2009.01.004

Pant, R., Thacker, S., Hall, J. W., Alderson, D., & Barr, S. (2018). Critical infrastructure impact assessment due to flood exposure. *Journal of Flood Risk Management, 11*(1), 22–33. https://doi.org/10.1111/jfr3.12288

Pasch, R. J., Penny, A. B., & Berg, R. (2023). National Hurricane center tropical cyclone report: Hurricane Maria (September 16–30, 2017). National Hurricane Center. https://www.nhc.noaa.gov/data/tcr/AL152017_Maria.pdf

Paté-Cornell, M. (1996). Uncertainties in risk analysis: Six levels of treatment. *Reliability Engineering and System Safety, 54*(2–3), 95–111. https://doi.org/10.1016/s0951-8320(96)00067-1

Paton, D. (2019). Disaster risk reduction: Psychological perspectives on preparedness. *Australian Journal of Psychology, 71*(4), 327–341. https://doi.org/10.1111/ajpy.12237

Pearson, L., & Pelling, M. (2015). The UN Sendai framework for disaster risk reduction 2015–2030: Negotiation process and prospects for science and practice. *Journal of Extreme Events, 2*(01), 1571001. https://doi.org/10.1142/S2345737615710013

Pine, J. C., Mashriqui, H., Pedro, S., & Meyer, J. (2005). Hazard mitigation planning utilizing the Hazus-MH flood model and hurricane wind model. *Journal of Emergency Management, 3*(2), 11–16. https://doi.org/10.5055/jem.2005.0014

Rosenzweig, C., & Solecki, W. (2014). Hurricane Sandy and adaptation pathways in New York: Lessons from a first-responder city. *Global Environmental Change, 28*, 395–408. https://doi.org/10.1016/j.gloenvcha.2014.05.003

Rosenzweig, C., Solecki, W., Romero-Lankao, P., Mehrotra, S., Dhakal, S., & Ali Ibrahim, S. (2019). *Climate change and cities: Second assessment report of the urban climate change research network.* Cambridge University Press.

Sahoo, B., & Bhaskaran, P. K. (2018). Multi-hazard risk assessment of coastal vulnerability from tropical cyclones – a GIS based approach for the Odisha coast. *Journal of Environmental Management, 206*, 1166–1178. https://doi.org/10.1016/j.jenvman.2017.10.075

Sastry, N., & Gregory, J. (2013). The effect of Hurricane Katrina on the prevalence of health impairments and disability among adults in New Orleans: Differences by age, race, and sex. *Social Science and Medicine, 80*, 121–129. https://doi.org/10.1016/j.socscimed.2012.12.009

Scawthorn, C., Flores, P., Blais, N., Seligson, H., Tate, E., Chang, S., Mifflin, E., Thomas, W., Murphy, J., Jones, C., & Lawrence, M. (2006). Hazus-MH flood loss estimation methodology. II. Damage and loss assessment. *Natural Hazards Review, 7*(2), 72–81. https://doi.org/10.1061/(asce)1527-6988(2006)7:2(72)

Schweizer, V. J., & Kriegler, E. (2012). Improving environmental change research with systematic techniques for qualitative scenarios. *Environmental Research Letters, 7*(4), Article 044011. https://doi.org/10.1088/1748-9326/7/4/044011

Shokry, G., Anguelovski, I., Connolly, J. J. T., Maroko, A., & Pearsall, H. (2022). "They didn't see it coming": Green resilience planning and vulnerability to future climate gentrification. *Housing Policy Debate, 32*(1), 211–245. https://doi.org/10.1080/10511482.2021.1944269

Suh, S. W., Lee, H. Y., Kim, H. J., & Fleming, J. G. (2015). An efficient early warning system for typhoon storm surge based on time-varying advisories by coupled ADCIRC and SWAN. *Ocean Dynamics, 65*(5), 617–646. https://doi.org/10.1007/s10236-015-0820-3

Tallapragada, V., Bernardet, L., Biswas, M. K., Gopalakrishnan, S., Kwon, Y., Liu, Q., & Zhang, X. (2014) Hurricane weather research and forecasting (HWRF) model: 2013 scientific documentation. *HWRF Development Testbed Center Tech. Rep, 99*. https://www.emc.ncep.noaa.gov/gc_wmb/vxt/pubs/HWRFScientificDocumentation2013.pdf

Texier, P. (2008). Floods in Jakarta: When the extreme reveals daily structural constraints and mismanagement. *Disaster Prevention and Management: An International Journal, 17*(3), 358–372. https://doi.org/10.1108/09653560810887284

Thakur, D. A., & Mohanty, M. P. (2023). A synergistic approach towards understanding flood risks over coastal multi-hazard environments: Appraisal of bivariate flood risk mapping through flood hazard, and socio-economic-cum-physical vulnerability dimensions. *Science of the Total Environment, 901*, Article 166423. https://doi.org/10.1016/j.scitotenv.2023.166423

Tonn, G. L., & Guikema, S. D. (2018). An agent-based model of evolving community flood risk. *Risk Analysis, 38*(6), 1258–1278. https://doi.org/10.1111/risa.12939

Tupper, A. C., & Fearnley, C. J. (2023). Mind the gaps in disaster early-warning systems—and fix them. *Nature, 623*, 478–482. https://www.nature.com/articles/d41586-023-03510-8

Unnikrishnan, V. U., & Barbato, M. (2016). Performance-based hurricane engineering: A multi-hazard approach. In P. Gardoni, & J. LaFave (Eds.), *Multi-hazard approaches to civil infrastructure engineering* (pp. 337–356). Springer. https://doi.org/10.1007/978-3-319-29713-2_16

van Oldenborgh, G. J., van der Wiel, K., Sebastian, A., Singh, R., Arrighi, J., Otto, F., Haustein, K., Li, S., Vecchi, G., & Cullen, H. (2017). Attribution of extreme rainfall from Hurricane Harvey, August 2017. *Environmental Research Letters, 12*(12), Article 124009. https://doi.org/10.1088/1748-9326/aa9ef2

Villarini, G., Zhang, W., Miller, P., Johnson, D. R., Grimley, L. E., & Roberts, H. J. (2022). Probabilistic rainfall generator for tropical cyclones affecting Louisiana. *International Journal of Climatology, 42*(3), 1789–1802. https://doi.org/10.1002/joc.7335

Wahl, T., Jain, S., Bender, J., Meyers, S. D., & Luther, M. E. (2015). Increasing risk of compound flooding from storm surge and rainfall for major US cities. *Nature Climate Change, 5*(12), 1093–1097. https://doi.org/10.1038/nclimate2736

Wei, Z., Pinelli, J. P., Gurley, K., & Hamid, S. (2024) Multi-hazard vulnerability modeling: An example of wind and rain vulnerability of mid/high-rise buildings during hurricane events. *Wind and Structures, 38*(5), 355–356. https://doi.org/10.12989/was.2024.38.5.355

Wing, O. E. J., Lehman, W., Bates, P. D., Sampson, C. C., Quinn, N., Smith, A. M., Neal, J. C., Porter, J. R., & Kousky, C. (2022). Inequitable patterns of us flood risk in the anthropocene. *Nature Climate Change, 12*(2), 156–162. https://doi.org/10.1038/s41558-021-01265-6

World Bank. (2021). *City resilience program: Building urban resilience through integrated modeling*. World Bank.

Zhang, Y. J., Ye, F., Stanev, E. V., & Grashorn, S. (2016). Seamless cross-scale modeling with SCHISM. *Ocean Modelling, 102*, 64–81. https://doi.org/10.1016/j.ocemod.2016.05.002

Zscheischler, J., Martius, O., Westra, S., Bevacqua, E., Raymond, C., Horton, R. M., van den Hurk, B., AghaKouchak, A., Jézéquel, A., Mahecha, M. D., Maraun, D., Ramos, A. M., Ridder, N. N., Thiery, W., & Vignotto, E. (2020). A typology of compound weather and climate events. *Nature Reviews Earth and Environment, 1*, 333–347. https://doi.org/10.1038/s43017-020-0060-z

Zscheischler, J., Westra, S., van den Hurk, B. J. J. M., Seneviratne, S. I., Ward, P. J., Pitman, A., AghaKouchak, A., Bresch, D. N., Leonard, M., Wahl, T., & Zhang, X. (2018). Future climate risk from compound events. *Nature Climate Change, 8*(6), 469–477. https://doi.org/10.1038/s41558-018-0156-3

Tropical Cyclone Risk in the Bay of Bengal: Advancing Multi-Hazard Planning and Resilience Strategies

Tanvir Hossain and Rubayet Bin Mostafiz

Abstract Communities along the Bay of Bengal face a deadly combination of hazards each cyclone season. Tropical cyclones in this region generate extreme winds, storm surges, and flooding that interact in complex, amplifying ways. This chapter advances multi-hazard planning theory by proposing an integrated risk governance framework that explicitly accounts for hazard interactions and institutional barriers. Drawing primarily on Bangladesh's experience the region's most cyclone-affected nation with comparative insights from India and Myanmar, we demonstrate how compound hazard events expose critical vulnerabilities in siloed disaster management systems. Recent cyclones (Sidr 2007, Amphan 2020, Mocha 2023) reveal that worst-case scenarios result not from individual hazard intensity but from their temporal and spatial convergence. We critically analyze why fragmented governance persists despite known risks, examining the political economy of institutional silos, bureaucratic path dependencies, and resource allocation conflicts that impede integrated planning.

Our novel contribution is a multi-scale resilience framework that integrates advanced hazard modeling with critical institutional analysis and participatory community engagement. We evaluate emerging tools storm surge models, composite risk atlases, and community-based early warning systems while assessing their feasibility, trade-offs, and scalability constraints. The chapter concludes with evidence-based recommendations for transforming disaster risk mitigation from reactive, single-hazard responses to proactive, systems-level resilience building.

Keywords Storm surge · Coastal flooding · Climate change adaptation · Disaster risk mitigation (DRR) · Vulnerability and equity · Integrated risk governance · Compound hazards · Community-based adaptation · Early warning systems · Participatory disaster management

T. Hossain (✉)
Louisiana State University, Baton Rouge, Louisiana, USA
e-mail: thossa4@lsu.edu

R. B. Mostafiz
Louisiana State University Agriculture Center, Baton Rouge, Louisiana, USA
e-mail: RMostafiz@agcenter.lsu.edu

J. Collins et al. (eds.), *Tropical Cyclone Risk and Resilience*,
https://doi.org/10.1007/978-3-032-28745-8_4

1 Introduction

1.1 Background and Regional Context

The Bay of Bengal, surrounded by India, Bangladesh, and Myanmar, is Earth's most lethal cyclone basin. The National Oceanic and Atmospheric Administration (NOAA) and Indian Meteorological Department (IMD) estimate that 80% of global cyclone-related deaths occur here, driven by storm frequency, dense coastal populations, and low-lying geography (Ali, 1999; Ritchie et al., 2022; Sahoo and Bhaskaran, 2018). Bangladesh bears disproportionate risk. The Ganges-Brahmaputra-Meghna delta, home to over 170 million people, forms the world's largest deltaic population cluster. Cyclones occur frequently: 5–6 major storms annually, with increasing intensity as measured by the Power Dissipation Index (Sahoo & Bhaskaran, 2018). The 1970 Bhola Cyclone, 1991 Bangladesh Cyclone, Cyclone Sidr (2007), Amphan (2020), and Mocha (2023) represent some of history's deadliest storms (Sahoo & Bhaskaran, 2018; Alam & Collins, 2010). India's eastern coast particularly Odisha and West Bengal and Myanmar's Rakhine State also face severe cyclone exposure. However, Bangladesh's unique vulnerability stems from its deltaic topography, where millions live at or below sea level, facing compound risks from storm surge, riverine flooding, and saltwater intrusion simultaneously (Ericson et al., 2006; Haque & Jakariya, 2023). This chapter therefore focuses primarily on Bangladesh while drawing comparative lessons from India and Myanmar where relevant. This approach allows deep analysis of multi-hazard dynamics in the region's most vulnerable context while acknowledging broader Bay of Bengal patterns (Fig. 1).

Cyclones in the Bay often intensify rapidly before landfall, coinciding with pre-monsoon (April–May) and post-monsoon (October–November) periods when rivers are swollen and estuarine conditions amplify surge risk (Mishra et al., 2021; United Nations in Bangladesh, 2021). Rather than discrete disasters, each major event comprises cascading, mutually reinforcing hazards what researchers now term "multi-hazard" or "compound" risk (Dastagir, 2015; Sahoo and Bhaskaran, 2018; United Nations in Bangladesh, 2021). Cyclone Amphan (2020) exemplifies this: $13 billion in economic damage, five million displaced, widespread infrastructure destruction, and months-long agricultural losses from saline intrusion (Hadi et al., 2021; Islam et al., 2019). Social vulnerability is inequitably distributed. Women, children, elderly, indigenous communities, and informal settlers disproportionately suffer due to geographic exposure and systematic exclusion from relief systems (Alam & Collins, 2010).

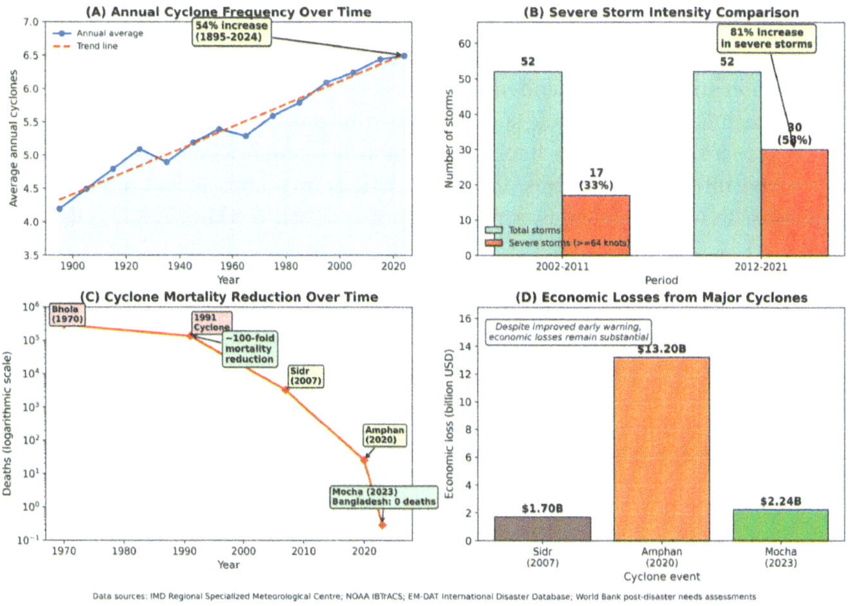

Fig. 1 Bay of Bengal cyclone trends (1895-2024) **a** Annual cyclone frequency; **b** Storm intensity comparison showing 81% increase in severe storms; **c** Mortality reduction; **d** Economic losses

1.2 Problem Statement: The Persistence of Fragmented Risk Management

Despite decades of evidence documenting compound hazard interactions, risk management in the Bay of Bengal remains fundamentally fragmented. This fragmentation is not merely technical inefficiency, it reflects deeper institutional, political, and economic structures that merit critical examination. Traditional approaches treat cyclone components separately: wind resistance standards, storm surge maps, and flood protocols developed by different agencies with distinct mandates and budgets (Brecht et al., 2012; Sahoo & Bhaskaran, 2018). Yet Bay of Bengal cyclones simultaneously generate winds exceeding 250 km/h, surges above 5 meters, and rainfall causing widespread flooding (Murty et al., 1986; Sahoo & Bhaskaran, 2018). These hazards interact nonlinearly: surge heights depend on tidal timing, rainfall saturates embankments increasing breach risk, and wind damage compromises flood defenses (Bernard et al., 2022).

Why does fragmentation persist? Three interrelated factors warrant analysis:

Political Economy of Institutional Silos: Bureaucratic structures reflect historical development patterns. In Bangladesh, separate ministries control meteorology, hydrology, and disaster management, each with distinct funding streams and political constituencies. Consolidation threatens established resource flows and administrative authority. India's federal system adds complexity, with state governments controlling disaster response while central agencies manage forecasting, creating coordination gaps (Brecht et al., 2012; NDMA India, 2022).

Bureaucratic Path Dependencies: Existing systems develop organizational cultures, technical expertise, and standard operating procedures optimized for single-hazard scenarios. Shifting to integrated approaches requires retraining personnel, redesigning information systems, and renegotiating costly inter-agency protocols that compete with immediate disaster response needs (Bhatt & Garge, 2023; UNDRR, 2023).

Knowledge-Action Gaps: Scientific understanding of compound hazards has advanced rapidly, yet translating research into operational policy faces barriers. Risk maps using integrated hazard modeling exist but are not systematically incorporated into land-use planning or infrastructure codes due to weak enforcement, informal construction practices, and political pressure for development in hazardous zones (United Nations in Bangladesh, 2021). Bangladesh's approach, documented in the National Early Action Protocol for Cyclone (Department of Disaster Management, Government of Bangladesh), represents regional best practice yet implementation varies by district due to capacity and resource constraints.

Recent cyclones demonstrate consequences: Amphan (2020) response was hampered by separate flood and cyclone warnings that confused evacuation timing (Hadi et al., 2021). Mocha (2023) exposed how Cox's Bazar refugee shelters, built without multi-hazard criteria, failed simultaneously from wind and flood (IFRC, 2023a). The core problem is thus not ignorance of hazard interactions but institutional inertia and structural barriers that prevent system-level resilience despite known vulnerabilities.

1.3 Contributions of the Chapter

This chapter advances multi-hazard disaster risk mitigation theory and practice through four interconnected contributions:

1.3.1 A Multi-Scale Integrated Risk Framework

We propose a novel analytical framework that simultaneously addresses: (a) physical hazard interactions (wind-surge-flood coupling); (b) institutional governance barriers (political economy analysis); (c) social vulnerability dimensions (equity and access); and (d) technological integration pathways (modeling tools to community warning systems). Unlike descriptive multi-hazard reviews, this framework provides actionable decision criteria for prioritizing interventions across scales (Fig. 2).

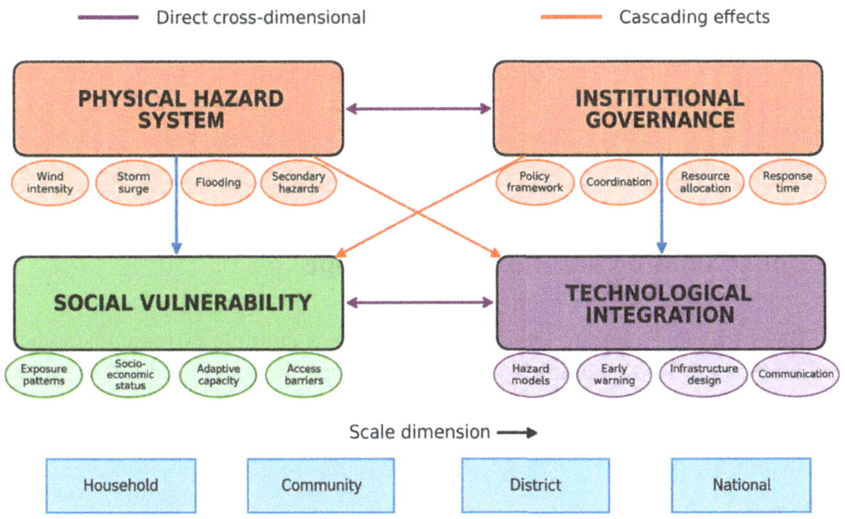

Fig. 2 Multi-scale integrated risk framework showing interactions between physical, institutional, social, and technological dimensions

1.3.2 Critical Institutional Analysis

Beyond documenting that silos exist, we examine "why" they persist through political economy and organizational theory lenses. This analysis reveals leverage points for governance reform, including funding mechanisms that incentivize coordination, regulatory changes that mandate integrated planning, and capacity-building strategies that overcome path dependencies.

1.3.3 Comparative Case Analysis with Feasibility Assessment

We analyze Sidr (2007), Amphan (2020), and Mocha (2023) not merely as disaster narratives but as natural experiments revealing what works, what fails, and under what conditions. For each adaptation measure (cyclone shelters, early warning systems, infrastructure standards), we assess implementation feasibility, cost-benefit trade-offs, and scalability constraints often absent from technical literature.

1.3.4 Evidence-Based, Context-Specific Recommendations

Drawing on Bangladesh's Cyclone Preparedness Program (76,000+ trained volunteers), India's coastal risk atlases, and Myanmar's capacity challenges, we provide differentiated recommendations recognizing that solutions must match local institutional capacity, resource availability, and socio-political contexts. Our approach combines system thinking with participatory principles, recognizing that sustainable resilience requires both top-down institutional reform and bottom-up community engagement. The chapter aims to transform multi-hazard planning from abstract concept to operational framework applicable across the Bay of Bengal and similar high-risk deltaic regions globally.

2 Understanding the Hazard Landscape

The Bay of Bengal's hazard environment comprises multiple extreme phenomena that interact in ways that amplify overall risk beyond the sum of individual threats. Understanding these dynamics is essential for evidence-based resilience strategies.

2.1 Primary Hazards in the Bay of Bengal

2.1.1 Cyclonic Winds

The India Meteorological Department (IMD) classifies Bay of Bengal cyclones by sustained wind speeds: depressions (<31 km/h), cyclonic storms (62–88 km/h), severe cyclonic storms (89–117 km/h), severe cyclonic storms (118–221 km/h), and super cyclonic storms (>221 km/h) (IMD, 2025). Rapid intensification near landfall limits evacuation time (Sahoo & Bhaskaran, 2018). Wind-driven damage includes structural collapse, agricultural destruction, and debris hazards.

2.1.2 Storm Surge

Storm surge abnormal sea level rise driven by cyclone winds and pressure represents the deadliest cyclone component. The Bay of Bengal's funnel-shaped northern basin amplifies surges, with heights exceeding 5–7 m recorded during Sidr and Nargis (Murty et al., 1986; Bernard et al., 2022). Surge height depends on cyclone intensity, track, coastline orientation, tidal phase at landfall, bathymetry, and coastal configuration (Antony & Unnikrishnan, 2013). Surges inundate low-lying deltas tens of kilometers inland, flooding homes, farmland, and infrastructure while causing long-term saltwater contamination of soil and groundwater (Ericson et al., 2006; Islam et al., 2019) (Table 1).

2.1.3 Inland and Coastal Flooding

Cyclones bring prolonged heavy rainfall that, combined with monsoonal hydrology, causes flash and riverine flooding. Rainfall saturates soil, raises river levels, and

Table 1 Historical storm surge heights (1970–2023)

Event	Year	Wind speed (km/h)	Surge height (m)	Tidal phase	Deaths
Bhola cyclone	1970	185–240	10.0+	High tide	>300,000
Cyclone Sidr	2007	240	5.0–6.0	High tide	3406
Cyclone Nargis	2008	215	5.0+	High tide	138,373
Cyclone Amphan	2020	250	3.0–5.0	Variable	26–130
Cyclone Mocha	2023	260+	Exceptional	Variable	145+

breaches embankments and polders. When storm surges block drainage outlets, inland flooding can equal or exceed surge impacts (Miyaji, 2020). Urban flash floods in Dhaka and Kolkata overwhelm drainage systems, while upstream riverine flooding can persist for weeks, delaying recovery (United Nations in Bangladesh, 2021; Hadi et al., 2021).

2.1.4 Secondary Hazards

Secondary effects include landslides in hilly areas (northeastern India, Myanmar) where saturated steep slopes fail; infrastructure collapse under combined wind-water-erosion stress; and waterborne disease outbreaks (cholera, dysentery, malaria) following floods and surges (Mishra et al., 2021; Islam et al., 2019).

2.2 *Hazard Interactions and Compounding Effects*

The Bay of Bengal hazardscape is defined by strong interdependence and nonlinear amplification of these hazards.

Temporal Sequencing: Tidal cycles significantly affect surge magnitude. Landfall during high tide can elevate surges by several meters (Sindhu and Unnikrishnan, 2013). Antecedent rainfall saturating soil before surge arrival alters drainage patterns and increases embankment overflow risk (NDMA India, 2025; Bernard et al., 2022).

Hydrological Saturation: Cyclones often follow or coincide with monsoon rains draining into shared river basins, compounding flood risk. Saturated soil lose stability, triggering landslides and embankment breaches in the Sundarbans and northeastern hilly areas (Mishra et al., 2021; Brecht et al., 2012).

Infrastructure Failure under Multiple Stressors: Infrastructure rarely withstands compound exposure. Wind weakens structures vulnerable to subsequent flooding; floodwaters undermine wind-damaged foundations (Islam et al., 2019). Cyclone shelters designed for wind but not flood risk inundation; structurally sound embankments fail under combined wave action and soil saturation (Miyaji, 2020;

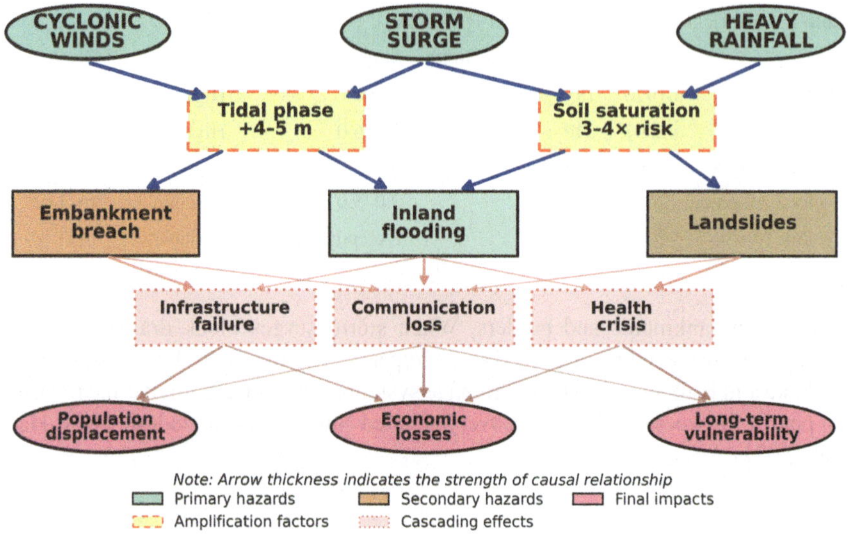

Fig. 3 Hazard interaction pathways and cascading effects

Bernard et al., 2022). Road networks critical for evacuation and relief are simultaneously blocked by fallen trees and floodwater, impeding emergency response (IFRC, 2023b) (Fig. 3).

Understanding these interactions is fundamental to the multi-hazard framework proposed in this chapter, as traditional single hazard approaches systematically underestimate actual risk.

3 Case Studies: Historical Events and Lessons Learned

The Bay of Bengal has experienced some of history's deadliest tropical cyclones. We analyze three recent events Sidr (2007), Amphan (2020), and Mocha (2023) to extract insights about persistent vulnerabilities, planning failures, and emerging adaptive innovations.

3.1 Cyclone Sidr (2007)

Cyclone Sidr struck Bangladesh on November 15, 2007, with sustained winds near 250 km/h and storm surges reaching 5.5 meters in coastal areas (Islam et al., 2019). The event starkly demonstrated compound hazard impacts in a densely populated delta system.

Multi-Hazard Impacts: Surges inundated coastal Bangladesh, submerging islands and villages while breaching embankments not designed for combined surge and wave action. Simultaneous heavy rainfall caused riverine flooding that extended impacts inland (Islam et al., 2019).

Consequences: Over 3000 deaths, millions displaced, billions in infrastructure damage. Many cyclone shelters were destroyed or isolated by floods with critical design failure. Early warning systems inadequately reached remote fishing communities most vulnerable to compound effects.

Institutional Failures: Post-event evaluations identified poor integration of hydrological and coastal surge data, lack of infrastructure designed for simultaneous wind and flood, and weak inter-agency coordination that delayed resource mobilization and shelter management (UNDP, 2008).

Adaptive Responses: Sidr catalyzed improvements in community-based early warning and shelter design standards, informing subsequent Bangladesh Cyclone Preparedness Program (CPP) enhancements (Miyaji, 2020).

Critical Assessment: While technical improvements followed, institutional fragmentation persisted. Separate agencies continue controlling surge forecasting, flood monitoring, and shelter management with limited real-time data integration illustrating how disaster-driven reforms often address symptoms rather than structural governance barriers.

3.2 Cyclone Amphan (2020)

Cyclone Amphan made landfall near the India-Bangladesh border on May 20, 2020, as a severe cyclonic storm with sustained winds of 250 km/h and 3–5-meter storm surges across the Sundarbans and West Bengal coast (Islam et al., 2019; Bernard et al., 2022).

Compound Hazard Dynamics: Amphan demonstrated urban vulnerability and compound exposure. Simultaneously with COVID-19, evacuation and sheltering faced social distancing constraints (Hadi et al., 2021). Urban flooding in Kolkata was severe due to overwhelmed drainage systems and tidal backflow in the Hooghly River amplified by surges (United Nations in Bangladesh, 2021). Infrastructure damage included widespread power outages, communication breakdowns, and transportation disruptions from both waterlogging and fallen trees (IFRC, 2023b). Storm surge caused prolonged saltwater contamination of groundwater and aquifers, impacting agriculture and drinking water for months (Islam et al., 2019).

Institutional Response: Despite early warnings and mass evacuations, absence of coordinated multi-hazard risk mapping in urban and rural areas limited resilience measure effectiveness (Bernard et al., 2022). However, integration of satellite-based precipitation monitoring and hydrological simulations into some district emergency management systems showed marked improvement (INCOIS, 2022).

Equity Dimensions: Amphan highlighted socioeconomic disparities. Marginalized populations informal settlers, forest-dependent indigenous communities, faced elevated risks from structural inequalities in shelter access and relief distribution (Shamsuddoha et al., 2024).

Feasibility Insights: Technological advances (satellite monitoring, coupled models) proved feasible and effective when implemented, but scalability remains constrained by district-level capacity and resources. The COVID-19 context revealed that multi-hazard planning must anticipate concurrent non-climatic crises, a consideration often absents from risk assessments.

Critical Analysis: Scientific assessment advocates continued development of integrated multi-hazard early warning systems and green infrastructure buffers like mangrove restoration (Islam et al., 2019). Yet implementation depends on overcoming resource allocation conflicts: funding for immediate relief consistently outcompetes investment in preventive infrastructure, reflecting short political cycles misaligned with long-term resilience needs.

3.3 *Cyclone Mocha (2023)*

Cyclone Mocha, a super cyclonic storm with winds exceeding 260 km/h and exceptional storm surges, made landfall in May 2023 along the Myanmar-Bangladesh coastline (IFRC, 2023b; Bernard et al., 2022). It affected nearly one million people through extreme winds, flooding surge, and landslides in hilly areas (Table 2).

Differential Outcomes: Bangladesh evacuated over 700,000 people, reflecting forecasting and mass evacuation improvements. However, Cox's Bazar refugee camps for displaced Rohingya populations suffered severe flood damage due to shelters not designed for multi-hazard exposure and absent multi-risk planning (IFRC, 2023a). Better integration of cyclone surge forecasts with hydrological inundation models enabled more effective road closures and resource distribution, though real-time data sharing gaps between agencies and communities still hindered rapid response (Bernard et al., 2022). Myanmar experienced higher casualty rates and prolonged recovery needs compared to Bangladesh due to infrastructural deficiencies and restricted early warning dissemination (UNDRR, 2023).

Intersectional Vulnerability: Mocha exposed how displacement, statelessness, and poverty create compounded vulnerabilities that standard disaster planning ignores.

Table 2 Governance comparison Mocha

Country	Deaths	Evacuated	Early warning quality	Infrastructure	Outcome
Bangladesh	0	>700,000	Excellent (96 hr. advance)	4000+ shelters	No casualties
Myanmar	145+ (possibly >400)	Limited	Poor (restricted)	Inadequate	High casualties

Refugee populations lack legal status to access government shelters, illustrating how social exclusion mechanisms amplify physical hazard exposure.

Governance Implications: Mocha accelerated calls for intersectional disaster governance combining meteorological, hydrological, engineering, and social science disciplines for coordinated real-time response (Brecht et al., 2012). Community-based adaptation measures participatory mapping, mobile alert systems proved effective were implemented and merit wider adoption (Hadi et al., 2021; Shamsuddoha et al., 2024).

Trade-Off Analysis: Bangladesh's evacuation success came at a significant economic cost (business closures, transportation shutdowns). As climate change increases cyclone frequency, balancing economic disruption against safety imperatives presents growing challenges. This trade-off is rarely quantified in disaster planning but fundamentally shapes community compliance with evacuation orders.

3.4 Comparative Synthesis

Analyzing these three cyclones comparatively reveals recurring patterns:

Compound Hazard Dynamics: Each storm demonstrates wind-surge-flood interactions creating vulnerabilities exceeding the sum of parts, validating the multi-hazard framework (Bernard et al., 2022; Brecht et al., 2012).

Institutional Fragmentation: Despite awareness of coordination needs, siloed governance and fragmented disaster management repeatedly emerge as barriers. This persistence suggests that technical knowledge is necessary, but insufficient deeper political economy reforms are required (NDMA India, 2022; UNDP, 2008).

Social Inequities: Vulnerability remains unevenly distributed, with informal, indigenous, and marginalized groups facing disproportionate risk and limited adaptive capacity (IFRC, 2023a; Shamsuddoha et al., 2024).

Technological Progress and Implementation Gaps: Increasing incorporation of real-time hazard modeling and multi-hazard platforms shows promise but requires greater interoperability and community-level accessibility (Bernard et al., 2022; INCOIS, 2022).

Emerging Resilience Practices: Adaptive infrastructure design, community early warning systems, and nature-based solutions like mangrove restoration offer pathways to enhanced multi-hazard resilience (Islam et al., 2019; Miyaji, 2020).

Critical Insight: Progress is evident but incremental. Transformative change requires addressing not just what needs to be done (integrated planning, better forecasting, stronger infrastructure) but why existing systems resist change (bureaucratic incentives, resource competition, political short-termism) (Table 3).

Table 3 Comparative case study analysis

Dimension	Cyclone Sidr (2007)	Cyclone Amphan (2020)	Cyclone Mocha (2023)
Wind speed (km/h)	240.00	250	260+
Storm surge (m)	5.0–6.0	3.0–5.0	Exceptional
Deaths	3406	26–130	0 (BD) / 145+ (MM)
Displaced	Millions	>five million	>700,000
Economic loss (USD)	$1.7 billion	$13.2 billion	$2.24 billion (MM)
Early warning	Improved but gaps	Excellent	Excellent (BD) / poor (MM)
Evacuation success	Partial	Very successful	Complete (BD) / failed (MM)

4 Barriers to Risk Management and Institutional Challenges

Despite scientific advances and technological improvements, effective multi-hazard cyclone planning in the Bay of Bengal remains impeded by persistent gaps and institutional failures. This section systematically examines these barriers within the socio-environmental contexts of Bangladesh, India, and Myanmar, identifying pathways for reform (Table 4).

Table 4 Mortality reduction through early warning systems (1970–2023)

Cyclone Event	Year	Wind speed (km/h)	Storm surge (m)	Deaths	Deaths per million exposed
Bhola Cyclone	1970	185–240	10 +	300,000–500,000	~ 5000
1991 Cyclone	1991	250	6–8	138,000	~ 2300
Cyclone Sidr	2007	240	5–6	3406	~ 38
Cyclone Amphan	2020	250	3–5	26	~ 0.1
Cyclone Mocha	2023	260+	Exceptional	0 (Bangladesh)	0

4.1 Fragmented Governance: A Political Economy Perspective

Fragmentation in disaster governance is the most significant obstacle to integrated multi-hazard risk management. Understanding why fragmentation persists despite known inefficiencies requires political-economic analysis.

Bureaucratic Silos and Resource Competition: In Bangladesh, the Department of Disaster Management, Bangladesh Meteorological Department, and Water Development Board monitor different hazards with separate rules, funding, and reporting structures, lacking seamless coordination or data sharing (United Nations in Bangladesh, 2021). This structure reflects historical development patterns where distinct ministries secured independent budgets and political constituencies. Consolidation threatens established resource flows and administrative authority, creating institutional resistance despite efficiency arguments (Bhatt & Garge, 2023). In India's eastern states, federal complexity compounds fragmentation. State governments control disaster response while central agencies manage forecasting, creating coordination gaps (NDMA India, 2022). Sectoral disaster risk mitigation policies at the state level often lack integration with flood control, cyclone warning, and urban planning departments (Brecht et al., 2012).

Path Dependencies and Organizational Culture: Existing systems develop organizational cultures, technical expertise, and standard operating procedures optimized for single-hazard scenarios. Personnel are trained in specialized domains; information systems are designed for specific data types; inter-agency protocols follow established patterns. Transitioning to integrated approaches requires retraining personnel across disciplines; redesigning information systems for interoperability; renegotiating inter-agency protocols and legal authorities. These are costly, time-intensive investments that compete with immediate disaster response needs. Bureaucracies prioritize maintaining operational capacity for known scenarios over uncertain benefits of systemic restructuring (Bhatt & Garge, 2023; UNDRR, 2023).

Political Economy of Short-Term Incentives: Political cycles incentivize visible, immediate interventions (post-disaster relief, infrastructure reconstruction) over long-term prevention (integrated planning, governance reform). Politicians gain credit for emergency response but face diffuse accountability for prevention failures. This misalignment perpetuates reactive rather than proactive risk management (World Bank, 2022a).

Regional Cooperation Barriers: At the Bay of Bengal regional level, cooperation remains inadequately institutionalized despite BIMSTEC (Bay of Bengal Initiative for Multi-Sectoral Technical and Economic Cooperation) and BOBLME (Bay of Bengal Large Marine Ecosystem) initiatives. Cross-border data sharing, shared modeling scenarios, and joint emergency drills are infrequent due to political sensitivities and capacity differences (World Bank, 2022b). Transboundary river basins and migrant communities face compounded risks from this fragmentation.

Leverage Points for Reform: Understanding these structural barriers suggests reform strategies: (1) funding mechanisms that incentivize coordination (e.g., budget supplements for inter-agency projects); (2) legal mandates for integrated planning in vulnerable districts; (3) capacity-building programs that create shared technical languages across agencies; (4) performance metrics that reward prevention and coordination rather than just response speed.

4.2 Data and Modeling Gaps

Robust risk-informed planning depends on high-resolution, interoperable hazard and vulnerability data systems. While cyclone track and surge forecasting have improved significantly, integrating meteorological, hydrological, socio-economic, and infrastructure data into unified platforms remains incomplete (INCOIS, 2022; Sahoo & Bhaskaran, 2018).

Data Accessibility and Interoperability: Bangladesh possesses substantial data, but much is siloed within government departments or inaccessible due to security or technical constraints (United Nations in Bangladesh, 2021). Divergent data collection methodologies across regions hinder comparability and real-time updating. Satellite-based remote sensing has improved spatial coverage of flood and surge impacts; however, insufficient ground validation and restricted community data integration limit localized accuracy (Bernard et al., 2022; United Nations in Bangladesh, 2021). India's disaster risk management frameworks employ advanced numerical models for predicting flooding, wind, and surge events, but tool interoperability and usability for local planners and responders remain under development (NDMA India, 2022). Local constraints in data analysis capacity and digital infrastructure hinder application effectiveness at subnational and community levels (Bhatt & Garge, 2023).

Integrated Scenario Analysis Platforms: Platforms capable of modeling compound hazard effects, including cascading infrastructure failures and social vulnerabilities, are emerging but not widely adopted or accessible to at-risk communities (Brecht et al., 2012; Bernard et al., 2022). This delays proactive risk mitigation in favor of a reactive response.

Technical Corrections: Notably, some references in disaster planning literature incorrectly describe HEC-RAS as a hydrologic model or satellite system. HEC-RAS is, in fact, a hydrodynamic model developed by the U.S. Army Corps of Engineers for modeling water flow in rivers and channels. When coupled with satellite-derived precipitation data, HEC-RAS can improve flood predictions following cyclones, but the model itself is a computational tool, not a data source (Bernard et al., 2022; Sahoo & Bhaskaran, 2018).

Feasibility Considerations: Developing comprehensive multi-hazard data platforms requires significant technical investment, institutional commitment to data sharing, and capacity building at local levels. Trade-offs include costs of maintaining real-time data systems versus accuracy benefits; open data access versus security concerns; and standardization versus local customization. These trade-offs are rarely explicitly evaluated in planning processes (Fig. 4).

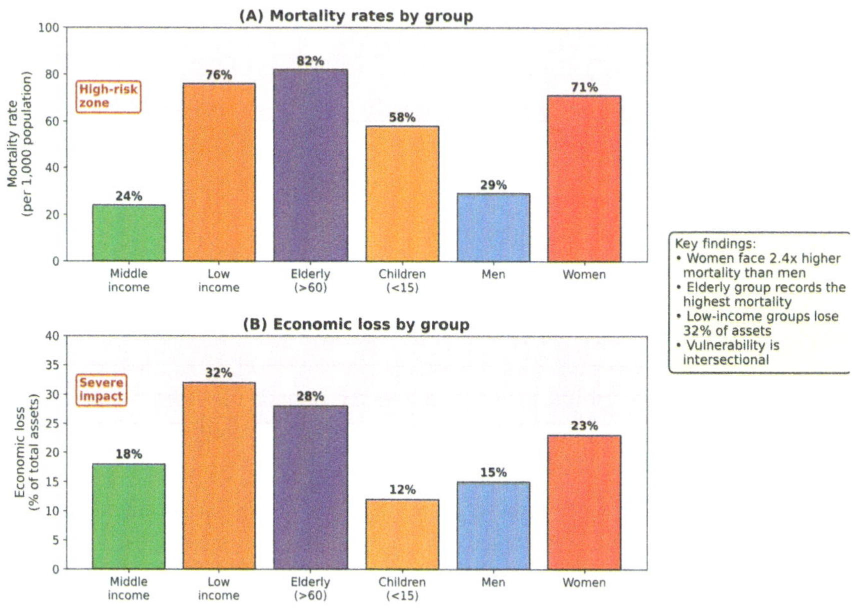

Fig. 4 Mortality rate and economic loss

4.3 *Socioeconomic Vulnerability*

Socioeconomic inequality significantly shapes differential risk exposure and adaptive capacity across Bay of Bengal cyclone-prone regions. Informal settlements, low-income agricultural communities, and displaced populations regularly inhabit areas highly vulnerable to storm surges and flooding yet face obstacles accessing risk information, evacuation assistance, and protective infrastructure (Hadi et al., 2021; Shamsuddoha et al., 2024).

Intersectional Vulnerabilities: Women, children, elderly, and individuals with disabilities face compounded vulnerabilities from socio-cultural norms, mobility constraints, and frequent exclusion from disaster governance participation (Shamsuddoha et al., 2024). Studies demonstrate that evacuation centers and shelters inadequately accommodate specific needs of these groups, compromising safety during cyclone events (IFRC, 2023a; UNDRR, 2023) (Table 5).

Livelihood and Economic Disparities: Coastal fishers and subsistence farmers depend significantly on natural resources compromised by cyclone damage and saltwater intrusion, while facing insufficient access to social safety nets or alternative income strategies (Islam et al., 2019). Structural poverty increases health risks and constrains investment in retrofitting or protective measures (World Bank, 2022b).

Critical Analysis of Perpetuation Mechanisms: Why do these inequities persist? Several reinforcing mechanisms operate: (1) Land use planning often prioritizes economic development over hazard exposure, pushing low-income populations to

Table 5 Differential mortality rates—sidr by demographics

Population Group	Percent of total population (%)	Percent of deaths (%)	Relative risk ratio
Coastal fishers	8	32	4.0x
Women	51	71	1.4x
Children < 15	35	42	1.2x
Elderly > 65	6	14	2.3x
Informal settlers	12	28	2.3x

riskier areas; (2) Informal settlements lack legal recognition, excluding residents from government services including disaster shelters; (3) Relief distribution systems favor populations with political voice and social capital; (4) Adaptation funding flows to visible infrastructure projects rather than social safety nets or livelihood diversification for vulnerable groups.

Pathways to Equity: Closing these gaps requires policies promoting inclusion, community participation, and targeted programs like women-led early warning volunteer groups and livelihood diversification programs (Bangladesh CPP, 2022; Shamsuddoha et al., 2024). However, equity-focused interventions face resource constraints and political resistance from elite interests benefiting from current allocation patterns.

5 Tools and Frameworks for Multi-Hazard Planning

Effective multi-hazard planning in the Bay of Bengal increasingly depends on integrating advanced scientific tools, innovative frameworks, and operational networks that can simulate, forecast, and communicate the compound risks cyclones pose. This section examines key tools and frameworks, assessing not only their technical capabilities but also their feasibility, trade-offs, and scalability.

5.1 Climate and Hazard Modeling Systems

Cyclone hazard modeling has advanced significantly due to improved computational power, satellite remote sensing, and integrated data assimilation. The Indian National Centre for Ocean Information Services (INCOIS) and other organizations operate numerical storm surge and cyclone track prediction models incorporating wind speed, atmospheric pressure, tidal phases, and bathymetric data to forecast storm surges up to 96 hours in advance (INCOIS, 2022; NDMA India, 2022). Hydrodynamic models such as HEC-RAS, when coupled with satellite-derived precipitation data, enhance predictions of urban and riverine flooding following cyclones

Table 6 Forecasting accuracy improvements (1990–2023)

Forecast lead time	1990s Error (km)	2020s Error (km)	Improvement
24 h	180	65	64% reduction
48 h	300	120	60% reduction
72 h	450	210	53% reduction

(Sahoo and Bhaskaran, 2018; Bernard et al., 2022). These models enable scenario analysis, allowing emergency managers to simulate compound hazard effects such as surges occurring simultaneously with blocked drainage or pre-existing soil saturation (Table 6).

Implementation Challenges: Difficulties persist in operationalizing these scientific advances. Interoperability among models, real-time data sharing protocols, and access for local authorities and communities remain constrained in many Bay of Bengal nations (United Nations in Bangladesh, 2021; Bhatt & Garge, 2023).

Emerging Solutions: Efforts to create unified composite risk atlases merging hazard probabilities with social vulnerability indices are ongoing. For instance, India's Web-based Dynamic Composite Risk Atlas integrates cyclone hazard maps with population density and infrastructure data for precise response planning (NDMA India, 2022).

Trade-Off Analysis: High-resolution modeling requires substantial computational resources and technical expertise. Trade-offs include accuracy versus operational speed (complex models may be too slow for real-time decision-making); spatial resolution versus computational cost; and model complexity versus user accessibility for local planners. Optimal approaches balance these factors based on specific decision contexts, coarser models for rapid initial response, and detailed models for infrastructure planning.

5.2 *Hazard Zoning and Resilience Indices*

Hazard zoning, spatially delineating areas by disaster probability and likely damage, is fundamental for disaster risk mitigation. In Bangladesh, coastal district-level hazard zoning employs multi-hazard indices based on historical cyclone paths, storm surge inundation depths, flood frequency, and soil salinity (United Nations in Bangladesh, 2021; Islam et al., 2019). Zoning informs land use regulations, evacuation route planning, and shelter siting. Complementing hazard zoning, composite resilience indices evaluate physical, social, and institutional capacities to withstand cyclonic events. These indices assess critical infrastructure coverage and condition, income diversification, healthcare accessibility, literacy rates, and governance quality (Sahoo and Bhaskaran, 2018). Indian coastal states have begun incorporating such indices into disaster risk management plans to prioritize infrastructure investments and community training programs (NDMA India, 2022).

Value Proposition: These multidimensional indices provide policymakers with frameworks to balance risk mitigation between hard infrastructure enhancements and intangible social investments, recognizing that resilience depends on both engineering solutions and social capacity.

Implementation Feasibility: Hazard zoning effectiveness depends on enforcement, often compromised by informal construction practices, political pressure for development in hazardous zones, and inadequate local governance capacity (Bhatt & Garge, 2023). Without enforcement, even accurate zoning maps have limited impact.

Scalability Considerations: Developing resilience indices requires significant data collection and analytical capacity. Replicating Bangladesh's district-level approach across the Bay of Bengal region would require substantial capacity building and resource investment, particularly in Myanmar where institutional capacity lags India and Bangladesh.

5.3 Infrastructure and Urban Applications

Urban and infrastructure planners increasingly use multi-hazard assessments to design resilient solutions. This section examines how risk assessments translate into practical applications.

5.3.1 Infrastructure Resilience

Building infrastructure capable of withstanding cyclonic winds, storm surges, and flooding is fundamental to resilience. In Bangladesh, where over 60% of the population inhabits coastal and deltaic cyclone-prone areas, cyclone shelter evolution demonstrates multi-hazard design progress. Following Sidr and subsequent storms, shelters were enhanced with raised plinths preventing flood ingress, reinforced concrete walls for high wind resistance, and multiple exits enabling rapid evacuation during compound hazards (IFRC, 2023a; Miyaji, 2020). Indian coastal states have similarly upgraded transportation systems by elevating roads, redesigning bridges for increased flood and debris load capacity, and hardening power and communication systems for operation during and after multi-hazard events (Bernard et al., 2022; NDMA India, 2022). Post-Cyclone Fani in Odisha, extensive electrical grid improvements reduced future outage probability, demonstrating proactive infrastructure planning benefits. Water supply and sanitation infrastructure design prioritizes cyclone resilience, emphasizing flood-proofing and rapid repairability to prevent post-disaster epidemics (Pramanik et al., 2021). Coastal embankments and polders are strengthened against both surge-induced erosion and riverine flood pressures, though maintenance challenges and climate change impacts threaten these initiatives' long-term durability (Islam et al., 2019).

Cost-Benefit Considerations: Analyses comparing proactive resilience investment to disaster reconstruction consistently show former economic superiority, justifying public funding and donor support for resilience corridors (Bernard et al., 2022). However, upfront costs are high, and benefits accrue over long-term horizons, creating financing challenges for resource-constrained governments.

Design Trade-Offs: Multi-hazard infrastructure design involves trade-offs between wind resistance (requiring rigid, reinforced structures) and flood resilience (requiring flexibility and rapid drainage). Optimal designs depend on local hazard profiles surge-dominant areas, prioritize elevation and drainage, while in wind-dominant areas, emphasize structural reinforcement.

5.3.2 Urban Planning and Land Use

Cities along the Bay of Bengal have experienced rapid growth, often encroaching on floodplains, wetlands, and hazard-prone areas. Multi-hazard urban design employs zoning regulations restricting development in the highest-risk coastal zones and flood-prone riverbanks (United Nations in Bangladesh, 2021). Natural infrastructure incorporation, mangrove belt restoration, and urban green space creation provides buffers, mitigating storm surge and reducing surface runoff (Islam et al., 2019; Bernard et al., 2022). Hazard zoning is integrated with satellite and GIS data, which informs land use planning and disaster management. Cities like Kolkata and Khulna use these tools for identifying at-risk neighborhoods and planning evacuation routes, accounting for compound cyclone effects (Sahoo and Bhaskaran, 2018). Participatory urban planning initiatives engage communities and local governments in co-designing flood-resilient neighborhoods, ensuring infrastructure upgrades align with residents' priorities (Bangladesh CPP, 2022).

Stormwater and wastewater infrastructure is being redesigned to accommodate sudden heavy rainfall and tidal flooding. Low-impact development principles permeable pavements, rain gardens, constructed wetlands are increasingly recognized for flood prevention and urban cooling roles (United Nations in Bangladesh, 2021; Islam et al., 2019).

Governance Challenges: Effective urban planning requires strong governance and enforcement. Informal settlements often develop outside regulatory frameworks, creating pockets of high vulnerability. Political economic factors land speculation, corruption, elite capture of planning processes frequently override technical hazard assessments.

5.3.3 Building Codes and Retrofitting

Updated building codes translate risk science into safer construction practices. Following Sidr, Bangladesh strengthened building codes with cyclone-resistant requirements: robust materials, anchoring methods, and multiple exits for emergency egress (Miyaji, 2020). India's national building code for cyclone-prone coastal states now includes stringent wind load resistance and flood-proofing standards (NDMA India, 2022). Retrofitting older buildings, particularly schools, hospitals, and residences in at-risk areas, remains challenging due to financial and technical constraints (Brecht et al., 2012). However, studies demonstrate retrofitting provides significant social and economic benefits through reduced disaster damage costs and accelerated recovery (World Bank, 2022b). Pilot projects in the Sundarbans and Odisha involve local masons and builders in training programs combining traditional building techniques with cyclone-resistant methods (Bangladesh CPP, 2022). Implementation Barriers: Widespread upgraded code implementation requires enhanced enforcement, frequently compromised by informal construction practices and inadequate governance (Bhatt & Garge, 2023). Many vulnerable populations cannot afford code-compliant construction, creating equity challenges. Financing Mechanisms: Innovative approaches to microfinance for retrofitting, building code subsidies for low-income households, and insurance premium discounts for compliant construction can improve adoption but require institutional development and sustained funding.

5.4 Early Warning Systems and Communication Strategies

Communication is critical for multi-hazard cyclone risk management in the Bay of Bengal. Rapidly delivering accurate, actionable information to at-risk populations enables timely evacuation and preparation. However, the region's challenges high population density, linguistic diversity, remote settlements, social inequalities complicate communication, necessitating innovative, culturally informed, trust-building approaches (Fig. 5).

Community-Based Early Warning: Bangladesh's Cyclone Preparedness Program (CPP), operating since 1972 as a government-Bangladesh Red Crescent Society partnership, exemplifies community-based early warning systems (Bangladesh CPP, 2022). The program trains and organizes thousands of local volunteers who disseminate cyclone warnings and preparedness messages using megaphones, mobile loudspeakers, radio, and door-to-door visits, reaching even the most remote coastal areas, including Cox's Bazar refugee camps (Communication and Community Engagement Strategy, 2021). The CPP communication strategy employs a recognized cyclone signal flag system indicating threat levels from "Flag 1" (warning stage) to "Flag 3" (great danger), triggering cascading preparedness actions. During the COVID-19 pandemic, this strategy incorporated health messaging to avoid compounding emergencies, demonstrating adaptability by integrating cyclone warnings with public

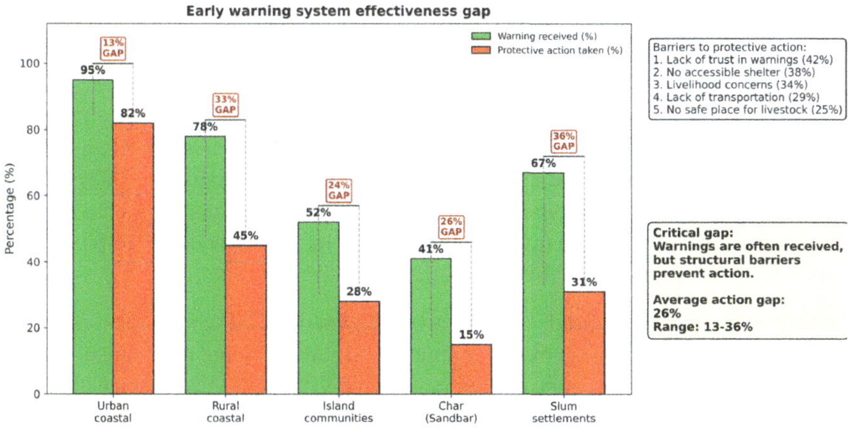

Fig. 5 Early warning system effectiveness gap

health measures like masking and distancing (Communication and Community Engagement Strategy, 2021).

Multi-Platform Dissemination: Digital and mass media platforms are increasingly important. Authorities use SMS alerts, social media (Facebook, Twitter, Instagram), interactive voice response (IVR) systems, community radio, and public announcement systems for rapid information dissemination (Bangladesh CPP, 2022; Bernard et al., 2022). During Cyclone Amphan (2020), Indian authorities extensively used social media for real-time updates, resource mobilization, and addressing public inquiries while using official channels to counter misinformation (World Bank, 2022a). In Bangladesh, during cyclone preparedness phases, local transportation modes (motorbikes, boats) rapidly mobilize volunteers for grassroots dissemination, shortening time between meteorological alerts and public response (Bangladesh CPP, 2022; Bernard et al., 2022). Using mosque loudspeakers and community leaders for culturally relevant warnings overcomes literacy barriers and social exclusion (Communication and Community Engagement Strategy, 2021) (Table 7).

Crisis Communication and Trust: Effective risk communication extends beyond technical warnings to include empathetic, transparent leadership. Kolkata power

Table 7 Technology access and preparedness

Technology	Urban access (%)	Rural access (%)	Remote island access (%)
Smartphone	78	45	18
Internet	85	52	12
Satellite TV	65	38	8
Radio	92	84	71
CPP volunteer network	98	96	93

sector experience during Cyclone Amphan demonstrates how communication failures erode safety and government trust. Early unclear messages during outages caused confusion and anger, hindering community cooperation, while clear, regular updates restored trust and accelerated recovery (PolicyCircle, 2021). Best practices emphasize visible leadership, messages including all vulnerable populations, and updates through trusted local and national channels (NDMA India, 2022; PolicyCircle, 2021). Crisis communication must also focus on rumor management and misinformation countering to maintain public trust and avoid panic (Communication and Community Engagement Strategy, 2021).

Participatory Mapping and Community Engagement: Technologies empower communities to participate in hazard mapping and early warning mechanism design. Participatory Geographic Information Systems (P-GIS) and citizen science enable real-time reporting of flooding, damage, and risk areas, incorporated into composite risk platforms accessible to disaster managers (United Nations in Bangladesh, 2021). Participatory approaches enhance situational awareness while fostering community ownership and compliance with evacuation protocols. Capacity-building workshops, regular volunteer training, and community knowledge-based simulation drills have improved early warning and evacuation system effectiveness and resilience (Bangladesh CPP, 2022; IFRC, 2023b). These initiatives have particularly benefited women, the elderly, and refugees, who are often excluded from formal planning.

Critical Assessment: Bangladesh's CPP represents global best practice, yet implementation quality varies by district due to capacity and resource constraints. Scaling similar systems across Myanmar and vulnerable districts in India requires sustained investment and political commitment. Additionally, evolving communication technologies (smartphones, social media) create digital divides; populations lacking access are increasingly marginalized from warning systems.

6 Toward Integrated and Inclusive Resilience

Building Bay of Bengal resilience to cyclonic multi-hazards requires transcending quick fixes toward transformative change encompassing science, institutions, and social inclusion. This chapter proposes pathways to resilience through cross-disciplinary collaboration, governance reform, and equity-centered planning.

6.1 Cross-Disciplinary Approaches

Bay of Bengal cyclone risk intersects atmospheric science, hydrology, engineering, urban planning, public health, and social psychology. Addressing such complexity requires collaborative research, policy design, and operational planning across these domains (United Nations in Bangladesh, 2021; Bernard et al., 2022). Integrating atmospheric modelers predicting cyclone paths, hydrologists modeling surge and flood propagation, civil engineers designing robust structures, and social scientists studying community vulnerability enables comprehensive risk understanding from storm genesis through post-disaster recovery (INCOIS, 2022; UNDRR, 2023). Cross-disciplinary innovation has produced multi-hazard risk atlases assisting local governments in optimizing evacuation routes and prioritizing shelter enhancements (Bangladesh CPP, 2022). Public health dimensions are critical as post-cyclone waterborne diseases and injuries require health authority planning. Behavioral scientists contribute to effective risk communication and evacuation compliance strategies (Hadi et al., 2021; Pramanik et al., 2021).

Implementation Mechanisms: Cross-disciplinary integration requires institutional structures facilitating collaboration joint research centers, integrated planning task forces, shared training programs. Academic institutions can play bridging roles through interdisciplinary disaster research programs training the next generation of resilience professionals.

6.2 Governance and Policy Reforms

Governance reform is essential for breaking down institutional silos and enabling coordinated multi-hazard risk management. Regional cooperation platforms such as BIMSTEC and SAARC can facilitate cross-border disaster data sharing, collaborative response exercises, and multi-hazard infrastructure funding; however, these initiatives require strong political backing and national policy alignment (Planetary Security Initiative, 2025; World Bank, 2022b). Nationally, creating institutional mandates integrating meteorology, disaster management, climate adaptation, urban planning, and social welfare improves preparedness and recovery (Brecht et al., 2012; NDMA India, 2022). Enhancing local government capacity for inclusive hazard mapping and response planning is critical, as frontline workers are closest to vulnerable populations (United Nations in Bangladesh, 2021). Innovative financial tools catastrophe bonds, pooled insurance schemes, climate resilience funds enable investment in sustainable infrastructure and community programs (World Bank, 2022a). Policy frameworks must incorporate multi-hazard risk assessments and encourage climate-smart development practices accounting for hazard evolution under climate change (Islam et al., 2019).

Political Economy Reforms: Beyond technical reforms, addressing political economy barriers requires: (1) performance metrics rewarding prevention and coordination rather than just response speed; (2) multi-year budget commitments insulating resilience investments from political cycles; (3) transparency mechanisms reducing corruption in disaster spending; (4) legal frameworks mandating integrated planning in vulnerable areas.

Critical Analysis: Governance reform is perhaps the most challenging component of resilience building, as it threatens entrenched interests and requires sustained political will. Success cases (Bangladesh CPP evolution, India's NDMA development) demonstrate that catastrophic disasters can create reform windows but capitalizing on these windows requires prepared reform proposals and advocacy coalitions ready to act.

6.3 *Inclusion and Equity in Planning*

Social equity must receive equal priority in resilience building. Women's leadership in early warning dissemination and community mobilization has demonstrated effectiveness in improving evacuation rates and reducing fatalities (Shamsuddoha et al., 2024). Programs targeting indigenous peoples, displaced communities, and informal settlers enhance adaptive capacity through participatory mapping, financial inclusion, and tailored risk communication (Bangladesh CPP, 2022; IFRC, 2023a). Technological tools, such as digital participatory GIS and mobile apps, enable citizen science and give voice to populations often excluded from disaster planning (United Nations in Bangladesh, 2021). Training programs emphasize cultural awareness and respect for local languages, knowledge systems, and social norms (Communication and Community Engagement Strategy, 2021). Equity-focused strategies also address specific vulnerabilities of individuals with disabilities and elderly populations through enhanced shelter accessibility, transportation assistance, and health service integration in disaster responses (Hadi et al., 2021).

Structural Barriers to Equity: Achieving meaningful equity requires confronting structural barriers: (1) Land tenure systems excluding informal settlers from legal protections; (2) Gender norms restricting women's mobility and decision-making; (3) Caste and ethnic discrimination affecting resource access; (4) Statelessness preventing refugee populations from accessing government services.

Leverage Points: Equity-oriented interventions should target multiple levels simultaneously: (1) National policies establishing disaster equity as explicit goal with measurable indicators; (2) Budget allocation mechanisms prioritizing vulnerable populations; (3) Community-level institutions (women's groups, indigenous organizations) as implementation partners; (4) Rights-based legal frameworks ensuring disaster assistance access regardless of social status.

7 Conclusions and Recommendations

7.1 Key Insights

The Bay of Bengal region, which experiences frequent and intense cyclones, exemplifies the complexity of multi-hazard risk. This chapter has demonstrated that regional vulnerability stems not only from cyclones but from intricate interactions among extreme winds, storm surges, river and urban flooding, and secondary hazards, including landslides and infrastructure failure. Conventional single-hazard approaches fail to capture these dynamics, producing underestimated risk and fragmented responses disproportionately impacting marginalized and low-income communities. Case studies of Sidr (2007), Amphan (2020), and Mocha (2023) reveal persistent challenges in governance, data integration, social equity, and infrastructure resilience. These events demonstrate that compound hazards amplify suffering and losses but can be mitigated through innovative forecasting, multi-hazard early warning, multi-purpose shelter design, and community engagement (Bernard et al., 2022; IFRC, 2023a; Miyaji, 2020). Our key theoretical contribution is demonstrating that effective risk management necessitates a transition from hazard identification to system-level resilience, which involves combining scientific modeling, institutional coordination, and inclusive social planning. The multi-scale integrated risk framework proposed here provides decision criteria for prioritizing interventions across physical, institutional, social, and technological dimensions simultaneously.

7.2 Strategic Recommendations

Invest in Integrated Multi-Hazard Modeling and Early Warning Platforms: Expand computational resources and data-sharing protocols to develop interoperable, real-time multi-hazard risk forecasting tools accessible to both policymakers and frontline responders (INCOIS, 2022; Bernard et al., 2022). Prioritize investments, balancing accuracy with operational usability.

Bridge Institutional Silos Through Incentive Restructuring: Strengthen coordination frameworks across sectors at local, national, and regional levels. Include: mechanisms for data sharing; joint emergency preparedness; resource mobility across health, cyclone, and flood systems; and budget incentives rewarding inter-agency collaboration (NDMA India, 2022; UNDRR, 2023).

Promote Inclusive Disaster Risk Mitigation: Ensure women, indigenous groups, informal settlers, the elderly, and individuals with disabilities participate meaningfully in risk communication, shelter design, and policy development through participatory governance and targeted capacity building (Shamsuddoha et al., 2024; Bangladesh CPP, 2022). Establish disaster equity indicators for monitoring progress.

Scale Up Resilient Infrastructure Investment with Equity Safeguards: Encourage large-scale adoption of cyclone-resilient building codes, multi-purpose cyclone shelters, green infrastructure (mangrove restoration), and adaptive urban planning guided by hazard zoning (Miyaji, 2020; Islam et al., 2019; Brecht et al., 2012). Ensure infrastructure investments include provisions for vulnerable populations rather than reinforcing existing inequalities.

Develop Innovative Financing Mechanisms: Create financial instruments, climate resilience bonds, insurance pools, and international aid linked to multi-hazard risk mitigation targets, ensuring long-term funding for infrastructure, early warning, and social safety nets (World Bank, 2022b). Design financing mechanisms explicitly addressing equity to prevent elite capture.

Foster Transboundary Collaboration: Support formalized regional cooperation through organizations like BIMSTEC to improve cross-border data sharing, joint risk assessment, and shared emergency response capabilities for threats transcending national borders (Planetary Security Initiative, 2025). Political commitment to overcoming sovereignty sensitivities is essential.

Conduct Explicit Trade-Off Analysis: Future planning should explicitly evaluate trade-offs inherent in resilience strategies accuracy versus speed, cost versus coverage, economic disruption versus safety, centralized versus distributed systems. Transparent trade-off analysis improves decision quality and public legitimacy.

Integrate Reference to Best Practice Protocols: Planning should explicitly reference and learn from documents like the National Early Action Protocol for Cyclone in Bangladesh (Department of Disaster Management, GoB), which represents regional best practice in integrated multi-hazard cyclone preparedness.

7.3 Final Reflection

The Bay of Bengal must lead in developing innovative multi-hazard resilience strategies integrating science, governance, and community action as climate change intensifies extreme weather and population growth continues. Moving forward requires not only technological innovation but a strengthened commitment to social equity and system thinking. Through sustained regional cooperation, cross-disciplinary innovation, and inclusive governance, cyclones' catastrophic impacts can be mitigated, enabling millions along this cyclone-prone coastline to thrive amid climate change rather than merely survive. However, transformation requires confronting uncomfortable truths about power, resources, and priorities. Technical solutions exist; political will to overcome entrenched interests and bureaucratic inertia remain the critical constraint. The question is not what should be done, which is increasingly clear, but whether political, economic, and social systems can be reformed to do it.

References

Alam, E., & Collins, A. E. (2010). Cyclone disaster vulnerability and response experiences in coastal Bangladesh. *Disasters, 34*(4), 931–954. https://doi.org/10.1111/j.1467-7717.2010.01176.x

Ali, A. (1999). Climate change impacts and adaptation assessment in Bangladesh. *Climate Research, 12*, 109–116.

Antony, C., & Unnikrishnan, A. S. (2013). Tide–surge interaction in the bay of Bengal. *Estuarine, Coastal and Shelf Science, 131*, 6–11. https://doi.org/10.1016/j.ecss.2013.08.004

Bangladesh Cyclone Preparedness Programme (CPP). (2022). *Annual report 2022*. Government of the People's Republic of Bangladesh.

Bernard, A., Long, N., Becker, M., Khan, J., & Fanchette, S. (2022). Bangladesh's vulnerability to cyclonic coastal flooding. *Natural Hazards and Earth System Sciences, 22*, 729–751. https://doi.org/10.5194/nhess-22-729-2022

Bhatt, R. M., & Garge, D. (2023). Efficiency in emergency: A perception study of hierarchical versus flat organizational structures for joint disaster response in the bay of Bengal region. *Strategic Analysis, 47*(1), 61–78.

Brecht, H., Dasgupta, S., Laplante, B., Murray, S., & Wheeler, D. (2012). Sea-level rise and storm surges: High stakes for a small number of developing countries. *Journal of Environment & Development, 21*(1), 120–138.

Communication and Community Engagement Strategy: Cyclone preparedness, awareness and community engagement in COVID-19 pandemic situation. (2021). Shongjog, http://www.shongjog.org.bd.

Dastagir, M. R. (2015). Modeling recent climate change induced extreme events in Bangladesh: A review. *Weather and Climate Extremes, 7*, 49–60. https://doi.org/10.1016/j.wace.2014.10.003

Ericson, J. P., Vörösmarty, C. J., Dingman, S. L., Ward, L. G., & Meybeck, M. (2006). Effective Sea-level rise and deltas: Causes of change and human dimension implications. *Global and Planetary Change, 50*, 63–82.

Hadi, T., Uddin, M. J., & others. (2021). The factors that influence refuge since Cyclone Gorky in Bangladesh. Progress in Disaster Science, 10, 100166.

Haque, C. E., & Jakariya, M. (2023). Bengal delta, Charland formation, and riparian hazards: Why is a flexible planning approach needed for deltaic systems? *Water, 15*(13), 2373. https://doi.org/10.3390/w15132373

Indian National Centre for Ocean Information Services (INCOIS). (2022). *Annual report 2022*. Ministry of Earth Sciences.

India Meteorological Department (IMD). (2025). Cyclone classification and intensity scales. In *Regional specialized meteorological Centre (RSMC)—Tropical cyclones*.

International Federation of Red Cross and Red Crescent Societies (IFRC). (2023a). Extreme weather and compound risks in South Asia

International Federation of Red Cross and Red Crescent Societies (IFRC). (2023b). Cyclone Mocha situation report

Islam, M. F., Brouwer, R., & others. (2019). Flood risk assessment due to cyclone-induced dike breaching in coastal Bangladesh. Natural Hazards and Earth System Sciences, 19, 353–374.

Mishra, M., et al. (2021). Geo-ecological impact assessment of severe cyclonic storm Amphan on Sundarban mangrove forest using geospatial technology. *Estuarine, Coastal and Shelf Science, 249*, 107115.

Miyaji, M. (2020). A study on the use of cyclone shelters in Bangladesh. *ALNAP*.

Murty, T. S., Flather, R. A., & Henry, R. F. (1986). The storm surge problem in the bay of Bengal. *Progress in Oceanography, 16*(4), 195–233.

National Disaster Management Authority (NDMA), Government of India. (2022). Cyclone risk mitigation project.

National Disaster Management Authority. (2025). Cyclone. https://ndma.gov.in/Natural-Hazards/Cyclone.

Policy Circle. (2021, June). A crisis communication playbook as Bay of Bengal turns into the world's cyclone epicenter.
Planetary Security Initiative. (2025). India's institutional cooperation on climate resilience in the Bay of Bengal region.
Pramanik, M., Szabo, S., Pal, I., Udmale, P., Pongsiri, M., & Chilton, S. (2021. often indexed as 2022). Population health risks in multi-hazard environments: Action needed in the Cyclone Amphan and COVID-19–hit Sundarbans region, India. *Climate and Development.*
Ritchie, H., Rosado, P., & Roser, M. (2022). Natural disasters. Our World in Data.
Sahoo, B., & Bhaskaran, P. K. (2018). Multi-hazard risk assessment of coastal vulnerability from tropical cyclones: A GIS-based approach for the Odisha coast. *Journal of Environmental Management, 206*, 1166–1178.
Sindhu, B., & Unnikrishnan, A. S. (2013). Characteristics of tides in the bay of Bengal. *Marine Geodesy, 36*(4), 377–399. https://doi.org/10.1080/01490419.2013.781088
Shamsuddoha, M., Jabed, M. A., Islam, M. S., Sultana, N., Imran, A., Rabbi, S. N. A., Jenat, T. U., Shams, S., & Sharif, M. M. (2024). Impacts of climate change-induced natural hazards on women and their human rights implications: A study in the southwest coast of Bangladesh. *Journal of Migration and Health, 9*, 100221. https://doi.org/10.1016/j.jmh.2024.100221
United Nations Development Programme (UNDP). (2008). *Post-Cyclone Sidr recovery review.* UNDP Bangladesh.
United Nations in Bangladesh. (2021). Multi-hazard risk analysis in Bangladesh: Final report.
United Nations Office for Disaster Risk Reduction (UNDRR). (2023). Disaster risk reduction in Bay of Bengal: Status report.
World Bank. (2022a). Bangladesh: Enhancing coastal resilience in a changing climate.
World Bank. (2022b). How Bangladesh can protect its development gains through coastal resilience.

Hurricane Impacts Compounded by Sea Level Rise: An Illustrative Case Study of Economic Impacts and Workforce Resilience Analysis

Juita-Elena Yusuf, Yin-Hsuen Chen, Robert M. McNab, and George M. McLeod

Abstract Expected sea level rise will increase the probability of extreme flooding events, such as those associated with hurricane surge and/or heavy rainfall. Projecting the economic impacts of a significant hurricane is necessary for weighing the costs and benefits of potential mitigation actions. Accounting for the compounding effects of sea level rise is critical, given the challenges elevated sea levels pose to coastal communities in the U.S. and worldwide. This chapter asks: How can we comprehensively estimate the economic impacts of a hurricane event that is compounded by sea level rise? To address this, we offer a methodology for conducting an economic impact analysis and apply it to the case study region of Hampton Roads (Virginia, U.S.). This chapter illustrates the methodology that accounts for compound hurricane and sea level rise, providing estimates of direct, indirect, and induced economic impacts. The economic impact estimates quantify uninsured building damages and reveal industry-specific variations in workforce disruptions.

Keywords Economic impact analysis · Sea level rise · Compound hazard · Induced economic impact · Indirect economic impact · Hurricane damages

---

J.-E. Yusuf (✉)
School of Public Service, Strome College of Business, Old Dominion University, Norfolk, VA, USA
e-mail: JYusuf@odu.edu

Y.-H. Chen · G. M. McLeod
Center for Geospatial Science, Education and Analytics, Old Dominion University, Norfolk, VA, USA

R. M. McNab
Department of Economics, Strome College of Business, Old Dominion University, Norfolk, VA, USA

J. Collins et al. (eds.), *Tropical Cyclone Risk and Resilience*,
https://doi.org/10.1007/978-3-032-28745-8_5

1 Introduction

While projections of future sea level rise are uncertain, there is increasing confidence that higher sea levels will increase the probability of extreme flooding events (Taherkhani et al., 2020), such as those associated with hurricane surge and/or heavy rainfall. Coastal regions, such as the Virginia Beach–Chesapeake–Norfolk metropolitan statistical area in the U.S. (also known as the Hampton Roads region of coastal Virginia), face increasing flood risks. Projections indicate that future moderate and major high tide flood events will occur with the same frequency as present-day minor and moderate flood events (Sweet et al., 2017).

Sea level rise (SLR) may amplify rainfall damages in coastal areas (Passeri et al., 2015) and increase stillwater elevations, raising the likelihood of moderate and severe flooding from storm events (Rahimi et al., 2020). Projecting the economic impacts of a significant storm event is essential to assessing flood risks and weighing the costs and benefits of mitigation, preparedness, and recovery efforts. Incorporating the compounding effects of SLR is critical given the challenges elevated sea levels pose to coastal communities in the U.S. and abroad. Given the amplifying effect of SLR on storm impacts, economic impact assessment should account for both SLR and storm impacts.

An economic impact analysis (EIA) estimates the impact of an event of a given duration and location on economic activity at local, regional, or national levels. It includes direct impacts on facilities, jobs, or payrolls, as well as the indirect and induced impacts that ripple through the economy. Indirect economic impacts are supply-chain related, reflecting how the direct impacts affect suppliers in the area. Induced economic impacts are generated by changes in compensation and household incomes. EIA methodologies typically estimate changes in industry production or output, value added, jobs, and employee compensation.

How can we comprehensively estimate the economic impacts of a hurricane event compounded by sea level rise? This chapter presents an EIA methodology for determining the economic impacts of a hurricane event compounded by SLR using a catastrophe model coupled with an input–output model. The assessment is based on a scenario involving a Category 3 hurricane combined with higher sea levels. We use the Hampton Roads region in coastal Virginia as the case study for applying this EIA methodology. Nevertheless, hurricanes and rising sea levels are not unique to Hampton Roads, but also pose increasing risks to coastal regions worldwide, from the U.S. Gulf Coast to island nations in the Caribbean and Atlantic to Southeast Asia. Thus, the methodology and insights in this chapter have broad applicability.

This chapter has three key contributions. First, the methodology offers insights into modeling the compounding effects of SLR on hurricane impacts, which is important as sea levels increase and more populations are at risk. Second, the methodology supports a comprehensive understanding of the compound event's impact by providing estimates of direct, indirect, and induced economic effects. Rather than focusing solely on direct damages and their associated costs, the methodology accounts for the ripple effects resulting from disruptions to physical infrastructure.

These include business interruptions, lower production of goods and services, and service losses. Additionally, it considers changes in economic activity arising from household consumption, driven by direct and indirect spending by industries in the affected region. Indirect and induced effects capture how a shock in one or several sectors propagates throughout the economy through interdependencies and intermediate consumption. Third, the methodology and illustrative case study provide information on variations in workforce impacts across industries, to determine which segments of the labor market would be most significantly affected. It highlights which elements of the labor market are likely to be the most (and least) resilient in the aftermath of a compound hurricane-SLR event in an urban area. This understanding of the economic resilience of the workforce and how the composition of the regional economy influences the pace and breadth of the recovery is important for planning and resource allocation for mitigation, preparedness, and recovery efforts.

2 Background

2.1 Sea Level Rise and Compounding of Hurricane Impacts

Hurricanes bring storm surges, strong winds, and heavy precipitation that can lead to significant coastal flooding (Goulart et al., 2024; Wahl et al., 2015; Wu et al., 2024). SLR compounds hurricane impacts by amplifying the extent and severity of coastal flooding through complex, non-linear interactions with the various storm-related processes (Deb et al., 2023; Rahimi et al., 2020; Wu et al., 2024).

By elevating the baseline water level, SLR increases the vulnerability of coastal areas to inundation (Castrucci & Tahvildari, 2018; Ghanbari et al., 2021; Sheng et al., 2022), directly raising relative sea level and reducing freeboard (the vertical distance between the water surface and coastal land elevations). This can result in existing flood thresholds being exceeded more frequently, with even minor storm surges leading to significant inundation (Castrucci & Tahvildari, 2018; Ghanbari et al., 2021). Additionally, with higher sea levels, hurricane storm surge can propagate further inland, inundating areas that would otherwise remain dry (Castrucci & Tahvildari, 2018; Sheng et al., 2022). SLR contributes to increases in both the depth and extent of storm flood inundation.

The combination of hurricanes and SLR amplifies flooding through storm surge, impaired drainage, and elevated groundwater levels. High water levels can create a "backwater effect" in which river water cannot freely discharge into the ocean, leading to inundation upstream and beyond the floodplain (Ghanbari et al., 2021; Wu et al., 2024). Low-elevation estuarine regions are particularly susceptible to this compounding effect (Ghanbari et al., 2021; Wu et al., 2024). Heavy precipitation during a hurricane can lead to direct runoff and pluvial flooding (Goulart et al., 2024; Wahl et al., 2015). SLR can cause groundwater levels to rise, which, when combined with hurricane-associated precipitation, leads to runoff that significantly exceeds the capacity of drainage infrastructure (Rahimi et al., 2020).

SLR is projected to account for a significant portion of total inundation associated with future hurricanes. For example, 1 m of SLR is projected to increase average flood volumes by 4.2 times during Hurricane Sandy scenarios in New York City (Goulart et al., 2024). Along the open coast, SLR and hurricane intensification contribute nearly equally to increased hurricane-related flooding, though SLR may be slightly more important for moderate to large surge events (Mousavi et al., 2011).

Thus, SLR cannot be ignored in mapping hurricane-associated flooding and in assessing future hurricane damage and costs. However, the relationship between SLR and hurricanes is not a simple linear addition (Castrucci & Tahvildari, 2018; Sheng et al., 2022; Wu et al., 2024). The interplay between the different drivers of flooding during a hurricane highlights the need to consider these dynamic interactions (Deb et al., 2023; Ghanbari et al., 2021; Wu et al., 2024). Modeling compound hurricane-SLR events should be part of assessing economic impacts to inform flood management and policy, mitigation actions, and recovery planning.

2.2 Economic Impacts of Disasters

Hazards cause significant economic disruptions and impacts. These impacts are broadly categorized into direct, indirect, and induced effects (Avelino & Dall'erba, 2019; Gc et al., 2024; Jin et al., 2020). Direct impacts generate indirect and induced impacts which, when summed with the direct impacts, comprise the total economic impact.

Direct impacts or stock damages are the immediate physical consequences of a hazard, occurring at or shortly after the event, including physical destruction of property and infrastructure, such as residential homes, businesses, manufacturing facilities, schools, roads, and utilities (Akhtar & Santos, 2013; Botzen et al., 2019; Hallegatte, 2008), and raw materials and natural resources (Botzen et al., 2019; Cavallo & Noy, 2011; Hallegatte, 2008). Direct impacts are typically estimated from the repair or replacement costs of damaged assets. Catastrophe models, insurance industry data, or disaster databases are often used to estimate damages (Avelino & Dall'erba, 2019; Botzen et al., 2019; Hallegatte, 2008).

Indirect impacts are changes in economic activity that follow, rather than being directly caused by, the hazard (Botzen et al., 2019; Cavallo & Noy, 2011; Hallegatte, 2008; Jin et al., 2020; Koks et al., 2016). These include business interruptions, lost production, utilities disruptions, and reductions in income, wages, retail sales, and tourism (Botzen et al., 2019; Cavallo & Noy, 2011; Okuyama & Santos, 2014). Also known as flow losses or higher-order effects (Koks et al., 2016), indirect costs are more complex and difficult to estimate due to their cascading nature and data collection challenges (Avelino & Dall'erba, 2019; Jin et al., 2020).

Indirect impacts capture the propagation of losses through the interconnected supply chains and industries. For instance, the 1993 Midwest flood caused $2 billion in direct losses by halting freight transportation and resulted in $3.6 billion in indirect losses due to crop damage (Avelino & Dall'erba, 2019). Hazard events can also lead to declines in labor market outcomes (Jerch et al., 2023; Meier et al., 2023), as labor-intensive services and manufacturing sectors can suffer significant losses in workforce productivity (Arabadzhyan et al., 2021; Avelino & Dall'erba, 2019; Meier et al., 2023).

Induced effects refer to the changes in economic activity resulting from reduced household spending and labor income generated by direct and indirect impacts (Gc et al., 2024). For example, when industries reduce production due to damage to manufacturing facilities (direct effects) or when their suppliers experience reduced demand (indirect effects), employee wages decline. Employees then have lower income to spend within the regional economy. This subsequent decline in economic activity is the induced effect.

Disasters generally have a negative short-term economic impact (Botzen et al., 2019; Cavallo & Noy, 2011). Furthermore, the total cost of a disaster, especially for large-scale events, increases nonlinearly relative to direct losses, emphasizing that direct impacts alone are insufficient measures of disaster consequences (Hallegatte, 2008). The overall economic impact of a hazard such as a compounded hurricane and SLR must account for direct, indirect, and induced effects.

2.3 Economic Impact Analysis

Economic impact analysis (EIA) is intended to answer "what if" questions related to changes in economic activity within a region resulting from an exogenous shock. EIA for disasters like hurricanes involves quantifying the multifaceted consequences and costs for the economy, including direct, indirect, and induced effects (Gc et al., 2024). EIA is crucial for planning recovery and reconstruction, evaluating preparedness and mitigation strategies, and assessing prevention investments (Botzen et al., 2019; Crowther et al., 2007; Okuyama & Santos, 2014).

Quantitative models and frameworks have been developed and employed to estimate the economic impacts of disasters (Avelino & Dall'erba, 2019; Jin et al., 2020; Okuyama & Santos, 2014). Input-output (I-O), computable general equilibrium, catastrophe, and econometric models comprise the most used frameworks. I-O models, widely used for EIA and assessing the consequences of hurricanes (Gc et al., 2024; Koks et al., 2016; Okuyama & Santos, 2014), structure the economy as a set of interconnected sectors that both produce and consume goods, allowing for the detailed understanding of economic interdependencies and how shocks propagate throughout the economic system (Botzen et al., 2019; Koks et al., 2016; Okuyama & Santos, 2014). The I-O model shows how goods and services produced by one industry are used by other industries and final consumers (Gc et al., 2024; McNab et al., 2024). They are especially useful for estimating the impacts of an initial shock and the flow losses that follow. Some commonly used I-O tools include Impact

Analysis for Planning (IMPLAN) and Regional Economic Models Inc. (REMI) models.

"The advantages of the I-O model for disaster impact analysis is based mainly on the ability to reflect the economic interdependencies … for deriving higher order effects, and partly on its simplicity" (Okuyama & Santos, 2014). However, I-O models have been criticized for their linear and rigid structure, which prevents them from accurately reflecting complex economic behaviors, leading to oversimplification and overestimation of impacts (Botzen et al., 2019).

I-O models have been applied to analyze the economic impacts of hurricane scenarios. For example, Hallegatte (2008) employed an I-O approach to estimate total costs associated withHurricane Katrina. Akhtar and Santos (2013) used I-O models for two hypothetical hurricanes in Virginia, focusing on workforce disruptions. In a non-US context, Yu et al. (2019) used an I-O model to estimate the economic impacts of Typhoon Fanapi (Taiwan).

This chapter contributes to research and practice by first modeling the economic impacts of hurricane damage on top of SLR, addressing a gap in the literature. Second, the EIA methodology underscores the importance of accounting for indirect and induced effects arising from the initial shock. Third, by decomposing the physical damage by building occupancy type (e.g., agriculture, commercial, education, government) and employment effects by industry (e.g., retail trade, utilities, accommodation and food service, manufacturing), the methodology and results can provide actionable information for adaptation and mitigation efforts. Fourth, by accounting for uninsured losses, the methodology and findings emphasize the important role of flood insurance and other risk transfer strategies, including federal and state assistance.

2.4 *Profile of the Hampton Roads Case Study Region*

Located in southeastern Virginia along the U.S. East Coast (Fig. 1a), the Hampton Roads region comprises 17 localities. The York and James Rivers flow into the Chesapeake Bay, and much of the region's elevation is low—over 30% of the land lies below 5 m, and another 19% is between 5 m and 10 m (Fig. 1b). This case study region had an estimated population of about 1.8 million in 2024 (U.S. Census Bureau, 2024b). The regional economy is anchored by three main sectors: federal spending, the Port of Virginia, and travel and tourism. In 2024, the region's Gross Domestic Product (GDP) exceeded $130 billion (Agarwal et al., 2024; Bureau of Economic Analysis, 2024). The median household income was $81,890, comparable to the national median of $78,538 (U.S. Census Bureau, 2024a).

Hurricanes are a recurring hazard for Hampton Roads. Category 1 and 2 hurricanes, as well as tropical storms, are more likely to affect the region in any given year. Since 1851, 76 significant storms have passed within 120 km of Hampton Roads (Hampton Roads Planning District Commission, 2022). While only two have

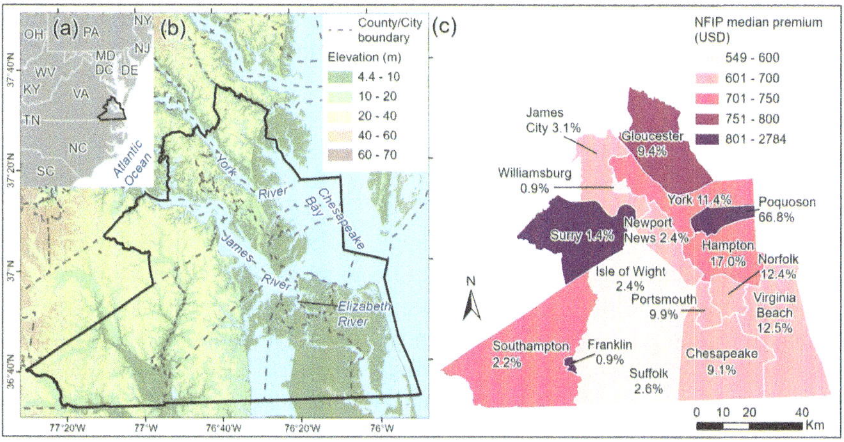

Fig. 1 (**a**) Inset map of the study area's location along the eastern coast of the U.S., (**b**) Elevation relief with major water bodies, and (**c**) National Flood Insurance Program (NFIP) uptake rate and median premium for 2022

been Category 3 hurricanes, modeling this higher-level threat yields a more conservative upper-limit estimate of economic impacts. Wind damage from these storms extends across large areas, leading to downed trees and widespread power outages. When combined with storm surge, particularly in low-lying and shorefront communities, the resulting damage can cause extensive business disruptions and substantial economic losses. Hurricane Isabel in 2003 caused an estimated $1.6 billion in regional damage (National Hurricane Center, 2004). Beach erosion was observed in several areas of Newport News and Hampton, and the Virginia Beach fishing pier sustained significant damage (National Weather Service, 2003).

Hampton Roads has several flood-prone areas due to its low elevations and extensive development. Numerous studies consistently show that the region bears the greatest share of future coastal-flooding impacts in Virginia. Analysis by McNab et al. (2024) found that Hampton Roads accounts for over 80% of projected storm-related economic losses in coastal Virginia. Watershed-scale analyses (Mitchell et al., 2023) identified urban areas across the region as particularly vulnerable. Additional work on ports (McLeod et al., 2018), transportation systems (Shen et al., 2022) and Natech disaster (Crawford et al., 2023) further underscores Hampton Roads as one of the most at-risk flood-prone regions in the state.

Figure 1c provides National Flood Insurance Program (NFIP) uptake rates and median premiums for localities in the region. Poquoson has the highest uptake rate, with 66.8% of occupied residences having an active NFIP policy. Four localities, Hampton, Virginia Beach, Norfolk, and York County, have uptake rates above 10% of occupied residences. However, other localities such as Franklin and Williamsburg have significantly lower percentages of households with flood insurance coverage. The pattern of flood insurance coverage is consistent with that of other communities along the U.S. East Coast (Kousky, 2017; McShane & Yusuf, 2019).

Note that the uptake rate is measured as the percentage of occupied households with an active policy (U.S. Government Accountability Office, 2023).

3 Methods

Our EIA methodology for a compound hurricane-SLR event impacting Hampton Roads includes three major components: (1) modeling of physical damages from wind and storm surge flooding; (2) estimating direct, indirect, and induced economic impacts from these damages using an I-O model; and (3) focusing on uninsured losses and damages. We incorporate SLR under the assumption of a 1.1 m tidal offset, which generates nuisance flooding that compounds the hurricane surge. We use Hazards U.S. Multi-Hazard (Hazus) to model and estimate direct physical damages, and IMPLAN to generate estimates of indirect, induced, and total economic impacts. The EIA methodology is summarized in Fig. 2.

3.1 Modeling the Physical Damages from Wind and Storm Surge Flooding

Estimating physical damage from a major hurricane is inherently uncertain. Each event is unique, and results depend on the model's underlying assumptions. We model a Category 3 hurricane scenario using a modified historical track of Hurricane Florence, which made landfall near Wrightsville Beach, North Carolina, on September 14, 2018. We selected this hurricane scenario because it is the most recent major storm to approach the Hampton Roads region. During this event, multiple tropical storm and storm surge watches and warnings were issued across coastal North Carolina and Virginia, highlighting its relevance as a representative high-impact scenario for the area (Stewart & Berg, 2018). We retain the storm's Category 3 intensity and shift its track northward by adding three degrees of latitude to each coordinate, resulting in landfall near the Virginia–North Carolina border (Fig. 3a).

Hazus (version 6.1) is applied to estimate the physical impacts and direct damages associated with the hurricane event. Hazus provides a standardized methodology for estimating losses from natural hazards, including hurricanes, earthquakes, and tsunamis (Schneider & Schauer, 2006). To assess the combined effects of wind and storm surge, we employ an integrated approach that uses both the Hazus Hurricane and Flood models. This combined method has been applied in prior studies to quantify hurricane impacts (Amini et al., 2023; Chen et al., 2021). Hazus operates as a catastrophe model, using Geographic Information Systems (GIS) to simulate the physical characteristics of a hazard at a specific location (Botzen et al., 2019), which includes defining physical features like potentially flooded areas and inundation depths.

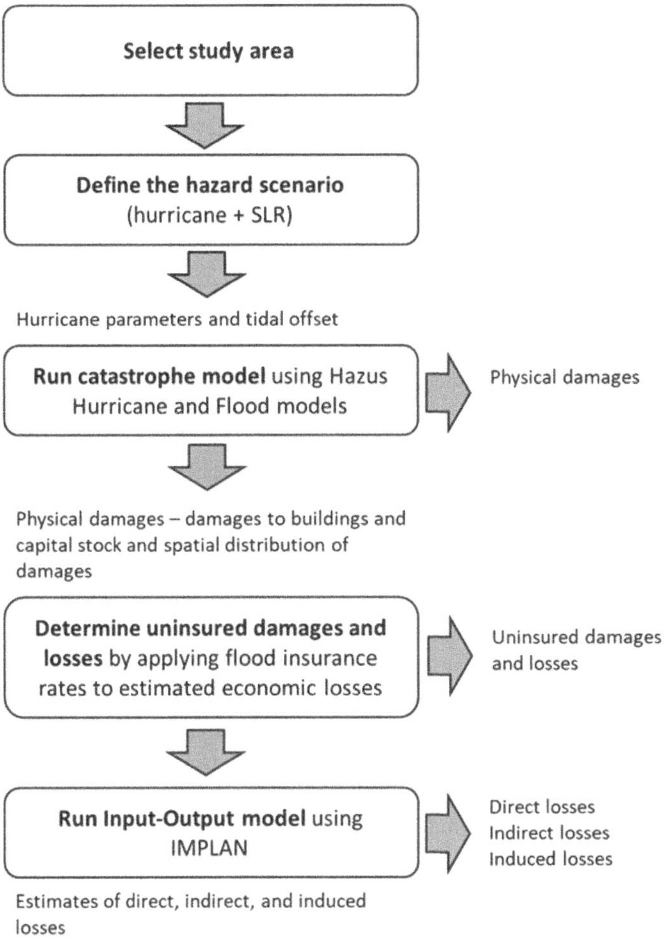

Fig. 2 Economic impact analysis (EIA) methodology

However, Hazus loss estimates have limitations in terms of validating the model's direct loss outputs. Prior work in the Upper Alabama River Watershed found that Hazus estimates were substantially higher than reported FEMA flood insurance claims (Gutenson et al., 2015). This discrepancy partly reflects the fact that the NFIP covers only a portion of total losses (Kousky & Michel-Kerjan, 2017), and that obtaining complete claims data from private insurers is challenging. We acknowledge the uncertainty inherent in Hazus-based direct economic loss estimates and interpret our results accordingly.

We apply the Level-1 Hazus model, which uses integrated FEMA databases (Federal Emergency Management Agency, 2025) that include estimated property characteristics at the census block level, such as building type, improvement values, and other structural attributes. A combined wind and storm surge analysis is

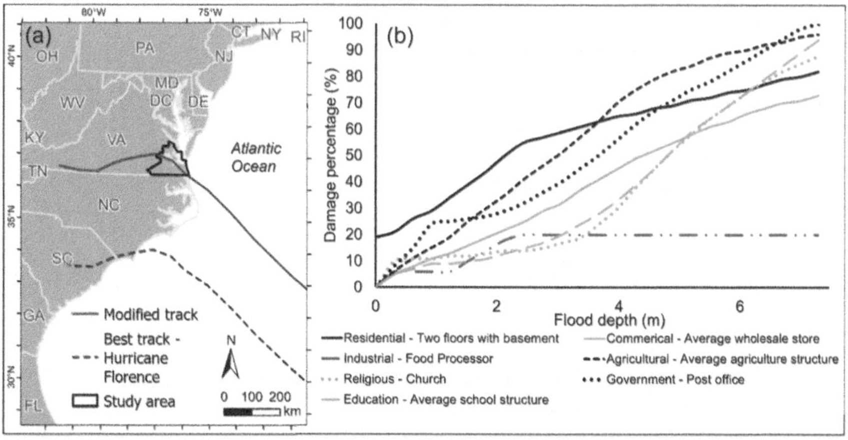

Fig. 3 (**a**) Map showing the observed track of Hurricane Florence and the modified track used for modeling, with the outline of the study area annotated to indicate its relative location, and (**b**) Line plot showing selected flood-depth damage curves used in the Hazus Flood Model to estimate property damage

conducted, starting with the Hurricane model. A custom hurricane scenario is developed based on key storm parameters, including latitude, longitude, forward speed, landfall time, radius to maximum wind, wind speed, and central pressure, derived from the best track data for Hurricane Florence (National Hurricane Center, 2019), with minor adjustments to latitude to better align the simulation with the study area. Hazus then generates a surface of peak gust wind speeds across the region. For each census block, wind-related damage is estimated using predefined loss functions that relate damage ratios to peak gust wind speeds, yielding spatially explicit estimates of wind damage throughout the study area.

A moderate flooding offset was used to reflect the influence of SLR on the potential damage from a hurricane. The moderate flooding threshold level for the Hampton Roads region was determined based on the Sewells Point tide gauge for Norfolk, Virginia, and estimated to be 0.82 m above mean higher high water (MHHW) datum (Sweet et al., 2021). To incorporate SLR, an accelerating trend of tidal flooding days per year is used (Sweet et al., 2017).

The difference between tidal datums mean sea level (MSL) and MHHW varies slightly depending upon location (0.42 m to 0.59 m), with higher differences being found along the Atlantic coastline in Virginia Beach. We add the moderate flooding value to the MHHW datum to obtain a 1.1 m tidal offset from the North American Vertical Datum of 1988 (NAVD88) datum which is used as the baseline for modeling higher sea levels. This initial water-level setting corresponds to a flood event with an approximate 2-year return interval (a 50% annual exceedance probability). This level also approximates projected sea levels around 2045 based on the NOAA intermediate-high forecast scenario (National Oceanic and Atmospheric Administration, 2025).

To simulate storm-surge flooding, the Hazus Wind Model incorporates the Sea, Lake, and Overland Surges from Hurricanes (SLOSH) model together with the hurricane wind-field module. We specified a tidal offset of 1.1 m as the Initial Water Level, representing the astronomical tide at the time of landfall. This Initial Water Level is coupled with the storm's wind field to drive the SLOSH model. Under the depth-limited wave assumption, Hazus then produces a still-water elevation surface that forms the basis for estimating flood depth (Federal Emergency Management Agency, 2009).

After completing the analysis in the Hurricane model, the study proceeds with the Flood model analysis. The stillwater elevation surface was used to calculate flood depth using a digital elevation model (DEM) and to delineate floodplain boundaries. Based on the resulting flood depth grid, property losses are estimated using depth-damage curves, which relate flood depth to the percentage of losses in building structure, contents, and inventory across various occupancy types, as illustrated in Fig. 3b. Finally, the combined wind and flood damage assessment ensures that total property losses do not exceed 100% of property value. The output of the Hazus catastrophe model is an estimate of damages to buildings and capital stock, expressed in 2022 dollars, reflecting FEMA's major inventory database update in 2022 (Federal Emergency Management Agency, 2022b). We convert the damages in 2022 dollars to 2024 dollars.

3.2 Estimating Direct, Indirect, and Induced Economic Impacts

We estimate the economic impacts of physical damage using regional I-O multipliers for regions as small as counties (Table 1). The core is based on the Bureau of Economic Analysis (BEA) national I-O accounts. The national I-O accounts illustrate how goods and services produced by an industry are used by other industries and end users (Horowitz & Planting, 2009). The BEA publishes the national I-O accounts and the Regional Input-Output Modeling System (RIMS II) to account for variations in regional supply conditions (Bureau of Economic Analysis, 2013).

We use IMPLAN, a commercially available economic impact software, to model the economic impacts of a hurricane making landfall in Hampton Roads. IMPLAN is comparable to the BEA's RIMS II multipliers but provides added detail on the workforce impacts associated with changes in household spending (Gc et al., 2024). However, the IMPLAN impacts are static and therefore fail to account for feedback effects from changes in direct expenditures or employment.

Table 1 The assumptions and equations for the economic model

Assumptions: • Producers are grouped into n industries. • Technology is homogenous within an industry but heterogeneous across industries. • Production is linear. • Purchase patterns are fixed. • Industries are homogeneous. • There are no supply constraints, regional feedback mechanisms, and time dimension. • Industry production is collectively exhaustive.
Each industry i produces gross output X_i. Industry i sells output X_i to other industries z_{ij} and to final users, Y_i. Equation 1 defines the gross output of industry i: $X_i = z_{i1} + z_{i2} + \ldots + z_{in} + Y_i$ (Eq. 1) Technical coefficients represent input requirements per unit of output for each commodity in the regional economy. They represent the amount of goods and services required to produce one unit of output while reflecting inter-industry relationships and flows of resources across industries. Eq. 2 defines the set of technical coefficients: $a_{ij} = \frac{z_{ij}}{X_i}$ (Eq. 2) The technical coefficients are used to construct the national make-and-use tables, which, in turn, are used to derive the domestic direct requirements, regional direct requirements, household use, and regional total requirement tables. From these regional total requirement tables, we obtain the indirect and induced multipliers to estimate total economic impact.

IMPLAN uses a social accounting matrix to extend the model to include industry-to-industry, industry-to-institution, and institution-to-institution transactions and transfers, thereby augmenting the model's estimates to encompass market and non-market monetary flows among industries, households, and government. Additionally, IMPLAN's Industry Schemes do not directly match NAICS industry codes, necessitating a manual bridge between the two.

Using IMPLAN and the estimates of flood and wind damages by building occupancy type, we estimate the impact of the compound hurricane-SLR event on economic output, compensation, payrolls, and value added. We aggregate the estimates of economic impact to derive our final projections of direct, indirect, and induced effects.

3.3 *Insurance Coverage and Uninsured Losses and Damages*

The physical damage from a hurricane is only part of the total economic impact. A major storm striking the region would not only damage structures and displace residents but also impact economic activity. While direct damage (to homes, contents, and cars) may be the most visible, households incur additional expenses in the aftermath of a disaster. Insured households may use their insurance to cover direct and additional disaster-related costs, such as evacuation and cleanup expenses, and to compensate for lost income. However, uninsured households may have to rely on support from their personal networks to reduce or manage financial losses (You & Kousky, 2024).

The transfer of damages from a disaster to insurers facilitates post-disaster recovery efforts. The voluntary purchase of insurance, including flood insurance, however, is low (McShane & Yusuf, 2019), with approximately 50% of homeowners in the 100-year floodplain participating in the NFIP (Kousky, 2017; Wagner, 2022). Uninsured losses from a disaster can lead to subsequent economic losses. For example, credit-constrained households residing outside the floodplain in Houston who experienced flooding after Hurricane Harvey had a 20% increase in bankruptcies and a 13% increase in the share of severely delinquent debt relative to households in non-flooded areas (Billings et al., 2022).

Our EIA methodology accounts for uninsured damage and losses. To estimate the proportion of residential buildings covered by flood insurance, we use the NFIP uptake rates by locality in the Hampton Roads region (see Fig. 1c), supplemented with survey data on private insurance uptake rates from the Insurance Information Institute. We assume that public and educational buildings are, in effect, fully insured against loss and that non-residential (commercial and industrial) buildings are covered at the same rate as residential buildings. We also assume that neither the agricultural nor the religious sector carries insurance, so their uninsured damages are treated as equal to their total wind and flood losses. Using the uptake rate and damages by industry, as well as these assumptions for public and educational buildings, we obtain the amount of insured damages by industry and locality, and subsequently calculate the uninsured damage. Our assumption that flood insurance fully offsets damages establishes a floor, rather than a ceiling, on uninsured damages, providing a conservative estimate.

4 Results

4.1 Direct Damages from Wind and Storm Surge

Table 2 presents the estimated wind and flood damages by locality from a hurricane making landfall in Hampton Roads for a 1.1 m tidal offset. Wind damages are assumed to be independent of the tidal offset. The Category 3 hurricane causes approximately \$4.4 billion in wind damage, with significant variability across the region. Estimated wind damages are highest in the Norfolk (\$794.6 million), Chesapeake (\$740.9 million), and Hampton (\$541.1 million) and lowest in the more inland localities—Franklin (\$6.3 million), Surry County (\$17.5 million), and Southampton County (\$18.3 million). Although Virginia Beach is the nearest modeled landfall location (Fig. 4a), its estimated wind losses (\$380.1 million) are lower than those of several nearby localities. This is primarily due to the hurricane's path crossing the less densely populated southern portion of Virginia Beach, which reduced the magnitude of wind-related structural damages.

Table 2 Expected wind and flood damages by locality

Locality	Wind damage ($000)	Flood damage ($000)	Total damage ($000)
Chesapeake	$740,868	$211,823	$952,691
Franklin	$6293	$0	$6293
Gloucester	$151,197	$739,194	$890,390
Hampton	$541,132	$2,619,021	$3,160,153
Isle of wight	$121,980	$44,901	$166,881
James city	$249,262	$21,829	$271,091
Newport news	$436,247	$26,198	$462,445
Norfolk	$794,595	$3,845,883	$4,640,478
Poquoson	$78,530	$1,800,744	$1,879,274
Portsmouth	$221,567	$955,066	$1,176,633
Southampton	$18,276	$0	$18,276
Suffolk	$206,288	$29,304	$235,592
Surry	$17,460	$5348	$22,808
Virginia Beach	$380,145	$177,135	$557,280
Williamsburg	$34,426	$0	$34,426
York	$385,099	$767,400	$1,152,499
Totals	$4,383,365	$11,243,846	$15,627,211

Note: Numbers may not sum to totals due to rounding

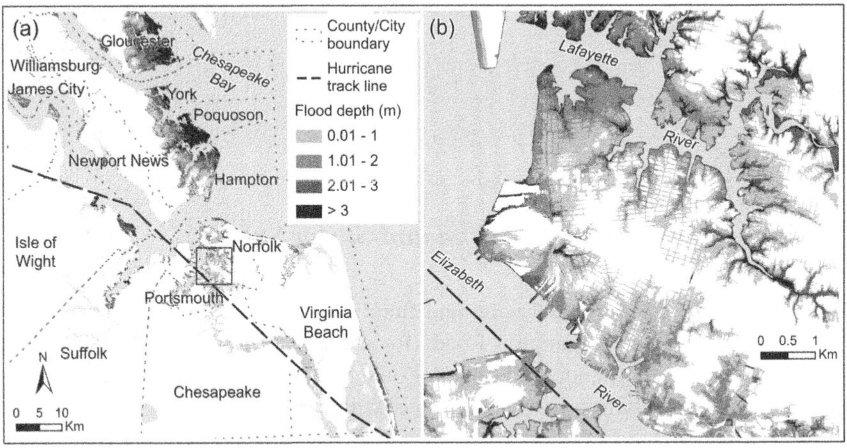

Fig. 4 (**a**) Modeled storm surge flood depths over the Hampton Roads region, shown with the hurricane track used in the analysis, and (**b**) zooms in on the City of Norfolk, illustrating flood impacts within highly populated neighborhoods along the Lafayette and Elizabeth Rivers

We estimate flood damage of approximately $11.2 billion for the region. Proximity to the flood and elevation influence the magnitude of direct flood damage. Norfolk ($3.9 billion), Hampton ($2.6 billion), and Poquoson ($1.8 billion) account for approximately 74.0% of projected flood damage in the region from a hurricane making landfall. In total, the combined direct damage given a Category 3 hurricane compounded by 1.1 m of tidal offset is approximately $15.6 billion, roughly equivalent to 13.3% of the region's 2022 regional GDP.

Similar to the wind-related impacts, flood losses in Virginia Beach are substantially lower ($177.1 million) than in nearby localities, again primarily due to the hurricane track used in the modeling. Aside from part of the oceanfront and communities around Lynnhaven Bay, most inundations occur in sparsely populated areas. In contrast, densely developed neighborhoods in Norfolk (Fig. 4b), Hampton, and Poquoson experience extensive flooding (Fig. 4a). Flood depths reach up to 3 m in some locations, resulting in significantly higher losses. The outcomes demonstrate that the hurricane's trajectory critically shapes regional damage patterns. They highlight the value of geospatial information for estimating the impacts of hazards.

Table 3 illustrates the projected damages by building occupancy type for the Category 3 hurricane with 1.1 m tidal offset that accounts for the higher sea levels due to SLR. Flood-related damage to buildings with residential occupancy is approximately $7.5 billion (67.1% of the $11.2 billion of flood-related damages). Residential damage also accounts for approximately $3.5 billion or 79% of the projected $4.4 billion in wind damage, followed by damage to buildings with commercial occupancy uses ($496.4 million or 11% of total wind damage). In other words, single and multi-family residences account for a substantial portion of projected total damage.

Table 4 presents estimated uninsured damages from the Category 3 hurricane under higher sea level conditions. Uninsured damages are estimated to be 69.0% of the total wind and flood damages. These results enable analysis of how damages are distributed among buildings with residential occupancy. We use the U.S. Census

Table 3 Expected wind and flood damage by occupancy type based on Hazus modeling

Occupancy type	Wind damage ($000)	Flood damage ($000)	Total damage ($000)
Agriculture	$25,267	$16,977	$42,244
Commercial	$496,429	$1,839,702	$2,336,131
Educational	$174,287	$672,752	$847,039
Government	$30,646	$222,959	$253,605
Industrial	$157,406	$653,509	$810,915
Religion	$33,497	$297,833	$331,330
Residential	$3,465,832	$7,540,114	$11,005,946
Totals	$4,383,365	$11,243,846	$15,627,211

Note: Numbers may not sum to totals due to rounding

Table 4 Expected uninsured damages

Occupancy type	Total wind and flood damages ($000)	Uninsured damages ($000)	Uninsured damages (% of Total Damages)
Agriculture	$42,244	$42,244	100.0
Commercial	$2,336,131	$1,765,225	75.6
Educational	$847,039	$0	0
Government	$253,605	$0	0
Industrial	$810,915	$614,392	75.8
Religion	$331,330	$331,330	100.0
Residential	$11,005,946	$8,031,052	73.0
Totals	$15,627,211	$10,784,244	69.0

Bureau's 2022 American Community Survey to estimate the distribution of households by household income and apportion the damage by household income distribution. Uninsured damages and losses from residential buildings are expected to be $8 billion, accounting for 73.0% of the total damages.

4.2 *Direct, Indirect, and Induced Economic Impacts*

Table 5 presents the estimated economic impacts in 2024 dollars from uninsured losses associated with a Category 3 hurricane making landfall in Hampton Roads. The total economic impact is the loss of 76,579 jobs and nearly $24 billion in economic losses. Direct impacts account for 22.7% of total economic impacts, whereas induced impacts account for 69.7% of total losses. The hurricane's direct damage will result in a loss of approximately 20,000 jobs (equivalent to $1.3 billion in labor income) and a reduction in regional GDP of approximately $1.4 billion. The indirect impacts are approximately 5100 jobs and a reduction in regional GDP of approximately $509.8 million in labor income. The induced economic impacts are the most significant, reflecting the predominance of damages to residential buildings. The impact is evident in a reduction of nearly 52,000 jobs and lower household incomes (equivalent to $2.4 billion) due to uninsured losses and decreased demand for goods and services, which, in turn, negatively affects businesses and employment in the region. The results show that the ripple effects of direct damages are substantial.

Table 6 presents the hurricane's impact on jobs. The largest impact on employment is for the Other Services (except Public Administration) sector. This sector comprises establishments primarily engaged in activities such as equipment and machinery repair, dry cleaning and laundry services, personal care and funeral services, and pet care services. This sector comprises jobs that provide services to residents who bear the majority of the costs associated with a disaster. We estimate that more than 23,000 jobs will be lost in the Other Services sector.

Table 5 Estimated economic impacts

Impact	Direct	Indirect	Induced	Total
Employment				
No. of jobs	19,797	5146	51,635	76,579
% of total	25.9%	6.7%	67.4%	100.0%
Labor income	$1,279,919,166	$293,288,982	$2,487,993,466	$4,061,201,615
Value added	$1,415,713,231	$509,807,018	$5,205,232,803	$7,130,753,053
Output	$2,715,170,000	$1,003,927,211	$8,900,187,302	$12,619,284,513
Total	$5,410,802,397	$1,807,023,211	$16,593,413,571	$23,811,239,181
% of total	22.7%	7.6%	69.7%	100.0%

Note: Numbers may not sum due to rounding

Table 6 Estimated employment impacts

Industry	Direct impact	Indirect impact	Induced impact	Total impact
Other services (except Public Administration)	17,495	154	5753	23,402
Retail trade	0	103	9327	9430
Health care and social assistance	0	1	9221	9222
Accommodation and food services	0	243	6774	7017
Real estate and rental and leasing	0	829	3066	3895
Transportation and warehousing	0	507	3168	3675
Administrative and support and waste management and remediation services	0	876	2540	3416
Finance and insurance	0	476	2894	3370
Professional, scientific, and technical services	0	700	1810	2510
Manufacturing	2117	69	264	2450
Arts, entertainment, and recreation	0	67	1697	1764
Wholesale trade	0	432	1101	1533
Educational services	0	24	1429	1453
Management of companies and enterprises	0	420	657	1077
Information	0	75	653	728
Government enterprises	0	87	624	711
Construction	0	61	377	438
Agriculture, forestry, fishing and hunting	185	3	102	290
Utilities	0	18	172	190
Mining, quarrying, and oil and gas extraction	0	1	7	8
Total	19,797	5146	51,636	76,579

Our analysis projects more than 9400 job losses in the Retail Trade sector and an additional 9200 job losses in the Health Care and Social Assistance sector. These job losses, like those in the Other Services sector, are closely tied to the consumption of personal services. Financial losses sustained by households reduce consumption of these services, inducing subsequent job losses. The Accommodation and Food Services sector is projected to lose more than 7000 jobs. These relate to losses incurred by households and a decline in tourism following the hurricane. Given the damage to buildings and infrastructure, tourism activities are expected to decline in the immediate aftermath of the hurricane.

The economic impacts in Tables 5 and 6 illustrate the critical role of flood insurance, which has benefits far greater than just to homeowners. Widespread flood insurance would have a preventive effect on household incomes, reducing financial losses associated with flood damage from a compound hurricane-SLR event. These financial damages reduce household consumption, which, in turn, reduces the demand for goods and services in the regional economy.

4.3 Limitations of the EIA Methodology

Limitations in the inventory dataset, building stock representation, and DEM resolution can significantly influence the accuracy of Hazus estimates. The inventory dataset, obtained from FEMA, is a compilation of multiple national datasets. For example, a survey by the Energy Information Administration is used to estimate the percentage of buildings by specific occupancy type (Federal Emergency Management Agency, 2022a). A more up-to-date approach would involve acquiring parcel data from local authorities and aggregating it to the census block level. Incorporating User-Defined Facilities to represent individual buildings could also substantially improve flood damage estimates (Chen et al., 2014; Ghimire & Sharma, 2021), as flood depth is often highly localized. Similarly, using higher-resolution DEMs would enhance the precision of flood damage estimates (Tate et al., 2015).

Another complementary approach would be to couple storms across different hurricane categories and conduct a sensitivity analysis to estimate the associated economic losses. This approach would provide a more detailed understanding of potential financial impacts and the probability of different storm scenarios, offering stronger support for regional mitigation planning. However, validating Hazus-based loss estimates remains inherently challenging, and NFIP claims often do not represent total property losses (Kousky & Michel-Kerjan, 2017). Future studies could undertake finer-scale local analyses combined with community engagement or questionnaire surveys to improve validation and refine the computation methods.

Additionally, SLR is likely to exacerbate health and income inequities and reduce the scale of public services in conjunction with projected declines in economic activity (Allaire, 2018). SLR may disrupt innovation and entrepreneurship, reduce economic efficiency, and create a reinforcing cycle of economic decline (Zhao et al.,

2022). These and other potential spillovers should be considered in assessments of economic impact.

Furthermore, our analysis does not incorporate potential changes resulting from future flood-mitigation measures, as the model relies on FEMA's default inventory dataset. While this approach is straightforward, it also omits adaptive responses that may alter future impacts. Some disaster-related losses can be reduced through strategies that increase economic resilience, as firms and regions adapt to mitigate potential damage (Rose & Liao, 2005). Modeling studies further emphasize that disaster impacts should account not only for direct effects but also for resilience and economic linkages. As these can substantially amplify indirect impacts when risks increase, systems are overwhelmed, and resilience is undermined (Rose, 2022a, 2022b; Rose, 2009). Previous work integrating land-use and land-cover data with storm-surge scenarios that include SLR projections shows that, in Hampton Roads, impacted residential areas could increase from 16% in 2000 to 24% and 31% by 2060 and 2090, respectively (Mitchell et al., 2023).

Future studies should consider updating building-stock inventories using simulated conditions that account for population shifts and development patterns influenced by increasing flood risk from SLR. In addition, economic modeling approaches, such as the computable general equilibrium approach used by Rose et al. (2017), could be applied to better capture adaptive behavior, sectoral interactions, and broader economic impacts that are not captured by I-O models such as those used in our methodology.

5 Conclusion

This chapter presents a methodology and illustrative case study that emphasize the critical need to account for the compounding effects of SLR on hurricane impacts and to comprehensively estimate not just direct damages, but also indirect and induced economic impacts. The application of the methodology project that a Category 3 hurricane, similar to Hurricane Florence but compounded by a 1.1 m tidal offset (accounting for higher sea levels), results in approximately $15.6 billion in combined direct wind and flood damages in the Hampton Roads region, equivalent to 13% of the regional GDP. Residential building damages comprise a substantial portion of these damages.

When considering uninsured losses, the total economic impact is estimated to exceed $12 billion, resulting in the loss of more than 76,000 jobs. Uninsured damages are estimated to account for 69% of total wind and flood damages, with uninsured residential losses alone projected at $8 billion, representing 73% of total residential buildings damages. These uninsured losses significantly reduce household incomes and consumption, affecting labor-dependent service sectors. Induced economic impacts are most significant for job losses, driven by reductions in household income. This highlights how initial direct damages ripple through the economy, affecting supply chains (indirect impacts) and household spending (induced impacts).

5.1 Contributions and Implications Beyond Hampton Roads

The EIA methodology developed and illustrated in this chapter is readily transferable to other coastal regions in the U.S. and beyond. The methodology combines (1) physical hazard modeling that combines wind, storm surge, and an SLR-induced tidal offset, (2) economic ripple-effect modeling that captures direct, indirect, and induced effects, and (3) a focus on uninsured damages and losses. It offers a standardized but customizable framework, since it uses widely available tools and datasets and can be replicated using location-specific data. Take Hazus, for instance. A growing body of literature demonstrates successful applications of Hazus for damage assessment worldwide (e.g., Fallah-Aliabadi et al., 2020; Frucht et al., 2021; Chantavilasvong & Guerrero, 2019), although many of these studies rely on more detailed and locally intensive data collection. For example, Chantavilasvong and Guerrero (2019) developed a building-stock database for Piura, Peru, delineated flood extents, and applied regionally appropriate damage curves to estimate impacts from riverine flooding.

Other localities or regions can substitute their own tide gauge and elevation data, building inventories, insurance coverage rates, and industry profiles. The emphasis on mapping uninsured physical damage to household income losses and business disruptions captures indirect and induced effects that are often overlooked. By quantifying job losses, reduced household spending, and supply chain impacts, the analysis can inform both emergency management planning and economic development strategies.

The explicit integration of SLR into hurricane loss estimates produces more realistic flood depths and damage footprints. This can reveal vulnerabilities in inland or higher-elevation zones and is critical for planning in areas where frequent "sunny day" flooding is already occurring. As such, this methodology is particularly relevant for U.S. regions such as Charleston, South Carolina, and Miami-Dade County, Florida. Applying the methodology to other coastal regions could reveal broader issues in economic vulnerability and resilience. Comparative analyses across regions—such as with Miami-Dade County and Charleston, as just noted—could strengthen the case for adoption of our approach and methodology. Future research should explore these cross-regional applications to identify common patterns and inform scalability.

Beyond economic impacts, disasters also have fiscal impacts. Hurricanes and other disasters can impose significant budgetary pressures on governments, including reduced tax revenues and increased expenditures for emergency response, debris cleanup, reconstruction, and welfare assistance. Our EIA methodology can support data-informed consideration of fiscal impacts by illuminating the extent of uninsured building damages that will require repair and reconstruction assistance, the sectors most affected in terms of employment and wages, which can impact tax revenues and welfare spending, and the reduction in production and output that can affect the revenue base.

5.2 Policy Recommendations

By highlighting the combined economic impacts, our findings underscore the need for policymakers to adopt a comprehensive policy approach. Although this analysis does not separately quantify the effects of SLR, the combined hurricane-SLR scenario produces markedly higher physical damage and regional economic losses, particularly through uninsured residential impacts that amplify induced effects on the workforce and household spending. This underscores the need for policies and programs that address both hazard exposure and the economic ripple effects of major storm events.

A first step is to recognize the importance of flood insurance as a loss mitigation strategy, as insured residences are less likely to sustain significant financial damage from a disaster. Improving flood insurance uptake is crucial for mitigating post-disaster financial hardship and stabilizing the local economy. Expanding insurance requirements to areas projected to experience future tidal or surge flooding, even if currently outside FEMA flood zones, could better align coverage with risk.

Second, incorporating compound hurricane-SLR scenarios into planning should become standard practice for hazard mitigation, capital planning, and land-use decision-making. For example, regional hazard mitigation planning should use updated modeling to guide investments and prioritize protection for high-multiplier economic sectors such as ports, tourism districts, logistics hubs, and healthcare facilities, where economic disruption can have outsized regional impacts. Proactive mitigation is cost effective, with $1 invested in hazard mitigation yielding $6 in avoided damages and losses (National Institute of Building Sciences, 2019). Mitigation actions, such as elevating structures, strengthening facilities, and improving drainage systems, not only reduce immediate physical damages but also minimize cascading indirect and induced economic impacts.

Third, workforce and business continuity activities that address economic resilience must also be considered. Policies that provide post-disaster housing assistance, facilitate employee relocation, and support payroll continuity can help maintain labor availability and consumer spending during recovery, limiting the prolonged economic downturns associated with workforce displacement. Beyond physical protection, resilience planning should include business continuity and adaptive measures that reduce economic losses. Employers, particularly in sectors sensitive to induced effects, should be supported in developing continuity-of-operations plans. Establishing contingency plans for rapid recovery of production capacity, such as through temporary facilities or shared-use spaces, can also help enhance business resilience. Incentivizing strategic facility relocation that moves critical business operations and high-value business assets out of high-risk areas can minimize post-disaster downtime and asset losses. Other strategies, such as inventory management (e.g., diversifying storage locations and reducing on-site inventories) and supply chain diversification (e.g., encouraging businesses to develop alternative suppliers and distribution channels) will help limit damages and losses while maintaining operations post-disaster.

References

Agarwal, V., Koch, J. V., Lian, F., Makhorkina, A., Nguyen, M., Parker, T., Rahoui, M., Walker, J., & Yuenyong, L. (2024). *State of the commonwealth report*. Dragas Center for Economic Analysis and Policy. https://digitalcommons.odu.edu/soc_reports/78

Akhtar, R., & Santos, J. R. (2013). Risk-based input–output analysis of hurricane impacts on interdependent regional workforce systems. *Natural Hazards, 65*(1), 391–405. https://doi.org/10.1007/s11069-012-0369-0

Allaire, M. (2018). Socio-economic impacts of flooding: A review of the empirical literature. *Water Security, 3*, 18–26. https://doi.org/10.1016/j.wasec.2018.09.002

Amini, M., Figueira, S. A., Cox, D. T., & Barbosa, A. R. (2023). Model-data validation and uncertainty quantification of the IN-CORE and HAZUS-MH damage models for buildings impacted by Hurricane Ian (2022) at Fort Myers Beach Florida. In *ASCE Inspire 2023: Infrastructure innovation and adaptation for a sustainable and resilient world*. https://doi.org/10.1061/9780784485163.090

Arabadzhyan, A., Figini, P., García, C., González, M. M., Lam-González, Y. E., & León, C. J. (2021). Climate change, coastal tourism, and impact chains—A literature review. *Current Issues in Tourism, 24*(16), 2233–2268. https://doi.org/10.1080/13683500.2020.1825351

Avelino, A. F. T., & Dall'erba, S. (2019). Comparing the Economic Impact of Natural Disasters Generated by Different Input–Output Models: An Application to the 2007 Chehalis River Flood (WA). *Risk Analysis, 39*(1), 85–104. https://doi.org/10.1111/risa.13006

Billings, S. B., Gallagher, E. A., & Ricketts, L. (2022). Let the rich be flooded: The distribution of financial aid and distress after Hurricane Harvey. *Journal of Financial Economics, 146*(2), 797–819. https://doi.org/10.1016/j.jfineco.2021.11.006

Botzen, W. J. W., Deschenes, O., & Sanders, M. (2019). The Economic Impacts of Natural Disasters: A Review of Models and Empirical Studies. *Review of Environmental Economics and Policy, 13*(2), 167–188. https://doi.org/10.1093/reep/rez004

Bureau of Economic Analysis. (2013). RIMS II: An essential tool for regional developers and planners. Bureau of Economic Analysis. https://www.bea.gov/sites/default/files/methodologies/RIMSII_User_Guide.pdf

Bureau of Economic Analysis. (2024). GDP by County, Metro, and Other Areas. Bureau of Economic Analysis. Retrieved August 23, 2025 from https://www.bea.gov/data/gdp/gdp-county-metro-and-other-areas

Castrucci, L., & Tahvildari, N. (2018). Modeling the impacts of sea level rise on storm surge inundation in flood-prone urban areas of Hampton Roads, Virginia. *Marine Technology Society Journal, 52*(2), 92–105.

Cavallo, E., & Noy, I. (2011). Natural Disasters and the Economy — A Survey. *International Review of Environmental and Resource Economics, 5*(1), 63–102. https://doi.org/10.1561/101.00000039

Chantavilasvong, W., & Guerrero, L. (2019). Application of HAZUS-MH Flood Model in Developing Countries: The Case of Piura, Peru. *Nakhara: Journal of Environmental Design and Planning, 16*(1), 33–52. https://digital.car.chula.ac.th/nakhara/vol16/iss1/3

Chen, Y.-H., Fik, T., & Mossa, J. (2014). Fine-scale vulnerability assessment of flooding an d storm surge hazard in coastal communities-A case study in Sarasota, Florida. In *Third international conference on urban disaster reduction-sustainable disaster recovery: Addressing risks and uncertainty*, Boulder CO.

Chen, Y.-H., Kim, J., & Mueller, N. (2021). Estimating the economic impact of natural hazards on shared accommodation in coastal tourism destinations. *Journal of Destination Marketing & Management, 21*, 100634. https://doi.org/10.1016/j.jdmm.2021.100634

Crawford, M. C., Bukvic, A., Rijal, S., & Gohlke, J. M. (2023). The exposure of vulnerable coastal populations to flood-induced Natech events in Hampton Roads, Virginia. *Natural Hazards, 119*(3), 1633–1663. https://doi.org/10.1007/s11069-023-06148-x

Crowther, K. G., Haimes, Y. Y., & Taub, G. (2007). Systemic Valuation of Strategic Preparedness Through Application of the Inoperability Input-Output Model with Lessons Learned from Hurricane Katrina. *Risk Analysis, 27*(5), 1345–1364. https://doi.org/10.1111/j.1539-6924.2007.00965.x

Deb, M., Sun, N., Yang, Z., Wang, T., Judi, D., Xiao, Z., & Wigmosta, M. S. (2023). Interacting Effects of Watershed and Coastal Processes on the Evolution of Compound Flooding During Hurricane Irene. *Earth's Future, 11*(3), e2022EF002947. https://doi.org/10.1029/2022EF002947

Fallah-Aliabadi, S., Ostadtaghizadeh, A., Ardalan, A., Eskandari, M., Fatemi, F., Mirjalili, M. R., & Khazai, B. (2020). Risk analysis of hospitals using GIS and HAZUS: A case study of Yazd County, Iran. *International Journal of Disaster Risk Reduction, 47*, 101552. https://doi.org/10.1016/j.ijdrr.2020.101552

Federal Emergency Management Agency. (2009). *Multi-hazard loss estimation methodology: Flood model - HAZUS-MH technical manual*. Federal Emergency Management Agency. https://www.fema.gov/sites/default/files/2020-09/fema_hazus_flood-model_technical-manual_2.1.pdf

Federal Emergency Management Agency. (2022a). *Hazus flood model user guidance: Hazus 5.1*. Federal Emergency Management Agency. https://www.fema.gov/sites/default/files/documents/fema_hazus-5.1-flood-model-user-guidance.pdf

Federal Emergency Management Agency. (2022b). *Hazus inventory technical manual-Hazus 6.0*. https://www.fema.gov/sites/default/files/documents/fema_hazus-6-inventory-technical-manual.pdf

Federal Emergency Management Agency. (2025). *FEMA flood map service center: Hazus*. U.S. Department of Homeland Security. Retrieved June 5, 2025 from https://msc.fema.gov/portal/resources/hazus

Frucht, E., Salamon, A., Rozelle, J., Levi, T., Calvo, R., Avirav, V., Burns, J. N., Zuzak, C., Gal, E., Trapper, P., Galanti, B., & Bausch, D. (2021). Tsunami loss assessment based on Hazus approach—The Bat Galim, Israel, case study. *Engineering Geology, 289*, 106175. https://doi.org/10.1016/j.enggeo.2021.106175

Gc, S., Thapa, I., Pokharel, R., Alward, G., Lamsal, B., Poudel, J., Dahal, R., Joshi, O., Parajuli, R., Wagner, J., & Leefers, L. (2024). A methodological framework for decomposing the value-chain economic contribution: A case of forest resource industries of the Lake States in the United States. *Forests, 15*(2), 305.

Ghanbari, M., Arabi, M., Kao, S.-C., Obeysekera, J., & Sweet, W. (2021). Climate Change and Changes in Compound Coastal-Riverine Flooding Hazard Along the U.S. Coasts. *Earth's Future, 9*(5), e2021EF002055. https://doi.org/10.1029/2021EF002055

Ghimire, E., & Sharma, S. (2021). Flood Damage Assessment in HAZUS Using Various Resolution of Data and One-Dimensional and Two-Dimensional HEC-RAS Depth Grids. *Natural Hazards Review, 22*(1), 04020054. https://doi.org/10.1061/(ASCE)NH.1527-6996.0000430

Goulart, H. M. D., Benito Lazaro, I., van Garderen, L., van der Wiel, K., Le Bars, D., Koks, E., & van den Hurk, B. (2024). Compound flood impacts from Hurricane Sandy on New York City in climate-driven storylines. *Natural Hazards and Earth System Sciences, 24*(1), 29–45. https://doi.org/10.5194/nhess-24-29-2024

Gutenson, J. L., Oubeidillah, A. A., Hicks, P., Durham, L., Ernest, A. N. S., Zhu, L., & Zhang, X. (2015). Using HAZUS-MH and HEC-RAS to Evaluate Real World Flooding Events in the Upper Alabama River Watershed. In K. Karvazy & V. L. Webster (Eds.), *World Environmental and Water Resources Congress 2015* (pp. 1607–1627). American Society of Civil Engineers. https://doi.org/10.1061/9780784479162.157

Hallegatte, S. (2008). An Adaptive Regional Input-Output Model and its Application to the Assessment of the Economic Cost of Katrina. *Risk Analysis, 28*(3), 779–799. https://doi.org/10.1111/j.1539-6924.2008.01046.x

Hampton Roads Planning District Commission. (2022). *Hampton roads hazard mitigation plan*. Hampton Roads Planning District Commission. https://www.hrpdcva.gov/DocumentCenter/View/3533/2022-Hampton-Roads-Hazard-Mitigation-Plan-PDF

Horowitz, K. J., & Planting, M. A. (2009). *Concepts and methods of the US input-output accounts*. Bureau of Economic Analysis. https://www.bea.gov/sites/default/files/methodologies/IOmanual_092906.pdf

Jerch, R., Kahn, M. E., & Lin, G. C. (2023). Local public finance dynamics and hurricane shocks. *Journal of Urban Economics, 134*, 103516. https://doi.org/10.1016/j.jue.2022.103516

Jin, X., Sumaila, U. R., & Yin, K. (2020). Direct and Indirect Loss Evaluation of Storm Surge Disaster Based on Static and Dynamic Input-Output Models. *Sustainability, 12*(18).

Koks, E. E., Carrera, L., Jonkeren, O., Aerts, J. C. J. H., Husby, T. G., Thissen, M., Standardi, G., & Mysiak, J. (2016). Regional disaster impact analysis: comparing input–output and computable general equilibrium models. *Natural Hazards and Earth System Sciences, 16*(8), 1911–1924. https://doi.org/10.5194/nhess-16-1911-2016

Kousky, C. (2017). Disasters as Learning Experiences or Disasters as Policy Opportunities? Examining Flood Insurance Purchases after Hurricanes. *Risk Analysis, 37*(3), 517–530.

Kousky, C., & Michel-Kerjan, E. (2017). Examining Flood Insurance Claims in the United States: Six Key Findings. *Journal of Risk and Insurance, 84*(3), 819–850. https://doi.org/10.1111/jori.12106

McLeod, G. M., Allen, T. R., & Behr, J. G. (2018). Geospatial Risk Assessment of Marine Terminal Infrastructure to Storm Surge Inundation and Sea Level Rise. *Transportation Research Record, 2672*(11), 19–29. https://doi.org/10.1177/0361198118774234

McNab, R. M., Chen, Y.-H., & McLeod, G. M. (2024). Estimating the economic vulnerability of 2020–2099 storm flooding with sea level rise on Coastal Virginia. *Ocean & Coastal Management, 259*, 107438. https://doi.org/10.1016/j.ocecoaman.2024.107438

McShane, M. K., & Yusuf, J. E. (2019). Toward Better Management of Flood Losses: Flood Insurance in a Wetter World. *Public Works Management & Policy, 24*(1), 88–109.

Meier, S., Elliott, R. J. R., & Strobl, E. (2023). The regional economic impact of wildfires: Evidence from Southern Europe. *Journal of Environmental Economics and Management, 118*, 102787. https://doi.org/10.1016/j.jeem.2023.102787

Mitchell, A., Bukvic, A., Shao, Y., Irish, J. L., & McLaughlin, D. L. (2023). Toward Collaborative Adaptation: Assessing Impacts of Coastal Flooding at the Watershed Scale. *Environmental Management, 71*(4), 741–754. https://doi.org/10.1007/s00267-022-01759-9

Mousavi, M. E., Irish, J. L., Frey, A. E., Olivera, F., & Edge, B. L. (2011). Global warming and hurricanes: The potential impact of hurricane intensification and sea level rise on coastal flooding. *Climatic Change, 104*(3), 575–597. https://doi.org/10.1007/s10584-009-9790-0

National Hurricane Center. (2004). *Tropical cyclone report: Hurricane Isabel, 6–19 September 2003*. National Hurricane Center. https://www.nhc.noaa.gov/data/tcr/AL132003_Isabel.pdf

National Hurricane Center. (2019). *Tropical cyclone reports: 2018 Atlantic Hurricane Season*. National Oceanic and Atmospheric Administration. https://www.nhc.noaa.gov/data/tcr/index.php?season=2018&basin=atl

National Institute of Building Sciences. (2019). *Natural hazard mitigation saves: 2019 report*. https://nibs.org/projects/natural-hazard-mitigation-saves-2019-report/

National Oceanic and Atmospheric Administration. (2025). *Sea level calculator*. NOAA Center for Operational Oceanographic Products and Services & NOAA Office for Coastal Management. Retrieved August 28, 2025 from https://coast.noaa.gov/sealevelcalculator

National Weather Service. (2003). *Hurricane Isabel Review—Wakefield, VA*. Retrieved August 10, 2025 from https://www.weather.gov/akq/Isabel_2003

Okuyama, Y., & Santos, J. R. (2014). DISASTER IMPACT AND INPUT–OUTPUT ANALYSIS. *Economic Systems Research, 26*(1), 1–12. https://doi.org/10.1080/09535314.2013.871505

Passeri, D. L., Hagen, S. C., Medeiros, S. C., Bilskie, M. V., Alizad, K., & Wang, D. (2015). The dynamic effects of sea level rise on low-gradient coastal landscapes: A review. *Earth's Future, 3*(6), 159–181. https://doi.org/10.1002/2015EF000298

Rahimi, R., Tavakol-Davani, H., Graves, C., Gomez, A., & Fazel Valipour, M. (2020). Compound inundation impacts of coastal climate change: Sea-level rise, groundwater rise, and coastal precipitation. *Water, 12*(10), 2776.

Rose, A. (2022a). Behavioral economic consequences of disasters. In M. Skidmore (Ed.), *Handbook on the Economics of Disasters* (pp. 50–71). Edward Elgar Publishing.
Rose, A. (2022b). Behavioral Economic Consequences of Disasters: A Basis for Inclusion in Benefit–Cost Analysis. *Economics of Disasters and Climate Change, 6*(2), 213–233. https://doi.org/10.1007/s41885-022-00107-9
Rose, A., & Liao, S.-Y. (2005). Modeling Regional Economic Resilience to Disasters: A Computable General Equilibrium Analysis of Water Service Disruptions. *Journal of Regional Science, 45*(1), 75–112. https://doi.org/10.1111/j.0022-4146.2005.00365.x
Rose, A., Prager, F., Chen, Z., Chatterjee, S., Wei, D., Heatwole, N., & Warren, E. (2017). *Economic consequence analysis of disasters: The E-CAT software tool.* Springer. https://link.springer.com/book/10.1007/978-981-10-2567-9
Rose, A. Z. (2009). A Framework for Analyzing the Total Economic Impacts of Terrorist Attacks and Natural Disasters. *Journal of Homeland Security & Emergency Management, 6*(1), 1–27.
Schneider, P. J., & Schauer, B. A. (2006). HAZUS—Its Development and Its Future. *Natural Hazards Review, 7*(2), 40–44. https://doi.org/10.1061/(ASCE)1527-6988(2006)7:2(40)
Shen, Y., Tahvildari, N., Morsy, M. M., Huxley, C., Chen, T. D., & Goodall, J. L. (2022). Dynamic Modeling of Inland Flooding and Storm Surge on Coastal Cities under Climate Change Scenarios: Transportation Infrastructure Impacts in Norfolk, Virginia USA as a Case Study. *Geosciences, 12*(6), 224. https://www.mdpi.com/2076-3263/12/6/224
Sheng, Y. P., Yang, K., & Paramygin, V. A. (2022). Predicting compound coastal inundation in 2100 by considering the joint probabilities of landfalling tropical cyclones and sea-level rise. *Environmental Research Letters, 17*(4), 044055. https://doi.org/10.1088/1748-9326/ac50d1
Stewart, S. R., & Berg, R. (2018). *National Hurricane Center tropical cyclone report, Hurricane Florence (AL062018), 31 August—17 September 2018.* National Hurricane Center. https://www.nhc.noaa.gov/data/tcr/AL062018_Florence.pdf
Sweet, W., Simon, S., Dusek, G., Marcy, D., Brooks, W., Pendleton, M., & Marra, J. (2021). *2021 State of High Tide Flooding and Annual Outlook* NOAA High Tide Flooding Report Issue,.
Sweet, W. V., Kopp, R., Weaver, C., Obeysekera, J., Horton, R., Thieler, E., & Zervas, C. (2017). *Global and regional sea level rise scenarios for the United States (NOAA Technical Report NOS CO-OPS 083).* National Oceanic and Atmospheric Administration. https://tidesandcurrents.noaa.gov/publications/techrpt83_Global_and_Regional_SLR_Scenarios_for_the_US_final.pdf
Taherkhani, M., Vitousek, S., Barnard, P. L., Frazer, N., Anderson, T. R., & Fletcher, C. H. (2020). Sea-level rise exponentially increases coastal flood frequency. *Scientific Reports, 10*(1), 6466. https://doi.org/10.1038/s41598-020-62188-4
Tate, E., Muñoz, C., & Suchan, J. (2015). Uncertainty and Sensitivity Analysis of the HAZUS-MH Flood Model. *Natural Hazards Review, 16*(3), 04014030. https://doi.org/10.1061/(ASCE)NH.1527-6996.0000167
U.S. Census Bureau. (2024a). *American community survey 5-year data (2009–2023).* U.S. Census Bureau. Retrieved August 22, 2025 from https://www.census.gov/data/developers/data-sets/acs-5year.html
U.S. Census Bureau. (2024b). *Population and housing unit estimates.* U.S. Census Bureau. Retrieved August 23, 2025 from https://www.census.gov/programs-surveys/popest.html
U.S. Government Accountability Office. (2023). *Flood insurance: FEMA's new rate-setting methodology improves actuarial soundness but highlights need for broader program reform.* U.S. Government Accountability Office. https://www.gao.gov/products/gao-23-105977
Wagner, K. R. H. (2022). Adaptation and Adverse Selection in Markets for Natural Disaster Insurance. *American Economic Journal: Economic Policy, 14*(3), 380–421. https://doi.org/10.1257/pol.20200378
Wahl, T., Jain, S., Bender, J., Meyers, S. D., & Luther, M. E. (2015). Increasing risk of compound flooding from storm surge and rainfall for major US cities. *Nature Climate Change, 5*(12), 1093–1097. https://doi.org/10.1038/nclimate2736

Wu, G., Liu, Q., Xu, H., & Wang, J. (2024). Modelling the combined impact of sea level rise, land subsidence, and tropical cyclones in compound flooding of coastal cities. *Ocean & Coastal Management, 252*, 107107. https://doi.org/10.1016/j.ocecoaman.2024.107107

You, X., & Kousky, C. (2024). Improving household and community disaster recovery: Evidence on the role of insurance. *Journal of Risk and Insurance, 91*(2), 299–338. https://doi.org/10.1111/jori.12466

Yu, J., Hu, P., & Chou, L. (2019). Apply Supply-Driven Input-Output Model to Evaluate Economic Losses from Coastal Typhoons. *Journal of Coastal Research, 96*(SI), 26–31. https://doi.org/10.2112/SI96-005.1

Zhao, X.-X., Zheng, M., & Fu, Q. (2022). How natural disasters affect energy innovation? The perspective of environmental sustainability. *Energy Economics, 109*, 105992. https://doi.org/10.1016/j.eneco.2022.105992

Fortifying the Gulf Coast: Adapting Infrastructure and Housing to Intensifying Storms

Rubayet Bin Mostafiz, Nazla Bushra, Robert V. Rohli, and Fahmida Mukarroma

Abstract U.S. Gulf Coast communities are increasingly vulnerable to amplifying storms and deteriorating infrastructure. More than 7000 mi^2 of land, including substantial parts of large cities such as Houston and New Orleans, are less than four feet in elevation, making crucial assets such as homes, transportation infrastructures, ports, and power grids exceedingly vulnerable to storm surge and flooding. Accelerating coastal erosion compounds these hazards by eroding natural defense structures and placing additional strain on older facilities. This chapter examines vulnerability across low-lying coastal Louisiana and bordering states, drawing upon both lessons from past failures, such as the devastation caused by Hurricane Katrina (2005), and post-Katrina successes that followed Louisiana's reforms that produced stronger building codes, stricter construction regulations, and roof-hardening design elements that reduce storm damage. Through analysis and engineering case studies, the chapter illustrates how public policy can operationalize new design ideas and retrofitting programs to reduce the storm impacts. It also points out the usefulness of incorporating technical solutions into public policy using updated codes, targeted economic incentives, and purposeful infrastructure investment. These measures are crucial in protecting Gulf Coast communities against the rising frequency and severity of tropical cyclones including the most intense hurricanes.

Keywords Wind hazards · Flood hazards · Risk management · FORTIFIED construction · Resilience

R. B. Mostafiz (✉)
Louisiana State University Agriculture Center, Baton Rouge, LA, USA
e-mail: RMostafiz@agcenter.lsu.edu

N. Bushra · R. V. Rohli · F. Mukarroma
Louisiana State University, Baton Rouge, LA, USA

J. Collins et al. (eds.), *Tropical Cyclone Risk and Resilience*,
https://doi.org/10.1007/978-3-032-28745-8_6

1 The Wind and Flood Hazard on the U.S. Gulf Coast

The U.S. Gulf Coast stands at the frontline of vulnerability to hurricanes and coastal flooding, a consequence of both its inherently low-lying terrain and the accelerating impacts of climate change. Extensive portions of the region's infrastructure and population centers are positioned precariously close to sea level. Notably, 27% of major roads, 72% of ports, and 9% of rail lines in the Gulf Coast region lie at or below four feet in elevation, making them acutely susceptible to storm surge (Savonis et al. 2008). This topography dramatically magnifies the impact of even modest sea-level rise and moderate storms, turning what were once routine weather events into significant flood threats (Bilskie et al. 2014, 2016). Over the past decade, a succession of powerful hurricanes—including Harvey (2017), Irma (2017), Michael (2018), Laura (2020), Ida (2021), Ian (2022), and the exceptionally active 2024 season featuring Beryl, Francine, Helene, and Milton—has caused widespread damage to energy systems, housing, transportation networks, and critical services across the Gulf states (NOAA 2025a). These events collectively highlight an unmistakable escalation in storm intensity and rainfall extremes that align with climate-model projections for the region (Salarieh et al. 2023). Under a 23-ft storm surge scenario, projections indicate inundation of 67% of the region's interstate highways, 57% of arterial roads, 48% of rail miles, nearly all ports, and numerous airports (NOAA 2025b).

Adding to this risk is rapid land subsidence, particularly across southeastern Texas and coastal Louisiana. Geophysical studies reveal that land in some areas is subsiding at rapid rates, driven by a combination of groundwater withdrawal, hydrocarbon extraction, and sediment compaction (Miller & Shirzaei 2021). Millennial-scale compaction rates that are dominantly peat-related can be 5 mm/year, which exceeds recent model projections locally, and on decadal to centennial timescales, the rates may exceed 10 mm/year. Compaction of Holocene strata is responsible for the very high rates of relative sea-level rise and coastal wetland loss in the Mississippi Delta and will cause subsidence in other organic-rich and often densely populated coastal plains (Törnqvist et al. 2008). In effect, once-dry parcels now routinely flood during high tides or minor storms, with the trend likely to intensify.

The region is also losing its natural wetlands that serve as storm buffers and have historically played a vital role in attenuating wave energy and absorbing floodwaters. Analyses indicate that coastal Louisiana has undergone a net change in land area of about 4833 km^2 (modeled estimate: 5197 ± 443 km^2) between 1932 and 2016. This net land area change equates to a loss of about 25% of the 1932 land area. To put these numbers into perspective, this is long-term average loss rate equivalent to an American football field of coastal wetlands in 34 minutes when losses were most rapid and within 100 minutes at more recent, slower loss rates (Couvillion et al. 2017). The ongoing degradation of these wetlands and barrier islands exposes inland communities to direct storm surge impacts.

This chapter explores how Gulf Coast communities can adapt to and mitigate these escalating threats. It begins by evaluating physical vulnerabilities in transportation, utilities, ports, and housing. Case studies from New Orleans, Houston, and Gulfport offer insight into local strategies for resilience, from retrofitting structures and revising zoning laws to constructing surge barriers and restoring wetlands. The chapter then outlines engineering and policy interventions, such as elevating infrastructure, hardening the electric grid, updating building codes, and expanding hazard disclosure. Considerations that ensure that resilience initiatives reach vulnerable populations disproportionately exposed to risk are emphasized. Data visualizations, maps of low-lying zones, and impact modeling help translate technical findings into accessible knowledge for policymakers, planners, and the public.

2 Physical Vulnerabilities of Gulf Coast Infrastructure and Housing

The Gulf Coast is characterized by an intricate network of infrastructure, including transportation systems, utilities, industrial hubs, and housing, much of which is aging and ill-equipped to confront the intensifying hazards of the twenty-first century. This section evaluates the physical vulnerabilities across key infrastructure categories considering increasing hurricane intensity, sea-level rise, and chronic environmental changes.

2.1 Roads and Bridges

Transportation infrastructure in the Gulf Coast, such as highways and bridges, often traverses low-lying coastal and wetland areas, making it highly susceptible to storm surge and flooding. Hurricanes Katrina (2005) and Ike (2008) highlighted the fragility of aging bridge infrastructure, as surge forces displaced entire spans, such as the Bay St. Louis and Biloxi Bay Bridges (Savonis et al. 2008). Many bridges built in the mid-twentieth century lack sufficient height above the base flood elevation (BFE; i.e., freeboard) and structural reinforcement, increasing vulnerability to strong winds and corrosion from saltwater intrusion. Despite some upgrades, many evacuation routes and smaller causeways remain structurally deficient and at risk during extreme weather events.

2.2 Electrical Power and Communications

The Gulf Coast power grid relies heavily on above-ground transmission lines and flood-prone substations, many of which were built decades ago and lack modern resilience standards. When Hurricane Ida made landfall near Port Fourchon, Louisiana, on 29 August 2021, as a Category 4 hurricane on the Saffir–Simpson scale, it left more than a million electric customers without power and caused the shutdown of water and wastewater treatment facilities as well as healthcare services (Tarabochia-Gast et al. 2022). Communication infrastructure proved equally vulnerable, as storm-induced power loss and physical damage to cell towers and fiber-optic networks disrupted critical information dissemination precisely when it was most needed.

2.3 Ports, Industrial Facilities, and Pipelines

The Gulf Coast is home to a high concentration of ports, refineries, and chemical plants that are critical to national logistics and energy supplies. Many of these facilities are located at or below four feet of elevation, making them particularly vulnerable to sea-level rise and hurricane storm surge (Abdelhafez et al. 2021). For instance, the Port of Gulfport suffered catastrophic damage from Katrina's 20–30-ft surge, causing prolonged operational losses. Beyond physical damage, disruptions ripple through national supply chains. Moreover, the risks of toxic chemical releases or fires pose additional environmental threats, as seen during Hurricane Harvey (Padgett et al. 2022).

2.4 Water Management and Flood Control Infrastructure

Many Gulf Coast communities depend on engineered levees, pumps, and drainage systems to remain habitable. However, these systems were often designed based on outdated flood benchmarks and now suffer from subsidence, erosion, and undercapacity during intense rainfall events (Yang et al. 2014). For example, during Hurricane Ida, areas outside the federal levee system such as LaPlace, Louisiana, faced catastrophic flooding, underscoring jurisdictional disparities in protection. Additionally, interior drainage systems fail when surge blocks outflow or reverses drainage, further highlighting systemic weaknesses.

2.5 *Housing and Buildings*

Southern Louisiana is experiencing severe shortages of low-cost, disaster-resilient housing, particularly in low-income communities, where impacts of flood and hurricane events are most disruptive. The majority of the regional housing remains inadequate for the current weather and long-term climate threats exacerbated by the growth in population and wealth in vulnerable zones, with insufficient elevation and structural elements to accommodate wind and water. This has led to long-term displacement and justifies enforcing housing solutions in accordance with modern resilience standards (Holton et al. 2018). Manufactured housing, especially those built before the 1994 U.S. Department of Housing and Urban Development (HUD) code update, faces extreme vulnerability even in moderate hurricanes. Additionally, flood risk is compounded by the prevalence of uninsured properties in flood-prone zones, leading to repetitive loss and slow recovery for residents with limited financial resilience.

3 Lessons from Major Storm Events: Failures and Successes

The Gulf Coast has been repeatedly tested by major hurricanes, each exposing critical weaknesses and, in some cases, yielding valuable lessons in resilience and recovery. Storms like Katrina (2005) have prompted critical debate on climate resilience, infrastructure planning, and transport policy (Savonis et al. 2008). Even more recent sea-level rise and flooding observations bring adaptation in the coastal cities to the fore (Kulp & Strauss 2017). These storms serve as real-world laboratories from which both systemic failures and adaptive successes can be analyzed.

3.1 *Hurricane Katrina (2005): Catastrophic Levee Failures and the Transformation of Risk Management*

Hurricane Katrina revealed the devastating consequences of inadequate flood protection infrastructure. Although it was a Category 3 hurricane at landfall on 29 August, Katrina's sheer size generated storm surges exceeding 25–28 ft. in coastal areas of Mississippi and Louisiana. New Orleans, much of which lies below sea level and was protected by a complex system of levees and floodwalls, became the epicenter of one of the worst infrastructure failures in U.S. history. Over 50 breaches occurred in New Orleans' levee and floodwall system, most notably along the Industrial Canal (IHNC) and the 17th Street Canal, allowing storm surge and inflows from Lake Pontchartrain to inundate roughly 80% of the city, with water depths in some neighborhoods reaching 10–15 ft. (ASCE, 2007; USACE, 2006, 2007).

Subsequent investigations revealed that New Orleans' flood protection infrastructure was an inconsistent patchwork of federal, state, and local efforts. For instance, the Mississippi River–Gulf Outlet (MR-GO), designed to facilitate navigation, channeled surge directly into vulnerable neighborhoods and exacerbated the failure of adjacent levees (Waugh 2009). The disaster highlighted not only technical failings in levee engineering and hydrodynamic modeling but also governance issues that left critical systems underfunded and mismanaged (Savonis et al. 2008). The toll was catastrophic: over 1800 lives were lost, and tens of thousands of homes were destroyed or rendered uninhabitable, particularly in low-income and minority neighborhoods already facing structural vulnerabilities (Elliott & Pais 2006). Hospitals, emergency services, and public utilities failed under floodwaters, highlighting the lack of resilience in essential infrastructure.

Yet, from this devastation came a wave of policy reforms and infrastructure investments that have improved resilience through stronger codes and standards. These measures are still not enough, and much more needs to be done, such as requiring FORTIFIED Home standards for new residential construction in vulnerable areas (Insurance Institute for Business and Home Safety (IBHS) 2025). In response, the federal government allocated approximately US$14.43 billion to design and construct the Hurricane & Storm Damage Risk Reduction System (HSDRRS), a comprehensive upgrade to the region's flood defenses (USACE, 2012). This system includes substantial levee enhancements, new floodwalls, pump stations and floodgates, and the construction of the 1.8-mi IHNC Lake Borgne Surge Barrier (LBSB), one of the largest surge barriers in the U.S., with features like a 150-ft sector gate and 26-ft wall height above sea level (Flood Protection Authority East 2025; Reynier 2025). Designed to reduce the risk from a 100-year storm surge (1% annual chance), these protections have greatly improved resilience in vulnerable coastal areas.

Rebuilding also extended to regulatory change. Post-Katrina, New Orleans and much of coastal Mississippi adopted updated building codes and floodplain ordinances, requiring new homes to be elevated several feet above revised BFEs. For example, in Pascagoula, Mississippi, new construction was required to be about 5 ft. higher than the mapped BFE. These changes, consistent with FEMA's Advisory BFE (ABFE) guidance, proved effective during subsequent events (FEMA 2006; U.S. Department of Energy 2008).

Katrina served as both a cautionary tale and a catalyst for change. The disaster underscored the imperative of designing infrastructure for extreme, low-probability events. It also emphasized the importance of coordinated governance, long-term maintenance, and equitable investment in community resilience.

3.2 Hurricane Ike (2008): Surge Devastation Spurs New Strategies

Hurricane Ike's landfall in September 2008 marked a pivotal moment in the Gulf Coast's confrontation with storm surge vulnerability. Though a Category 2 storm at landfall, Hurricane Ike's unusually huge size produced a 15–20 ft. (4.6–6.1 m) storm surge along parts of the Texas Gulf Coast. The surge caused catastrophic damage to the Bolivar Peninsula and Galveston Island, where entire neighborhoods were washed away (Yao et al. 2023).

Among the most emblematic images of the storm's aftermath was that of a lone elevated house left standing amid widespread devastation, underscoring the value of resilient construction practices. In total, Ike caused an estimated $24.9 billion in damage and dozens of deaths, making it one of the costliest hurricanes in the U.S. history at the time (Jafarzadegan & Moradkhani 2023).

The lessons from Ike extend beyond immediate damage to long-term risk mitigation and regional planning. The storm revealed the consequences of continued development in surge-prone areas without sufficient elevation or defense. Despite Galveston's deadly history of catastrophic hurricanes, including the 1900 storm that remains the deadliest in U.S. history, with an estimated 8000 deaths, urban development has since extended farther into severely exposed coastal regions. Ike illustrated how moderate-intensity storms can generate apocalyptic surge if they are large, slow-propagating, and interact with shallow bay shelves (Musinguzi & Akbar 2021; Peters 2024).

In response, Texas officials and engineers proposed a sweeping coastal barrier concept, colloquially known as the "Ike Dike." Initially conceived in academic and policy circles, this idea evolved into the Coastal Texas Protection and Restoration Project. By 2021, the U.S. Army Corps of Engineers (USACE) completed a comprehensive feasibility study, and in 2022, Congress formally authorized the estimated $31 billion project under the Water Resources Development Act (USACE 2025). If fully funded and constructed, this "coastal spine" will integrate large-scale surge gates, elevated dunes, and extended seawalls to protect the economically vital Houston-Galveston region from future surge threats (Greer et al. 2023).

Beyond megaprojects, Hurricane Ike spurred localized resilience efforts. Galveston Island revised building codes to require elevated construction in flood-prone areas. Key facilities such as the University of Texas Medical Branch (UTMB), which suffered extensive flood damage, have since relocated critical infrastructure to higher floors or implemented flood-proofing strategies (Woodruff et al. 2022). The storm also revealed logistical vulnerabilities: the surge cut off road access to the Bolivar Peninsula, stranding residents and impeding emergency services. The failure of evacuation system in Hurricane Katrina revealed severe gaps in regional preparedness, which prompted greater investment in evacuating route planning and led to the consideration of major infrastructure improvements, such as elevated causeways (Wolshon 2006).

Notably, natural systems offered some defense. Preserved wetlands and dune complexes in Galveston's West Bay absorbed portions of the surge energy, although they could not stop the storm's full force. These observations strengthened support for hybrid approaches, combining engineered defenses with the restoration of ecological buffers such as marshes, dunes, reefs (Yao et al. 2023).

In summary, Ike catalyzed a twofold response: the advancement of a major federal infrastructure initiative (the Coastal Texas Protection and Restoration Project) and the reinforcement of best practices in resilient urban design and land use. Ike's legacy endures as a case study in how surge risk can drive transformational change in coastal resilience.

3.3 Hurricane Harvey (2017): Unprecedented Rainfall and Urban Flooding

Hurricane Harvey, which struck southeast Texas in August 2017, redefined the understanding of urban flood risk. Unlike previous Gulf Coast hurricanes, which caused major destruction through wind and surge, Harvey's primary weapon was rain. The storm stalled over the Houston metropolitan area for several days, dumping over 50 in of rainfall in some locations, making it the most extreme rainstorm in U.S. history for an area of its size (Zhang et al. 2018). This deluge overwhelmed Houston's drainage systems, reservoirs, and bayous, submerging major highways and inundating approximately more than 150,000 homes, many of which were outside the FEMA-designated 100-year flood zones (Karaye et al. 2019).

Harvey's devastation exposed severe vulnerabilities in urban design and floodplain governance. Houston's history of rapid, largely unzoned development resulted in vast expanses of impervious surfaces and construction within natural floodplains. These land-use patterns exacerbated runoff and limited water absorption. Neighborhoods near the Addicks and Barker reservoirs were inundated by both uncontrolled overtopping and managed releases, flooding areas that were never officially designated as high-risk, leaving residents uninformed and unprepared (Mobley et al. 2019).

In response to widespread public concern, local governments acted swiftly. In 2018, the City of Houston passed one of its more stringent floodplain regulations (Ordinance No. 2018–258), mandating that new and substantially improved structures in the 500-year floodplain be built at least two feet above the FEMA-mapped 500-year flood elevation. According to a 2023 Baker Institute analysis, this regulatory standard now serves as Houston's adopted norm (Blackburn & Borski 2023; City of Houston, 2018). Furthermore, Harris County voters approved a $2.5 billion bond for flood control initiatives, including new detention basins, bayou improvements, and voluntary buyouts of repeatedly flooded properties (Newman & Li 2022).

Harvey also catalyzed a paradigm shift in climate adaptation thinking. Some have said that human-induced climate change increased the amount of rainfall associated with Hurricane Harvey by approximately 15% (Zhang et al. 2018). This concern has led planners to incorporate "Houston's future climate" into urban infrastructure, emphasizing larger drainage capacities, raised roadways, and watershed-scale planning. One major proposal that is currently under study is the creation of a "third reservoir" in northwest Harris County to alleviate stress on the aging Addicks and Barker systems (Juan et al. 2020).

On a grassroots level, Harvey galvanized voluntary buyout programs aimed at reducing exposure. In flood-prone neighborhoods like Greenspoint and Brays Bayou, buyouts have transformed residential blocks into green spaces that function as floodwater buffers. Though demand outpaces funding, these efforts represent a long-term strategy for equitable resilience (Jiang et al. 2023).

Harvey was a wake-up call: even in the absence of high winds, water alone can inflict hurricane-level devastation. The storm's legacy includes a suite of adaptive responses from revised building codes and infrastructure funding regional cooperation and climate-aware planning that now guide urban flood resilience in Houston and beyond.

3.4 Hurricane Ida (2021): Testing Post-Katrina Protections

Hurricane Ida made landfall in Louisiana on 29 August 2021, the 16th anniversary of Katrina, with sustained winds of 150 mph and storm surges reaching 10–15 ft. in some coastal regions. As a high-end Category 4 storm, Ida represented the most formidable test of the $14.5 billion flood protection system installed around New Orleans in Katrina's aftermath. The outcome was a pivotal validation of resilient infrastructure: the upgraded levees, floodwalls, and surge gates functioned as designed, preventing catastrophic flooding in New Orleans (Jafarzadegan & Moradkhani 2023). Officials described the performance as the system's "first major test," and it "passed" by keeping much of the city dry from surge (Xu et al. 2024).

However, the success in New Orleans revealed deeper regional vulnerabilities. Communities outside the federal levee system suffered significant flooding, particularly in LaPlace and other parts of St. John the Baptist Parish, as Ida's surge pushed through Lake Pontchartrain and adjacent lowlands. Construction on a long-delayed levee for these towns had only recently begun, and residents lamented that while New Orleans was protected, surrounding areas bore the brunt of redirected water, illustrating the phenomenon of flood risk displacement (Mansur et al. 2024). As noted in resilience literature, levee systems that protect one area can inadvertently increase hazards in others if not coordinated regionally (Mashrur & Kameshwar 2024).

Another critical failure exposed by Ida was the fragility of power infrastructure. Despite the success of flood defenses, the storm destroyed transmission towers and left the entire city of New Orleans without power for days, endangering hospitals, disrupting communications, and exacerbating heat exposure (Feng et al. 2025). The incident emphasized that resilience must be multi-dimensional: robust storm surge defenses alone are insufficient if power, transportation, and communication systems remain vulnerable (Coleman et al. 2023).

Structural performance varied as well. Post-2005 building codes generally supported improved structural resilience, but many older homes suffered severe wind damage. Roof failures led to interior water intrusions, rendering buildings uninhabitable. In response, Louisiana has increasingly promoted "FORTIFIED" roofing standards, which involve enhanced fastening systems and water barriers (Shen et al. 2024).

Ida also devastated other smaller, more isolated communities, particularly in Terrebonne and Lafourche Parishes, where traditional infrastructure offers limited protection. In these areas, strategic adaptation options such as voluntary relocation, home elevation, or nature-based defenses (e.g., wetlands restoration) are gaining attention (Day et al. 2023).

Ultimately, Ida confirmed that infrastructure investments can yield protective dividends. The successful performance of New Orleans' levee system stands as a testament to post-Katrina reforms. At the same time, the storm exposed the unfinished work of equitable and comprehensive resilience planning, especially regarding regional coordination, energy infrastructure, and at-risk rural populations.

4 Case Studies: Adapting in Practice in New Orleans, Houston, and Gulfport

Every community along the Gulf Coast has its own unique challenges and approaches to resilience. In this section, we spotlight three case studies—New Orleans, Louisiana; Houston, Texas; and Gulfport, Mississippi—to illustrate how different locales are implementing adaptation strategies in infrastructure and housing. These cases show a spectrum of measures: from massive engineered systems (like New Orleans' levees) to policy shifts (like Houston's floodplain regulations) to combined economic redevelopment and resilience projects (like Gulfport's port restoration). Each case studies details specific projects or reforms and examines outcomes in recent storms to discern what is working.

4.1 New Orleans, Louisiana: Rebuilding a Safer City Below Sea Level

New Orleans, long emblematic of coastal vulnerability, has transformed into a model for integrated urban resilience. Situated largely below sea level and flanked by the Mississippi River and Lake Pontchartrain, the city's geography leaves it acutely exposed to both storm surge and rainfall flooding. The catastrophic levee failures during Katrina in 2005 catalyzed one of the most ambitious flood defense overhauls in American history.

The centerpiece of this transformation is the federally funded HSDRRS completed by USACE in the early 2010s. The system includes 133 miles of levees and floodwalls, sector gates, massive outfall pump stations, and the IHNC-LBSB, a 1.8-mile, 26-foot-tall structure with a 150-ft-wide sector gate, designed to block surge from entering the IHNC. Built in 2011 at the site of fatal failures during Katrina, the LBSB operated as designed during Hurricane Ida in 2021, preventing surge flooding into the city (USACE 2015; Reynier 2025).

Beyond physical defenses, New Orleans instituted sweeping policy reforms, including stricter building codes and mandatory elevation of new or rebuilt homes above BFE, (i.e., freeboard). In New Orleans, elevated houses have long provided site-specific flood protection, with buildings having been traditionally built on pilings or piers. The centuries-long tradition, common for over 200 years, was largely abandoned after World War II for more cost-effective slab-on-grade construction that placed too much reliance on levees and drainage systems. The city has also embraced green infrastructure and water-sensitive urbanism, influenced by Dutch resilience models. A flagship initiative is the Gentilly Resilience District, supported by a $141 million HUD grant. Projects within the district include restored urban wetlands such as the Mirabeau Water Garden, along with bioswales and permeable pavements that all aimed at promoting stormwater infiltration, adding storage space for standing water, and reducing reliance on pumps, thereby reducing flood risk in the neighborhood while also recharging groundwater and slowing subsidence and land loss concurrently (City of New Orleans 2025; Stantec 2025).

Cultural resilience and land use adaptation are equally critical. In the Lower Ninth Ward, the Bayou Bienvenue Wetland Triangle was preserved as green space rather than redeveloped, acknowledging the unsuitability of rebuilding in the area's most flood-prone zones. The city also initiated voluntary home buyouts and partnered with non-profits to fund elevations for low-income homeowners. Programs like Louisiana's Restore Program further support mitigation by offering relocation or retrofitting grants to flood-prone households.

Forward-looking efforts are underway to address risk escalation. Proposed strategies include restoration of coastal barrier islands, expansion of tidal gates at Lake Pontchartrain's mouth, and enhanced subsidence monitoring. City planners acknowledge that current defenses, while robust, may be outpaced by future sea-level rise and intensifying storms.

New Orleans illustrates the power of combining engineered (grey) infrastructure, natural (green) solutions, and regulatory reforms into a cohesive adaptation strategy. However, it also serves as a reminder that resilience is not static; continued investment in maintenance, fairness, and innovation remains vital to sustaining gains.

4.2 Houston, Texas: Tackling Urban Flooding and Future Surge

As one of the largest metropolitan regions in the U.S., greater Houston, which includes Galveston, the Bolivar Peninsula, and petrochemical facilities along the Houston Ship Channel, represents a uniquely complex environment for climate adaptation. Unlike coastal cities such as New Orleans, Houston's primary threat is not storm surge but extreme rainfall and riverine flooding, particularly evident during Hurricane Harvey (2017), which produced over 50 in of rain and flooded more than 150,000 structures across the metro area. Yet, the Houston Ship Channel and neighboring coastal communities also face substantial surge risks.

In the wake of Harvey, the City of Houston and Harris County enacted landmark changes in floodplain regulations. New construction is now required to be elevated two feet above the 500-year flood level, significantly exceeding FEMA's minimum standards. This policy aims to reduce future flood damage and has already reshaped development practices, prompting redesigns in new subdivisions and the installation of retention ponds (Hooimeijer et al. 2022). While developers initially pushed back on cost grounds, city officials emphasized long-term savings and public safety. These changes reflect a growing recognition that urban flood exposure cannot be managed without land-use reform.

The Harris County Flood Control District has installed flood mitigation infrastructure throughout Houston's vast Brays Bayou watershed. Project Brays, a $480 million project, includes four stormwater detention basins that can store 3.5 billion gallons of stormwater, widening 21 mi of Brays Bayou, and the replacement of 32 bridges (Harris County Flood Control District 2022). These basins have already prevented repeat flooding in neighborhoods such as Meyerland during subsequent storms. Similar improvements are underway on White Oak Bayou and Hunting Bayou, often facilitated by buyout programs converting flood-prone properties into green space (Merrell et al. 2011).

The Coastal Spine Project, often referred to as the Ike Dike (Davlasheridze et al. 2019) authorized by Congress in 2022, addresses the surge risk with a comprehensive suite of structural defenses designed to protect the region's infrastructure and population. The system would include massive vertical lift gates at the entrance of Galveston Bay between Galveston Island and Bolivar, construction of new, extensive levees along the Bolivar Peninsula and Galveston Island, and extension of the Galveston Seawall with supplementary floodwalls designed to safeguard critical industrial assets, including those along the Houston Ship Channel (Merrell et al. 2017). If realized, this multibillion-dollar engineering initiative would become one of the largest civil infrastructure projects in U.S. history and serve to safeguard critical components of the national energy sector from catastrophic surge impacts (Davlasheridze & Atoba 2022). While full funding has not yet been secured, the state of Texas has proactively created a cost-sharing mechanism and a dedicated coastal resiliency fund to facilitate the eventual implementation of the project.

Beyond structural defenses, Houston has embraced green infrastructure and community-centered resilience. The Bayou Greenways initiative has converted large segments of floodplain into linear parks, trails, and natural detention zones. These greenways double as recreational amenities and functional flood buffers, absorbing overflow during storm events such as Harvey. Their performance validated the concept: in several cases, green corridors flooded while adjacent residential zones remained dry (Hooimeijer et al. 2022).

Houston has also enhanced flood monitoring and warning systems, installing hundreds of new gauges that feed real-time data to the public through the Flood Warning System managed by HCFCD. This system helps inform evacuation decisions and provides visibility into changing flood conditions.

In 2018, Harris County voters approved a $2.5 billion flood control bond that included a plan to prioritize historically underserved neighborhoods in flood-prone areas for project funding (Harris County Flood Control District, 2018; Bayou City Waterkeeper, 2021). Complementary initiatives supported by HUD's Community Development Block Grant–Disaster Recovery (CDBG-DR) program have provided resources for home elevations, resilient retrofits, and voluntary buyouts in neighborhoods such as Kashmere Gardens and Greenspoint, with reconstruction efforts emphasizing elevated housing, flood-resistant materials, and reinforced utilities to improve long-term resilience (Koslan 2023).

Houston's adaptation efforts span both internal resilience through drainage, detention, policy, and external defense, via the proposed Ike Dike. While challenges remain, especially around funding and implementation, the city exemplifies a pragmatic, multi-dimensional strategy that combines engineering, green space, zoning reform, and justice. It represents a living case study of how a major urban center can translate disaster into systematic reform and forward-looking climate resilience.

4.3 Gulfport, Mississippi: Resilient Port and Community Renewal

The port, commercial districts, and residential neighborhoods of Gulfport, Mississippi, suffered the effects of 25 ft. of storm surge from Katrina on 29 August 2025. Yet Gulfport offers a compelling example of post-disaster recovery that leverages infrastructure reconstruction as a catalyst for long-term resilience. The city's response, led by the Port of Gulfport Restoration Program (PGRP), highlights how federal recovery dollars can be deployed to rebuild not just what was lost, but what is needed for the future.

At the heart of Gulfport's recovery was the $567 million PGRP, funded through HUD's CDBG-DR program and administered by the Mississippi Development Authority. The program aimed not only to restore port operations after Katrina but also to incorporate mitigation measures against future storms. A central feature was the elevation of the West Pier, originally planned for 25 ft. but later adjusted to approximately 14 ft. above sea level to balance resilience goals with project costs and timelines. The program also included reconstruction of wharves, bulkheads, container yards, and warehouses, with design standards that sought to reduce vulnerability to future hurricane damage and support rapid recovery of critical operations (Mississippi Joint Legislative Committee on Performance Evaluation and Expenditure Review 2014).

This deliberate elevation strategy of lifting infrastructure physically out of the flood zone rather than merely reinforcing it in place reflects best practice in resilient design. Models suggest that, in a surge event comparable to Katrina, the rebuilt port would experience minimal service disruption. The port's reconstruction was aligned with broader resilience planning for Gulfport. Coastal buffer zones were expanded where feasible, including the restoration of wetlands and oyster reefs to attenuate wave energy. Additionally, elevated land around the port was designated for mixed-use development, effectively creating a "resilience platform" that co-locates economic activity and protective elevation.

In parallel, Mississippi updated its building codes to align with high-wind design standards like those in Florida. Gulfport enforced elevation mandates in FEMA flood zones, and many homes were rebuilt on tall piers or with breakaway ground levels. The Mississippi Cottages program, an alternative to FEMA trailers, offered residents safer, semi-permanent modular homes with hurricane-rated features. Some of these cottages remain in use today as affordable, resilient housing.

Further bolstering community safety, Gulfport Memorial Hospital was retrofitted with floodwalls and elevated mechanical systems. Schools and public buildings were similarly reconstructed with flood elevations in mind. The city also expanded its drainage canal network and installed pumps in flood-prone areas to manage heavy rainfall investments that have reduced the severity of flooding in more recent storms.

In recent years, Gulfport has participated in the IBHS FORTIFIED Home program, particularly after the damage wrought by Hurricane Zeta in 2020. Dozens of homes have been re-roofed using FORTIFIED Roof techniques, such as ring-shank nails and secondary water barriers such as sealed roof decks, that have proven highly effective at preventing water intrusion during wind-driven rain events.

The PGRP stands as a model of multi-level governance coordination. The port is state-controlled, the funding federal, and the surrounding land-use policies locally administered. Achieving alignment across these levels was complex but ultimately productive. The port's expansion catalyzed further investment in resilience, including transportation upgrades (e.g., discussions around elevating U.S. Highway 90) and ecosystem restoration in partnership with organizations like The Nature Conservancy.

Gulfport exemplifies how recovery from catastrophic loss can seed long-term resilience. The city's integrated approach, combining elevated hard infrastructure, revised codes, modular housing innovation, and ecosystem restoration offers a blueprint for other small-to-midsize Gulf communities. While future hurricanes will test these systems, Gulfport has effectively leveraged disaster lessons learned to build back smarter and stronger.

5 Engineering and Design Solutions for Resilience

Adapting Gulf Coast infrastructure to a future of intensified storms requires a layered toolkit of engineering strategies, ranging from house-level retrofits to region-wide surge defense systems. The goal is not only to prevent catastrophic failure but also to ensure rapid recovery and reduce the social and economic costs of extreme weather events.

5.1 Elevating Structures and Critical Equipment

Elevation is the most effective flood mitigation strategy. Raising structures typically 1–3 ft. above the BFE can greatly reduce the risk of structural loss. Following Katrina, higher elevation requirements in Louisiana and Mississippi caused many homes to be raised by as much as 10–15 ft. (Subasinghe 2021). Such measures proved effective during subsequent events, such as Hurricane Zeta (2020), which caused less damage to elevated structures.

Elevation also applies to critical infrastructure. Wastewater plants in Texas have been redesigned with elevated control rooms, and hospitals in Louisiana and Texas have moved generators to higher floors following failures during past floods. The elevation of transportation arteries is another example: Louisiana's LA-1 project features raised road segments to preserve access to Port Fourchon, which is a critical facility for the national oil supply chain.

However, lower-income residents may lack the resources to elevate. This has prompted innovations like forgivable loans, grant-funded elevations, and insurance incentives, aiming to make elevation more accessible across income brackets (Wilson & Lazarus 2020).

5.2 Wind-Resilient Buildings and Roofing

High winds are often as destructive as storm surge, and mitigating their impact requires deliberate engineering interventions. The IBHS FORTIFIED program offers detailed guidance to strengthen residential structures, particularly roofs and framing systems. The Fortified Roof requirement of ring-shank nails provides superior holding strength, while the secondary water barriers beneath shingles prevent interior damage if shingles are lost. Even higher standards include impact-rated windows or storm shutters to protect openings (for FORTIFIED Silver) and continuous load paths through roof-to-wall and wall-to-foundation strapping (for FORTIFIED Gold). These components work in concert to substantially improve a building's ability to withstand hurricane-force winds. A study of Hurricane Sally (2020) insurance claims revealed that the houses built to the FORTIFIED standard experienced much less damage than their non-FORTIFIED counterparts; it was estimated that more than $106 million in insurance claims could have been avoided had all houses affected been built to the FORTIFIED Roof standard (Subasinghe 2021). Reflecting growing adoption, more than 35,000 FORTIFIED designations had been issued across the U.S. by 2025, with a significant concentration in Gulf Coast states (Feng et al. 2025).

Beyond construction standards, building shape and material selection also significantly influence wind resilience: for example, hip roofs—which slope on all sides—reduce wind pressures and outperform gable roofs under high-wind stress (Building America Solution Center 2020; Meecham 1992). Building materials such as insulated concrete forms (ICFs) and reinforced masonry provide superior structural integrity and wind resistance compared to lighter or less continuous wall systems (ICFmag 2009; BuildBlock 2025). In commercial settings, glazing systems rated for hurricane wind loads and missile impact are increasingly standard. For example, after Ike (2008), which had shattered windows on many downtown Houston high-rises due to windblown debris, several high-rises were retrofitted with reinforced window systems and impact-resistant glazing to better resist subsequent storms (Konz & Vega 2009).

5.3 Structural Retrofitting and Reinforcement

Retrofitting legacy structures is a critical component of coastal resilience, particularly for communities with aging infrastructure exposed to increasing storm threats. For residential buildings, retrofits may involve FORTIFIED Roof principles like additional roof deck fasteners, FORTIFIED Silver impact-resistant features such as strengthened windows and doors, and FORTIFIED Gold principles like the installation of hurricane straps, all of which significantly enhance a structure's ability to withstand high winds. In the case of older bridges, structural reinforcements such as column jacketing, scour protection, and deck tie-downs have become standard retrofitting techniques, especially post-Katrina.

Both dry and wet floodproofing are also commonly employed. Dry flood proofing is done by sealing the walls and installing flood barriers at the building entrances (Bignami et al. 2019). In New Orleans' historic French Quarter, several buildings now incorporate automatic flood barriers that activate as water level rise. Larger-scale dry floodproofing surge protection systems such as seawalls, floodgates, and levees are undergoing modernization across the Gulf Coast. The Seabrook Floodgate in New Orleans, for example, was constructed to block storm surge from entering Lake Pontchartrain, while the seawall in Port Arthur, Texas, has recently been upgraded to higher elevations and supplemented with new pump stations to improve its flood-handling capacity. Similarly, the Texas City levee system was fortified after Hurricane Ike to enhance its structural integrity and reduce overtopping risks during future storm events. By contrast, wet floodproofing involves designing ground-level spaces, such as parking garages, to safely admit water using water-resistant materials. Wet floodproofing can be effective as a means of reducing pressure on walls from rushing or standing water, while also holding water in less-vulnerable places rather than risking much more damage if it were to accumulate elsewhere.

5.4 *Hybrid and Nature-Based Infrastructure*

The Gulf Coast's engineering response to climate risk illustrates a continuum from hard infrastructure (e.g., elevated roads, levees) to structural retrofits (e.g., roof reinforcements, flood barriers) to hybrid solutions (e.g., oyster reefs, bioswales) to nature-based solutions (e.g., wetlands, dunes, oyster reefs, green urban infrastructure). As storms intensify and the lives and property value in harm's way continue to increase, widespread implementation of these proven methods will be vital. The challenge is not technological; it is economic, political, and social. Ensuring access to these measures is just as important as their technical robustness.

Living shorelines and geotextile tubes represent hybrid solutions that integrate structural and ecological benefits. While not substitutes for engineered defenses in high-risk areas, these strategies are valuable complements that can reduce erosion, attenuate surge impact, and enhance biodiversity. For example, low-lying breakwaters seeded with oyster reefs not only attenuate waves but also provide ecosystem services. These designs are increasingly incorporated in Gulf restoration plans, including The Nature Conservancy's initiatives in Mississippi and Louisiana. Nature-based solutions are increasingly central to Gulf Coast resilience planning, offering synergistic benefits for both storm protection/risk reduction and ecosystem restoration such as enhancing biodiversity, water quality, and adaptive capacity in a changing climate.

5.5 *Wetland Restoration as Natural Surge Buffers*

In Hurricane Ike in 2008, coastal wetlands on Galveston Bay, Texas, prevented property damage by $934 million, a 13% decrease, compared to no-wetlands scenarios, by absorbing wave energy and tidal flows (Sutton-Grier et al. 2015). The attenuation effect varies by storm and landscape, but healthy marshes can reduce surge heights by several inches to a foot per linear mile.

The *2023 Master Plan for the Coast* prioritizes the building of marshes via sediment diversions and dredged material placement and subsequent re-vegetation. These projects will start to build habitat and reduce stress on levees during major storms (Day et al. 2023). However, by late July 2025, CPRA determined that construction of the project as constructed in the Final RP 3.2 is no longer viable at this point based on a totality of the circumstances, such as costs, permitting, and current litigation (Coastal Protection and Restoration Authority [CPRA] 2025).

5.6 Barrier Island and Dune Rebuilding

Barrier islands and dunes are the first line of defense along much of the Gulf. Past hurricanes, including Katrina and Camille (1969), severely eroded these features, increasing vulnerability inland. Restoration efforts, such as sand nourishment and planting dune grasses, have been implemented on Dauphin Island, the Chandeleur Islands, and Mississippi's Ship Island, where a major breach was repaired with dredged sediment.

These features have proven effective. For instance, restored dunes along the Florida Panhandle, rebuilt after Hurricanes Ivan (2004) and Dennis (2005), helped mitigate impacts from later storms like Michael in 2018 (Powell et al. 2019). Barrier island restoration requires continual investment, as erosion and sea-level rise can outpace passive recovery.

5.7 Oyster Reef Restoration: Living Breakwaters

Oyster reefs provide biogenic protection, functioning as natural breakwaters that reduce wave energy and stabilize shorelines. In Mobile Bay, Alabama, restored oyster reefs constructed of limestone or concrete reduced wave height by as much as 90% and wave energy by as much as 99%, providing effective protection to adjacent marsh edges (Scyphers et al. 2011). Oyster reefs also enhance water quality, stabilize and buffer shorelines from erosion, and protect billions of dollars in infrastructure and the livelihoods of individuals who live and work in the coastal zone. In coastal Alabama, reef breakwaters made of seeded oysters or oyster shell have been shown to reduce shoreline erosion by more than 40% while preserving fisheries and mobile macro-invertebrate habitats compared to control sites (Scyphers et al., 2011). Such reefs are highly effective at reducing small- to medium-scale wave action such as boat wake and wind chops. Other research suggests that oyster reefs reduce wave height by up to 68%, which can reduce shoreline erosion and retreat substantially (Morris et al. 2021).

5.8 Urban Green Infrastructure for Rainfall Flooding

While coastal features mitigate surge, urban green infrastructure plays a critical role in managing pluvial (rainfall-induced) flooding. In cities such as Houston and New Orleans, strategies like green roofs, rain gardens, permeable pavements, and bioswales integrated into streets and parking lots are increasingly used to reduce stormwater runoff. These interventions are particularly valuable during intense thunderstorms, as they help to absorb rainfall, slow runoff, and reduce pressure on aging drainage systems. A prime example is the Gentilly Resilience District in New Orleans described previously.

5.9 Long-Term co-Benefits and Limitations

Nature-based solutions can also provide a number of co-benefits such as carbon sequestration and recreation (Sutton-Grier et al. 2015). However, their effectiveness often hinges on spatial requirements. Wetlands, for example, need room to migrate inland as sea levels rise, while dune systems require space for lateral expansion and natural reshaping. Oyster reefs, another key natural defense, depend on the presence of appropriate salinity gradients and stable substrates to thrive. Therefore, strategic land acquisition and adaptive management, such as periodic sediment augmentation or the restoration of tidal connectivity, are critical to sustain and enhance these nature-based defenses over time.

5.10 Gray-Green Integration

The ideal approach combines natural buffers with engineered structures. For example, wetlands fronting levees reduce incoming wave energy, prolonging levee lifespan. Louisiana's Master Plan supports this "gray-green" model, citing hybrid systems as cost-effective and adaptive. Sutton-Grier et al. (2015) found that wetlands in the Northeast reduced damages incurred by Hurricane Sandy by $625 million, underscoring the fiscal and protective value of natural defenses in reducing surge, controlling runoff, and building ecological health. While not a panacea, when paired with gray infrastructure, nature-based solutions offer a flexible, scalable path forward; one that blends engineering with environmental stewardship.

6 Modernizing Power Grids and Smart Infrastructure

In the wake of increasingly intense Gulf Coast storms, ensuring power continuity has become a central challenge for infrastructure resilience. Grid modernization involves hardening vulnerable components, embedding intelligence in network systems, and decentralizing power generation to reduce single-point failures. These strategies together form a blueprint for a more durable, adaptable, and equitable energy future.

6.1 Physical Hardening of Grid Components

Post-disaster assessments following hurricanes such as Laura (2020) have prompted regional utilities like Entergy to invest in comprehensive upgrades to transmission infrastructure. As part of a roughly $15 billion grid resilience plan in the next decade, Entergy has installed transmission towers, such as the Mississippi River Avondale to Harahan (both in the New Orleans area) crossing, that are designed to sustain wind gusts of up to 175 mph, beyond Category 5 hurricane levels (Entergy Corporation 2022; Entergy Louisiana 2022). The initiative encompasses several key measures, including replacing traditional wooden utility poles with more durable steel or composite alternatives, reinforcing exposed infrastructure with guy-wires to enhance wind resistance, elevating substations to prevent flood damage, and configuring transmission lines into looped systems to facilitate power rerouting and minimize the scope of outages during severe weather events.

6.2 Undergrounding and Distributed Energy Systems

Undergrounding distribution lines is a widely-discussed solution, particularly in high-density or historic districts like downtown Galveston, where lines were buried after Ike. While underground cables are naturally protected from wind damage, they must still be sealed against floodwater ingress using water-blocking elements, moisture-barrier jackets, and pressure-resistant duct seals to prevent system failures. Industry standards, including ANSI/ICEA S-94-649 and S-97-682, as well as AEIC specifications, require longitudinal water-blocking and water-penetration resistance testing to ensure cable integrity in wet or submerged environments (Insulated Cable Engineers Association (ICEA), 2017; Jacksonville Electric Authority (JEA), 2020; U.S. Nuclear Regulatory Commission (NRC), 2009). Unfortunately, cost considerations typically limit this strategy to new developments or high-value corridors.

Simultaneously, distributed energy resources and microgrids are being deployed to improve energy autonomy for critical facilities. Microgrids use local generation (e.g., solar + battery or combined heat and power) to operate independently from the main grid during outages. This model proved effective during Hurricane Ida and winter storm Uri (2021) at facilities like Texas Medical Center, which operated continuously, thanks to an internal thermal energy system (Chandak & Rout 2021).

6.3 Smart Grid and Automation Technologies

Modern grid resilience increasingly hinges on smart grid technologies, which use sensors, data analytics, and automation to reduce downtime by improving efficiency, responsiveness, and adaptability of energy systems. These technologies include automated switches that reroute power to isolate faults, sensors and fiber optic cables that provide real-time diagnostics, and Supervisory Control and Data Acquisition (SCADA) systems that enable remote control of equipment such as valves and pumps. In cities like New Orleans and Houston, the rollout of smart grid infrastructure has significantly enhanced outage tracking and reduced recovery time after extreme weather events (Center for Climate and Energy Solutions (C2ES) 2018; U.S. Department of Energy, 2014). Furthermore, smart meters and embedded analytics have been instrumental in supporting predictive maintenance and improving communication with customers, allowing utilities to respond more effectively to service disruptions and potential failures (Koukouvinos et al., 2025).

6.4 Water and Sewer Infrastructure Modernization

Utilities are also integrating SCADA systems and backup generators for water and sewer infrastructure. During Katrina, the failure of electrically powered pump stations led to uncontrolled urban flooding. Since then, New Orleans and other Gulf cities have installed redundant systems, including automated sluice gates for drainage canals and remote pump controls to manage flash flood events more dynamically.

6.5 Passive Survivability and Building-Scale Resilience

Another frontier is passive survivability, which calls for designing buildings to remain habitable during extended power outages. Design experience and research (e.g., U.S. Department of Energy, 2024) show that resilient construction often focuses on architectural and material strategies that reduce dependence on mechanical systems. Key features include enhanced insulation, natural ventilation, solar thermal systems, and daylighting, which together help critical facilities like hospitals and community centers continue to function as safe havens. Increasingly across the Gulf Coast, such facilities are incorporating passive survivability into their design, ensuring they can serve their communities even when conventional energy systems fail.

The modernization of Gulf Coast energy and utility systems is moving from centralized vulnerability to distributed resilience. Whether through wind-hardened towers, neighborhood microgrids, smart sensors, or passive building design, the emphasis is on avoiding cascading failures and ensuring lifeline services during disasters. As climate extremes intensify, this systems-based approach to resilience will be increasingly indispensable.

7 Surge Barriers and Coastal Megaprojects

As storm intensity and sea-level rise increase risk to coastal communities, Gulf Coast planners are advancing megaproject engineering interventions like surge barriers and comprehensive levee systems. These large-scale defenses are intended to mitigate catastrophic flooding during hurricanes and are designed to complement smaller-scale protective measures such as resilient building codes and wetlands restoration. Among the most prominent forms of such infrastructure are surge barriers, movable gates, or fixed walls. These structures are constructed at key water entry points such as estuaries, inlets, or river mouths, to prevent hurricane-driven storm surge from flooding inland areas.

The most established example in the Gulf Coast region is the 2011-completed IHNC–LBSB in New Orleans as described earlier, which was a major component of the HSDRRS. It has proven effective in reducing flood risk in eastern New Orleans neighborhoods that were disproportionately impacted by Katrina (USACE 2011). But the success of the system in the long run relies on the frequent maintenance and inspections (Jonkman & van Berchum 2022). Also in Louisiana, expansion of the levee network continues, including a planned ring levee for Houma and the long-delayed levee system for LaPlace, which had been scheduled for completion in 2024. It is now expected to be completed by early 2030. Meanwhile the Houma Region's "Morganza to the Gulf" system is making progress. The Houma Navigation Canal lock started construction in August 2024 and is estimated to be finished by 2028. Additional levee and gate systems are also being built.

Perhaps the most ambitious surge barrier proposal in the U.S. is the Galveston Coastal Spine (i.e., the Ike Dike), as described previously. Drawing on the Netherlands' sophisticated flood defense strategies, the Ike Dike is being engineered not only for structural strength and operability during extreme weather events but also to maintain ecological functionality under normal conditions. The gates are designed to withstand intense wave action and can be closed on short notice in response to storm threats. If completed, the Ike Dike would be among the largest and most ambitious civil engineering projects ever attempted in North America, designed to protect the Houston-Galveston region, one of the nation's most densely populated and economically important coastal areas (Merrell et al. 2011).

Beyond the Houston-Galveston area, other Gulf Coast regions are exploring similar protective strategies. Tampa Bay and Mobile Bay have initiated preliminary discussions about surge barrier installations, particularly following recent storm events like Hurricane Ian (2022). Additionally, smaller-scale, community-specific barriers are being considered to protect individual neighborhoods or assets while maintaining the broader floodplain's ecological functions. Internationally, notable examples include the Maeslantkering in the Netherlands and the Thames Barrier in London. A persistent engineering challenge in levee design is the "end-of-levee problem," where water pours through the unconnected ends of protective walls. To address this issue, modern levee systems are increasingly designed to fully encircle at-risk areas or tie into natural elevations to close the protection loop.

The introduction of large-scale flood barriers brings a secondary engineering challenge: managing interior drainage. When levees or surge gates are closed to block external water, rain that falls inside the protected area must still be evacuated via robust pumping infrastructure to avoid pluvial flooding, with the urgency of such a system highlighted by Harvey. New Orleans, for instance, relies on a network of massive outfall pump stations to discharge rainwater during flood events. Similarly, the Ike Dike concept includes high-capacity pumps to serve the Clear Lake and NASA Johnson Space Center area, ensuring that rainwater does not accumulate dangerously behind closed barriers.

While megastructures like the IHNC barrier and the Ike Dike offer extraordinary protective potential, safeguarding billions or even trillions in infrastructure and countless lives, they are not without trade-offs. These include their high construction and maintenance cost, extended timelines for design and construction, and potential environmental impacts. Moreover, even the most well-designed structure carries residual risk: gates can fail, systems can be overwhelmed, and unanticipated ecological impacts may emerge over time. Consequently, these hard-engineered defenses are being integrated into broader, layered resilience strategies that include evacuation protocols, land-use planning, and nature-based solutions such as wetland restoration and oyster reef construction.

Surge barriers and levee systems represent the upper echelon of coastal engineering, constructed not for routine conditions but for the most extreme and destructive events. They are not standalone answers but rather critical elements in a multifaceted strategy that also incorporates elevated buildings, wind-resistant codes, decentralized energy systems, and ecological restoration. The future resilience of the Gulf Coast depends not only on these monumental interventions but on how effectively they are integrated with smaller-scale, community-driven, and adaptable measures that address both present and evolving risks.

8 Policy Tools and Strategies for Resilience

Resilience on the Gulf Coast is not just a product of engineering; it is deeply shaped by policy. From strengthened building codes to financial incentives and community engagement, the region's ability to adapt to disaster threats hinges on regulatory innovation, fiscal investment, and inclusive governance. This section synthesizes policy mechanisms into four key domains: codes and standards, financial incentives, disclosure and risk communication, and public infrastructure funding.

8.1 Strengthening Building Codes and Infrastructure Standards

Building codes form the foundation of hazard mitigation. After disasters like Katrina and Harvey, Gulf states have progressively strengthened construction codes, especially for wind and flood. After Hurricane Andrew devastated South Florida in 1992, the state pioneered adoption of impact-resistant building design standards through the 1994 South Florida Building Code and, later, the statewide Florida Building Code in 2002. Over time, other Gulf Coast states incorporated similar hurricane-resistant provisions into their codes as well (FEMA, 2019; IBHS, 2021). Louisiana adopted a statewide code in 2005, aligning wind and flood resistance standards with those of Florida. The IBHS FORTIFIED program enhances building codes by offering voluntary, science-based standards for construction and retrofit, and has seen growing adoption in states such as Alabama and Mississippi, where insurance incentives, legislative backing, and homeowner interest have all increased markedly (Awondo et al. 2020; Insurance Institute for Business and Home Safety (IBHS) 2025; Urban Land Institute (ULI) 2025).

Infrastructure codes are also evolving. The American Society of Civil Engineers flood standards and the Federal Highway Administration's coastal road guidelines now recommend designing for climate-adjusted loads, such as higher storm surge and increased precipitation intensities (Thomas 2024). Projects like the Sabine Pass Bridge and Southwest Louisiana levee systems now incorporate future risk into design.

Yet adoption of codes alone is insufficient, and enforcement remains a challenge. Many local jurisdictions lack trained inspectors, and political pressure can hinder compliance. Moreover, older buildings remain vulnerable unless incentivized to retrofit. Complementary policies, such as retrofit grants and insurance discounts, voluntary standards like FORTIFIED, and other approaches described below are essential to bridge this gap.

8.2 Incentive Programs and Insurance Reforms

States have launched grant programs to subsidize retrofits. Louisiana's Fortify Homes Program, started in 2022, provides up to $10,000 for FORTIFIED roof upgrades. As of 2025, over 1800 homes have participated, with 22% reductions in insurance premiums observed (Feng et al. 2025). Alabama's similar Strengthen Alabama Homes program, funded through insurance surcharges, has retrofitted thousands of homes.

Insurance reforms are also underway. FEMA's Risk Rating 2.0 for the National Flood Insurance Program ties premiums more closely to actual risk, which encourages mitigation but can price out lower-income households if not balanced by subsidies. Some states mandate insurance discounts for mitigation, while others provide state reinsurance to stabilize markets strained by storm losses.

Tax incentives and sales tax exemptions for hurricane-resilient building materials, as used in Mississippi, offer another tool. Additionally, federal-level tax credits for climate mitigation have been proposed.

At the community scale, FEMA's Building Resilient Infrastructure and Communities (BRIC) program and the Hazard Mitigation Grant Program (HMGP) fund projects like drainage upgrades, buyouts, and resilient housing development; states like Texas and Louisiana have aggressively competed for these funds and secured large awards to support long-term resilience transformation (FEMA, 2023a; FEMA, 2023b; North Dallas Gazette, 2023).

8.3 Hazard Disclosure and Risk Communication

One of the most cost-effective resilience strategies is accurate risk assessment and transparency in disclosing risk. Florida, Texas, and Louisiana have updated real estate disclosure laws, now requiring sellers to inform buyers about flood history, insurance claims, and zone designations. Florida's new law, which took effect on 1 October 2024, is part of House Bill 1049 (Sect. 689.302, Florida Statutes). It mandates these disclosures and represents a significant step toward giving consumers more information about risks (Florida Realtors 2024; Florida House of Representatives 2024).

Public tools like Flood Factor used on platforms like Realtor.com (First Street Foundation 2025) are democratizing risk awareness. Meanwhile, evacuation planning and public awareness has improved dramatically since Hurricane Rita (2005), with contraflow routes, tiered evacuation zones, and real-time communication strategies.

Cities like New Orleans and Houston have implemented localized alerts and sensor-based flood warning systems. In Miami, street-level tide markers and automated pumps help visualize risk. Some jurisdictions are also beginning to include future conditions, such as sea-level rise in official planning documents.

8.4 Public Investment and Infrastructure Funding for Resilience

Federal, state, and local funds are now proactively allocated for resilience. The Infrastructure Investment and Jobs Act and FEMA's BRIC program provide billions for coastal defense, grid hardening, and adaptive planning. For example, Louisiana used HUD's post-Harvey funds to launch LA SAFE, a program that supports community-designed adaptations, including elevating homes, developing high-ground housing, and restoring wetlands (State of Louisiana 2019).

States like Louisiana, Texas, and Florida have created dedicated resilience offices and funding mechanisms. Florida's Resilient Florida program has allocated hundreds of millions to local adaptation projects. Meanwhile, infrastructure upgrades, like the I-10 Twin Span bridge near Slidell, Louisiana (rebuilt post-Katrina), showcase how transportation spending can yield dual benefits for mobility and resilience.

A policy frontier is emerging around benefit-cost analysis (BCA) reform. Traditional BCA methods prioritize property value over social need, disadvantaging rural or low-income areas. Agencies are now incorporating environmental justice considerations to justify projects in high-need communities (Feng et al. 2025).

9 Conclusion

Policy shapes the U.S. Gulf Coast's resilience as much as engineering. Strong building codes, transparent disclosures, targeted incentives, equitable funding, and community partnerships collectively drive adaptation. As climate risks accelerate, policy must continue to evolve, moving from reaction to anticipation, from uniformity to inclusion, and from short-term recovery to long-term transformation.

The Gulf Coast's escalating confrontation with atmospheric- and environmentally-driven threats including intensifying hurricanes, accelerating sea-level rise, subsiding lands, and entrenched socio-economic vulnerabilities present a formidable yet navigable challenge. As perhaps the most hazard-prone U.S. region, the U.S. Gulf Coast has emerged as a testing ground for multi-faceted resilience. The chapter has detailed a broad array of vulnerabilities like aging infrastructure, flood-prone housing, and regional inequities, and also a growing arsenal of solutions such as engineered protections, ecological buffers, forward-thinking policies, and inclusive governance mechanisms. Taken together, these reflect a regional shift toward comprehensive, layered adaptation.

A central insight that emerges is the necessity of integrated, systems-based planning. Resilience cannot be constructed in fragments. Strengthening a levee or elevating a building has little value if surrounding systems such as roads, utilities, and/or neighborhoods remain exposed. Effective adaptation strategies are now being conceived not as isolated interventions but as interlinked networks that span from the household to the watershed, and from city zoning plans to federal flood defense megaprojects. States like Louisiana, Texas, and Florida have developed forward-looking adaptation frameworks, such as Louisiana's 50-year Coastal Master Plan, which is updated regularly and grounded in community participation. The Dutch Delta Program is built on adaptive delta management, with a focus on flexibility in decision-making, broad stakeholder engagement, and cross-sectoral cooperation in order to meet future uncertainties about climate and socio-economic developments (Zevenbergen et al. 2015).

Another key takeaway is that the effort to achieve resilience must remain dynamic. Climate projections indicate that the Gulf Coast could experience approximately 14–18 in (0.35–0.45 m) of relative sea-level rise by 2050 (U.S. Climate Resilience Toolkit 2025). Accompanying this threat are warnings of more frequent high-intensity hurricanes, faster storm intensification, and more extreme rainfall events. These evolving threats necessitate infrastructure and policy that are designed with flexibility in mind. For example, global-scale application of CoastalDEM identifies far more land and people below projected flood heights than previously estimated, including in areas such as New Orleans, where levees are designed with wide bases to allow future raising and in places where roadway reconstructions are increasingly accounting for expected sea-level rise (Kulp & Strauss 2019). Adaptive building codes, which are updated on three-year cycles, further embed climate flexibility into local governance. These foresighted design principles are critical, as today's "future-proofed" infrastructure may still need enhancement in the coming decades.

Yet even the most technically sound systems can falter without sustained monitoring and maintenance. Maintaining stormwater systems, upgrading pumps, inspecting levees, and funding repairs are as essential as the initial construction. Vertical ground deformation is derived from Interferometric Synthetic Aperture Radar (InSAR), allowing for accurate measures of subsidence at the community scale and providing necessary information for proactive risk management (Navarro-Hernández et al. 2023). Without such feedback mechanisms, resilience may be short-lived. Accordingly, Gulf states are developing stormwater utilities and resilience funding mechanisms to institutionalize maintenance, often linking these to broader fairness frameworks.

The Gulf Coast's trajectory offers valuable lessons for coastal regions worldwide. It starkly illustrates the cost of under preparedness, as seen in Katrina's devastation, and the rewards of sustained investment, exemplified by Ida's comparatively mitigated impacts. It also demonstrates the power of nature-based strategies like marsh and barrier island restoration, and oyster reef construction as low-carbon, multifunctional defenses that provide co-benefits for biodiversity and fisheries. Just as importantly, the Gulf's experience highlights the critical role of social resilience including community planning, housing support, and transparent risk communication. Adaptation efforts that neglect the most vulnerable not only leave people behind but also undermine the legitimacy and effectiveness of resilience strategies for everyone (Neumann et al. 2021).

As of 2025, the Gulf Coast is demonstrably more resilient than it was a generation ago. New homes are elevated, public infrastructure is being built to more rigorous standards, and awareness of flood and wind risk has increased significantly. But climate risks have also intensified. The 2020 hurricane season, which exhausted the alphabetical naming system with 30 named storms, was likely a harbinger of more volatile years ahead. The combination of physical vulnerability and growing exposure makes it clear that resilience is not a destination but an ongoing process. Maintaining this progress will require not only sustained political and financial commitment but also deeper integration of scientific data and community engagement.

The story of the Gulf Coast adaptation is still unfolding. The jury remains out regarding whether the resilience improvements are outpacing the escalating threats. Yet progress to date offers at least some confidence for a promising trajectory. With continued learning, investment, and inclusive planning, the region can move from being a symbol of climate vulnerability to a model of proactive adaptation and resilience.

References

Abdelhafez, M. A., Ellingwood, B., & Mahmoud, H. (2021). Vulnerability of seaports to hurricanes and sea level rise in a changing climate: A case study for Mobile, AL. *Coastal Engineering, 166*, 103884. https://doi.org/10.1016/j.coastaleng.2021.103884

ASCE. (2007, May). *The New Orleans hurricane protection system: What went wrong and why*. American Society of Civil Engineers. https://doi.org/10.1061/9780784408933

Awondo, S. N., Hollans, H., Powell, E., & Wade, K. (2020). *Estimating the effect of FORTIFIED home™ construction on home resale value*. University of Alabama & University of Mississippi. https://www.smarthomeamerica.org/assets/uploads/UniversityofAL_Value-Study_FORTIFIEDReport_V2__2.pdf. Last accessed: 15 September 2025

Bayou City Waterkeeper. (2021). Advocating for equity-based flood prioritization in Harris County. *Bayou City Waterkeeper.*. https://bayoucitywaterkeeper.org/advocating-for-equity-based-flood-prioritization-in-harris-county/

Bignami, D. F., Rosso, R., & Sanfilippo, U. (2019). *Flood proofing in urban areas*. Springer Nature Switzerland. https://doi.org/10.1007/978-3-030-05934-7

Bilskie, M. V., Hagen, S. C., Medeiros, S. C., & Passeri, D. L. (2014). Dynamics of sea level rise and coastal flooding on a changing landscape. *Geophysical Research Letters, 41*(3), 927–934. https://doi.org/10.1002/2013GL058759

Bilskie, M. V., Hagen, S. C., Alizad, K., & Medeiros, S. C. (2016). Dynamic simulation and numerical analysis of hurricane storm surge under sea level rise with geomorphologic changes along the northern Gulf of Mexico. *Earth's Future, 4*(5), 177–193. https://doi.org/10.1002/2015EF000347

Blackburn, J., & Borski, J. (2023, August 22). Assessing Houston's flood vulnerability 6 years after Harvey. Rice University's Baker Institute for Public Policy. https://www.bakerinstitute.org/research/assessing-houstons-flood-vulnerability-6-years-after-harvey. Last accessed 16 September 2025.

BuildBlock. (2025). High-winds resistance of BuildBlock ICFs. Indicates the strength and durability of concrete formed with ICFs under severe wind conditions. https://buildblock.com/insulating-concrete-forms-icfs/icf-disaster-resistant/high-winds/. Last accessed 16 September 2025.

Building America Solution Center. (2020). *Hip roof vs gable roof*. PNNL. Shows that hip roofs experience smaller wind pressures than gable roofs. https://basc.pnnl.gov/resource-guides/hip-roof-vs-gable-roof. Last accessed: 16 September 2025

Chandak, S., & Rout, P. K. (2021). The implementation framework of a microgrid: A review. *International Journal of Energy Research, 45*(6), 8626–8647. doi: 10.1002/er.6064

Center for Climate and Energy Solutions (C2ES). (2018). Resilience strategies for power outages. Center for Climate and Energy Solutions. https://www.c2es.org/wp-content/uploads/2018/08/resilience-strategies-power-outages.pdf. Last accessed 16 September 2025.

City of Houston. (2018, April 10). Ordinance No. 2018-258: An ordinance amending chapter 19 of the code of ordinances relating to floodplains. (Houston Permitting Center).

City of New Orleans. (2025). *Gentilly resilience district*. City of New Orleans. https://www.nola.gov/next/stormwater-green-infrastructure/topics/gentilly-resilience-district/. Last accessed 15 September 2025

Couvillion, B. R., Beck, H., Schoolmaster, D., & Fischer, M. (2017). Land area change in coastal Louisiana (1932 to 2016). U.S. Geological Survey scientific investigations map 3381. U.S. Department of the Interior. doi: https://doi.org/10.3133/sim3381.

Coleman, N., Esmalian, A., & Lee, C. C. (2023). Energy inequality in climate hazards: Empirical evidence of social and spatial disparities in managed and hazard-induced power outages. *Sustainable Cities and Society, 94*, 104644. https://doi.org/10.1016/j.scs.2023.104644

Coastal Protection and Restoration Authority (CPRA). (2025, July 17). State, Louisiana trustee implementation group announce termination of mid-Barataria sediment diversion. Louisiana Coastal Protection and Restoration Authority. https://coastal.la.gov/news/state-louisiana-trustee-implementation-group-announce-termination-of-mid-barataria-sediment-diversion/. Last accessed 15 August 2025

Davlasheridze, M., & Atoba, K. O. (2022). Cost-benefit analysis of a proposed coastal infrastructure for reducing storm surge-induced impact in the upper Texas coast. In *Coastal flood risk reduction* (pp. 141–162). Elsevier. https://doi.org/10.1016/B978-0-323-85251-7.00008-1

Davlasheridze, M., Atoba, K. O., & Brody, S. D. (2019). Economic impacts of storm surge and the cost-benefit analysis of a coastal spine as the surge mitigation strategy in Houston-Galveston area in the USA. *Climatic Change, 152*, 395–413. https://doi.org/10.1007/s11027-018-9814-z

Day, J. W., Twilley, R. R., Freeman, A., & Couvillion, B. (2023). The concept of land bridge marshes in the Mississippi River Delta and implications for coastal restoration. *Nature-Based Solutions, 3*, 100090. https://doi.org/10.1016/j.nbsj.2023.100090

Elliott, J. R., & Pais, J. (2006). Race, class, and hurricane Katrina: Social differences in human responses to disaster. *Social Science Research, 35*(2), 295–321. https://doi.org/10.1016/j.ssresearch.2006.02.003

Entergy Corporation. (2022). Integrated Report 2022. https://s201.q4cdn.com/714390239/files/doc_financials/2022/ar/entergy-2022-integrated-report.pdf. Last accessed 9 August 2025.

Entergy Louisiana. (2022, August 24). Entergy completes transmission line segment over Mississippi River built to withstand winds of up to 175 mph. Entergy Newsroom. https://www.entergy.com/news/entergy-completes-transmission-line-segment-over-mississippi-river. Last accessed 9 August 2025.

FEMA. (2006). Hurricane Katrina in the Gulf coast: Mitigation assessment team report, building Performance observations, recommendations, and technical guidance. FEMA 549. *Federal Emergency Management Agency.*. https://www.preventionweb.net/files/1543_549cvrtoc1.pdf. Last accessed 17 September 2025

FEMA. (2019). The role of Florida's building codes in 2018 Hurricane Michael. https://www.fema.gov/case-study/role-floridas-building-codes-2018-hurricane-michael. Last accessed 15 September 2025.

FEMA. (2023a). Hazard mitigation Grant program (HMGP). U.S. Department of Homeland Security. https://www.fema.gov/grants/mitigation/learn/hazard-mitigation. Last accessed 15 September 2025.

FEMA. (2023b). Building resilient infrastructure and communities (BRIC). U.S. Department of Homeland Security. https://www.fema.gov/grants/mitigation/learn/building-resilient-infrastructure-communities. Last accessed 15 September 2025.

Feng, K., Lin, N., Gori, A., Xi, D., Ouyang, M., & Oppenheimer, M. (2025). Hurricane Ida's blackout-heatwave compound risk in a changing climate. *Nature Communications, 16*(1), 4533. https://doi.org/10.1038/s41467-025-59737-8

First Street Foundation. (2025). https://firststreet.org/methodology/flood. Last accessed 15 September 2025.

Flood Protection Authority East. (2025). Lake Borgne Surge Barrier: Great Wall of Louisiana. https://www.floodauthority.org/the-system/lake-borgne-surge-barrier/. Last accessed 17 September 2025.

Florida Realtors. (2024, August 29). New flood disclosure requirements coming. Florida Realtors.. https://www.floridarealtors.org/news-media/news-articles/2024/08/new-florida-flood-disclosure-requirements-coming. Last accessed: 15 September 2025.

Florida House of Representatives. (2024, February 21). CS/CS/HB 1049 flood disclosure in the sale of real property (staff analysis). *Florida House of Representatives.*. https://www.flsenate.gov/Session/Bill/2024/1049/Analyses/h1049b.JDC.PDF. Last accessed: 15 September 2025

Greer, A., Huntsman, D., Wu, H. C., Murphy, H., & Clay, L. (2023). Household hurricane evacuation during a dual-threat event: Hurricane Laura and COVID-19. *Transportation Research Part D: Transport and Environment, 120*, 103820. https://doi.org/10.1016/j.trd.2023.103820

Harris County Flood Control District. (2018). 2018 bond program. Harris County, Texas. https://www.hcfcd.org/2018-bond-program. Last accessed 15 September 2025.

Harris County Flood Control District. (2022, May 26). 30+-year flood damage reduction effort "project brays" now complete [press release]. *Harris County Flood Control District.*. https://www.hcfcd.org/Community/Press-Room?post=30%2B+Year+Flood+Damage+Reduction+Effort+%E2%80%9CProject+Brays%E2%80%9D+Now+Complete+. Last accessed 17 September 2025

Holton, R., Barbato, M., & Kumar, N. (2018). Resilient & affordable housing for the US Gulf Coast: Earthen building materials re-appropriated for use in hot wet climates. *Louisiana State University College of Art & Design/College of Engineering.*. Available at https://www.arcc-journal.org/index.php/repository/article/view/559. Last accessed 9 August 2025

Hooimeijer, F., Yoshida, Y., Bortolotti, A., & Iuorio, L. (2022). Integrated urban flood design in the United States and The Netherlands. In *Coastal flood risk reduction* (pp. 255–266). Elsevier. https://doi.org/10.1016/B978-0-323-85251-7.00018-4

Insurance Institute for Business and Home Safety (IBHS). (2025). FORTIFIED Home™: Frequently asked questions. https://fortifiedhome.org/frequently-asked-questions/. Last accessed 15 September 2025

IBHS. (2021). Rating the states: 2021 hurricane coast building code performance. https://ibhs.org/wp-content/uploads/RatingtheStates_2021.pdf. Last accessed 15 September 2025.

ICFmag. (2009, August). Wind resistance and ICFs. Discusses how ICF walls are highly resistant to wind damage. https://www.icfmag.com/2009/08/wind-resistance-and-icfs/. Last accessed 15 September 2025.

Insulated Cable Engineers Association (ICEA). (2017). ICEA T-34-664: Test method for conducting longitudinal water penetration resistance tests on longitudinal water-blocked cables, ICEA. Retrieved from https://www.icea.net/docs. Last accessed 15 September 2025.

Jacksonville Electric Authority (JEA). (2020). Appendix A—Technical specifications for medium-voltage power cables (5–46 kV). Formal Procurement Opportunities, 032-20 ITN. https://www.jea.com/About/Procurement/Formal_Procurement_Opportunities/032-20_ITN_-_Appendix_A_-_Technical_Specifications.aspx. Last accessed 15 September 2025

Jafarzadegan, K., & Moradkhani, H. (2023). Recent advances and new frontiers in riverine and coastal flood modeling. *Reviews of Geophysics, 61*(1), e2022RG000788. https://doi.org/10.1029/2022RG000788

Jiang, H., Zhang, J., Liu, Y., Li, J., & Fang, Z. N. (2023). Does flooding get worse with subsiding land? Investigating the impacts of land subsidence on flood inundation from hurricane Harvey. *Science of the Total Environment, 885*, 163964. https://doi.org/10.1016/j.scitotenv.2023.163964

Jonkman, S. N., & van Berchum, E. C. (2022). Flood risk reduction for Galveston Bay: Preliminary design of a coastal barrier system. In *Coastal flood risk reduction* (pp. 313–336). Elsevier. https://doi.org/10.1016/B978-0-323-85251-7.00019-6

Juan, A., Gori, A., & Sebastian, A. (2020). Comparing floodplain evolution in channelized and unchannelized urban watersheds in Houston, Texas. *Journal of Flood Risk Management, 13*(4), e12604. https://doi.org/10.1111/jfr3.12604

Karaye, I., Stone, K. W., Casillas, G. A., & Newman, G. (2019). A spatial analysis of possible environmental exposures in recreational areas impacted by hurricane Harvey flooding, Harris

County, Texas. *Environmental Management, 64*(4), 456–468. https://doi.org/10.1007/s00267-019-01204-4

Koslan, K. L. (2023). Comparative analysis of Houston, Texas and new York City, New York's incorporation of climate adaptive strategies in long term disaster recovery plans after large-scale hurricanes (Master's professional report). The University of Texas at Austin. https://repositories.lib.utexas.edu/server/api/core/bitstreams/ffea4bbe-d883-4221-b70d-ec770dc0594e/content.

Konz, R. C., & Vega, R. E. (2009). Cladding performance of high-rise buildings in the Houston central Business District during hurricane Ike. In *Proceedings of the forensic engineering 2009 conference: Pathology of the built environment, American Society of Civil Engineers* (pp. 739–748) https://iawe.org/Proceedings/11ACWE/11ACWE-Vega4.pdf

Koukouvinos, K. G., Koukouvinos, G. K., Chalkiadakis, P., Kaminaris, S. D., Orfanos, V. A., & Rimpas, D. (2025). Evaluating the performance of smart meters: Insights into energy management, dynamic pricing and consumer behavior. *Applied Sciences, 15*(2), 960. https://doi.org/10.3390/app15020960

Kulp, S., & Strauss, B. H. (2017). Rapid escalation of coastal flood exposure in US municipalities from sea level rise. *Climatic Change, 142*(3–4), 477–489. https://doi.org/10.1007/s10584-017-1963-7

Kulp, S. A., & Strauss, B. H. (2019). New elevation data triple estimates of global vulnerability to sea-level rise and coastal flooding. *Nature Communications, 10*(4844), 1–12. https://doi.org/10.1038/s41467-019-12808-z

Mansur, A. V., Langhorn, G., & Nelson, D. R. (2024). Exploring social contracts of disaster risk through twitter narratives during a major storm. *Nature-Based Solutions, 3*, 100090. https://doi.org/10.1016/j.nbsj.2024.100197

Mashrur, N., & Kameshwar, S. (2024). Effectiveness of regional risk mitigation policies in equitably improving connectivity to essential services during hurricane-induced floods. *International Journal of Disaster Risk Reduction, 102*, 104078. https://doi.org/10.1016/j.ijdrr.2024.104078

Meecham, D. (1992). The improved performance of hip roofs in extreme winds—A case study. *Journal of Wind Engineering and Industrial Aerodynamics, 43*(1–3), 1717–1726. https://doi.org/10.1016/0167-6105(92)90583-V

Merrell, W. J., Reynolds, L. G., Cardenas, A., & Gunn, J. R. (2011). The Ike Dike: A coastal barrier protecting the Houston/Galveston region from hurricane storm surge. In *Natural hazards* (pp. 189–198). Springer. https://doi.org/10.1007/978-3-642-14779-1_31

Merrell, W., Waterworth, L., Brody, S., & Coffman, C. (2017). The status of the Ike Dike/coastal spine [Technical paper]. Texas A&M University,. https://hurricane.egr.uh.edu/sites/hurricane.egr.uh.edu/files/files/2017/9-THE-STATUS-OF-THE-IKE-DIKE.pdf. Last accessed 15 September 2025.

Miller, M. M., & Shirzaei, M. (2021). Assessment of future flood hazards for southeastern Texas: Synthesizing subsidence, sea-level rise, and storm surge scenarios. *Geophysical Research Letters, 48*(8), e2021GL092544. https://doi.org/10.1029/2021GL092544

Mississippi Joint Legislative Committee on Performance Evaluation and Expenditure Review. (2014). The status of the port of Gulfport restoration program. Report #575. https://www.jea.com/About/Procurement/Formal_Procurement_Opportunities/032-20_ITN_-_Appendix_A_-_Technical_Specifications.aspx. Last accessed 15 September 2025

Mobley, W., Sebastian, A., & Highfield, W. (2019). Estimating flood extent during hurricane Harvey using maximum entropy to build a hazard distribution model. *Journal of Flood Risk Management, 12*(4), e12549. https://doi.org/10.1111/jfr3.12549

Morris, R. L., La Peyre, M. K., Webb, B. M., Marshall, D. A., Bilkovic, D. M., Cebrian, J., McClenachan, G., Kibler, K. M., Walters, L. J., Bushek, D., Sparks, E. L., Temple, N. A., Moody, J., Angstadt, K., Goff, J., Boswell, M., Sacks, P., & Swearer, S. E. (2021). Large-scale variation in wave attenuation of oyster reef living shorelines and the influence of inundation duration. *Ecological Applications, 31*(6). https://doi.org/10.1002/eap.2382

Musinguzi, A., & Akbar, M. K. (2021). Effect of varying wind intensity, forward speed, and surface pressure on storm surges of hurricane Rita. *Journal of Marine Science and Engineering, 9*(2), 128. https://doi.org/10.3390/jmse9020128

Navarro-Hernández, M. I., Valdes-Abellan, J., Tomás, R., Tessitore, S., & Ezquerro, P. (2023). Analysing the impact of land subsidence on flooding risk: Evaluation through InSAR and hydraulic modeling. *Water Resources Management, 37*(23), 9123–9143. https://doi.org/10.1007/s11269-023-03561-6

Neumann, J. E., Chinowsky, P., Helman, J., Black, M., & Fant, C. (2021). Climate effects on US infrastructure: The economics of adaptation for rail, roads, and coastal development. *Climatic Change, 167*, 5. https://doi.org/10.1007/s10584-021-03179-w

Newman, G., & Li, D. (2022). Green infrastructure-based design in Texas coastal communities. In *Coastal flood risk reduction* (pp. 267–288). Elsevier. https://doi.org/10.1016/B978-0-323-85251-7.00017-2

NOAA. (2025a). National Centers for environmental information (NCEI). U.S. Billion-Dollar Weather and Climate Disasters. https://www.ncei.noaa.gov/access/billions/. Last accessed 26 November 2025.

NOAA. (2025b). Storm Surge Overview. National Weather Service. https://www.weather.gov/phi/stormsurge. Last accessed 15 September 2025.

North Dallas Gazette. (2023, September 6). FEMA announces nearly $3 billion in funding selections to drive resilience to climate change, extreme weather. North Dallas Gazette. https://northdallasgazette.com/2023/09/06/fema-announces-nearly-3-billion-in-funding-selections-to-drive-resilience-to-climate-change-extreme-weather/. Last accessed 15 September 2025.

Padgett, J. E., Panakkal, P., & González-Dueñas, C. (2022). Infrastructure impacts and vulnerability to coastal flood events. In *Coastal flood risk management* (pp. 321–342). Elsevier. https://doi.org/10.1016/B978-0-323-85251-7.00012-3

Peters, X. (2024). Galveston's Texas-size plan to stop the next big storm. Smithsonian Magazine.. https://www.smithsonianmag.com/science-nature/galveston-texas-plan-stop-next-big-storm-hurricane-ike-180984487/. Last accessed 6 August 2025.

Powell, E. J., Tyrrell, M. C., Milliken, A., Tirpak, J. M., & Staudinger, M. D. (2019). A review of coastal management approaches to support the integration of ecological and human planning for climate adaptation. *Journal of Coastal Conservation, 23*, 1–13. https://doi.org/10.1007/s11852-018-0632-y

Reynier, W. (2025). Inner Harbor Navigation Canal Lake Borgne Surge Barrier: Resilient Storm Surge Protection for New Orleans. https://www.cakex.org/case-studies/inner-harbor-navigation-canal-lake-borgne-surge-barrier-resilient-storm-surge-protection-new-orleans. Last accessed: 17 September 2025.

Salarieh, B., Ugwu, I. A., & Salman, A. M. (2023). Impact of changes in sea surface temperature due to climate change on hurricane wind and storm surge hazards across US Atlantic and gulf coast regions. *SN Applied Sciences, 5*, 62. https://doi.org/10.1007/s42452-023-05423-7

Savonis, M., Burkett, V. R., & Potter, J. R. (2008). Impacts of climate change and variability on transportation systems and infrastructure: Gulf Coast study, phase I. https://rosap.ntl.bts.gov/view/dot/17351. Last accessed 15 September 2025.

Scyphers, S. B., Powers, S. P., Heck, K. L., & Byron, D. (2011). Oyster reefs as natural breakwaters mitigate shoreline loss and facilitate fisheries. *PLoS One, 6*(8), e22396. https://doi.org/10.1371/journal.pone.0022396

Shen, J., Murray-Tuite, P., & Wernstedt, K. (2024). Estimating pre-impact and post-impact evacuation behaviors: An empirical study of hurricane Ida in coastal Louisiana and Mississippi. *Journal of Transport Geography, 117*, 103630. https://doi.org/10.1016/j.jtrangeo.2024.103630

Stantec. (2025). Blue and green corridors project. Stantec.. https://www.stantec.com/en/projects/united-states-projects/b/blue-green-corridor. Last accessed 15 September 2025.

State of Louisiana. (2019). LA-SAFE. https://lasafe.la.gov/. Last accessed: 9 August 2025.

Subasinghe, C. (2021). Sustainability-resiliency status quo. In *Rebuilding for resilience: A Barrier Island case* (pp. 45–62). Springer. https://doi.org/10.1007/978-3-030-65532-7_3

Sutton-Grier, A. E., Wowk, K., & Bamford, H. (2015). Future of our coasts: The potential for natural and hybrid infrastructure to enhance the resilience of our coastal communities, economies, and ecosystems. *Environmental Science & Policy, 51*, 137–148. https://doi.org/10.1016/j.envsci.2015.04.006

Tarabochia-Gast, A. T., Michanowicz, D. R., & Bernstein, A. S. (2022). Flood risk to hospitals on the United States Atlantic and gulf coasts from hurricanes and sea level rise. *GeoHealth, 6*(1), e2022GH000651. https://doi.org/10.1029/2022GH000651

Thomas, V. (2024). Climate mitigation and adaptation in policymaking. In *Risk and resilience in the era of climate change* (pp. 121–144). Springer. https://doi.org/10.1007/978-981-97-2769-8_8

Törnqvist, T. E., Wallace, D. J., Storms, J. E. A., Wallinga, J., van Dam, R. L., Blaauw, M., Derksen, M. S., Klerks, C. J. W., Meijneken, C., & Snijders, E. M. A. (2008). Mississippi Delta subsidence primarily caused by compaction of Holocene strata. *Nature Geoscience, 1*(3), 173–176. https://doi.org/10.1038/ngeo129

USACE. (2006). Performance evaluation of the New Orleans and Southeast Louisiana hurricane Protection system: Volume I—Executive summary and overview. *Interagency Performance Evaluation Task Force (IPET)*.. https://biotech.law.lsu.edu/katrina/ipet/gpo/Vol%20I%20ExecutiveSummaryandOverview.pdf?utm_source. Last accessed 16 September 2025

USACE. (2007). Performance evaluation of the New Orleans and Southeast Louisiana hurricane Protection system: Volume V—The Performance: Levees and floodwalls. Interagency Performance Evaluation Task Force (IPET). https://biotech.law.lsu.edu/katrina/ipet/FINAL%20Vol%20V%20The%20Performance%20-%20Levees%20and%20Floodwalls%20-%20maintext%20-%2001Jun07.pdf?utm_source. Last accessed 16 September 2025.

USACE. (2011). Fact sheet: Inner Harbor Navigation Canal–Lake Borgne Surge Barrier (Hurricane and Storm Damage Risk Reduction System). U.S. Army Corps of Engineers, New Orleans District. https://www.mvn.usace.army.mil/Missions/Engineering/Hurricane-Design-Guidelines/. Last accessed: 15 September 2025.

USACE. (2012). Greater New Orleans Hurricane and Storm Damage Risk Reduction System-Peer Review Plan. https://www.mvn.usace.army.mil/Portals/56/docs/PD/PeerReview/GNOHSDRRSReviewPlan.pdf. Last accessed: 15 September 2025.

USACE. (2015). HSDRRS Fact Sheets—includes dimensions and features of IHNC Surge Barrier. https://www.mvn.usace.army.mil/Portals/56/docs/HSDRRS/Hurricane%20%20Storm%20Damage%20Risk%20Reduction%20System%20Fact%20Sheets.pdf. Last accessed: 17 September 2025.

USACE. (2025). Hurricane & Storm Damage Risk Reduction System (HSDRRS). *New Orleans District*.. https://www.mvn.usace.army.mil/Missions/HSDRRS/. Last accessed: 15 September 2025

U.S. Climate Resilience Toolkit. (2025). Sea level rise. https://toolkit.climate.gov/sea-level-rise. Last accessed: 15 September 2025.

U.S. Department of Energy. (2008). Building America best practices series: High-performance home solutions for hurricane Katrina reconstruction. In *Office of Energy Efficiency and Renewable Energy*. https://www1.eere.energy.gov/buildings/publications/pdfs/building_america/building_post_katrina.pdf. Last accessed: 15 September 2025

U.S. Department of Energy. (2014). Smart grid investment Grant program: Progress report highlights. *Office of Electricity Delivery & Energy Reliability*.. https://energy.gov/sites/prod/files/2014/12/f19/SGIG-SGDP-Highlights-October2014.pdf

U.S. Department of Energy. (2024). The nexus of building energy codes and resilience. https://www.energycodes.gov/sites/default/files/2024-12/The%20Nexus%20of%20Building%20Energy%20Codes%20and%20Resilience.pdf. Last accessed: 15 September 2025.

U.S. Nuclear Regulatory Commission (NRC). (2009). Suitability of medium voltage cables for wet or submerged applications (report no. ML092810125). Washington, DC: U.S. NRC. Retrieved from https://www.nrc.gov/docs/ml0928/ml092810125.pd. Last accessed 15 September 2025.

Urban Land Institute (ULI). (2025). FORTIFIED homes: Developing resilience. https://developingresilience.uli.org/case/fortified-homes/. Last accessed 15 September 2025.

Wilson, A., & Lazarus, M. A. (2020). Resilient design. In *Sustainable Built Environments* (pp. 389–398). Springer. https://doi.org/10.1007/978-1-0716-0684-1_1031

Waugh, W. L. (2009). *Living with hazards, Dealing with Disasters: An Introduction to Emergency Management.* M.E. Sharpe. https://doi.org/10.4324/9781315702810

Wolshon, B. (2006). Evacuation planning and engineering for hurricane Katrina. *The Bridge, 36*(1), 27–34.

Woodruff, S., Bae, J., Sohn, W., Newman, G., Tran, T., & Lee, J. (2022). Planning, development pressure, and change in green infrastructure quantity and configuration in coastal Texas. *Land Use Policy, 112*, 105816. https://doi.org/10.1016/j.landusepol.2021.105816

Xu, Y. J., Lam, N. S. N., Liu, K., & Xu, K. (2024). Delta sustainability: A report to the mega delta community on Mississippi River Delta. In *Delta Sustainability* (pp. 227–248). Springer. https://doi.org/10.1007/978-981-97-7259-9_11

Yang, Z., Wang, T., Leung, R., Hibbard, K., & Janetos, T. (2014). A modeling study of coastal inundation induced by storm surge, sea-level rise, and subsidence in the Gulf of Mexico. *Natural Hazards, 73*(2), 1055–1074. https://doi.org/10.1007/s11069-013-0974-6

Yao, Q., Liu, K., Fan, D., & Cohen, M. C. L. (2023). Eco-morphological evolution of the bolivar peninsula (Texas, USA) during the last 2,000 years: A multi-proxy record of coastal environmental changes. *Quaternary Science Reviews, 301*, 107933. https://doi.org/10.1016/j.quascirev.2023.107933

Zevenbergen, C., Rijke, J., van Herk, S., & Bloemen, P. J. T. M. (2015). Room for the river: A stepping stone in adaptive delta management. *International Journal of Water Governance, 3*(3), 121–140. https://journals.open.tudelft.nl/ijwg/article/view/5881

Zhang, W., Villarini, G., Vecchi, G. A., & Smith, J. A. (2018). Urbanization exacerbated the rainfall and flooding caused by hurricane Harvey in Houston. *Nature, 563*(7731), 384–388. https://doi.org/10.1038/s41586-018-0676-z

Nature's Barrier: Strategic Wetland Restoration for Tropical Cyclone Flood Risk Mitigation

Rose E. Noggle and Rubayet Bin Mostafiz

Abstract Flooding from storm surge and rain is one of the threats tropical cyclones pose to the built environment. As climate change progresses, the severity of tropical cyclones and of their resulting flooding will increase. Risk management measures which target the structural resilience of building stock, like freeboard elevation ordinance, can be combined with measures that target the severity of flooding to provide a greater margin of safety for future larger storms. Natural ecosystems like marshes and fluvial floodplains provide a natural defense against tropical cyclone-related flooding by storing flood water and attenuating wave energy and height. Those ecosystems are disappearing due to factors like sea level rise, urban development, and harmful land use practices. Already, larger-scale efforts to restore lost wetlands and pre-development river states are underway to correct this issue, like the Coastal Master Plan of southern Louisiana. However, attempts by local governments to follow suit will be complicated by the geographical and temporal dependence of land use intervention utility in flood peak attenuation, the difficulties in imposing development restrictions on privately-owned land, and the demands of settlement growth. The authors provide an analysis of these and other considerations for identifying locations for useful wetland protection and restoration zoning. We propose that a sandbox tool with a cost–benefit analysis for implementation would make natural mitigation measures aiming to attenuate floods more accessible for local planners and floodplain managers.

Keywords Flood risk · Tropical cyclone · Wetland restoration · Cost–benefit analysis · Risk mitigation

---

R. E. Noggle (✉) · R. B. Mostafiz
LaHouse Research and Education Center, Louisiana State University Agricultural Center, Baton Rouge, USA
e-mail: rnoggle@agcenter.lsu.edu

J. Collins et al. (eds.), *Tropical Cyclone Risk and Resilience*,
https://doi.org/10.1007/978-3-032-28745-8_7

1 Resilient Systems and Flood Resilience

Flood and wave damage from tropical cyclones present a major and increasing threat to life and infrastructure. Due to climate change, risk from both is projected to increase over the next fifty years. Resilience is a major topic of concern when it comes to preparation for current and increasing flood hazard. Resilience is the ability of a system to absorb shocks without impeding function. Figure 1 provides a visual representation of this concept. A system's operation is variable, but the system remains functional when it operates within a safety boundary. The safety boundary changes depending on the size of shock, or disruption, that the system can absorb and remain functional. By increasing the area enclosed by the safety boundary, one can increase the margin of safety and the tolerance of the functionality of the system to shocks (Furuta, 2015).

In our case, the system is the built environment and the shock is the wind and flood loads of a tropical cyclone. We can increase the size of the safety boundary and the margin of safety through a plethora of means. Interventions to buildings, like raising freeboard elevation, increase the resistance capacity of building stock to flood loads. Structural and nature-based solutions, like bulkheads and living shorelines (Smith et al., 2017), reduce the flood wave loads themselves. Extensive research has been performed on the effectiveness of these solutions (Boutwell & Westra, 2015; Hoang Tri & Adger, 1998; Justine & Seenath, 2025; Kreibich et al., 2015; Rezaie et al., 2020; Smith et al., 2017; Van Rees et al., 2024).

To discuss strategy in implementing these defenses, we can borrow from the nuclear engineering field. The International Atomic Energy Agency introduced the concept of the five-layered Defense-in-Depth for nuclear energy (International Nuclear Safety Advisory Group, 1996). The first layer is prevention of abnormal operation and failures via conservative and high-quality design and construction. The second layer is control of abnormal operation and detection of failures through various surveillance features. Next is control of accidents within the design basis with engineered safety features. The fourth layer is control of accident progression

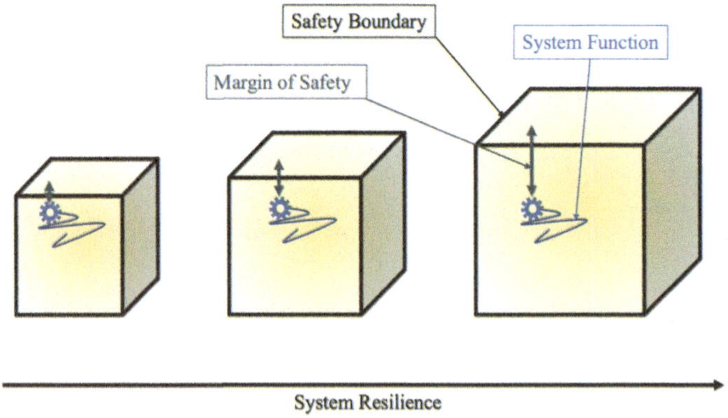

Fig. 1 Visual representation of the systems resilience concept described in Furuta (2015)

and mitigation of accident consequences. The fifth and final layer is mitigation of radiological consequences via off-site emergency response (International Nuclear Safety Advisory Group, 1996).

While these defenses are specific to nuclear events, a similar strategy can be assumed to increase the margin of safety for the operation of the built environment during tropical cyclone events. The Defense-in-Depth layers are intended to be independent and redundant such that no single layer is relied upon alone (International Nuclear Safety Advisory Group, 1996). The combination of independent and redundant defenses would create a larger margin of safety for the build environment as well. A larger margin of safety would better protect the built environment from tropical cyclones of unanticipated magnitude, or "black swan" cyclone events. As climate change has varied and complex effects on the global weather system, it will be difficult to predict the magnitude of future storms with a high degree of accuracy. And so, a wider margin of safety will be an advantage to maintaining an appropriately low level of cyclone and flood damage risk in the future.

Historically, flood risk mitigation has been achieved with hard engineered structures, relocation of critical infrastructure to lower risk areas, and raising buildings above the 100 year flood level (Juan-Diego et al., 2025; "Mississippi Coastal Improvement Program (MsCIP) Hancock, Harrison, and Jackson Counties, MS: Comprehensive Plan and Integrated Programmatic Environmental Impact Statement," 2009; Smith et al., 2017). Hard engineered structures, like levees and breakwaters, reduce the flood peak of fluvial and coastal storm surge flooding to reduce risk. Relocation of infrastructure reduces potential damage expenses by removing high-value buildings and infrastructure to lower risk zones. Raising buildings lowers potential damage by removing buildings from the flood path vertically. These strategies are independent in that the first two reduce the anticipated flood load the building stock and critical infrastructure *must* withstand and the third increases the anticipated flood load that the building stock *can* withstand.

However, hard engineered structures like breakwaters and levees have some major issues associated with them. Hard engineered structures in coastal systems, most commonly bulkheads, breakwaters, and riprap, have been shown to have a deleterious effect on the surrounding ecosystems and shore erosion patterns (Smith et al., 2017). In fluvial systems, levees disconnect rivers from the riparian ecosystems of their floodplains, which reduces the water, sediment, and nutrient exchange processes that keep these ecosystems healthy (Juan-Diego et al., 2025). Levee construction also can incentivize development in the floodplain protected by the levee, as levee presence enhances the perception of safety from flood risk. This is known as the levee effect, and generally increases tropical cyclone-related flood damage risk (Juan-Diego et al., 2025; Kreibich et al., 2015; Serra-Llobet et al., 2022).

Nature-based solutions have risen in recent literature as a potential solution to the problems generated by hard engineered structures. Nature-based solutions are active designs which leverage, manage, and restore natural ecosystems and processes to address societal challenges in an environmentally, socially, and economically conscious way (Cohen-Shacham et al., 2016; Ruangpan et al., 2020). Nature-based solutions are often more cost-effective and socially and aesthetically pleasing than their hard-engineered counterparts (Gedan et al., 2011; Juan-Diego et al., 2025; Smith et al., 2017). Among the nature-based solutions for storm surge and fluvial flood mitigation is wetland restoration. This chapter will discuss the role wetlands play in flood mitigation, their loss and restoration, and considerations for the implementation of wetland restoration as a tropical cyclone-related flood mitigation measure on the local scale. This review primarily draws on literature and case studies from the United States. While the hydrological mechanisms discussed are generally applicable across cyclone-prone regions, implementation pathways for wetland restoration and nature-based solutions are context-dependent and may differ in other regions of the tropical cyclone impact zone. Finally, recommendations for implementation through a cost–benefit analysis framework will be discussed.

2 Wetlands and Tropical Cyclones

Tropical cyclones cause flooding by several mechanisms, including pluvial and fluvial flooding, and storm surge and wave action-related flooding. In pluvial flooding, rainfall from the tropical cyclone overwhelms the drainage capabilities of creeks, streams, and urban drainage systems. This results in ponding over land and, in some cases, flash flooding. Fluvial flooding results from tropical cyclone rainfall runoff being introduced into a river stream such that water levels are raised in a section of the river and the flow capacity of that section of the river channel is exceeded (Acreman & Holden, 2013; Juan-Diego et al., 2025). Storm surge is caused by low atmospheric pressure and high winds, which raise the sea level and cause the coast to flood. Similarly, large waves are generated from cyclone winds. Their setup, an accumulation of water near shore, and runup, travel distance up shore, can raise water levels on land (Gedan et al., 2011; Justine & Seenath, 2025; Rappaport, 2014; Rezaie et al., 2020).

2.1 Coastal Wetlands

Wetlands provide damage mitigation services for United States coastal communities to the tune of 23.2 billion US dollars per year (Costanza et al., 2008). Coastal wetlands mitigate flooding from storm surge and wave action, as well as reduce the fetch and drag from wind on water surfaces that are necessary to create large waves (Costanza et al., 2008; Justine & Seenath, 2025). Wetlands attenuate flood risk through various mechanisms, including vegetation interaction with waves, sediment retention, and nearshore bathymetry change. These mechanisms are described in detail in this section and summarized in Table 1.

Wetlands attenuate wave energy and amplitude primarily via frictional forces between wave water and wetland vegetation. These are illustrated in the equation

Table 1 Summary of coastal wetland characteristics contributing to storm risk mitigation

Factor	Risk mitigation effect	Mechanism
Vegetation stem diameter	Attenuates wave energy and amplitude	Increased drag on storm surge waves, decreased bending under large wave force
Vegetation growth density	Attenuates wave energy and amplitude	Increased drag and turbulence on storm surge waves
Vegetation root density	Retains and stabilizes sediment	Decreased erosion during storm events, decreased coastal land loss
Bathymetry	Attenuates wave amplitude	High bottom elevation gradient reduces wave amplitude over short landward distances

$$F_x = \frac{1}{2}\rho C_d b_v N u|u|$$

(Abdolali et al., 2022; Dalrymple et al., 1984), where F_x is the drag force on the water, ρ is water density, C_d is the drag coefficient, b_v is the stem diameter of the wetland vegetation, N is the plant density in number of stems per meter squared, and u is the horizontal wave water velocity.

As shown in the equation above, the friction force on the water is modified by the stem diameter and the stem density, or growth density, of wetland vegetation. Stems of greater diameter present a larger frontal area in the direction of floodwater flow, creating greater drag resistance. Higher stem density also presents a larger frontal area, but directs wave energy into turbulence, thereby dissipating wave energy (Justine & Seenath, 2025). This modification means that the wave energy and amplitude attenuation services provided by wetlands varies with seasonal changes. Many plants drop leaves or otherwise reduce growth in autumn and winter seasons, which would make them less effective for storm surge mitigation (Justine & Seenath, 2025).

Modification of frictional force by the frontal area presented by vegetation stems also reduces when vegetation stems are bent by wave velocity or when inundation depth exceeds vegetation height (Justine & Seenath, 2025). However, bending and other movements of vegetation add complexity to the flow regime of floodwaters in a wetland. By creating turbulence in the flow, flexible vegetation can increase dissipation of wave energy. Rigid vegetation with irregular geometry from branches, leaves, etc., can also add flow complexity and dissipate wave energy (Arkema et al., 2017). Large storms with very high wave energy and rapid unidirectional flows can break and scour wetland vegetation. So, the storm surge mitigation effects of wetlands are more significant in lower-intensity recurring storms (Arkema et al., 2017; Rezaie et al., 2020). As these lower-intensity recurring storms contribute to a high proportion of flood damages, wetlands can still make a significant impact on reducing those damages (Gourevitch et al., 2022).

Coastal wetlands can also contribute to storm surge mitigation by trapping sediment in stagnation zones behind the vegetation and by creating new soil from decaying vegetation (Gedan et al., 2011). Wetlands, saltmarshes, mangroves, and other coastal habitats trap and build sediment over time, which increases the land area and topographical height of the coast (Arkema et al., 2017; Gedan et al., 2011; Justine & Seenath, 2025). This retention of sediment also prevents large scale erosion over the course of a tropical cyclone event, improving retention of coastal lands (Arkema et al., 2017). Vegetation roots stabilize the soil by enhancing cohesion and tensile strength. However, this strengthening only takes effect over the length of the roots, such that depths below the roots are still subject to slump (Gedan et al., 2011).

Wave attenuation effects of coastal wetlands are nonlinear across space, and depend on bottom elevation gradient, vegetation species and morphology, and the storm duration and track across wetlands (Gedan et al., 2011). Wave attenuation is greater with higher bottom elevation gradient, making attenuation effects stronger over short distances at the seaward edge of wetlands than longer distances over the landward area (Gedan et al., 2011). As mentioned earlier in the section, vegetation drag on wave water increases with higher growth density and changes with stem stiffness (Arkema et al., 2017; Gedan et al., 2011; Justine & Seenath, 2025), each of which vary across vegetation species. Storms with steady winds tracking over wetlands can overcome drag from the vegetation and generate greater wave heights. Attenuation effects are reduced, but still present, at small and large ratios of water depth to vegetation height, and greater at intermediate ratios (Gedan et al., 2011).

2.2 *Inland Wetlands*

The degree to which inland wetlands contribute to fluvial flood attenuation is still a subject in need of further investigation. The prevalent method of study of the effects of wetlands on flooding are hydrological models of wetland loss scenarios (Ameli & Creed, 2019; Fossey et al., 2015; Wu et al., 2020, 2023). Wetlands are generally expected to attenuate flood volume and peak flow due to their water storage capacity and their surface roughness (Acreman & Holden, 2013; Ameli & Creed, 2019; Wu et al., 2020, 2023). Wetlands catch runoff and rainfall and redirect it slowly into the ground water table (Wu et al., 2023). The degree to which inland wetlands attenuate flood peak flow and volume varies depending on the location of the wetland within the catchment system, the vegetation present, and the water levels within the wetland before a tropical cyclone or other rainstorm event (Acreman & Holden, 2013; Ameli & Creed, 2019; Wu et al., 2023).

Wetlands in the headwater or upstream portion of a river basin may enhance flood risk in lower regions under saturation conditions (Acreman & Holden, 2013; Wu et al., 2023). The upstream portion of the basin is at a higher elevation, where rains fall more often. This means that upstream wetlands are fully or mostly saturated more often before a tropical cyclone event, and have far less capacity for water storage during the event (Acreman & Holden, 2013). Wetlands at full capacity generate overland runoff at a high flow rate during rain, which contributes to flooding further down the river basin (Acreman & Holden, 2013; Wu et al., 2023). A hydrological study of wetland loss in the Nenjiang River Basin by Wu et al. (2023) showed that wetland loss in the upstream portion of the basin caused an increase in peak flow and volume in the upstream portion of the basin and a decreased flow in the middle and lower portions of the basin.

These results imply that the presence of upstream wetlands reduces flood risk locally in headwater regions while redistributing and enhancing flood risk downstream through delayed runoff contributions. This leads to Wu et al. concluding that "upstream wetlands have a stronger role in enhancing flooding than in reducing it," (Wu et al., 2023). However, it should be noted that an attenuation effect was observed in the upstream region by upstream wetlands, as well as in the midstream region by midstream wetlands. This implies that wetlands maintain an attenuative effect on flood risk near to their location, though wetland nature-based solution designers should account for downstream effects. The key results of Wu et al. (2023) are summarized in Table 2 for clarity.

Wetlands in downstream portions of the river basin have greater attenuative effects on peak flow and volume. Downstream wetlands have less contributing rainfall, and therefore larger storage capacity during cyclone and other rainstorm events (Acreman & Holden, 2013). Similar to coastal wetlands, the vegetation present in inland wetlands creates higher surface roughness or drag on floodwater, slowing its flow rate and dissipating its energy (Acreman & Holden, 2013). The same hydrological study by Wu et al. (2023) showed wetland loss in the middle river basin region increased peak flow in the middle and downstream regions of the basin but decreased flood volume in those regions. Wetland loss in the downstream region caused an increase in both peak flow and volume in that region (Wu et al., 2023).

So, upstream wetlands tend to attenuate flooding upstream and enhance flooding downstream, and downstream wetlands tend to attenuate flooding in their area. However, the degree to which each wetland effects fluvial flooding depends on its connectivity to the river. Wetlands in close proximity to the river, meaning 100 m away or closer, have a greater attenuation effect on peak flow and volume than those farther away (Ameli & Creed, 2019). These are generally riparian wetlands, hydrologically characterized by periodic flooding from the river and fed by flooding, hydrological connectivity to the river, and the groundwater table (Fossey et al., 2015).

Table 2 Summary of flood attenuation effects of wetlands by basin position (Wu et al., 2023)

Wetland position	Upstream effect	Midstream effect	Downstream effect
Upstream	Decreased peak flow and flood volume	Increased peak flow and flood volume	Increased peak flow and flood volume
Midstream	–	Decreased peak flow and increased flood volume	Decreased peak flow and increased flood volume
Downstream	–	–	Decreased peak flow and flood volume

In downstream regions, riparian wetlands decrease peak flow in fluvial floods in their area and reduce flow volume downstream (Wu et al., 2023).

Isolated wetlands, those disconnected from surface waters and fed by the groundwater table and the precipitation and evaporation processes of the watershed (Fossey et al., 2015), decrease both peak volume and flow in their area by a smaller margin (Ameli & Creed, 2019; Wu et al., 2023). However, loss of isolated wetlands beyond a 50% benchmark still created a disproportionate increase in both flow peak and volume in simulations of the Nose Creek watershed in the Prairie Pothole region of North America (Ameli & Creed, 2019). Both isolated and riparian wetlands can contribute to flood risk attenuation if retained or restored.

3 Wetland Loss and Restoration

3.1 Why Wetlands Are Disappearing

Wetland loss has occurred across the world due to various anthropogenic and environmental factors. In areas impacted by tropical cyclones, the main anthropogenic causes are urbanization, agriculture, and hydrologic intervention to river flow. The main environmental causes in these regions are repeated strong tropical cyclone impacts, subsidence, and sea level rise (Ballut-Dajud et al., 2022; Khalil et al., 2011; McCorvie & Lant, 1993; Ramachandra et al., 2012).

One land use practice that has been detrimental to wetlands in the United States and other countries in Asia, the Caribbean, and Latin America is drainage of wetlands for agricultural use (Acreman & Holden, 2013). In the United States, drainage of wetlands for the purpose of converting the land into agricultural land was still subsidized by the USDA Agricultural Stabilization and Conservation Service until the middle 1950s. At that time and in the century prior, drainage of wetlands was viewed as a method of reclaiming valueless land. It was only when priorities shifted to conservation of wildlife and their ecosystems with the National Environmental Policy Act of 1969 that the prevention of wetland loss became a supported practice (McCorvie & Lant, 1993). Agricultural drainage was also identified as an incident factor in wetland loss in Tanzania, South Africa, Mexico, Belize, Guatemala, Costa Rica, Trinidad, Tobago, China, Thailand, and India (Ballut-Dajud et al., 2022).

Urbanization is another anthropogenic cause of wetland loss across countries and continents (Ballut-Dajud et al., 2022; Ramachandra et al., 2012). Urbanization, especially unplanned urbanization, causes wetland loss by means of encroachment onto wetlands via land filling, drainage to provide water utilities, and introduction of pollutants through effluent waste and other means. In Greater Bangalore, India, a 632% increase in built area from 1973 to 2009 caused a loss of 79% area of water bodies in the region (Ramachandra et al., 2012). In this area and other urbanized regions, wetland and vegetation conversion to impermeable surfaces has resulted in an increase in urban flooding and flood risk (Ramachandra et al., 2012; Serra-Llobet et al., 2022).

Engineered hydrological structures alter the courses of water, nutrient, and sediment flow in river channels and deltas (Ballut-Dajud et al., 2022; Batker et al., 2010). The construction of dams and levees contributed to wetland loss in South Africa, coastal Louisiana and northern California, United States, Korea, and Australia (Ballut-Dajud et al., 2022). Levees direct sediment flow downstream in major rivers rather than out into wetlands along the coast and in riparian floodplains, where sediment would contribute to wetland nourishment and soil elevation levels (Batker et al., 2010; Edwards, 2023; Gedan et al., 2011). In the Mississippi River delta, for example, sediments are directed out into the deeper waters of the Gulf of Mexico (i.e. Gulf of America) rather than building up land area along the coasts of Mississippi and Louisiana (Batker et al., 2010). Channelization in that delta for oil and gas industrial services has converted interior wetlands into open water and contributed to the erosion of barrier islands (Khalil et al., 2011).

Sediment starved wetlands are more vulnerable to erosion caused by large tropical cyclones, as they are less able to recover sediment eroded in those storms (Batker et al., 2010). A series of four major hurricanes inside of a three-year period caused a net loss of 328 square miles, or 850 km^2, of land in coastal Louisiana, United States (Barras, 2009; Khalil et al., 2011). Hurricanes and other tropical storms can tear away vegetation and vegetation roots which stabilize soil and sediment, leaving soil and sediment vulnerable to erosion in the flooding and high winds of these storms (Justine & Seenath, 2025; Khalil et al., 2011). As the frequency of severe tropical cyclones continues to increase, this mode of wetland loss may be a persistent issue.

The combined effects of wetland subsidence and sea level rise also serve to degrade coastal wetlands. Rising sea levels have contributed to vegetation loss in wetlands in estuaries from Delaware Bay and the Indian River Florida and along the Georgia coast, United States, and in the Sian Ka'an Biosphere Reserve of Yucatán Peninsula, Mexico (Ballut-Dajud et al., 2022). Dramatic subsidence in the Cheneir plane and the Mississippi River Delta in combination with sea level rise has increased the rate of erosion of coastal wetlands there (Ballut-Dajud et al., 2022; Khalil et al., 2011). Intrusion of sea water also increases the salinity of wetland waters, which, if not countered with freshwater from riverine sources, causes degradation of wetland vegetation and fauna (Khalil et al., 2011).

3.2 Efforts Towards Restoration

Restoration projects concerning rivers tend to take the form of total floodplain restoration, motivated by conservation goals, flood mitigation goals, or a combination of the two (Serra-Llobet et al., 2022; Widis et al., 2015). Though wetland restoration does play a role in fluvial floodplain restoration, especially where conservation is a critical goal in the project, it is superseded by levee setbacks and flood bypasses in the flood risk mitigation role. Some examples of this include the Bear River Levee Setback

Project and the Yolo Bypass Project of California, United States, and the Elbe River Setback Project and Isar River Restoration Project of Germany (Serra-Llobet et al., 2022). The Bear River, Elbe River, and Isar River projects all used levee setbacks to obtain flood peak reductions by providing a higher volume of flooded area contained by the levees. The Yolo Bypass project provides another kind of flood volume storage for the Sacramento River by rerouting flood waters into its area, creating an artificial and optimally placed floodplain to store flood waters (Serra-Llobet et al., 2022).

Due to the distortion of risk perception that traditional grey infrastructure for flood mitigation generates, often referred to as the "levee effect," there has been a push in recent literature for natural flood management approaches that allow flooding to occur in certain regions of the flood plain to mitigate peak flow and volume downstream (Juan-Diego et al., 2025). These approaches use nature-based solutions to alter flood flow. Wetland restoration typically falls into the runoff attenuative features category of nature-based solutions for fluvial flood risk mitigation. Along with forest and sediment restoration, wetland restoration aims to store and redirect surface flow by increasing the permeability of ground surfaces within the catchment system (Juan-Diego et al., 2025).

One restoration project in progress that is using wetland restoration as a flood mitigation measure is the Elizabeth River Floodplain and Wetland Restoration project of the City of Virginia Beach Public Works. This project is still underway as of 2025 and was funded by the Stormwater Green Infrastructure Master Project of the City's Flood Protection Program and the Virginia Department of Conservation & Recreation Community Flood Preparedness Fund grant. The project's aims are to reverse wetland and shoreline erosion and improve floodplain connectivity in the Eastern Branch of the Elizabeth River to improve the flood resilience of the City of Virginia Beach. Restoration of upland marshes in this project will be achieved by installing coir logs at the edges of existing marshes paired with hardwood logs below the high tide line, and marsh vegetation planting in between the log pairs (Searles, n.d.).

Coastal restoration projects lean heavily on wetland restoration as a strategy to attenuate storm surge flooding, reverse coastal land loss, and provide other recreational and ecosystem services to coastal areas. There are several such projects in the United States. For example, the Coastal Master Plan of the state of Louisiana and the United States Army Corps of Engineers Coastal Improvement Program (CIP), which covers the coasts of Mississippi, Alabama, and the Florida Panhandle. The coasts of these states were damaged and heavily eroded by successive hurricanes between 2005 and 2008, and by the Deepwater Horizon Oil Spill in the Gulf of Mexico in 2010, as well as erosion from a combination of delta subsidence and sea level rise (Khalil et al., 2011; Widis et al., 2015).

The aim of coastal restoration in Louisiana is to reverse land and ecosystem loss to sediment deprivation effects along the Mississippi River Delta and Louisiana coast. Sediment deprivation in this region is a result of the management decisions of the previous century, successive strong hurricanes including Hurricane Katrina, and the Deepwater Horizon Oil Spill event (Khalil et al., 2011). Restoration is achieved through four main activities: barrier island restoration, marsh creation and nourishment, shoreline protection, and river diversion.

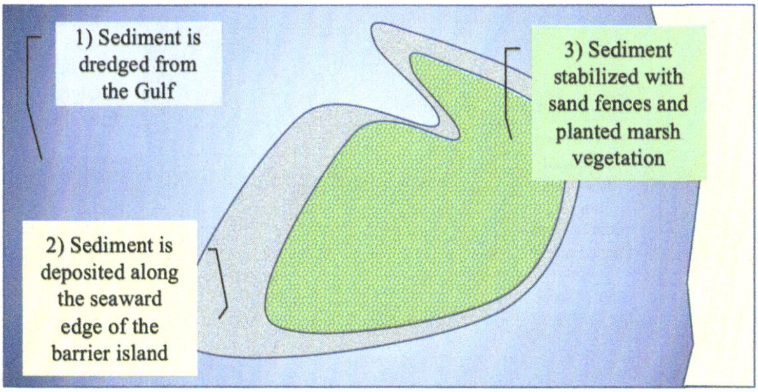

Fig. 2 Barrier Island restoration illustrated in the 2023 Comprehensive Master Plan (Edwards, 2023)

Barrier islands are the first line of defense against tropical cyclones. They also prevent the higher salinity water of the Gulf from entering and damaging lower salinity wetlands along the coast. Barrier islands are subject to landward retreat as storm surge from tropical cyclones and sea level rise erode their seaward soil, vegetation, and sediment. Landward retreat is prevented by placing dredged sediments around beaches, dunes, and marshes along the seaward edge of the islands, and stabilizing them with sand fences and planted vegetation. This process is illustrated in Fig. 2. One example of a barrier island restoration project within the Coastal Master Plan of Louisiana is the Barrier Island Comprehensive Monitoring (BCIM) program (Khalil et al., 2011).

Marshes provide the flood attenuation services discussed in the Wetlands and Tropical Cyclones section of this chapter. They act as "horizontal levees" which prevent wave formation by reducing fetch and attenuate wave energy by increasing drag on and turbulence in wave water. In Louisiana, coastal wetlands also provide recreational and agricultural services to residents and support the state's seafood industry (Khalil et al., 2011). Marsh restoration is achieved by placing suitable sediment in the lost area and planting vegetation over it. The sediment provides the increase in soil surface elevation necessary to support vegetation (Hardy & Wu, 2021), and the vegetation both stabilizes the sediment and increases surface roughness to attenuate wave energy. Marshes are then maintained by the enhancement of connectivity to adjacent water bodies, which provide water and nutrient exchange. These processes are illustrated in Fig. 3. Some example projects within the Coastal Master Plan include the Bayou LaBranche Wetland Creation Project and the Bayou Perot and Bayou Rigoulet sediment dredging and marsh creation projects (Khalil et al., 2011).

Shorelines, as exist around the shores of the Gulf, bayous, lakes, and bays, are lost to erosion from waves breaking on them and dislodging plant roots and soil. Shoreline loss is the primary mode of land loss in the coastal Louisiana region. Shoreline protection is achieved with the installation of rock dikes and terraces.

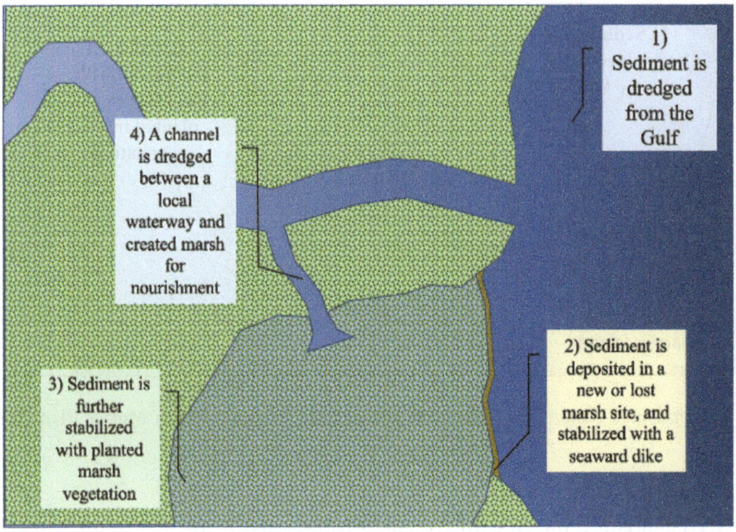

Fig. 3 Marsh creation and nourishment processes illustrated in the 2023 Comprehensive Master Plan (Edwards, 2023)

Rock dikes are made by piling large rocks along a depth contour parallel to the shoreline. Terraces are built parallel to one another and to the shoreline to reduce the fetch and amplitude of wind-generated waves. They are made from vegetated sediments. Wetlands can be created between dikes and the shore by natural retention of, and sometimes addition of, sediment in the calmer waters of that area (Khalil et al., 2011).

River diversions, connections between the river and the lands and wetlands of the delta, help maintain delta wetlands with the introduction of fresh water, sediments, and nutrients (Khalil et al., 2011). As illustrated in step 4 of Fig. 3, sediment and water in river channel are redirected away from channel flow and into a built conveyance channel. This redirected flow deposits sediment into the wetland at the opposite mouth of the conveyance channel (Edwards, 2023). Sediments from the river provide nutrients and material for rebuilding to the wetland, and the addition of fresh water maintains the moderate salinity level of the wetland waters. This salinity level is essential to maintaining the health of the vegetation and fauna in the wetland (Khalil et al., 2011).

Wetland restoration efforts in the Louisiana Coastal Master Plan serve the dual function of ecosystem restoration and flood risk mitigation. In contrast, wetland restoration efforts in the 2009 Mississippi Coastal Improvement Program (MsCIP) plan mainly serve the purpose of ecosystem restoration on restored barrier islands. In the MsCIP plan, flood risk mitigation is achieved through interventions like restoration of barrier islands, elevation of at-risk buildings and the construction of levees, seawalls, embankments, gates, berms, and breakwaters, and through the removal of critical infrastructure from high-risk zones by means of zoning and land acquisition ("Mississippi Coastal Improvement Program (MsCIP) Hancock, Harrison, and Jackson Counties, MS: Comprehensive Plan and Integrated Programmatic Environmental Impact Statement," 2009).

However, the site selection tool for locating wetland restoration sites as part of the MsCIP project does include variables related to likelihood of flood risk mitigation utility. These include *topographic depressions*, *storm surge category*, and *stream buffer* (Lin & Kleiss, 2007; Widis et al., 2015). *Stream buffer* refers to the distance between the potential restoration site and nearby streams, and relates to the utility of wetlands in storing runoff and exceeded flow capacity from such streams in a pluvial flooding event (Lin & Kleiss, 2007). *Topographic depressions* within a wetland serve to increase their storage capacity for runoff in pluvial and fluvial flooding events. *Storm surge category* addresses storm surge flooding by incorporating the location of a potential restoration site within the storm surge area of different predicted tropical cyclones (Lin & Kleiss, 2007). The 2009 MsCIP was expanded to projects in coastal Alabama and the Florida Panhandle under the Coastal Infrastructure Program (CIP) by the USACE, and now includes marsh restoration for the purposes of flood risk reduction in addition to ecosystem restoration (United States Army Corps of Engineers, Mobile District, 2018).

Wetland restoration has also been applied in several countries of the East Asia and Western Pacific region with greater emphasis on nature-based solution co-benefits. For example, the Hong Kong Drainage Services Department (DSD), in the Yuen Long Bypass Floodway Project, converted a 70 ha section of land adjacent to the bypass to address ecosystem degradation caused by the bypass. The constructed wetland provides habitat area for birds, butterflies, dragonflies, and assorted amphibians and reptiles (Chan et al., 2018). In this case, the wetland does not provide the primary flood control function but alleviates the ecological consequences of the grey infrastructure flood risk measure to achieve both flood risk mitigation and environmental goals.

As another example, in Osaki City, Japan, the Ministry of the Environment of Japan and several other entities collaborated to convert fallow fields adjacent to the Kabukuri-numa freshwater marsh into wetland to fortify the flood control function of the marsh. The expanded marsh and rice paddy system provides flood risk mitigation, food production, biodiversity and habitat conservation, and recreation services to Osaki City (The Ministry of the Environment of Japan, 2016). In the context of Miyagi Prefecture, the system also reduces public health risks by providing roosting habitat for migratory geese that would otherwise over-concentrate in a small number of habitat sites. Together, the Kabukuri-numa marsh and surrounding rice paddies illustrate a locally tailored nature-based solution approach that integrates flood risk reduction with ecological and societal co-benefits under conditions of limited land availability.

4 Considerations for Local Restoration Efforts

4.1 Hydrological Considerations

Unlike intervention strategies which can be effectively applied and scaled with consideration mainly to cost, like freeboard elevation ordinance or levee construction, blanket land use and restoration policies cannot be used to reduce fluvial flooding. Wetland restoration and other nature-based solutions have variable effectiveness depending on where they are placed geographically within a river basin (Acreman & Holden, 2013; Pattison & Lane, 2012; Reaney, 2022; Wu et al., 2023). This is mainly due to the complexities of the hydrological problem of fluvial flooding caused by rainfall from tropical cyclones and other rainstorms.

Fluvial flooding is dependent on several factors within a catchment system that make placement of nature-based solutions like wetland restoration a complex problem. The main factors are local floodwater generation, hydrological connectivity, and travel times for floodwaters within a catchment, and the rainfall patterns inherent to the tropical storm (Pattison & Lane, 2012; Reaney, 2022). Local floodwater generation occurs where rainfall is converted into fast flows, and depends on the soil type and land cover of the land that the runoff flows over (Bornschein & Pohl, 2018; Reaney, 2022). For example, soils found in woodlands generate slower flows than those in areas where compaction from livestock activity or heavy machinery occurs, as they absorb more rainfall (Reaney, 2022). In hydrological modeling, this variable is frequently represented by surface roughness and has a significant impact on the accuracy of the model (Bornschein & Pohl, 2018).

In our context, hydrological connectivity of the catchment is the ease with which rainwater runoff can reach rivers and lakes, and depends on the presence of disconnects and trackways and their spatial locations within the catchment (Reaney, 2022). Travel times relates to the timing of when flows reach different locations within the catchment. When tributaries contribute to a flood at several different times, the flood peak can be attenuated, whereas when they contribute at similar times, they build up flood peak (Pattison & Lane, 2012). Positioning a wetland in a location where its flow contributes to the receding limb of the flood will reduce the peak of the flood, but positioning a wetland where its flow contributes to the rising limb will increase the peak (Reaney, 2022).

Timing also is impacted by the rainfall pattern, i.e. the path and rainfall rate, of the tropical cyclone or other rainstorm. The rainfall patterns, along with land management in the catchment and pre-storm hydrological conditions, dictate whether the storm will cause flooding or not under unaltered conditions. The importance of each of these factors on the occurrence of a flood also changes with the storm events (Reaney, 2022). This is typically where modelling efforts for this problem incorporate uncertainty, as a range of events must be considered for thorough flood risk management (Reaney, 2022).

As discussed in the Wetlands and Tropical Cyclones section, both riparian and isolated wetlands positioned at the headwater or upstream can exacerbate flood peak flow and volume. It is tempting to remove or allow the loss of those wetlands through drainage and other means, but this can have the opposite of the intended effect on flooding. Drainage of wetland water creates preferential flow pathways, or pipes, through wetland soil, which increases the drainage flow velocity and drainage runoff volume (Acreman & Holden, 2013). Drainage velocity and volume increases fluvial flood peak and flow in turn. Upstream wetlands also play a major role in contributing to baseflow and preventing droughts in river basins (Wu et al., 2023), so drainage in these wetlands may increase drought risk in addition to harming other features like biodiversity (Acreman & Holden, 2013; Wu et al., 2023).

The major factors contributing to the attenuative effects of middle- and downstream wetlands, both riparian and isolated, on fluvial flooding are their antecedent water storage conditions and their vegetation characteristics. The volume of water stored in a wetland before a tropical cyclone or other rainstorm event dictates the remaining storage capacity available to detain runoff during the event (Acreman & Holden, 2013). One potential strategy to ensuring middle- and downstream wetlands have high storage capacity for mitigation of flooding in a cyclone event is to drain them during the storm season (Acreman & Holden, 2013). Drainage has a second effect relevant to the flood attenuation capacity of wetlands. This effect is the creation of drainage channels which alter the flow of surface water such that the groundwater table is lowered or subsurface flow within the wetland increases over surface flow. Depending on the synchronicity of wetland and drainage channel flows reaching the river channel, draining downstream wetlands can either increase or decrease peak flow (Acreman & Holden, 2013).

An alternate strategy may be to encourage vegetation growth in upstream and downstream wetlands. As mentioned in the Wetlands and Tropical Cyclones section, vegetation slows runoff drainage velocity by enhancing drag forces on the water, usually modeled as higher surface roughness in hydrological simulation (Acreman & Holden, 2013). The presence of vegetation in both upstream and downstream wetlands causes a decrease in flood magnitude (Acreman & Holden, 2013). Hydrological factors influencing wetlands' effect on flooding are summarized in Table 3. Hydrological modeling should be used to compare the impacts of these strategies on flood magnitude within the river basin of interest to best incorporate all relevant factors in decision making.

Unlike inland wetlands, coastal wetlands will not exacerbate flooding, in their case due to storm surge, if they are placed in the wrong location. The degree to which they attenuate storm surge related flooding does depend on several factors, including vegetation species and growth density, area, coastal bathymetry, and storm characteristics (Arkema et al., 2017; Costanza et al., 2008; Gedan et al., 2011; Justine & Seenath, 2025). Multiple studies have demonstrated that the flood-attenuation benefits of coastal wetlands scale measurably with wetland width, vegetation density, and structural complexity. Empirical and modeling studies indicate that storm surge and wave heights are reduced by approximately 3–9 cm per kilometer of marsh, with attenuation rates depending on vegetation type, stem density, bathymetric slope, and storm characteristics (Costanza et al., 2008; Gedan et al., 2011). Vegetation drag plays a dominant role in this process. Increases in stem diameter and stem density increase

Table 3 Summary of hydrological factors for wetlands mitigating fluvial flood risk

Factor	Effect	Design implication
Basin hydrological connectivity	Changes timing and synchronicity of floodwater peaks	Employ hydrological modeling to ensure intended attenuation will be realized
Wetland location in the flood basin	Can attenuate or enhance flood peak and flow velocity	Employ hydrological modeling to ensure intended attenuation will be realized
Water storage capacity	Greater storage capacity enhances floodwater detention	Drain wetlands only during flood seasons or before anticipated storms, and depending on wetland location in the basin
Vegetation	Slows floodwaters and retains sediment	Prefer planted wetlands in all feasible cases

hydraulic resistance and turbulence, leading to 30–50% greater wave energy dissipation under moderate storm conditions when vegetation density is doubled (Abdolali et al., 2022; Dalrymple et al., 1984).

Greater wetland area increases the attenuation of storm surge up to a point due to the associated increase in drag applied and wave fetch reduced (Costanza et al., 2008). However, these benefits are nonlinear. The marginal reduction in surge height and damage decreases beyond wetland widths of roughly 5–10 km, indicating diminishing returns for very large restoration footprints (Boutwell & Westra, 2015). This is in part due to the strong effect of the seaward bathymetry change on wave attenuation (Gedan et al., 2011). The storm characteristics also impact the utility of coastal wetlands in mitigating storm surge flood damage. The storm wind path must track over coastal wetland area for the reduction in fetch and increase in drag to have a pronounced effect (Gedan et al., 2011). This contributes some variability in valuation of the utility of wetlands in storm surge flood damage mitigation.

It has been recognized in the Louisiana restoration projects of the Coastal Master Plan that local project-by project solutions are insufficient to maintain the landmass and storm surge flood risk reduction utility of coastal Louisiana (Khalil et al., 2011). However, project-by-project solutions tend to be the only feasible solutions for governments at the local scale (Burby et al., 2000). So, managers at the local scale must balance the conditions of their own location with the needs of the wider catchment, and sometimes an even wider region when planning local wetland restoration projects. When wider region projects like the Coastal Master Plan in Louisiana are available, they can provide context for local-scale policy decisions (Burby et al., 2000). In turn, local scale projects may be feasible for local scale governments where they are not for the regional project, and so can provide additional support for flood risk mitigation in the region (Edwards, 2023).

Event-based damage analyses further show that intact coastal wetlands reduce storm-surge-related property losses by tens of thousands of dollars per hectare per storm, translating physical attenuation into economically meaningful risk reduction (Rezaie et al., 2020). Empirical and modeling studies from recent major hurricanes provide concrete evidence of the flood-mitigation performance of wetlands under extreme conditions. Post-event analyses of Hurricane Katrina (2005) demonstrated that coastal wetlands in southeast Louisiana attenuated storm surge and wave energy, while areas experiencing substantial wetland loss exhibited higher surge penetration and greater structural damage (Barbier et al., 2013; Costanza et al., 2008). These simulation-based studies estimate that continuous wetland areas reduced surge heights by several centimeters per kilometer, translating into meaningful reductions in property losses during Katrina-scale events (Barbier et al., 2013; Costanza et al., 2008). Similarly, hydrodynamic modeling and damage assessments following Hurricane Sandy (2012) showed that coastal marshes and wetlands along the U.S. Northeast significantly reduced flood depths and delayed surge propagation, avoiding an estimated $600+ million in property damages across affected regions and reducing surge-related losses by 10–40% in some locations, depending on wetland width and continuity (Arkema et al., 2017; Rezaie et al., 2020; Sheng et al., 2021). Together, these case studies confirm that wetland design parameters have quantifiable and decision-relevant impacts on flood mitigation effectiveness.

Wetland restoration is most effective when implemented as part of a hybrid flood mitigation strategy that combines nature-based solutions with traditional engineered defences. Rather than serving as substitutes for levees, seawalls, or floodwalls, wetlands can complement grey infrastructure by reducing hydrodynamic loads, delaying flood wave propagation, and increasing system redundancy (Arkema et al., 2017; Temmerman et al., 2013).

In coastal settings, wetlands placed seaward of levees or seawalls can attenuate wave energy and storm surge heights before they reach engineered structures, thereby reducing overtopping risk, structural stress, and long-term maintenance requirements (Gedan et al., 2011; Temmerman et al., 2013). Similarly, inland and riparian wetlands integrated upstream or adjacent to levee systems can store floodwaters, alter runoff timing, and reduce peak flows, enhancing the effectiveness of engineered conveyance and protection systems (Acreman & Holden, 2013). These complementary functions increase the overall margin of safety by distributing flood mitigation across multiple, partially independent layers of defense.

Hybrid approaches also improve adaptability under future climate change. While gray infrastructure is typically designed for fixed design thresholds, wetlands can adjust dynamically through vegetation growth and sediment accretion, providing adaptive capacity that helps offset uncertainties in future flood magnitude and frequency (Arkema et al., 2017; Temmerman et al., 2013). As a result, integrated gray–green systems are increasingly recognized as more robust, resilient, and cost-effective than reliance on single-strategy flood defenses alone.

4.2 Human Considerations

With considerations of the hydrological impacts of land use on surface roughness and runoff behavior comes considerations of the demands of human settlement growth. Expected population and settlement growth should be incorporated into hydrodynamic modeling iteratively, as flood risk impacts where settlements arise and the location of settlements that contribute to impervious land will impact flood risk. For example, the development of a new subdivision for housing on previously undeveloped land will add impervious surfaces to that tract of land. This increases the runoff and drainage flow during a tropical storm (Huang et al., 2020), and which locations are ideal for wetland restoration. Restoration of wetlands for fluvial flood control impacts where such development can occur. Due to this dependency, accurate estimation of wetland restoration impact on flooding should include population expansion estimation.

Wetland restoration for mitigation of fluvial flooding demands that rivers be allowed to extend flooding into the floodplain. This flood process helps to maintain the water and nutrient levels within riparian wetlands to keep them healthy, and gives a higher volume of floodwater capacity along the adjoining sections of the river channel (Serra-Llobet et al., 2022). However, populations are expanding into river floodplains due to a combination of factors, including benefits associated with proximity to river courses, urban and population expansion, and artificially increased perception of safety via the levee effect (Kreibich et al., 2015). This expansion pressure may be felt more greatly in countries in East Asia and Western Pacific regions that have rapidly expanding populations, high population density cities, and less formal settlement planning under rapid urbanization (Molnar-Tanaka & Surminski, 2024; Ramachandra et al., 2012).

As a result, wetland restoration as a flood risk mitigation measure may be limited in highly urbanized settings where development already exists close to flood sources. In some contexts, governance-supported interventions such as levee setbacks, e.g. in the Isar River Restoration Project in Munich, have demonstrated that partial floodplain reconnection is possible despite urban constraints (Kreibich et al., 2015; Serra-Llobet et al., 2022). For areas with very high biodiversity, like those in Southeast Asia, further difficulty may arise from human-wildlife conflicts after restoration is completed (Molnar-Tanaka & Surminski, 2024).

Acquisition of lands using public funds is one avenue toward limiting development in high risk areas and restoring wetlands in hydrologically and developmentally optimal locations within a watershed. Some considerations for this method are the cost of parcels acquired and the status of development on the desired land (Burby et al., 2000; Widis et al., 2015). Flood risk mitigation and conservation projects run by local governments have a need to be cost-effective, so land value should be factored in to the selection of sites for acquisition and wetland restoration (Widis et al., 2015). While disaster-related damages to buildings and other land features may draw the value down (Widis et al., 2015), waiting for these damages to occur before acquiring the land may be more costly due to incurred litigation costs (Burby et al.,

2000). If one of the goals of land acquisition is to prevent development in a high-risk area, such as would be the case in wetland restoration within a fluvial flood plain, land should ideally be purchased before it is subdivided and sold. Once sold, there is very little local governments in the United States can do to prevent development (Burby et al., 2000). However, the governance style of the country where wetland restoration occurs will have an impact on land ownership rights and the historical provenance of potential restoration sites (Terrisse et al., 2025), so other countries may face different obstacles to land acquisition.

Another consideration for wetland restoration land acquisition cost is land availability. In the river restoration projects described in the Wetland Loss and Restoration section, wide tracts of land were available at lower cost to the governments responsible for restoration activities. In the case of the Yolo Bypass in California, United States, the required land had been set aside before widespread development for the purposes of flood conveyance. In the case of the Elbe River Levee Setback Project in Germany, the land for the setbacks became available briefly at the end of the Cold War. It had been relatively undeveloped before this point because of its use under the socialist government for large scale agricultural purposes (Serra-Llobet et al., 2022). Similar efforts towards floodplain restoration, and wetland restoration within floodplains for flood risk management, may be hindered by the prohibitive cost of land acquisition on that spatial scale (Serra-Llobet et al., 2022). Floodplain land considerations are summarized in Table 4.

Proximity to critical road networks is another variable that must be taken into consideration when allowing expansion of floodplains. Coastal and other rural communities may rely on only a few roads and bridges to facilitate evacuation from tropical cyclones. If these are inundated or collapsed under floodwaters, evacuations can be heavily impeded and evacuation shelters isolated (Frazier et al., 2010). However in more walkable regions and dense urban areas like cities in Japan, a greater emphasis may be placed on vertical evacuation shelters and the dynamics of evacuation progress change significantly (Noggle, 2022). Road networks and shelter locations should therefore be included as a factor in the selection of optimal wetland restoration sites, especially for inland restoration projects.

Planners should also consider the distribution of resulting flood risk and flood risk mitigation to property owners in their community when selecting sites for wetland restoration. The impacts of current and future storms are not evenly distributed

Table 4 Land-related human considerations

Consideration	Parameters
Population expansion	Urbanization and land development, water body proximity benefits to population, risk and safety perception of floodplain areas
Land acquisition	Cost of land, ownership of land, development status of land, quantity of land needed, land availability
Restoration siting	Proximity to road networks, proximity to flood and storm evacuation shelters, socially equitable distribution in resulting flood inundation risk

across space, such that economically vulnerable property owners can be more heavily impacted. Modeling flood risk from floods of various return periods in the Vermont portion of the Lake Champlain Basin revealed that owners of mobile homes and other low value properties were disproportionately exposed to flood inundation, but evaluation based on property damage monetary value showed higher damage to higher value property (Gourevitch et al., 2022). Planners must take care when evaluating flood risk exposure in different restoration scenarios to ensure that economically vulnerable community members benefit equitably from restoration projects.

Timing and time are another issue. Restoration of wetlands and riparian ecosystems can take a long time to deliver flood risk mitigation and other benefits. Riparian forest type wetlands may offer flood resistance by increasing surface roughness and soil infiltration capacity, but it takes several decades before newly planted forests can offer further resistance via logjams and increases in water channel complexity (Dixon et al., 2019). Resilience activities that offer benefits only in future scenarios can be less attractive to stakeholders, especially those who have not had experience with a major storm in recent history (Frazier et al., 2010). Mobilization of the affected community through education and other efforts may be necessary to increase the viability of wetland restoration as a resilience measure (Frazier et al., 2010). Factors which drive stakeholder acceptance of wetland restoration projects in flood risk mitigation contexts are summarized in Table 5.

Nature-based solutions and wetlands in particular tend to have positive stakeholder acceptance due to aesthetic appeal and recreational and economic benefit (Khalil et al., 2011; Serra-Llobet et al., 2022). Though ranked below economic cost and durability, aesthetics were a contributing factor in the decisions of coastal homeowners in North Carolina, United States in favor of installing natural shorelines over bulkheads on their properties (Smith et al., 2017). Also in the case of the Isar River Restoration Project in Munich, Germany, recreational access to the river and its banks was listed as a main goal of the project and increased its popularity among the Munich population (Serra-Llobet et al., 2022). Coastal wetland restoration in Louisiana, United States is viewed as a support to the seafood industry that relies on coastal wetland ecosystems in that state (Khalil et al., 2011). The additional

Table 5 Stakeholder acceptance factors for wetland nature-based solutions (NBS) and traditional flood mitigation infrastructure

Factor	Wetland NBS	Traditional Infrastructure
Time-to-effect (after construction)	Long time required to develop full capabilities	Immediately at flood risk mitigation capacity
Aesthetic appeal	More aesthetically appealing	Less aesthetically appealing
Recreational benefit	Recreational activities supported	No recreational opportunities
Economic cost	Support for industries relying on habitat, property value increase, low cost SLR adaptation and construction	Higher cost for SLR adaptation and initial construction, but better-understood costs and benefit values are available

recreational, economic, and other services provided by wetlands beyond flood risk mitigation may be promoted by local planners to enhance restoration project appeal.

By far the biggest motivator for stakeholder acceptance of nature-based solutions is cost (Smith et al., 2017). Hard engineered structures for attenuating floods tend to be more costly than nature-based solutions because they are static. This means they cannot be adapted easily and inexpensively to rising flood peaks due to sea level rise and other future meteorological changes (Temmerman et al., 2013). In contrast, nature-based solutions are easily adapted or keep up with sea level rise at lower costs. Mangrove restoration projects are between three and five times cheaper than submerged breakwater projects of the same protection value in Vietnam (Narayan et al., 2016). Salt marsh restoration projects are between one and three times cheaper than their equivalent breakwater projects in the United States and United Kingdom (Narayan et al., 2016).

Specifically in the United States, local governments which implement wetland restoration as a flood risk mitigation measure may reduce their community members' National Flood Insurance Program (NFIP) flood insurance premiums through the Community Rating System (CRS) incentive program. This may be another valuable incentive for stakeholder and political support to consider when evaluating wetland restoration projects (Li & Landry, 2018). However, the Community Rating System is a points-based system that gives community members premium discounts in amounts based on the total number of points the community gains for their risk reduction and other activities. For communities in areas of very high population density, like the waterfront community of New York City, this program is less viable as a way to reduce insurance costs because of the limited opportunity for implementation of point-earning measures (Aerts & Wouter Botzen, 2011).

The NFIP CRS program may also be less valuable as an incentive for stakeholder and political support due to the low rate at which homeowners purchase flood insurance and the exclusion of many at-risk households from NFIP hazard maps. Roughly half to one fifth of homeowners in United States Federal Emergency Management Agency (FEMA)-designated flood hazard zones purchase flood insurance (Li & Landry, 2018). NFIP only dictates minimum standards for the current 100-year floodplain, which is expected to increase in area over the next 50 years and does not reflect the risk of households exposed to flooding outside of the 100-year mean recurrence interval (Aerts & Wouter Botzen, 2011). For those who do not purchase flood insurance, or who are not in FEMA-designated flood hazard areas, insurance premium reductions will not be a salient motivator for supporting wetland restoration or other types of flood risk mitigation projects.

Damage mitigation cost benefits are not limited to flood insurance premium reductions, and damage avoidance estimates make for strong statements in wetland restoration projects' favor. Modeling of storm surge damage in coastal counties under different scenarios of wetland loss under sea level rise in the Jacques Cousteau National Estuarine Research Reserve, New Jersey, United States showed the value of the protection provided by the current wetlands is 13.1 million US dollars in avoided damages (Rezaie et al., 2020). Another valuation of the flood risk benefits of wetlands using data from twenty-four storms in the Gulf of Mexico estimates that wetlands provided 26,410 US dollars in damage mitigation benefits per hectare of wetland per kilometer of coastline per tropical cyclone (Boutwell & Westra, 2015).

4.3 Wetland Management and Long-Term Performance Considerations

Wetland restoration for tropical cyclone flood risk mitigation depends not only on initial placement and design, but also on post-storm recovery, ongoing maintenance, and long-term resilience under cumulative environmental and anthropogenic pressures. Wetlands are dynamic systems whose flood attenuation capacity evolves over time in response to storm disturbance, vegetation condition, sediment availability, and hydrologic connectivity (Acreman & Holden, 2013; Gedan et al., 2011).

Recovery times following major storm events vary widely across wetland types. Coastal marshes may regain vegetation structure within years under favorable sediment and salinity conditions, whereas riparian and forested wetlands often require decades to recover hydraulic roughness and floodplain storage capacity (Acreman & Holden, 2013; Dixon et al., 2016). During these recovery periods, flood mitigation performance may be temporarily reduced, particularly following high-energy events that damage vegetation or erode wetland surfaces (Arkema et al., 2017; Gedan et al., 2011).

Sustained flood mitigation benefits further depend on active management. In coastal systems, sediment supply is essential for maintaining wetland elevation and vegetation productivity relative to sea-level rise (Batker et al., 2010; Temmerman et al., 2013). Vegetation degradation caused by salinity intrusion or prolonged inundation can reduce wave attenuation and soil stabilization, while altered drainage and hydrologic disconnection in fluvial systems can diminish flood storage capacity (Acreman & Holden, 2013; Gedan et al., 2011).

Long-term resilience is increasingly constrained by sea-level rise, erosion, subsidence, and human encroachment. These pressures can limit wetland recovery, disrupt sediment and freshwater pathways, and reduce protective function over time, even where wetland area is preserved (Arkema et al., 2017; Temmerman et al., 2013). Recent synthesis work further highlights that wetland restoration outcomes must be evaluated over multi-decadal timescales and across multiple ecosystem service dimensions, including flood mitigation and carbon dynamics (Misteli et al., 2025).

Overall, wetland restoration should be treated as a long-term, actively managed strategy rather than a one-time intervention. Accounting for recovery timelines, maintenance needs, and long-term resilience is essential to avoid overestimating near-term flood protection benefits and to sustain wetland flood attenuation services under future climate and land-use change.

5 A CBA Framework for Implementation

We recommend assembling the considerations in the Considerations for Local Restoration Efforts section into a sandbox tool with a cost-based output to aid local planners in evaluating wetland restoration as a tropical cyclone-related flood risk mitigation measure.

5.1 *Illustrative Cost–Benefit Framework for Wetland-Based Flood Mitigation*

While a full cost–benefit analysis (CBA) is beyond the scope of this review chapter, a simplified framework can illustrate how wetland restoration can be evaluated alongside traditional flood mitigation strategies. At its core, the net present value (NPV) of a wetland restoration project can be expressed as the discounted difference between expected avoided flood damages and total lifecycle costs:

$$NPV = \sum_{t=1}^{T} \frac{\left(D_{a,t} - C_{m,t}\right)}{(1+r)^t} - C_0$$

where $D_{a,t}$ represents expected annual avoided flood damages attributable to the wetland, $C_{m,t}$ represents annual maintenance and management costs, C_0 is the initial restoration cost, r is the discount rate, and T is the project lifespan.

Avoided damages $D_{a,t}$ can be estimated by comparing modeled flood losses under baseline conditions with losses under scenarios that include wetland restoration. Empirical and modeling studies have demonstrated that coastal wetlands reduce storm surge heights and wave energy, translating into measurable reductions in property damage during tropical cyclones (Arkema et al., 2017; Batker et al., 2010). These reductions are often expressed as avoided damages per hectare of wetland per storm or per unit coastline length, providing inputs that can be scaled to local restoration footprints. In applied settings, the avoided damages term $D_{a,t}$ may be evaluated under multiple climate scenarios, for example by using projections consistent with SSP–RCP pathways as described in IPCC AR6 (Intergovernmental Panel on Climate Change (IPCC), 2023), to capture uncertainty in future storm frequency, intensity, and sea-level rise.

Costs include not only initial construction or sediment placement, but also long-term management expenditures associated with sediment nourishment, vegetation maintenance, and monitoring. Importantly, wetlands differ from hard infrastructure in that their protective capacity can evolve over time, potentially increasing benefits as vegetation matures or declining if sediment supply is insufficient (Gedan et al., 2011; Temmerman et al., 2013). As such, CBAs should explicitly incorporate recovery timelines and performance variability rather than assuming constant annual benefits.

Hybrid gray-green strategies can be evaluated within the same framework by accounting for reduced lifecycle costs of engineered defenses. For example, wetlands positioned seaward of levees or seawalls can lower wave loads and overtopping frequency, extending infrastructure service life and reducing maintenance and retrofit costs (Arkema et al., 2017; Temmerman et al., 2013). Even when wetlands do not replace engineered structures, these avoided costs can materially improve project benefit–cost ratios.

Traditional cost–benefit analyses have drawn criticism for neglecting social equity and distributional impacts (McGee, 2021). While aggregate cost–benefit analysis provides a useful basis for comparing flood mitigation strategies, it may obscure the distribution of costs and benefits across different social groups. Wetland restoration can alter land values, tax burdens, and access to livelihoods, resulting in uneven benefit capture even when total avoided damages are positive. Environmental justice scholarship has shown that disaster mitigation and recovery efforts can disproportionately burden marginalized communities while delivering greater protection to more advantaged populations (Bullard & Wright, 2012; Environmental Protection Agency, 2021). To address these concerns, the simple framework presented here may be incorporated into frameworks like CBA+ which extend the capabilities of CBA to account for social equity (Cross et al., 2025). Indicators like the Atkinson index may also be integrated into the CBA framework (Atkinson, 1970). In the context of this review, such approaches are noted as important extensions for equitable decision-making, though their formal integration lies beyond the scope of the simplified framework presented here.

This simplified framework demonstrates how wetland restoration can be assessed using standard CBA principles while accounting for the dynamic and adaptive nature of nature-based solutions. Such an approach is well suited for implementation within decision-support tools that allow local planners to compare alternative restoration locations, social equity impacts, hybrid strategies, and investment timelines under uncertainty.

5.2 *Benefits of a CBA Tool*

Tools which frame the hazard of tropical cyclones in financial terms to help local planners and other decision makers already exist. For example, FEMA built HAZUS, which simulates storm surge flooding from hurricanes and estimates the damage in monetary terms (Boutwell & Westra, 2015). Other tools, like FloodSafeHome (Friedland et al., 2023) and HazardAware (Mostafiz et al., 2024), support individual homeowner decisions on flood and wind risk mitigation strategies based on cost–benefit information. A similar tool using cost-benefit information to support decisions on a community scale would assist in effective implementation of wetland restoration as a risk mitigation measure.

Tools like these support decision makers by allowing them to trial potential policies and projects virtually and assess their effects on damage totals. The tools take the burden of hydrodynamic modelling and risk assessment off decision makers so that they can elect strategies for damage and risk mitigation more efficiently while staying accurate to their local context. Cost-based framing allows stakeholders from all sectors, not just academic, to easily understand their flood and other cyclone-related risks. It also allows for clear justification of mitigation actions where upfront costs are higher than stakeholders (e.g. property developers) would like to accept.

Tools which provide cost–benefit analysis of different mitigation measures also reduce the burden of proof for hazard mitigation planners in the United States who want to implement mitigation actions which have not been previously funded or implemented by FEMA. This becomes especially relevant for the construction of community hazard mitigation plans, which are a requirement for eligibility for federal disaster aid in the United States under the Disaster Mitigation Act of 2000 (Gall et al., 2024). These plans favor projects which have been successfully implemented and funded in the past over new projects, regardless of risk mitigation utility, because of the burden of proof element required in their justification (Gall et al., 2024). Having the cost–benefit assessment performed as part of a sandbox tool removes that burden, making newer mitigation actions like nature-based solutions more actionable.

A sandbox tool that simulates the impacts of wetland restoration projects with variable restoration sites and frames those impacts in a cost–benefit framework would be ideal for encouraging the implementation of wetland restoration as a cyclone-related flood risk mitigation strategy. Such a tool would simplify the considerations discussed in the Considerations for Local Restoration Efforts section for local planners and offer ease of justification for considered wetland restoration projects. If integrated into a tool which targets structural interventions or interventions to building structural resilience, it would further the use of the Defense-in-Depth strategy for increasing community resilience and increase safety margins for at-risk communities in future storms.

Data Availability Statement No new data were generated or analyzed in this chapter. Data sharing is not applicable.

References

Abdolali, A., Hesser, T. J., Anderson Bryant, M., Roland, A., Khalid, A., Smith, J., Ferreira, C., Mehra, A., & Sikiric, M. D. (2022). Wave attenuation by vegetation: Model implementation and validation study. *Frontiers in Built Environment, 8*, Article 891612. https://doi.org/10.3389/fbuil.2022.891612

Acreman, M., & Holden, J. (2013). How wetlands affect floods. *Wetlands, 33*, 773–786. https://doi.org/10.1007/s13157-013-0473-2

Aerts, J. C. J. H., & Wouter Botzen, W. J. (2011). Flood-resilient waterfront development in New York City: Bridging flood insurance, building codes, and flood zoning. *Annals of the New York Academy of Sciences, 1227*, 1–82. https://doi.org/10.1111/j.1749-6632.2011.06074.x

Ameli, A. A., & Creed, I. F. (2019). Does wetland location matter when managing wetlands for watershed-scale flood and drought resilience? *JAWRA Journal of the American Water Resources Association, 55*, 529–542. https://doi.org/10.1111/1752-1688.12737

Arkema, K. K., Griffin, R., Maldonado, S., Silver, J., Suckale, J., & Guerry, A. D. (2017). Linking social, ecological, and physical science to advance natural and nature-based protection for coastal communities. *Annals of the New York Academy of Sciences, 1399*, 5–26. https://doi.org/10.1111/nyas.13322

Atkinson, A. B. (1970). On the measurement of inequality. *Journal of Economic Theory, 2*, 244–263. https://doi.org/10.1016/0022-0531(70)90039-6

Ballut-Dajud, G. A., Sandoval Herazo, L. C., Fernández-Lambert, G., Marín-Muñiz, J. L., López Méndez, M. C., & Betanzo-Torres, E. A. (2022). Factors affecting wetland loss: A review. *Land, 11*, 434. https://doi.org/10.3390/land11030434

Barbier, E. B., Georgiou, I. Y., Enchelmeyer, B., & Reed, D. J. (2013). The value of wetlands in protecting Southeast Louisiana from hurricane storm surges. *PLoS ONE, 8*, Article e58715. https://doi.org/10.1371/journal.pone.0058715

Barras, J. A. (2009). Land area change and overview of major hurricane impacts in Coastal Louisiana, 2004–08, Scientific Investigations Map. U.S. Department of the Interior, U.S. Geological Survey.

Batker, D. D. L. T., Costanza, R. S., & Day, J. B. (2010). Gaining ground: Wetlands, hurricanes and the economy; The value of restoring the Mississippi River Delta. https://doi.org/10.3411/col.08230102

Bornschein, A., & Pohl, R. (2018). Land use influence on flood routing and retention from the viewpoint of hydromechanics. *Journal of Flood Risk Management, 11*, 6–14. https://doi.org/10.1111/jfr3.12289

Boutwell, J. L., & Westra, J. V. (2015). Evidence of diminishing marginal product of wetlands for damage mitigation. *Natural Resources, 06*, 48–55. https://doi.org/10.4236/nr.2015.61006

Bullard, R. D., & Wright, B. (2012). *The wrong complexion for protection: How the government response to disaster endangers African American communities*. NYU Press.

Burby, R. J., Deyle, R. E., Godschalk, D. R., & Olshansky, R. B. (2000). Creating hazard resilient communities through land-use planning. *Natural Hazards Review, 1*, 99–106. https://doi.org/10.1061/(ASCE)1527-6988(2000)1:2(99)

Chan, F. K. S., Chuah, C. J., Ziegler, A. D., Dąbrowski, M., & Varis, O. (2018). Towards resilient flood risk management for Asian coastal cities: Lessons learned from Hong Kong and Singapore. *Journal of Cleaner Production, 187*, 576–589. https://doi.org/10.1016/j.jclepro.2018.03.217

Cohen-Shacham, E., Walters, G., Janzen, C., & Maginnis, S. (Eds.). (2016). *Nature-based solutions to address global societal challenges*. IUCN International Union for Conservation of Nature. https://doi.org/10.2305/IUCN.CH.2016.13.en

Costanza, R., Pérez-Maqueo, O., Martinez, M. L., Sutton, P., Anderson, S. J., & Mulder, K. (2008). The value of coastal wetlands for hurricane protection. *Ambio: A Journal of the Human Environment, 37*, 241–248. https://doi.org/10.1579/0044-7447(2008)37[241:TVOCWF]2.0.CO;2

Cross, B., Doberstein, B., & Lueck, V. (2025). Improving economic assessment and decision-making for managed retreat through CBA+: A targeted literature review. *Frontiers in Climate, 7*. https://doi.org/10.3389/fclim.2025.1481824

Dalrymple, R. A., Kirby, J. T., & Hwang, P. A. (1984). Wave diffraction due to areas of energy dissipation. *Journal of Waterway, Port, Coastal and Ocean Engineering. American Society of Civil Engineers, 110*, 67–79. https://doi.org/10.1061/(ASCE)0733-950X(1984)110:1(67)

Dixon, S. J., Sear, D. A., & Nislow, K. H. (2019). A conceptual model of riparian forest restoration for natural flood management. *Water Environment Journal, 33*, 329–341. https://doi.org/10.1111/wej.12425

Dixon, S. J., Sear, D. A., Odoni, N. A., Sykes, T., & Lane, S. N. (2016). The effects of river restoration on catchment scale flood risk and flood hydrology. *Earth Surface Processes and Landforms, 41*, 997–1008. https://doi.org/10.1002/esp.3919

Edwards, J. B. (2023). Louisiana's Comprehensive Master Plan for a Sustainable Coast, Executive Summary.

Environmental Protection Agency. (2021). *Climate change and social vulnerability in the United States: A focus on six impacts* (No. EPA 430-R-21-003). U.S. Environmental Protection Agency.

Fossey, M., Rousseau, A. N., Bensalma, F., Savary, S., & Royer, A. (2015). Integrating isolated and riparian wetland modules in the PHYSITEL/HYDROTEL modelling platform: Model performance and diagnosis: Integrating wetland-specific modules in hydrologic model. *Hydrological Processes, 29*, 4683–4702. https://doi.org/10.1002/hyp.10534

Frazier, T. G., Wood, N., & Yarnal, B. (2010). Stakeholder perspectives on land-use strategies for adapting to climate-change-enhanced coastal hazards: Sarasota, Florida. *Applied Geography, 30*, 506–517. https://doi.org/10.1016/j.apgeog.2010.05.007

Friedland, C. J., Lee, Y.-C., Mostafiz, R. B., Lee, J., Mithila, S. P., Rohli, R. V., Rahim, M. A., Gnan, E., & Farris, M. T. (2023). FloodSafeHome: Evaluating financial benefits and savings of freeboard for improved decision-making in flood risk mitigation. *Frontiers in Communication, 8*, 1060901. https://doi.org/10.3389/fcomm.2023.1060901

Furuta, K. (2015). Resilience engineering. In J. Ahn, C. Carson, M. Jensen, K. Juraku, S. Nagasaki, & S. Tanaka (Eds.), *Reflections on the Fukushima Daiichi Nuclear Accident: Toward social-scientific literacy and engineering resilience* (pp. 435–454). Springer. https://doi.org/10.1007/978-3-319-12090-4_24

Gall, M., Li, P., & Friedland, C. J. (2024). Strategic hazard mitigation planning. *International Journal of Disaster Risk Reduction, 114*, Article 104923. https://doi.org/10.1016/j.ijdrr.2024.104923

Gedan, K. B., Kirwan, M. L., Wolanski, E., Barbier, E. B., & Silliman, B. R. (2011). The present and future role of coastal wetland vegetation in protecting shorelines: Answering recent challenges to the paradigm. *Climate Change, 106*, 7–29. https://doi.org/10.1007/s10584-010-0003-7

Gourevitch, J. D., Diehl, R. M., Wemple, B. C., & Ricketts, T. H. (2022). Inequities in the distribution of flood risk under floodplain restoration and climate change scenarios. *People and Nature, 4*, 415–427. https://doi.org/10.1002/pan3.10290

Hardy, T., & Wu, W. (2021). Impact of different restoration methods on coastal wetland loss in Louisiana: Bayesian analysis. *Environmental Monitoring and Assessment, 193*, 1. https://doi.org/10.1007/s10661-020-08746-9

Hoang Tri, N., & Adger, W. (1998). Natural resource management in mitigating climate impacts: The example of mangrove restoration in Vietnam. *Global Environmental Change, 8*, 49–61. https://doi.org/10.1016/S0959-3780(97)00023-X

Huang, Y., Tian, Z., Ke, Q., Liu, J., Irannezhad, M., Fan, D., Hou, M., & Sun, L. (2020). Nature-based solutions for urban pluvial flood risk management. *Wires Water, 7*, Article e1421. https://doi.org/10.1002/wat2.1421

Intergovernmental Panel on Climate Change (IPCC). (2023). *Climate change 2021—The physical science basis: Working Group I Contribution to the Sixth Assessment Report of the Intergovernmental Panel on Climate Change.* Cambridge University Press. https://doi.org/10.1017/9781009157896

International Nuclear Safety Advisory Group (Ed.). (1996). *Defence in depth in nuclear safety: A report by the International Nuclear Safety Advisory Group, INSAG series.* International Atomic Energy Agency.

Juan-Diego, E., Mendoza, A., Arganis-Juárez, M. L., & Berezowsky-Verduzco, M. (2025). Alteration of catchments and rivers, and the effect on floods: An overview of processes and restoration actions. *Water, 17*, 1177. https://doi.org/10.3390/w17081177

Justine, Y. E. D., & Seenath, A. (2025). Vegetative nature-based solutions for coastal flood risk management: Benefits, challenges, and uncertainties. *Ocean and Coastal Management, 261*, Article 107520. https://doi.org/10.1016/j.ocecoaman.2024.107520

Khalil, S. M., Raynie, R. C., Muhammad, Z., & Killebrew, C. (2011). Overview of coastal restoration in Louisiana 79.

Kreibich, H., Bubeck, P., Van Vliet, M., & De Moel, H. (2015). A review of damage-reducing measures to manage fluvial flood risks in a changing climate. *Mitigation and Adaptation Strategies for Global Change, 20*, 967–989. https://doi.org/10.1007/s11027-014-9629-5

Li, J., & Landry, C. E. (2018). Flood risk, local hazard mitigation, and the community rating system of the National Flood Insurance Program. *Land Economics, 94*, 175–198. https://doi.org/10.3368/le.94.2.175

Lin, J. P., & Kleiss, B. A. (2007). ERDC/EL TR-07-12, a wetland restoration spatial decision support system for the Mississippi Gulf Coast.

McCorvie, M. R., & Lant, C. L. (1993). Drainage District Formation and the loss of Midwestern Wetlands, 1850–1930. *Agricultural History, 67*, 13–39.

McGee, K. (2021). A place worth protecting: Rethinking cost-benefit analysis under FEMA's flood-mitigation programs comments. *University of Chicago Law Review, 88*, 1925–1970.

Mississippi Coastal Improvement Program (MsCIP) Hancock, Harrison, and Jackson Counties, MS: Comprehensive Plan and Integrated Programmatic Environmental Impact Statement. (2009).

Misteli, B., Morant, D., Camacho, A., Adamo, M., Bachi, G., Bègue, N., Bučas, M., Cabrera-Brufau, M., Carballeira, R., Cavalcante, L., Cazacu, C., Coelho, J., Doebke, C., Dinu, V., Guelmami, A., Giuca, R., Kataržytė, M., Lillebø, A., Marin, A., …, Vaičiūtė, D. (2025). *Coastal wetland restoration and greenhouse gas pathways: A global meta-analysis.* https://doi.org/10.31223/X51B39

Molnar-Tanaka, K., & Surminski, S. (2024). Nature-based solutions for flood management in Asia and the Pacific. OECD Dev. Cent. Work. Pap. https://doi.org/10.1787/f4c7bcbe-en

Mostafiz, R. B., Al Assi, A., Rahim, M. A., & Friedland, C. J. (2024). HawardAware: A comprehensive flood and wind risk assessment and mitigation tool for enhancing resilience in gulf of Mexico Coastal Communities. Presented at the EGU General Assembly 2024, Vienna, Austria. https://doi.org/10.5194/egusphere-egu24-13553

Narayan, S., Beck, M. W., Reguero, B. G., Losada, I. J., Van Wesenbeeck, B., Pontee, N., Sanchirico, J. N., Ingram, J. C., Lange, G.-M., & Burks-Copes, K. A. (2016). The effectiveness, costs and coastal protection benefits of natural and nature-based defences. *PLoS ONE, 11*, Article e0154735. https://doi.org/10.1371/journal.pone.0154735

Noggle, R. (2022). *Developing a spatial decision support tool for planning tsunami evacuation shelter locations* (Major Qualifying Project Report). Worcester Polytechnic Institute, Shibaura Institute of Technology.

Pattison, I., & Lane, S. N. (2012). The link between land-use management and fluvial flood risk: A chaotic conception? *Progress in Physical Geography: Earth and Environment, 36*, 72–92. https://doi.org/10.1177/0309133311425398

Ramachandra, T. V., Aithal, B. H., & Kumar, U. (2012). Conservation of wetlands to mitigate urban floods. *Journal of Resources, Energy and Development, 9*, 1–22. https://doi.org/10.3233/RED-120001

Rappaport, E. N. (2014). Fatalities in the United States from Atlantic tropical cyclones: New data and interpretation. *Bulletin of the American Meteorological Society, 95*, 341–346. https://doi.org/10.1175/BAMS-D-12-00074.1

Reaney, S. M. (2022). Spatial targeting of nature-based solutions for flood risk management within river catchments. *Journal of Flood Risk Management, 15*, Article e12803. https://doi.org/10.1111/jfr3.12803

Rezaie, A. M., Loerzel, J., & Ferreira, C. M. (2020). Valuing natural habitats for enhancing coastal resilience: Wetlands reduce property damage from storm surge and sea level rise. *PLoS ONE, 15*, Article e0226275. https://doi.org/10.1371/journal.pone.0226275

Ruangpan, L., Vojinovic, Z., Di Sabatino, S., Leo, L. S., Capobianco, V., Oen, A. M. P., McClain, M. E., & Lopez-Gunn, E. (2020). Nature-based solutions for hydro-meteorological risk reduction: A state-of-the-art review of the research area. *Natural Hazards and Earth Systems Sciences, 20*, 243–270. https://doi.org/10.5194/nhess-20-243-2020

Searles, K. (n.d.). Elizabeth River Wetland and Floodplain Restoration. https://pw.virginiabeach.gov/stormwater/flood-protection-program/stormwater-green-infrastructure/elizabeth-river

Serra-Llobet, A., Jähnig, S. C., Geist, J., Kondolf, G. M., Damm, C., Scholz, M., Lund, J., Opperman, J. J., Yarnell, S. M., Pawley, A., Shader, E., Cain, J., Zingraff-Hamed, A., Grantham, T. E., Eisenstein, W., & Schmitt, R. (2022). Restoring rivers and floodplains for habitat and flood risk reduction: Experiences in multi-benefit floodplain management from California and Germany. *Frontiers in Environmental Science, 9*, Article 778568. https://doi.org/10.3389/fenvs.2021.778568

Sheng, Y. P., Rivera-Nieves, A. A., Zou, R., & Paramygin, V. A. (2021). Role of wetlands in reducing structural loss is highly dependent on characteristics of storms and local wetland and structure conditions. *Science and Reports, 11*, 5237. https://doi.org/10.1038/s41598-021-84701-z

Smith, C. S., Gittman, R. K., Neylan, I. P., Scyphers, S. B., Morton, J. P., Joel Fodrie, F., Grabowski, J. H., & Peterson, C. H. (2017). Hurricane damage along natural and hardened estuarine shorelines: Using homeowner experiences to promote nature-based coastal protection. *Marine Policy, 81*, 350–358. https://doi.org/10.1016/j.marpol.2017.04.013

Temmerman, S., Meire, P., Bouma, T. J., Herman, P. M. J., Ysebaert, T., & De Vriend, H. J. (2013). Ecosystem-based coastal defence in the face of global change. *Nature, 504*, 79–83. https://doi.org/10.1038/nature12859

Terrisse, A., Karner, M., Kaufmann, J., & Ernoul, L. (2025). Characterizing governance models for upscaling wetland restoration. *Environmental Management, 75*, 1155–1167. https://doi.org/10.1007/s00267-025-02132-2

The Ministry of the Environment of Japan. (2016). *Ecosystem-based disaster risk reduction in Japan. A handbook for practitioners*. The Ministry of the Environment of Japan.

United States Army Corps of Engineers, Mobile District. (2018). Mississippi Coastal Improvements Program Comprehensive Barrier Island Restoration Monitoring and Adaptive Management Plan.

Van Rees, C. B., Chambers, M. L., Catalano, A. J., Buhr, D. X., Mansur, A. V., Hall, D. M., Nelson, A., Suedel, B., Hawley, R. J., Bledsoe, B., & Nibbelink, N. (2024). An interdisciplinary overview of levee setback benefits: Supporting spatial planning and implementation of riverine nature-based solutions. *Wires Water, 11*, Article e1750. https://doi.org/10.1002/wat2.1750

Widis, D. C., BenDor, T. K., & Deegan, M. (2015). Prioritizing wetland restoration sites: A review and application to a large-scale coastal restoration program. *Ecological Restoration, 33*, 358–377. https://doi.org/10.3368/er.33.4.358

Wu, Y., Sun, J., Hu, B., Zhang, G., & Rousseau, A. N. (2023). Wetland-based solutions against extreme flood and severe drought: Efficiency evaluation of risk mitigation. *Climate Risk Management, 40*, Article 100505. https://doi.org/10.1016/j.crm.2023.100505

Wu, Y., Zhang, G., Rousseau, A. N., Xu, Y. J., & Foulon, É. (2020). On how wetlands can provide flood resilience in a large river basin: A case study in Nenjiang river Basin, China. *Journal of Hydrology, 587*, Article 125012. https://doi.org/10.1016/j.jhydrol.2020.125012

Tropical Cyclones in Systems Thinking: Testing Perspectives in Resilience Connections

Justin Nguyen and Kesley J. Richardson

Abstract There is a correlation between resilience connections and systems thinking when facing the threat of tropical cyclones. This correlation focuses on the systems thinking approach when organizing and processing changes before, during, and after a tropical cyclone. This paradigm reflects a systems thinking approach that encompasses multiple case studies on how organizations can further enhance their social, economic, and environmental contexts. The findings in this chapter aim to identify tropical cyclone risk through the lens of climate change and introduce its impacts on emergency management professionals and managers operating in the ever-evolving disaster management landscape. This understanding of the complexities of disaster and emergency decision-making not only has to improve for the practitioner continuously but also be translatable and understood by the elected officials who grant delegation of authority (DOA), who provide resources and personnel, and finally, impacts the ability and capability of Emergency Management Departments (EMDs) and organizations within their jurisdiction. Improving the decision-making process and supporting the whole community and its stakeholders promotes safety, reduces loss of life, and enhances EOC operations at various levels.

Keywords Tropical disruptions · Decision-making · System thinking · Emergency management

1 Introduction

Systems thinking involves viewing multiple components and factors and how they interact and function with one another. This paradigm, when applied to the topic of tropical cyclones and the associated risks, has connections that impact communities

J. Nguyen (✉)
Voinovich School of Leadership and Public Service, Ohio University, Athens, OH, USA
e-mail: nguyenjustin850@gmail.com

K. J. Richardson
School for Continuing and Professional Studies, Loyola University of Chicago, Chicago, IL, USA
e-mail: Krichardson8@luc.edu

J. Collins et al. (eds.), *Tropical Cyclone Risk and Resilience*,
https://doi.org/10.1007/978-3-032-28745-8_8

in various ways. This differs from the most common view on linear structured interaction, but rather a dynamic and interconnected array of relationships and feedback loops. Every organizational system will vary from one another, as they contain unique parts that interact with them. For example, a business company will have a different system compared to a non-profit organization, a public sector organization, or a private organization. The business system will contain components such as dealing with other companies, international clients, and the supply chain, while the non-profit interacts with volunteers, local organizations, and participates in fundraising events. For systems thinking, the focus is on observing and analyzing the various components, such as social, economic, or environmental factors, that interact with an organization. By observing these interactions, we can determine the emergence or outcome of synergies with different components and gain a better understanding of the system (Acaroglu, 2017).

Communities and cities are constantly experiencing rapid growth as people move into communities or as modern services become accessible. With new developments and interactions, many systems are becoming increasingly complex as society evolves and grows. However, the systems thinking approach will enable us to understand challenges better and create the desired outcomes through the complex development of communities and jurisdictions. According to Arnold and Wade (Arnold & Wade, 2015), systems thinking can be best defined as a system that identifies its goal and then elaborates its elements and the interconnections between these elements. The review of Parkland, FL, will be used as a case study to demonstrate systems thinking and its application to societal disruptions.

2 Background of Parkland

The city of Parkland, Florida, is located in Broward County, Florida, and was founded in 1963. As a city, it consists primarily of residential housing developments and communities in the present day. It features several locations and areas that provide economic activity, including business and shopping plazas; however, it also boasts a range of recreational areas and outdoor green spaces. The city supports residents and local businesses through its governmental departments and staff. These departments help ensure the city is operational and supporting residents' livelihoods within its jurisdiction. During times of extreme weather events, such as a tropical cyclone, the city shifts its role and day-to-day operations to prepare for, respond to, and recover from the event. In the past decade, six major tropical cyclones have impacted the state of Florida. Although some tropical cyclones were not in the direct path of the city, the impact of high winds and heavy precipitation still occurred in the city.

3 Methods

This chapter focuses on analyzing how the systems thinking approach can be applied to local governments and organizations. The local government of the city of Parkland, located in South Florida, is the focus of this case study. The approach to understanding the organization and its existing structure was to analyze materials available publicly on the local city website (Parkland, 2025). Within the city's resources, key departments were identified as to how they could contribute to the city's function and continuity. The identification of key departments, such as City Clerk, City Manager, Communications, Development Services, Finance, Human Resources, Information Technology, Library, Parks and Recreation, Public Safety, Public Works, and Purchasing, is linked to the components that comprise the Parkland governmental system. Each department had its resource page that covered its responsibilities for the city. Internal documents, such as the city code and ordinances, were evaluated to understand the policies and procedures in place for the city's activities. These documents helped to understand the interconnectedness between city departments and other parties, such as businesses, city residents, and other local entities. Lastly, news media outlets, including online media postings, new articles, and reports, were also utilized to highlight the city's activities.

4 The Impact of Climate Change on Tropical Cyclones

The climate impacts on Tropical Cyclones can be tracked chronologically over time, and as the world has changed, so has its climate, with its relationships to tropical cyclones. Over time, there have been significant impacts of global warming, particularly over the last decade (2015–2025). With these changes to tropical cyclone strength, even in weaker storms with limited global warming impacts, can still lead to the cascading effects of these same storms on a worldwide scale, particularly in the southern hemisphere, where notable increases have been recorded. The stronger tropical cyclones reported in the Intergovernmental Panel on Climate Change's (IPCC) Sixth Assessment Report (AR6) confirm that climate change has a direct effect on the increased intensity of the strongest tropical cyclones, particularly in terms of rainfall and wind speed (IPCC, 2023). Throughout this report, work has informed policymakers with regular scientific data on climate science to support global governmental positions on climate and its intersecting impacts on weather. As locations for reliable and valid data shrink in the West, sources like reports help paint an objective picture of science, connected impacts from our environment, and its cross-sectional impacts on disasters.

These impacts of climate change on tropical cyclones can be linked to rising sea levels, which have amplified storm surge impacts generally where coastal communities are impacted more, the heavier rainfalls during tropical cyclones which have been observed to be connected with continued global warming (IPPC, 2023), and connections to higher proportion of intense storms (category 4–5) with increased rainfall (NOAA, 2024). These impacts and cross-sectional increases in storm winds and rain can also be observed from the reflection of hurricanes in the Atlantic, where a recorded increase in storm strength and impact has been attributed to the effects of climate change and global warming (Richardson, 2024). These impacts have not only cascading effects on decision-making for policymakers, city managers, government officials, and emergency managers, but also are interconnected through the lens of systems thinking dynamics. The connections between climate change, tropical cyclone risk, and systems thinking are interlinked at a fundamental level, which have been shown to lead to impacts on social, economic, and environmental paradigms, among others.

The cascading impacts connected to tropical cyclone risk can be identified as follows: wind damage, storm surge, flooding, heavy rainfall, erosion (CISA, 2025), infrastructure damage, injuries and fatalities, psychological trauma to responders and communities (Done et al., 2025), exacerbation of existing conditions, displacement, economic losses (Australian Climate Service, 2025), food security issues, increased community tensions, and long-term impact on ecosystems. These impacts are just some of the problems identified, connected not only to tropical cyclone risk and impacts, but also to the increasing disruptions when tropical cyclones come into contact with communities within the historical hazard exposure area. Even if there were a halt to climate change and global warming now, it would take around 2050, or 25 years, to approach a net-zero state, according to the World Economic Forum (World Economic Forum, 2022). This leads humanity to operate in a deficient way if there is any hope of combating global warming, tropical cyclone risk, and its impacts on climate change, as well as other global phenomena.

5 Tropical Cyclones and Their Increasing Impacts

Tropical Cyclones' Impacts on Communities

The destructive nature of tropical cyclones will continue to impact communities. Local communities in Florida are all vulnerable to the effects of tropical cyclones, as the state is prone to hurricanes. Hurricane season occurs annually, lasting approximately six months. During this time, the hurricane season reaches its peak from mid-August to late October, due to the increased warm waters in the Atlantic and Gulf of Mexico, which support the development of tropical waves (FSU, 2025). The physical impacts of tropical cyclones will continue to affect individuals, businesses, organizations, and local governments as shown in Fig. 1. Tropical cyclones pose three hazards that can have a severe impact on local communities: wind damage, storm surge, and flooding. Tropical cyclones can cause wind damage that destroys buildings and homes and create flying debris that can inflict additional harm to people and their property. In addition, tropical cyclones generate storm surges that generate higher than typical waves, leading to flooding, especially in coastal areas. These

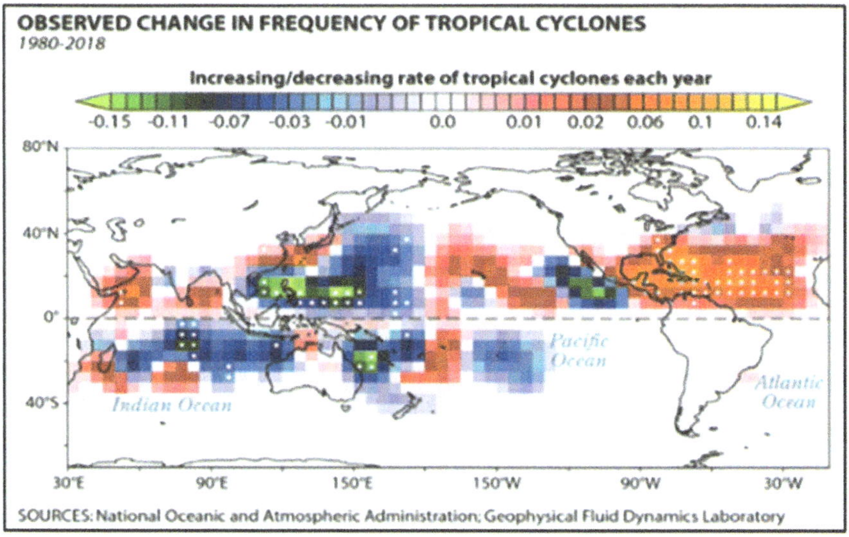

Fig. 1 CISA 2025 tropical cyclone map—*Source* NOAA

severe hazards will continue to impact communities and organizations in three key areas: social, economic, and environmental.

The wind damage from tropical cyclones has historically intensified over time, driven by the increasing temperatures of the sea surface, which have led to more intense storms, including stronger winds (Wu et al., 2022). With this increase in cyclone strength leading to more severe wind damage, stronger sustained winds are impacting coastal communities, which challenges existing community jurisdictions. Urban infrastructure is challenged with aging buildings with dated or insufficient wind-resistant design from the original construction. This wind damage, of course, can damage and dislodge power lines, buildings, debris, plant life, streetlights/signs, and other detachable items that can become airborne. These airborne items lead to further impact damage throughout communities where HVAC units, boats, and other heavier items can be thrown across cities, counties, and communities. The wind damage can lead to more than just structural damage. Still, it can also result in significant agricultural damage, resulting in substantial crop loss due to the damage caused by high winds (Kunze, 2021).

The storm surge accompanying tropical cyclones can cause coastal flooding, which has been recognized as an increasingly dominant hazard in the southeastern United States (Song & Chung, 2024). Storm surge can pull sand from both man-made and natural beaches and push boats and ocean equipment up onto land, while flooding neighborhoods, streets, and communities. As sea levels rise, amplifying storm surge impacts tied to prolonged rainfall can lead to cascading economic, social, and environmental effects on communities. These disruptions not only destroy livelihoods but also incur severe long-term recovery costs, which can displace individuals from their homes and communities, redirecting their lives for generations to come (Schleypen et al., 2024).

The flooding associated with tropical cyclones poses a significant risk to the United States in areas where coastal communities are located. Although the city of Parkland is not directly on the coast, the likelihood of flooding can occur and impact the city due to the increased rainfall and rising groundwater. These flood-related events are one of many concerns that can compromise and overwhelm the city's stormwater infrastructure, leading to flooding (Reilly, 2024). Climate change impacts are estimated to increase flood exposure and lead to more cascading economic, social, and environmental impacts, as well as pose a significant public health risk in the U.S. (Anderson et al., 2022). The financial losses from tropical cyclones continue to rise, even with adaptation efforts that are lagging behind the future flood vulnerability impacts (Bakkensen & Mendelsohn, 2019). As these damages rise sharply without proven flood mitigation methods, the approach to secure funding for mitigation projects has been in flux with the 2025 shift in political climate and support from federal funding. The types of damage that can come from tropical cyclones can vary depending on a variety of factors, but they can also impact communities in various ways, as shown in Fig. 2. Without existing mitigation funding, which often fails to reach the most at-risk communities, and the historical existing funding allocation and misalignment with actual vulnerability hotspots, an unstable source of federal funding support (Tate et al., 2021). While the Southeast is particularly vulnerable to compound flooding events, especially during hurricane seasons, the flood risk is amplified by social vulnerability and infrastructure gaps (Song & Chung, 2024).

The Social Impact of Tropical Cyclones

The threat of tropical cyclones does more than create physical damage; it disrupts and affects the livelihood of communities and individuals. It is important to be aware of the potential danger of disruption and take proactive approaches to address it. An example would be creating strategic plans and strategies. Parkland created a strategic plan to better align the needs of the community and address key areas such as improving and enhancing the city's infrastructure, which will increase its resilience towards the disruption of tropical cyclones (City of Parkland, 2025). Displacement is one of the primary social impacts of tropical cyclones, which can have ongoing effects on all communities (Ramachandran et al., 2025). When intense tropical cyclones are forecast to hit a region, many individuals are advised to evacuate their homes for safety. In addition, the destructive wind speeds from tropical cyclones pose another threat and reason for evacuation, as issued by local NOAA offices, emergency management agencies, and government outlets. Often, the follow-through on evacuation can be hindered by government mistrust, confirmation bias, and a lack of resources or transportation to evacuate. There are several reasons why individuals will evacuate their homes, one of which is if their area is vulnerable to flooding from either heavy rain or storm surges. This is particularly true for areas located in floodplains or near bodies of water, or where there is a lack of water relocation infrastructure. Unsafe sustained wind speeds of over 45 mph will disrupt and prevent normal day-to-day

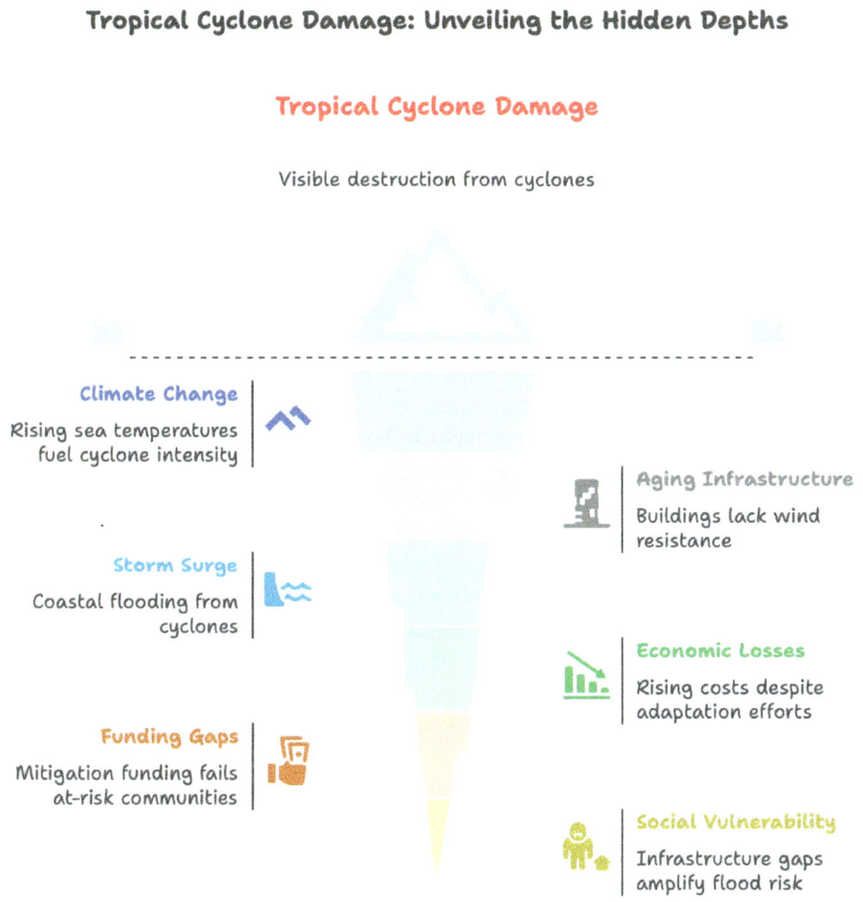

Fig. 2 Tropical cyclone damage: unveiling the hidden depths

activities from functioning and limit transportation. These challenges of displacement create a social problem for many communities, with cascading impacts being frequent. Normal day-to-day activities, such as grocery shopping, and even special events, like planned gatherings, cannot occur under unsafe conditions caused by tropical cyclones.

Another social impact of tropical cyclones is their effect on food security in communities where this insecurity may already be prevalent, such as those experiencing food deserts. Tropical cyclones can destroy crops, farmland, terrestrial, and marine ecosystems, which makes it more challenging to feed communities. Harvest failures and infrastructure damage disrupt the food supply chain due to the impacts of extreme weather events, such as tropical cyclones (Mirzabaev et al., 2023). With these agricultural systems damaged, it creates a strain on the supply chain and food production for both coastal and non-coastal communities. The impact has been felt by businesses, nonprofits, individuals, and families throughout an affected community or jurisdiction. These impacts will be observed through a shortage in food supply and increased costs for food and certain limited items. Communities whose livelihoods depend on agricultural land and fishing areas will face the most significant impact, as it is not only part of their lifestyle but also their primary means of income and financial support. Lastly, civil tension and unrest are a direct result of the food shortages and increased costs persisting for an extended period. The possibility of civil unrest, including physical altercations, rioting, and violent looting, can arise in the event of a food shortage. With insufficient food availability or access to food due to distribution barriers, this can increase the strain on social impacts and exacerbate social disruptions.

One of the biggest threats and impacts of tropical cyclones is their effect on the health and well-being of communities, especially those considered vulnerable populations. Vulnerable populations include those residing in flood-prone areas, as well as individuals at higher risk due to underlying health conditions, such as older adults and those with chronic illnesses. The direct impact of tropical cyclones can be caused by physical impact. For example, tropical cyclones can have a direct impact on communities through physical destruction, such as wind and flood damage. According to an article by Young and Hsiang (2024), it is estimated that roughly 500 tropical cyclones have directly killed 24 individuals on average within the United States. Beyond direct harm and death, tropical cyclones can lead to indirect harm and deaths. A study estimated that the average number of indirect deaths between 1930 and 2015 from tropical cyclones was between 7,000 and 11,400 individuals (Young & Hsiang, 2024).

Indirect harm and deaths from tropical cyclones are often caused by damage to infrastructure and power outages. During these disruptions and damages, vulnerable community members, such as older adults and people with disabilities, are at a higher risk than any other population. Community members rely on reliable and secure electricity sources to power their homes, medical devices, and vehicles. Without it, they become vulnerable to unsafe conditions due to the lack of ventilation and air conditioning in their homes, which increases the chances of heat-related illnesses. In Florida, it is a requirement for all health care facilities to be equipped with a backup generator. This was mandated after Hurricane Irma in 2017, which knocked out the air-conditioning, oxygen, and dialysis systems at a nursing homes and eventually led to the deaths of 12 patients (Sexton, 2021). Other areas outside the state have recognized the impact that tropical cyclones can have on communities' power sources and the cascading impacts to power failure for homes, services, and systems. For example, Harris County in Texas recently mandated that assisted living facilities and nursing homes install backup generators to maintain operational heating and cooling systems during disruptions from hazardous events, such as tropical cyclones (Oxenden, 2025). Mandates and policy reforms are among the initiatives that many local governments and organizations have implemented in response to the health impacts and risks associated with tropical cyclone events.

The Economic Impact of Tropical Cyclones

The long-term effects of tropical cyclones on the economies of affected communities are exacerbated when communities face exposure to disaster risk. This is essentially the sentiment expressed by NPR, an independent, non-profit media organization, as highlighted in an article that highlights how federal money disproportionately flows to the rich and inequitably boosts the recovery of fiscally affluent communities, while disregarding the more vulnerable and underfunded communities (Hersher, 2019). Exposure to tropical cyclones further exacerbates the compound economic issues faced by all impacted communities. However, closing down places of employment, destroying homes, and the disproportionate allocation of aid within these communities lead to generational disproportionality of impacts on vulnerable families and communities. When considered alone, these fiscal impacts cause significant losses for communities, and the problem worsens as these financial pressures interact with other systems.

The loss of property and housing from tropical cyclones can occur in every impact, especially when inadequate mitigation measures are in place. Parkland has begun investing and improving local drainage systems in some neighborhoods to prevent failures from heavy rain events like tropical cyclones (White, 2025). The investment was feasible thanks to past federal funding and local contributions; however, it will not always be the case, especially when there are difficulties and challenges in obtaining federal support with the current federal administration. With the inconsistent funding and lack of localized support to bolster protective measures, an increase in vulnerability is inevitable for exposed communities (Castro et al., 2025). The loss of property and housing is especially damaging to socially vulnerable populations, who are more likely to be exposed to flood-prone areas, where a lower cost of living is often available. With the continued misalignment of federal funds in communities and regions with the highest risk, the disproportionate effects on people will persist in the context of storm impacts and tropical cyclone risk.

Infrastructure damage to communities can leave them without essential services and resources, creating a dire need for sustainability. The broken continuity of infrastructure can also leave communities functionally exposed. The increase in vulnerability and damage to infrastructure continues to rise due to the connection to climate-related hazards, such as flooding and storm surges, which accompany tropical cyclones (Neumann et al., 2021). The limitations of federal aid and inconsistent funding often leave sustainable infrastructure unattainable for most communities, with long-term resilience remaining elusive without stable investment (Badr et al., 2024). This includes, but is not limited to, the disruption of businesses and other organizational resources that support the physical, social, or environmental infrastructure of a community. The interruption of companies is only amplified by inconsistent federal support, with domestic firms/organizations being more vulnerable to funding gaps and continuity gaps in the hazard mitigation (Mithani et al., 2022). These disruptions can also be reflected in an international context, where businesses are exposed to high-risk zones (flood zones), where strategic planning is hindered by inconsistent support both internally to businesses and externally by local regulations.

The Environmental Impact of Tropical Cyclones

The environmental impacts of tropical cyclones not only disrupt the society that is impacted but also impact the surrounding ecosystem. The effect of a tropical cyclone can lead to the loss of biodiversity within surrounding ecosystems and environments exposed to tropical cyclone conditions. This loss of biodiversity is driven by climate change, habitat modifications that occur when nature comes into contact with human society, and underfunded conservation efforts (Wudu et al., 2023). This issue is also linked to the lack of federal and organizational support for protecting and preserving native ecosystems, as well as protecting species. The loss of biodiversity has a cascading impact on both ecological and social systems, in which damage to coastal wetlands, reefs, and mangroves can exacerbate the effects of future cyclones where communities are exposed to tropical cyclones (Mori et al., 2022).

Systems or policies that have been established to address the environmental impacts of tropical cyclones and other climate-related storms have manifested in reactionary policies as shown in Fig. 3, such as the Post-Katrina Emergency Management Reform Act of 2006, the Pets Evacuation and Transportation Standards Act of 2006, the Sandy Recovery Improvement Act of 2013, and the Disaster Recovery Reform Act of 2018 (Syed, 2024). These policies and actions encompass changes to zoning and land-use regulations, building codes, and resilience standards, as well as ecological restoration, nature-based solutions, and integration with the Sustainable Development Goals (SDGs). Even with these policies, the substantial effects of tropical cyclones impacted a myriad of conditions for human existence and well-being within communities and populations (Ramachandran et al., 2025). These policies, intended to improve operational responses to disasters and climate-related hazards such as tropical cyclones, often still fall short of the intent and nature behind reactionary policies. Reactionary policies are policies that were created in response to an impact or social disruption that established policies or acts as a conceptual framework for governments. These policies are also supported by American and international initiatives, such as nature-based solutions by the United Nations Institute for Environment and Human Security (UNU-EHS), which is pushing cutting-edge research to support tomorrow's solutions in climate mitigation, including but not limited to tropical cyclones.

As environmental effects of cyclones are taken into account, community-wide approaches have been considered to support local neighbors through collaboration with multiple organizations, aiming to mitigate the loss resulting from environmental impacts associated with historically exposed hazards. A major community-wide approach that many local governments utilize is mitigation planning. Mitigation planning involves engaging with both local government officials and the public to develop strategies that mitigate the loss from environmental impacts, such as flooding, on property and reduce their impact on the population's health and wellbeing. When it comes to mitigation planning, the primary goal is to develop a comprehensive mitigation plan that encompasses a range of mitigation action items to further enhance a community's resilience capabilities. Local communities follow the FEMA Local

Fig. 3 Key U.S. disaster management policies and initiatives

Mitigation Planning Handbook. As recommended within the handbook, local governments should involve their community members as part of the planning process to ensure a community-wide approach is met when developing mitigation activities and projects (*Community Rating System | FEMA.Gov*, 2025). This is intended to ensure that the mitigation strategies are aligned with the needs of community members. One example of an environmental support system that can benefit local jurisdictions and communities is FEMA's Community Rating System.

The Community Rating System (CRS) is a voluntary incentive program that promotes community floodplain management practices that exceed the minimum requirements of the National Flood Insurance Program. The CRS program is a nationally recognized initiative that many communities participate in across the country to support their local jurisdictions. There are numerous benefits to participating in this program, including helping to reduce and prevent flood damage to insurable property, strengthening and supporting the insurance aspect of the National Flood Insurance Program, and promoting comprehensive floodplain management. This system can enhance the mitigation approaches to communities if leveraged properly with adequate support from local leadership. The highlight of this program is that it offers a premium reduction for flood insurance based on the CRS classification. The CRS classification ranges from 1 (highest) to 10 (lowest) (*Community Rating System | FEMA.Gov*, 2025). Depending on the CRS class rating, communities can obtain a specific CRS discount or premium reduction for flood insurance. These ratings serve as incentives for communities to implement preparedness initiatives and policies as outlined in the CRS manual. The following table (Fig. 4) from FEMA outlines the CRS classification classes and the premium discount associated with them:

CRS Credit Points, Classes and Premium Discounts

CRS Credit Points	CRS Class	CRS Discount (Premium Reduction)
4,500+	1	45%
4,000 – 4,499	2	40%
3,500 – 3,999	3	35%
3,000 – 3,499	4	30%
2,500 – 2,999	5	25%
2,000 – 2,499	6	20%
1,500 – 1,999	7	15%
1,000 – 1,499	8	10%
500 – 999	9	5%
0 – 499	10	0

Fig. 4 CRS credit points—*Source* FEMA

6 Cascading After Effects of Tropical Cyclones and Systematic Impacts

Among the previously stated impacts in this chapter associated with tropical cyclones across the United States, there are interconnected and cross-sectional impacts that compound when they interact. At this nexus and intersection of the cascading effects of tropical cyclones, systematic impacts would take place, examining the intersections of social and economic effects, social and environmental impacts, and economic and environmental impacts. These intersections create cascading issues within social disruptions that exacerbate the strain on communities, families, and individuals who experience the effects of tropical cyclones. These impacts can range from displacement and its social and economic implications, for example. One way to measure this intersected impact would be to leverage the Innovative Structural Theory of Emergency Management (IST-EM) to identify interconnected metrics, evaluate the impact against the mitigation efforts or emergency management capabilities, and support areas of expansion with objective, peer-reviewed evidence in practice (Richardson, 2023).

The intersection of social and economic impacts when facing tropical cyclone risk has been reviewed to reveal interconnected effects on ecological degradation, exacerbation of public health crises, and contribution to infrastructure failure (Bahlai et al., 2021), as shown in Fig. 5. Regarding environmental degradation, the connection between tropical cyclones and land results in the loss and damage of wetlands, forests, and biodiversity, reducing natural storm buffers and increasing future vulnerability from various perspectives, including the physical infrastructure in communities integrated into natural or rural areas. This intersection can also exacerbate the amplification of social inequity, where vulnerable populations are disproportionately affected and further strained by disaster impacts (Ramachandran et al., 2025). The intersections of disruptions would be identified as well, but not limited to the educational disruption, where schools are closed for extended periods beyond what was anticipated, due to infrastructure damage or other cascading impacts, which can have long-term developmental effects on students, as seen during the COVID-19 pandemic (Zhou, 2025). The economic implications of schools being closed limit parents' ability to go to work, as parents' jobs are essential for nuclear family survival. In addition, another economic drain can also be reflected in the cost of schools' physical repair or unanticipated costs for shifts to the educational environment to support the continuity of instruction for students. These agile shifts to return educational continuity can be limited by a lack of power, internet, and the cost associated with maintaining an alternative learning environment or extending the learning environment outside of the traditional timeline.

The intersection of social and environmental effects of cyclones operates more within the context of the human environment than that of the nature-based environment. From this perspective, one significant intersectional impact of housing displacement is a recurring outcome of climate-based disasters, including tropical cyclones. This has a social aspect, as societies are often faced with homelessness or displacement, resulting in the creation of internally displaced persons, while also altering the typical human environment in terms of living circumstances (Zhou, 2025). Even with housing damage and displacement, doubling down on the concept of infrastructure damage and its impact on community and societal stability can be damaging through the social and environmental paradigm. This displacement can lead to additional stressors on the well-being and public health exposures of families, which increases the risk of diseases, mental health issues, and strains on the health system within communities (Bahlai et al., 2021). The connection between social and environmental problems may also be closely related to social and economic impacts, as it is also closely connected to economic and environmental impacts.

A very long-standing intersection between these two impacts is tourism. Many local communities, particularly those in coastal areas, rely on the natural environment and its attractions to sustain and expand their economies. In the state of Florida, tourism is one of its most significant economic drivers (Macdonald et al., 2023). According to Macdonald et al. (2023), a considerable portion of the tourism industry is directly reliant on outdoor activities and the environment. This major factor is what influences the importance for many communities to proactively protect these environmental attractions, as it plays a significant role in contributing to their economy.

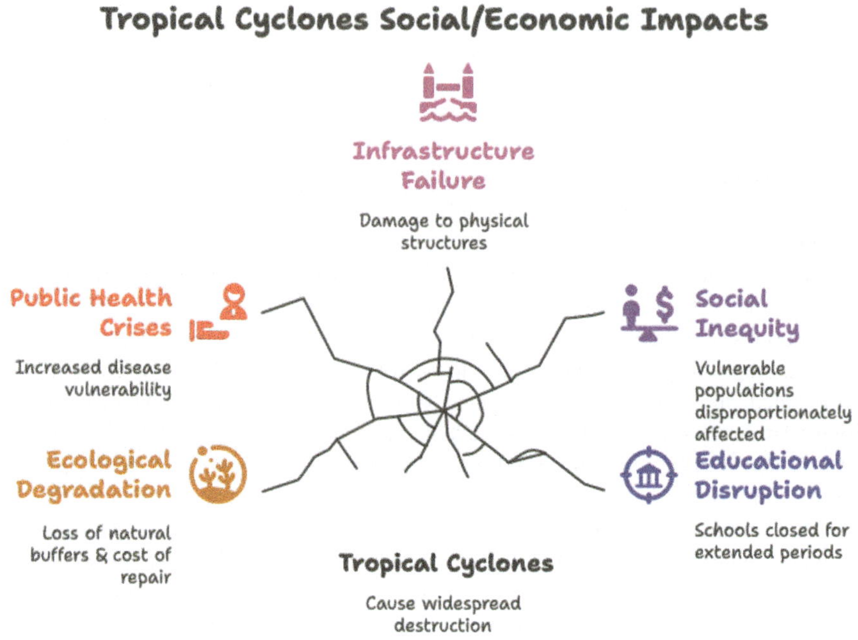

Fig. 5 Tropical cyclone social and economic impacts

The local government and community incur costs and make investments to ensure the natural attractions remain protected. However, this makes it challenging when massive physical damage from tropical cyclones occurs. Communities would be forced to enter a lengthy recovery period to restore natural environments to their original conditions. The post-recovery period would require significant funding for communities, as it entails a substantial amount of labor, materials, and costs to restore environmental conditions and properties. In addition, the damage would force many tourism-based operations and businesses to shut down, potentially for weeks or even months (Brown et al., 2021).

7 Systems Thinking Framework

When it comes to understanding communities and tackling challenges, leaders must adopt an approach that analyzes all interconnected parts to find the right solution. In today's world, many organizations and local communities interact with each other. Parkland has adopted the systems thinking framework through its strategic plan. One of their key strategies focuses on intergovernmental engagement and how they can work with other public bodies, beyond the city boundaries, to further improve and enhance the quality of life for the city and its residents (City of Parkland, 2025).

Each organization has its unique system and considerations when seeking solutions. As mentioned earlier, the impact of tropical cyclones crosses multiple dimensions. These impacts are numerous and often overlap, as one tropical cyclone event can create various social and environmental challenges. Many communities could benefit from using the system's thinking process. According to Dr. Jennifer Teague, "Systems thinking is a holistic approach to analysis that focuses on the way different parts of a system interact and how they influence one another within a whole" (Morganelli, 2024). The benefits of a holistic understanding, enhanced collaboration, informed decision-making, and adaptability support government officials in mitigating the risks associated with tropical cyclones. The holistic mindset of systems thinking will enable local officials to better understand the various components of their community, whether social, economic, or environmental, and see how each interacts and influences the others in developing potential solutions.

There are many benefits to systems thinking, but the core benefit is that it promotes and enhances better decision-making. Systems thinking allows us to surface hidden dynamics by revealing how different parts of a system interact over time. This helps reduce the risk of solving one problem while creating another elsewhere as a side effect. Several types of tools are available to apply systems thinking. One tool is the RACI method, which is an acronym that stands for Responsible, Accountable, Consulted, and Informed. The RACI is a matrix that captures all activities and decision-making authorities within an organization, aligned with all the people and/or roles. An example would be policy development. Elected officials and senior leadership staff would fill the responsible role in developing policy, while the public information staff would fill the informed role to ensure the organization and community are aware of the new policy. The intersection of activity and role makes it possible to assign somebody responsible, accountable, consulted, or informed for that activity or decision (Farrell et al., 2021). Another system thinking tool is causal loop diagrams (CLD). CLD is a systems thinking tool that allows users to approach problems from a feedback perspective. CLD creates a visual hypothesis of how components of a system are interconnected in a way that generates the system behavior we see (Farrel et al., 2021). For example, a CLD example could be based on the activity of engaging the public. Different loops can stem from this one activity. Some will be positive loops, such as ensuring public satisfaction and meeting expectations, while others will be considered negative loops, like the public voting against mitigation projects for their community. The CLD can help navigate the system's behavior and understand the effects of one loop on another.

8 Summary and Discussion

Throughout the writing of this chapter, the discussion focused on how to connect best thinking on systems in a manner that would not only convey the cascading impacts of tropical cyclones but also align with the reader. Concerning the impacts of disasters on communities, the focus is on the intersectional connections between social, environmental, and economic impacts. In these intersections, there were similar or oftentimes the same impacts, like business disruption, that spoke to all three connected categories. These interconnected impacts create ripple effects through the communities impacted by tropical cyclones, in which additional supportive and mitigation measures are needed with stable funding support to avoid social disruptions. As shown in Figs. 2 and 5, highlight key points that connect systems thinking in its connection to tropical cyclones, which have and will continue to impact communities.

Within the writing of this chapter, certain areas of focus were honed to concentrate on the social, economic, and environmental factors through a systems thinking lens. This lens focuses on the interconnected nature of concepts that are influenced by a range of variables, often disrupting or changing multiple capabilities within communities. Through this chapter, the identification of increased tropical cyclone strength and similar climate-related hazards has been enhanced with a direct connection to climate change. As these enhanced hazards continue to cause social disruptions, the interconnected effects and cascading issues continue to compound and generate larger and larger impacts for communities impacted storm after storm. The risk of tropical cyclones has increased over time in frequency, strength, and sometimes even in duration of the associated (rain, surge, flooding) hazards and vulnerabilities that communities face. The lens of systems thinking helps us understand the interconnected nature of societal elements and their impact on those elements.

As we close the writing of this chapter there have been several shifts of the political landscape within the United States. This constant shift may have limited the access to some related sources within this chapter from digital access. During the creation of this chapter for example FEMA has changed rapidly and is still changing throughout publication. This is not limited to other information access sources from federal agencies, changes in data retention practices, and contextual data collection and value around areas of equity, disadvantaged communities, and functions of agency resources. Even after publication the access and availability of some of the data mentioned here may range from having difficulty accessing to no longer being available.

References

Acaroglu, L. (2017, September 7). *Tools for systems thinkers: The 6 fundamental concepts of systems thinking*. Medium. https://medium.com/disruptive-design/tools-for-systems-thinkers-the-6-fundamental-concepts-of-systems-thinking-379cdac3dc6a

Anderson, G. B., Schumacher, A., Done, J. M., & Hurrell, J. W. (2022). Projecting the impacts of a changing climate: Tropical cyclones and flooding. *Current Environmental Health Reports, 9*(2), 244–262. https://doi.org/10.1007/s40572-022-00340-0

Arnold, R. D., & Wade, J. P. (2015). A definition of systems thinking: A systems approach. *Procedia Computer Science, 44*, 669–678. https://doi.org/10.1016/j.procs.2015.03.050

Australian Climate Service. (2025). *Hazards—Cyclones*. https://www.acs.gov.au/pages/hazards-cyclones

Badr, M., El-Badawy, S., & Elgawady, M. (2024). *Innovative infrastructure solutions*. SpringerLink. https://link.springer.com/journal/41062

Bahlai, C. A., Hart, C., Kavanaugh, M. T., White, J. D., Ruess, R. W., Brinkman, T. J., Ducklow, H. W., Foster, D. R., Fraser, W. R., Genet, H., Groffman, P. M., Hamilton, S. K., Johnstone, J. F., Kielland, K., Landis, D. A., Mack, M. C., Sarnelle, O., & Thompson, J. R. (2021). Cascading effects: Insights from the U.S. long term ecological research network. *Ecosphere, 12*(5), e03430. https://doi.org/10.1002/ecs2.3430

Bakkensen, L. A., & Mendelsohn, R. O. (2019). Global tropical cyclone damages and fatalities under climate change: An updated assessment. In J. M. Collins & K. Walsh (Eds.), *Hurricane risk* (pp. 179–197). Springer International Publishing. https://doi.org/10.1007/978-3-030-02402-4_9

Brown, C. E., Alvarez, S., Eluru, N., & Huang, A. (2021). The economic impacts of tropical cyclones on a mature destination, Florida, USA. *Journal of Destination Marketing and Management, 20*, Article 100562. https://doi.org/10.1016/j.jdmm.2021.100562

Castro, D. R., Shahi, K. R., Sairam, N., Fischer, M., Mohor, G. S., Thieken, A., Dewals, B., & Kreibich, H. (2025). Key drivers of flash flood damage to private households. *Journal of Flood Risk Management*.

City of Parkland. (2025, November 10). Strategic planning | Parkland, FL. https://www.cityofparkland.org/1540/Strategic-Planning

Community Rating System | FEMA.gov. (2025, June 16). https://www.fema.gov/floodplain-management/community-rating-system

Done, J. M., Fussell, E., Anderson, G. B., & Schumacher, A. (2025). Chapter 9—Understanding human health impacts of tropical cyclones across scales. In G. Villarini, G. A. Vecchi, & E. Scoccimarro (Eds.), *Tropical cyclones and associated impacts* (pp. 179–188). Elsevier. https://doi.org/10.1016/B978-0-323-95390-0.00009-1

Farrell, A., Hu, M., Chin, L., Evbuoma, E., Liem, W., & Ballard, E. (2021). Understanding systems from a feedback perspective. *Social System Design Lab*. https://doi.org/10.7936/t4k1-8g36

FSU. (2025). *Hurricanes—Florida Climate Center*. https://climatecenter.fsu.edu/topics/hurricanes

Hersher, R. (2019, March 5). How federal disaster money favors the rich. *NPR*. https://www.npr.org/2019/03/05/688786177/how-federal-disaster-money-favors-the-rich

IPCC. (2023). *Chapter 11: Weather and climate extreme events in a changing climate*. https://www.ipcc.ch/report/ar6/wg1/chapter/chapter-11/

Kunze, S. (2021). Unraveling the effects of tropical cyclones on economic sectors worldwide: Direct and indirect impacts. *Environmental and Resource Economics, 78*(4), 545–569. https://doi.org/10.1007/s10640-021-00541-5

Macdonald, C., Turffs, D., McEntee, K., Elliot, J., & Wester, J. (2023). The relationship between tourism and the environment in Florida, USA: A media content analysis. *Annals of Tourism Research Empirical Insights, 4*(1), Article 100092. https://doi.org/10.1016/j.annale.2023.100092

Mirzabaev, A., Bezner Kerr, R., Hasegawa, T., Pradhan, P., Wreford, A., Cristina Tirado Von Der Pahlen, M., & Gurney-Smith, H. (2023). Severe climate change risks to food security and nutrition. *Climate Risk Management, 39*, 100473. https://doi.org/10.1016/j.crm.2022.100473

Mithani, M. A., Narula, R., Surdu, I., & Verbeke, A. (2022). Unique implications of crises and disruptions for international business: How established MNEs are exposed to crises and mitigate their effects. In M. A. Mithani, R. Narula, I. Surdu, & A. Verbeke (Eds.), *Crises and disruptions in international business: How multinational enterprises respond to crises* (pp. 3–18). Springer International Publishing. https://doi.org/10.1007/978-3-030-80383-4_1

Morganelli, D. M. (2024, May 31). *What is systems thinking?* https://www.snhu.edu/about-us/newsroom/business/what-is-systems-thinking

Mori, A. S., Sasaki, T., Kagami, M., Miki, T., & Yasuhara, M. (2022). Feedbacks between biodiversity and climate change. In *The ecological and societal consequences of biodiversity loss* (pp. 281–304). Wiley & Sons, Ltd. https://doi.org/10.1002/9781119902911.ch13

Neumann, J. E., Chinowsky, P., Helman, J., Black, M., Fant, C., Strzepek, K., & Martinich, J. (2021). Climate effects on US infrastructure: The economics of adaptation for rail, roads, and coastal development. *Climatic Change, 167*(3), 44. https://doi.org/10.1007/s10584-021-03179-w

NOAA. (n.d.). *Global warming and hurricanes*. https://www.gfdl.noaa.gov/global-warming-and-hurricanes/

Oxenden, M. (2025, January 9). Harris County unveils backup power mandate for nursing homes, assisted living facilities. *Houston Landing*. http://houstonlanding.org/harris-county-unveils-backup-power-mandate-for-nursing-homes-assisted-living-facilities/

Parkland. (n.d.). *Departments | Parkland, FL*. Retrieved July 31, 2025, from https://www.cityofparkland.org/8/Departments

Ramachandran, A. K., Panneer, S., Rice, L., Rathnayake, U., & Kantamaneni, K. (2025). Addressing tropical cyclone risks: Stratified for wellbeing and global policy coordination. *Natural Hazards, 121*(6), 6441–6466. https://doi.org/10.1007/s11069-024-07097-9

Reilly, I. (2024, August 29). *Flooding in parkland: A growing concern*. parklandnews.net. https://parklandnews.net/flooding-in-parkland-a-growing-concern-p2016-175.htm#:~:text=The%20City%20of%20Parkland%20is,runoff%20and%20maintain%20water%20quality.

Richardson, K. (2023). *Quality check: How good is your emergency management program?* [Dissertation, West Chester University]. https://digitalcommons.wcupa.edu/all_doctoral/229/

Richardson, K. J. (2024). A climate change paradigm and its effects on emergency operations center decision-making process. In J. Collins, J. Done, Y.-J. Zhu, & P. Wilson (Eds.), *Advances in hurricane risk in a changing climate* (pp. 281–294). Springer Nature. https://doi.org/10.1007/978-3-031-63186-3_11

Schleypen, J. R., Plinke, C., & Geiger, T. (2024). *The impacts of multiple tropical cyclone events and associated precipitation on household income and expenditures | economics of disasters and climate change*. https://link.springer.com/article/10.1007/s41885-024-00149-1

Sexton, C. (2021, May 11). *Most Florida nursing homes in compliance with new backup power regulations*. https://www.mysuncoast.com. https://www.mysuncoast.com/2021/05/11/most-nursing-homes-compliance-with-new-backup-power-regulations/

Song, J. Y., & Chung, E.-S. (2024). Increasing wind threat of Atlantic tropical cyclones based on a comprehensive risk analysis using multi-hazard hurricane index and social vulnerability index. *Earth Systems and Environment, 8*(4), 951–962. https://doi.org/10.1007/s41748-024-00471-4

Syed, Z. (2024). Hurricane Katrina 19 year later, what policies have changed. *National Center for Disaster Preparedness | NCDP*. https://ncdp.columbia.edu/ncdp-perspectives/hurricane-katrina-19-years-later-with-policies/

Tate, E., Rahman, M. A., Emrich, C. T., & Sampson, C. C. (2021). Flood exposure and social vulnerability in the United States. *Natural Hazards, 106*(1), 435–457. https://doi.org/10.1007/s11069-020-04470-2

White, A. (2025, October 28). *Residents press for noise controls, youth programs, and flood relief*. parklandnews.net. https://parklandnews.net/residents-press-for-noise-controls-youth-programs-and-flood-relief-p2493-170.htm

World Economic Forum. (2022, June 15). *World has 42% chance of breaching 1.5 degrees if emissions stopped today*. World Economic Forum. https://www.weforum.org/stories/2022/06/world-has-42-chance-of-breaching-1-5-degrees-if-emissions-stopped-today-new-study-finds/

Wu, L., Zhao, H., Wang, C., Cao, J., & Liang, J. (2022). Understanding of the effect of climate change on tropical cyclone intensity: A review. *Advances in Atmospheric Sciences, 39*(2), 205–221. https://doi.org/10.1007/s00376-021-1026-x

Wudu, K., Abegaz, A., Ayele, L., & Ybabe, M. (2023). The impacts of climate change on biodiversity loss and its remedial measures using nature based conservation approach: A global perspective. *Biodiversity and Conservation, 32*(12), 3681–3701. https://doi.org/10.1007/s10531-023-02656-1

Young, R., & Hsiang, S. (2024). Mortality caused by tropical cyclones in the United States. *Nature, 635*(8037), 121–128. https://doi.org/10.1038/s41586-024-07945-5

Zhou, T. (2025). *A bibliometric review of climate change cascading effects: Past focus and future prospects*. CHANS-Net: The International Network of Research on Coupled Human and Natural Systems. https://www.canr.msu.edu/resources/A-bibliometric-review-of-climate-change-cascading-effects-past-focus-and-future-prospects

Hurricane Season in the Caribbean: The Case of Preparing for the Known Unknown

Nicola Williams

Abstract The Atlantic Hurricane Season is a known and relatively anticipated phenomenon in the Caribbean. However, the increasing frequency and severity of hurricane events have exacerbated known and unknown risks for vulnerable sectors and people in the Caribbean. This chapter examines the challenges and opportunities faced by Small Island Developing States (SIDS) in their efforts to prepare for the annual hurricane season in the Caribbean. It discusses the intersection of local knowledge and communication from technical stakeholders for hurricane preparation and disaster mitigation. Using case examples, the chapter elucidates lessons learnt for improving the hurricane preparation communication ecosystem in the Caribbean. The chapter concludes by proposing a multicomponent, interactive risk communication approach that facilitates local action in alignment with national disaster mitigation strategies.

Keywords Disaster communication · Hurricane preparedness · Tropical cyclones · Atlantic hurricane season · Climate change · Caribbean adaptation

1 Hurricane Season in the Caribbean: The Case of Preparing for the Known Unknown

> "The Caribbean is considered one of the most disaster-prone regions in the world. According to the Long-Term Climate Risk Index (CRI), three out of the ten countries most affected by extreme weather events in the last 20 years are located in the Caribbean" (United Nations Office for Disaster Risk Reduction (UNDRR)-Regional Office for the Americas and the Caribbean and Economic Commission for Latin America and the Caribbean (ECLAC/CEPAL), 2021, p. 15)

N. Williams (✉)
University of Calgary, Calgary, Canada
e-mail: nicola.williams1@ucalgary.ca

J. Collins et al. (eds.), *Tropical Cyclone Risk and Resilience*,
https://doi.org/10.1007/978-3-032-28745-8_9

Meteorological predictions for "above normal" activity are becoming the new normal in the annual opening forecasts for the Atlantic Hurricane Season (Government of Canada and Global Affairs Canada, 2025; "NOAA Predicts Above-Normal, 2025 Atlantic hurricane season | National Oceanic and Atmospheric Administration," 2025). Furthermore, the potential for hurricane-related disasters is inextricably linked to decades of socio-political and economic processes shaped by colonial governance structures that reproduce social inequalities (Bonilla, 2020; Rhiney, 2020).

The Atlantic Hurricane Season is a well-documented and anticipated phenomenon that affects millions of people in the Caribbean every year (Dookie & Spence-Hemmings, 2022). Hurricanes are strong tropical cyclones that bring strong winds, rain and can trigger secondary hazards like floods and landslides. Hurricanes are typically categorized and reported using the Saffir-Simpson scale, with a range from category 1 to 5 based on the wind speed (Whitmer & Sims, 2021). The Saffir-Simpson scale has been critiqued as a system for communicating hurricane risks because it does not capture the nonlinear association of hurricane category levels and the potential deleterious effects of an impending system (Whitmer & Sims, 2021). Hurricanes are both a hydrometeorological hazard and a phenomenon that transforms local cultures as they affect infrastructure, livelihoods and ecological systems. The devastation caused by hurricanes such as Hurricane Gilbert in 1988, in the Caribbean has been anthropomorphized (Barker & Miller, 1990), as generational markers of the risks, transformation, and resilience in the region. Increasingly severe tropical storm events transform the cultural landscape as their impacts contribute to changes in local traditions, government policies, sustainable livelihood prospects and land-use practices. The increasing frequency of more severe tropical storms creates a new imperative for adaptation in both the natural and the social world of the people living in the Caribbean. This transformation is being accelerated by the adoption of new technologies and the repositioning of local knowledges for addressing the potentially deleterious impacts of a rapidly changing climate.

As climate-related hazards intensify, there is growing recognition that an integration of context-specific local knowledge and new technologies may be an effective macro strategy for honing local resilience (Hao & Lun, 2024; Prasad & Nigam, 2023). Improvements in meteorological technologies have enhanced predictive models for estimating the number of tropical systems expected in each season. However, accelerating climate change has tested the efficacy of these models as systems occur more frequently and the rate of escalation to devastating intensity is increasingly unpredictable. The capacity for real-time on the ground communication and response mobilization is increasingly vital for mitigating disasters.

To improve the efforts to mitigate the potential adverse effects of a hurricane on local communities, communications from meteorological experts and state entities must transcend formal communication focused on the meteorological event, to a more holistic approach that extends into local knowledge and preparedness capacity. Government-led communication for hurricane awareness has been top down, passive, one directional and reactive. Typically, a standard generic public broadcast like the one below is the go-to mechanism for communication shared via traditional and social media platforms (Fig. 1).

FOR IMMEDIATE RELEASE: Thursday, October 23, 2025 –2:00 p.m.

BULLETIN No: 17

***** HURRICANE WATCH AND TROPICAL STORM WARNING REMAIN IN EFFECT FOR JAMAICA…AS MELISSA CONTINUES TO SLOWLY DRIFT NORTH-NORTHWEST… *****

A TROPICAL STORM WARNING, and a HURRICANE WATCH remain in effect as Tropical Storm Melissa moves slowly across the Caribbean Sea south-southeast of Jamaica. This means tropical storm conditions pose a possible threat to Jamaica within 36 hours and Hurricane conditions are likely to pose a threat within 48 hours.

At 1:00 p.m. the centre of Tropical Storm Melissa was located near latitude 15.5 degrees North, longitude 75.3 degrees West. This is about 280 kilometres (175 miles) south-southeast of Morant Point, Jamaica or about 320 kilometres (200 miles) south-southeast of the capital Kingston.

Tropical Storm Melissa is moving very slowly toward the north-northwest near 4 km/h (2 mph). A slow forward speed and northward motion is forecast during the next couple of days, followed by a westward turn over the weekend.

Maximum sustained winds are near 75 km/h (45 mph) with higher gusts. Gradual strengthening is forecast during the next day or so, and Melissa could become a hurricane in a couple of days, and a major hurricane by Sunday.

On its forecast track, **Tropical Storm Melissa** is expected to continue moving toward Jamaica before turning westward on Saturday and passing over the waters south of the island. As the system approaches, **showers and thunderstorms are forecast to begin affecting the island late this evening or by Friday morning**, with **periods of moderate to heavy rainfall expected to continue through the weekend resulting in flooding and possible landslides.**

Strong, gusty winds, possibly reaching tropical storm strength, are expected to gradually develop on Friday, continuing into the weekend. These winds will initially affect eastern and southern parishes, before gradually spreading across the remainder of the island. Winds reaching hurricane strength are possible over sections of eastern parishes as early as Saturday, as Tropical Storm Melissa continues to move closer to Jamaica.

Marine interests should expect widespread heavy showers and thunderstorms, accompanied by tropical-storm-force winds today and continuing through the weekend. Winds reaching hurricane strength are likely on Saturday, particularly over southern and eastern offshore waters, as Tropical Storm Melissa moves closer to the island. **All small craft operators including fishers from the cays and banks should by now be in port and are urged to complete their necessary safety precautions without delay.**

Tropical Storm Warning - A public release from the Meteorological Service indicating tropical storm conditions, including possible sustained wind speeds of 63–118 km/h (39–73 m/h), are expected to affect Jamaica in 36 hours or less.

Hurricane Watch - A public release from Meteorological Service indicating that hurricane conditions including sustained winds of 119 km/h or higher (74 m/h or more) pose a possible threat to Jamaica within 48 hours.

The Meteorological Service continues to closely monitor the progress of Tropical Storm Melissa, and all interests are encouraged to pay special attention to further releases.

The next Bulletin on this system will be issued at 5:00 p.m.

For recorded updates, please dial 116.
See also our website: http://metservice.gov.jm
END.

Fig. 1 Tropical storm warning and hurricane watch communication example from Jamaica (Meteorological Service Jamaica, 2025)

The Meteorological Service's communications are among the primary public facing alerts that guide government and citizen weather response in Jamaica. However, the language used in these communications tends to heavily emphasize the technical descriptions of the impending hazard event and assessments of risks. It is at this juncture that people interpret the information presented in relation to their situation and determine their course of action (Morss et al., 2024). Therefore, it is essential that people's experience, situational knowledge and contextual dynamics are taken into account in a broader hurricane preparedness communication strategy.

The Caribbean Disaster Management Agency (CDEMA) undertook significant institutional reform towards improved adaptation efforts and disaster risk mitigation through regional collaboration and training. However, on-the-ground implementation is often fragmented and divorced from the local contexts (Dookie & Spence-Hemmings, 2022). Dookie and Spence- Hemmings (2022) found that the communication timing of awareness information to the citizenry is a major vulnerability in hurricane awareness and preparedness efforts. Communication timing about an impending tropical storm event in the region ranged from within 12 h to impact to 24 h before impact across a sample of 14 countries. The qualitative findings of the same study revealed that "information awareness and dissemination is generally slower on a weekend due to capacity limitations" (Dookie & Spence-Hemmings, 2022, p. s103). Added to that, social factors such as knowledge gaps and social roles, as well as economic and infrastructural vulnerabilities are important variables that can impact risk perception (Morss et al., 2024). Consequently, there is a need for improved risk assessment and communication that is responsive to the differential risk perception among the different demographics of the population and its evolution over time (Derbyshire & Aven, 2025). Relying on a review of the literature and cases from the Caribbean, this chapter examines the lessons learnt from hurricane experiences in the Caribbean and articulates important considerations for hurricane preparedness. The chapter also presents an option for adapting the hurricane awareness communication strategy to facilitate timely, context-appropriate and effective support at all phases of the disaster risk mitigation cycle before the start of and during the Atlantic Hurricane Season.

2 Locating the Caribbean

The Caribbean region is comprised of independent states and dependent (overseas) territories with similar yet diverse historical, geological, social and economic characteristics (Meeks & Lindahl, 2001) that shape their experience of tropical cyclones. The shared experience of colonization and subsequent imperialist hegemonies played a significant role in the patterns of infrastructural development in coastal zones, economic dependency and the marginalization of traditional and local knowledge in the Caribbean. Climate injustices, including the unequal burden of climate change impacts, is a product of these colonial influences on Caribbean societies resulting in climate responses being both a social and a political phenomenon. Consequently, "marginalized groups experience climate change effects differently from the wealthy and privileged" (Baptiste & Rhiney, 2016, p. 17). Global geopolitical relationships continue to dictate the developmental and adaptation agenda of the world (Bishop et al., 2025). Located in the Tropical Cyclone (hurricane) Belt, the Caribbean is affected by multiple hazards concurrently, including, volcanoes, earthquakes, wildfires, and hydrometeorological hazards such as tropical cyclones and secondary hazards including flooding, landslides, large forest and crop destruction and storm surges. This multi-hazard reality has a systemic impact on the region's development due to the limited resources available to adequately address loss and damage caused by recurring hurricane events. Furthermore, the capacity of the Caribbean to effectively adapt to exacerbating climate change is constrained by its "subordinate position within the global capitalist system [which] is in part a function of centuries of colonial exploitation" (Baptiste & Rhiney, 2016, p. 18). This reality makes community-based capacity to prepare for and respond to hurricane hazards a critical prong in regional adaptation.

3 The Legacy of Knowledge Hegemony

The relationship between a state, its systems of governance and its citizens is a constantly evolving process that dictates the allocation and use of resources as well as the different priorities and responsibilities in the society. Historical legacies that reproduce the systemic oppression of local people and their knowledge systems through coercive tactics, paved the path for imperialist Eurocentric mechanisms to dominate knowledge production and communication in the Caribbean (Alatas, 2022). This stratification of knowledge is institutionalized through various systems presented as "intellectual and moral leadership, which takes the form of orientations circulated via the dominant educational, religious and associational institutions" (Alatas, 2022, p. 3). Local knowledge systems in the Caribbean are attuned to both the content, and the context of communication transmitted through dynamic, relational and sensory processes (Nakhid-Chatoor et al., 2018). However, the Westernized practice prioritizes knowledge that subscribes to prescribed standards of scientific rigour or espoused by accredited knowledge producers over "knowledge forms that are informed, established and passed on within distinct communities, cultures, and/or contexts in other (non-academic) ways" (Benz et al., 2022, p. 5). This contributes to the marginalization of knowledges that have developed locally, including community-based knowledge about hurricanes and hurricane preparedness.

Another critical yet often overlooked element that upholds knowledge hegemony in the Caribbean, is the prioritization of standardization of languages, like French, Spanish, Dutch and English -embedded by colonial powers, and are now the official language of a post-colonial state in the Caribbean. Oral traditions and dialogic communications, an essential aspect of Caribbean cultural identities (Sindoni, 2010) have been sidelined in favour of one way, top down sensitization communication for hazard management and disaster mitigation. Today, while many Caribbean countries utilize some creolized version of the imperialist language in day-to-day life, mainstream communication from the state and other technical stakeholders are often conveyed in the official language of the country. This is further compounded by the use of technical jargons when presenting climate related information. The result is a sustained system that not only delegitimises local knowledge production and dissemination but also reifies knowledge that has come to be known as scientific, expert knowledge that can be divorced of contextual nuances and relevance. The mismatch between the technical knowledge that often guides state action and the local knowledge that guides citizen, will evoke different concerns and priorities for actions taken towards preparedness. Citizens aim to ensure their personal safety and secure their property, which is directly linked to the state's ability to maintain critical infrastructure and reduce large scale vulnerabilities for all. Both levels of action are needed, however communication gaps undermine collaboration and erodes the efficacy of hurricane preparedness interventions.

4 Hydrometeorological Disaster Risks and Vulnerabilities in the Caribbean

The economies of many Caribbean countries are heavily dependent on the tourism and agriculture sectors, which are vulnerable to increasing climate change risks and various hydrometeorological hazards including droughts, floods, and hurricanes. While physically and economically vulnerable states, including those in the Caribbean, are on the front line for the deleterious effects of accelerating climate change they have limited power in the global governance decision making processes that dictate the broader responses to climate risks (Alatas, 2022). Therefore, a targeted local and regional disaster risk reduction strategy is essential to safeguarding the welfare of Caribbean people. Despite contributing the least to accelerated climate change, Small Island Developing States (SIDS) in the Caribbean continue to be increasingly vulnerable to the adverse effects of more frequent and severe tropical cyclones.

Tropical cyclones in and of themselves pose significant risks to Caribbean lives and livelihoods; they are multi-hazard phenomena (strong winds and rain) that often trigger secondary hazards such as storm surges, flooding, and landslides. Due to their recurring nature and destructive impact, tropical cyclones also have a compounding effect. In any given hurricane season as well as every year, the occurrence of multiple tropical systems interacts with other human-made and environmental hazards which elevate the risk of disasters in the Caribbean (UN Migration, 2024). For example, a tropical system impacting communities built in hydrologically sensitive areas prone to flooding and landslides or communities that did not have the time or resources to completely recover from a previous system is likely to have elevated vulnerabilities in their infrastructures and capacity to ensure protective measures if they never had effective measures in place.

Caribbean countries experience compounding loss and damages from hydrometeorological hazards year over year (Jones et al., 2022). Over the last decade, tropical cyclones have consistently resulted in significant disruptions in the economies and day-to-day life in the Caribbean. In September 2017, hurricanes Irma and Maria wreaked havoc in the Caribbean in rapid succession, causing widescale infrastructural damage, displacement of people, loss of lives, livelihoods, and homes as well as significant macro-economic disruptions in the affected countries. The impact of these and subsequent storm events highlights that, in addition to infrastructural considerations, knowledge production and communication play an important role in mitigating hurricane induced disasters in the region.

5 The Interplay Among Knowledges and Preparedness Communication

At the crux of a good disaster mitigation plan is high-quality, up-to-date data and how it is interpreted by those who are expected to act on the information provided. The power dynamics shaping knowledge production and dissemination in the Caribbean have produced two primary strands of information sources that influence decision-making and communication about and throughout the hurricane season. Communication via traditional media (radio, television, print news) and new media (social media, blogs and online databases) is often used to provide updates on impending tropical cyclone events. Collectively, media communications have emerged as an important avenue for disseminating information on hurricane awareness and disaster mitigation.

Scientific modelling and communication are essential components of forecasting and tracking tropical cyclone development and potential impact (Machen & Nost, 2021). This information is usually developed over longer periods and updated as new theories and ideas are tested. This information is often filtered by different stakeholders using different priority lenses and transformed for presentation to a wider audience (Bostrom et al., 2016). Hurricane preparedness communication through state-led initiatives tends to take a passive, top-down approach, while media entities highlight what they perceive to be relevant for their audience. When the intensity and risk of direct impact by different tropical cyclones become apparent to the technical experts, advisories are issued to communicate the level of urgency and possible risks associated with the hazard, again using various media. The top-down approach utilized by governance institutions and experts when communicating with the public focuses on information dissemination through a unidirectional knowledge transfer, rather than an interactive knowledge-sharing process. However, unidirectional approaches tend to be passive and less engaging to their intended audience, which reduces the prospects of being translated to action (Fig. 2).

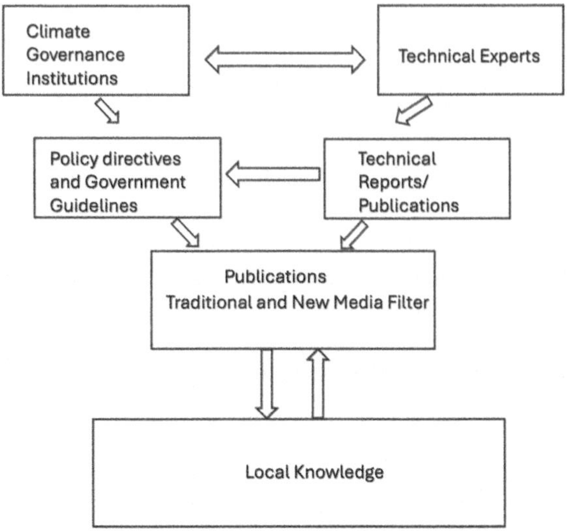

Fig. 2 Conventional knowledge transfer among stakeholders for Hurricane preparedness communication

Given the increased risks posed by more severe hurricane events in the region, hurricane preparedness communication in the Caribbean must employ strategies that will provoke proactivity among all stakeholders and especially at the level of local communities. The information sources accessed at the local level are critical to how people respond to hurricane risks. Added to this, hurricane preparedness response is inextricably linked to people's risk perception which intersects with the risk information they receive from a myriad of sources including state communication, informal channels and their lived experience (Bostrom et al., 2016). For example, in 2024, Hurricane Beryl affected the Caribbean leading to damages in Grenada, St. Vincent and the Grenadines and Jamaica (UN Migration, 2024). Hurricane-affected persons in Jamaica were reluctant to relocate even after their homes were damaged or destroyed because they did not perceive shelters to be adequately resourced to accommodate them and feared looting of their properties if they did not stay to secure them (UN Migration, 2024). This indicates that efforts to improve available information and support facilities may be insufficient or not effectively communicated to communities, and thus, result in a reluctance on the part of the community to utilize these facilities when these recommendations are made by the state.

Local communities, particularly those with severely limited resources and little to no technology-based warning systems, rely heavily on their lived experiences, generational knowledge shared primarily through oral traditions, and their informal networks to navigate hurricane risks. In some instances, local knowledge is the primary source of information for decision making and action to mitigate potential hurricane related disasters. Of note, the efficacy of these ways of knowing is built on a strong connection to the natural environment, heeding lessons from the past and camaraderie (Fugarino, 2015). This information is often not formally written anywhere and is adapted with every new experience and awareness. Sensory cues, including the changing sound of the river or noticeable absence of bird sounds, the visible changes in the colour of the sea or tidal movements, the stillness of the air and changes in animal behaviours will override technical and government communication and act as early warning systems for local communities.

While local understanding and experience of climate hazards including tropical cyclones are frequently documented in research outputs from various stakeholders (Prasad & Nigam, 2023; Rivosecchi & Singh, 2023; UN Migration, 2024; United Nations Development programme, 2018) engaged in developing proactive plans for mitigating disasters, it is often not given prominence in technical plans or frameworks. This is in part due to the multiuse nature of these reports and greater emphasis on using the language and terminologies shared within the technical communities, perhaps with the assumption that different organizations at different levels will add the relevant context and nuances for knowledge translation. This has left a communication gap that is being filled by informal communication and increases the risk of local communities being inundated with information or being exposed to misinformation, potentially undermining the trust of technical data. Consequently, local understandings become even more critical as people use a combination of technical data, their traditional knowledge and other media resources to generate an understanding of the phenomenon and potential risks for themselves, their property and the wider region. This interactive relationship among different knowledge sources also actively contributes to the evolution of local knowledge with the benefit of the addition of on-the-ground realities.

The preceding contextual information highlights the need for a shift from knowledge production and dissemination practices that minimise the value of local knowledge to one that facilitates the dialogic exchange between local knowledge and technical (scientific) knowledge.

Successful hurricane preparedness in the Caribbean relies on multiple stakeholders at all levels within and outside the region to ensure high-quality and effective communication as well as adequate resources. Climate governance institutions tend to operate globally, regionally and at the country level, with government officials and international partners often at the forefront of information and communication. Despite being the major power brokers in the discursive domain, they are often less prominent on the ground when a country is negatively affected by a hurricane. Other sociopolitical factors, including low trust of government officials, the increasing risk of disinformation and misinformation through new media and the increasing occurrence of hurricanes in the region, have different implications for different aspects of disaster risk reduction and management. It therefore becomes necessary to develop communication systems that are responsive to local changes while making technical knowledge more accessible through various media.

6 Facing the Unknown Potential of the Known Hazard

The last two decades have been replete with more severe and frequent hurricane activity in the Caribbean. The devastating impact of Hurricane Ivan in the region in September 2004 laid bare the fact that, as hurricanes become more severe in the region, lessons learnt from the past do not always translate into effective planning for the present or the future. Even with the benefit of new experiences and technologies, it is impossible to predict the total possible effect of a hurricane. While technical assessments of hurricane events tend to focus on injuries, loss of lives and property damage, hurricanes can also have a psycho-emotional impact on the people affected. Therefore, on the ground, embodied experiences and risk perception are the first warning systems for people and their communities. However, "when people lack knowledge about a hazard, their risk judgments are based on the degree to which they trust the responsible risk managers. As such, knowledge is conceptually related to trust" (Kellens et al., 2013, p. 42). Risk communicators, therefore, must be transparent and accountable to the communities they serve and hone strong trust-based relationships. They also have to actively manage the expectations that they can provide precise information that captures all the possible outcomes for a hurricane event. The complex interaction between technical facts and lived realities is among the major obstacles to building and maintaining public trust in technical information for hurricane preparedness.

7 The Case of Hurricane Ivan

In 2004, Hurricane Ivan, a Category 5 system, devastated several Caribbean Islands, including the Cayman Islands. Twenty years after Hurricane Ivan devastated the Cayman Islands, a chilling fact about the prospect of more severe hurricanes in the region has become apparent. There is no way to know what is going to happen or how bad things can get when a country is impacted by a tropical cyclone (*Hurricane Ivan Documentary: Past. Present. Resilient*, 2024). In a retrospective interview the Hon. Franz Manderson, Deputy Governor of the Cayman Islands in speaking about the devastation stated that, "I was not prepared for that. I didn't understand the type of devastation that a hurricane could bring" (*Hurricane Ivan Documentary: Past. Present. Resilient*, 2024, 1:48). This inability to comprehensively assess and anticipate the risks posed by Ivan manifested as a real challenge to what was thought to be a good hurricane preparedness plan. Dr. Gelia van Genderen, Director of the Water Authority Cayman Islands, shared that the Water Authority had what was thought to be a good plan. "We followed our plan, we put a lot of everything in place, but what we didn't really plan for was storm surge. That is a lesson learned" (*Hurricane Ivan Documentary: Past. Present. Resilient*, 2024, 3:48). The reflections on the Hurricane Ivan experience in the Cayman Islands also highlights the limitations of technology in monitoring the potential risks of impending hurricanes which can create a false sense of security. In the case of the Cayman Islands, the original path of the hurricane was expected to have minimal impact on the islands, however, Hurricane Ivan changed its course and made a more direct approach. Richard Hew, CEO and President Caribbean Utilities Company Ltd, stated

> I think, we tend to look at the computer screens and the television reports, and we see this line that's drawn, the projected track of the hurricane and we somehow believe that it's going to magically follow that. The forecasters are getting better at forecasting the track but we can't, um [sic], these hurricanes can be very unpredictable (*Hurricane Ivan Documentary: Past. Present. Resilient.*, 2024, 4:20).

The valuable insights gained from the reflections on the Cayman Islands' experience underscore the fact that, as the region prepares for the known aspects of the potential impacts of hurricanes, there are many unknown variables that will manifest during and post-hurricane events.

Another important area to consider, in addition to the impact of hurricanes on critical infrastructure and the wider populace, is the impact of hurricane events on the people working to implement disaster management plans during an active hurricane event. When essential service workers are activated to keep communication going, they provide logistical, medical and evacuation services, among many other critical functions during hazard events. They are often expected to have all the solutions ready and provide leadership for the communities affected. Concurrently, they also have close personal connections, such as their own families, properties, and communities that are in imminent danger. When their capacity, including material resources and their psychological capacity to support the community, is overwhelmed, their inherent vulnerabilities are heightened. This makes an already complex and tenuous situation more challenging as life-altering decisions are made to preserve life and property to the best of their abilities. John Tibbetts, Director General of the Cayman Islands National Water Service, whose family was with him at the Fire Department where some members of the wider community sought refuge, recounts the moment the storm surge breached the building from which he was managing community response. He recalls,

> we could see the storm surge coming in right, and I go around and I said, I went to my family and I go like guys I think it's time we go upstairs and stuff like that and my son looked at me and he says 'Dad I think you are a little bit too late' (*Hurricane Ivan Documentary: Past. Present. Resilient.*, 2024, 11:15).

In that moment, he had to embrace his role in his family even while navigating his professional responsibilities to help his family and the wider community navigate the oncoming hazard and trying to stay safe. This is the experience of many frontline responders working in an active hurricane event. Therefore, capacity building for hurricane preparedness and the necessary tactical and psychosocial support skills must transcend equipping those expected to be professionals on the frontline to include the wider community network so that mutual aid is readily available for all despite professional designations. Community members will be able to step into support roles when the designated professionals are overstretched or unavailable. This point was made clear with the example shared by Dr. Hazel Brown the Chief

Nursing Officer with the Health services Authority (HSA), who shared that at the time the entire island including the hospital was

> without power for about 23 hours and we had a man at the time on the ventilator, so we had to manually ventilate this person using an AMBU bag, and everybody, housekeepers, porters, physicians, nurses, physio, anyone who was in the hospital, administrators... Anyone who was available to squeeze that bag helped to keep that gentleman ventilated for those 23 hours" (*Hurricane Ivan Documentary: Past. Present. Resilient.*, 2024, 3:59)

Noteworthy is that in the face of adversity, people will work together to respond to the challenges; however, a good strategy is to equip people with varied skills before a crisis-level event. This will make it easier for them to step into response roles when needed and enhance their understanding of the limitations faced by the professionally designated team.

8 The Case of Hurricanes Irma and Maria

The Caribbean region continues to experience varied levels of tropical cyclones annually. Thirteen years after Hurricane Ivan, on September 6, 2017, Hurricane Irma, the 9th named tropical cyclone of the season with a Category 5 classification, hit Barbuda. Irma left the island of Barbuda uninhabitable with approximately 95% of the houses damaged and the entire population displaced. Hurricane Irma proceeded to the flood affected (August 7, 2017) British Virgin Islands (BVI), damaging major infrastructure including roads and telecommunications (United Nations Development programme, 2018). On September 18, 2017, Category 5 Hurricane Maria impacted Dominica affecting the entire population, damaging 90% of houses as well as power and water infrastructure (United Nations Development Programme, 2018). While, the region has a strong culture of multi-country mobilisation to support affected islands in a time of crisis, when telecommunication went down in Dominica and help began to move in, disaster managers were overwhelmed with human resources that they had no way of effectively deploying because teams from other countries were unfamiliar with the island's terrain (Personal Communication, May 26, 2025). Upon reflection, it became apparent that a major risk to effective communication post-hazard impact is the absence of up-to-date baseline data on the countries before catastrophic events (ACAPS, 2018). Another critical insight garnered from the experience was the realization that existing capacities on the ground and in the region, even with increased training initiatives prior, were overwhelmed by the hurricane. Furthermore, it was determined that "[B]etter communication and collaboration between UN [the United Nations] agencies, actors, and CDEMA [the Caribbean Disaster Emergency Management Agency] was needed, to avoid considerable delays in assessments and overlap in activities" (ACAPS, 2018, p. 1). This underscores the fact that "context-specific communication mechanisms need to be tailored in order to ensure two-way communication with affected communities" (ACAPS, 2018, p. 1).

9 The Case of Hurricane Beryl

On June 29, 2024, Hurricane Beryl made the historical transition to a Category 5 hurricane, earlier than any other Atlantic hurricane on record. Hurricane Beryl affected multiple islands, damaging communication infrastructure, buildings, and water resources. While response agencies reported that the felt improved technologies and communication among partner organization was more efficient leading to a more effective warning systems and coordinated response post event (UN Office for Disaster Risk Reduction (UNDRR), 2025) reports on the ground indicated that communities were taken by surprise and may not have prepared as much as they could have for the impending hazards (Villani, 2024). Added to this is the recognition of the limited understanding of the psychological and social dimensions that impact individuals and communities' capacity to respond to, and adapt for future hurricane risks because post-event assessments tend to focus primarily on infrastructure, loss of lives and economic gains (Smith, 2025). Smith's (2025) exploration of the coping strategies and support systems available to people in the aftermath of Hurricane Beryl highlights the intersecting experiences of displacement, loss of livelihood, shelter, and property while facing the need to rebuild their lives and community. This demonstrates the differential impact for different demographic segments of the population including, children, the elderly, and people living with disabilities. The study also reiterated the importance of strengthening community capacity as they are the first responders, securing shelter and property before and during hurricane events. Added to that, they provide psychological and other recovery support in the critical post event period when outside support is not yet on the ground and networked communication systems are inactive (Smith, 2025).

10 A Caribbean Framework for Disaster Management

A regional inter-governmental agency, the Caribbean Disaster Emergency Response Agency (CDERA), now the Caribbean Disaster Management Agency (CDEMA) was established in 1991 to institutionalize the cooperation among Caribbean Community (CARICOM) states and strengthen the region's ability to mitigate disasters in the region and strategically respond to major hazard and disasters events. In 2001, the Caribbean Disaster Management (CDM) Strategy was adopted by CARICOM. The strategy aims to "embrace and integrate relevant initiatives underway, whilst identifying gaps which needed addressing" (Caribbean Disaster Emergency Management Agency, 2014, p. 21). Subsequent revisions and consultations culminated in various iterations of the framework, with the most current being the Regional Comprehensive Disaster Management (CDM) Strategy and Results Framework 2014–2024. The CDM Strategy and Results Framework 2014–2024 prioritizes institutional strengthening, increased knowledge development and training, integrating CDM in different sectors, and honing sustainable community disaster resilience. The strategy has focused on building technical capacity and leveraging community capacity for tasks-based activities when there is a major disaster event. However, as demonstrated by the cases presented in this chapter, communities are first in line for the impact of hurricane hazards and the first responders on the ground before additional help can be mobilized. Therefore, their capacity to understand the hurricane risks they face annually, as well as to communicate their resource needs and capacity is essential to an effective disaster mitigation strategy. Consequently, there is a need for a stronger integrated knowledge exchange framework that explicitly considers local knowledge and experiences in the region while building community capacity.

11 Rebuilding a Culture of Hurricane Preparedness in the Caribbean

The macro discourse on Climate Change Risks (CCRs), including more intense hurricane seasons have largely remained esoteric in practice, often engaging leaders in subsectors, researchers, academicians and technical practitioners, leaving the translation on the ground to what gets prioritized in media spaces and local organizations working with communities. Added to this, the knowledge production and dissemination mechanisms, such as meetings and written reports utilized by institutional organizations are proving inefficient for enhancing resource allocation and awareness across countries. In 2025, the United Nations (UN) found that the despite the UN secretariat spending "more than 10% of the regular budget" (p. 18) on meetings (27,000 in 2024) and reports (more than 1100 in 2024), a significant increase in the number of reports produced since 1990, duplication of efforts and underutilization of reports is a major limitation of these mechanisms. Generally, existing cultural norms in the day-to-day affairs of Caribbean societies have weakened stakeholder relationships and eroded trust in the decision makers and power brokers in society. However, trust is an important component for effective communication (Haynes et al., 2008).

The World Meteorological Organization's (WMO) Deputy Secretary-General, Ko Barrett, reiterated that hurricane hazards cause significant setbacks to the socioeconomic development of Caribbean nations when they are affected. Barrett (2024) notes that,

> It takes just one landfalling hurricane to set back years of socio-economic development. For example, Hurricane Maria in 2017 cost Dominica 800% of its Gross Domestic Product. Early warnings by the WMO community and improved disaster risk management have dramatically reduced fatalities, but Small Island Developing States in the Caribbean still suffer disproportionately (World Meteorological Organization, 2024, para 7.)

Effective communication is a critical element of disaster mitigation and capacity building in the Caribbean. While there are, on paper, laudable progress in regional and national hurricane preparedness strategies as a result of lessons learnt over the last three decades, there is still the need for improvement, particularly in communication that enhances the culture of preparedness on the ground and strengthens

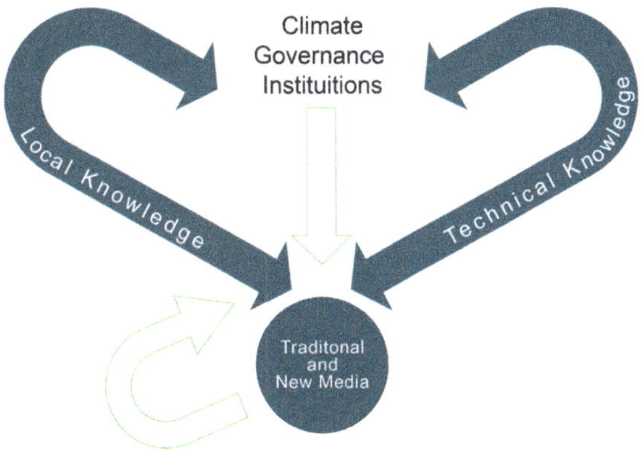

Fig. 3 Ideal knowledge exchange among stakeholders for hurricane preparedness communication

collaboration among local communities and technical experts. It is essential that the communication strategies employed are interactive with technical assessments integrating community-based knowledge and improved technical knowledge. A variety of media should be incorporated to ensure that different demographics are receiving information put out by the technical stakeholders, as well as being engaged for feedback based on their lived realities and experiences (Fig. 3).

12 A Multiphase Communication Framework

The cases of Hurricanes Ivan, Irma, Maria, and Beryl highlight the reality that effective communication for hurricane preparedness must transcend one way information sharing to incorporate active capacity building, resource mobilization and integration of local knowledge. The nature of the Atlantic Hurricane Season, known and recurring, necessitates four distinct components for an effective communication strategy to enhance hurricane preparedness and effective disaster mitigation in the Caribbean. These communication measures can be implemented sequentially or concurrently depending on the existing dynamics in each country.

I. *Hurricane Preparation Communication:* This communication should take place year-round and intensified at the onset of the Atlantic Hurricane Season. This phase includes active information sharing among communities, government entities and other technical stakeholders about the state of infrastructure, resources and community mobilization plans in the advent of a hurricane event, as well as proactively addressing gaps as they are identified. This will facilitate effective planning and implementation of strategic capacity building, collection of baseline data on areas of vulnerability as well as the improvement of infrastructure and resource management. This will also ensure that when communication infrastructure is damaged the communities are able to self-organize until further support can be provided.

Hurricane preparation communication should be multisectoral and supported by the mobilization of adequate resources to strengthen the capacity of communities to shelter in place or relocate when necessary. During this phase of communication, efforts should be made to help the public adjust their approach to hurricane risk assessment to include the understanding that a hurricane's potential to trigger a disaster is not solely based on the category level assigned. Added to this, the emerging trend of rapid development of systems over short periods as they strengthen to higher levels means that communication with the public should help them to understand how quickly the outlook can change, as well as the importance of preparing for the risk of the rapid intensification before it happens. Individual and community capacity to accurately assess risks and resources are honed before a hurricane event. However, it is important to note that communication preparation without the attending resources will be insufficient to ensure community based capacity to respond to the communication guidelines.

II. *Hurricane Warning Communication:* Hurricane warning typically progresses along a continuum progressing from a hurricane watch when there is a system that can potentially impact within 48 h to a hurricane warning a system that is likely to make an impact within 36 h (US Department of Commerce, n.d.). Added to the hurricane-specific communications are communication about secondary hazards such as storm surges, extreme wind and flood risks. It is usually at this stage that on the ground understanding is critical for disaster risk mitigations as people activate their safeguarding measures and ensure that the community understands how to access resources and support. Community based disaster mitigation plans are activated and people are on stand by for the post hazard response.

III. *Post-Hazard Communication*: This communication occurs in the immediate aftermath of a hurricane event. It aims to facilitate information exchange between local communities and technical stakeholders for the assessment of the hurricane impact and the response activities necessary. This necessarily relies on strong relationships between community and technical stakeholders before a hurricane event and a clear communication strategy that considers both mainstream, technology-assisted communication as well as informal communication when communication technologies are impacted.

IV. *Recovery Communication:* Hurricane events invariably produce a continuum of hazardous situation and recovery timelines. Typically, there is a communication blitz in the immediate aftermath of a hurricane, particularly if it triggered disaster outcomes. Recovery communication should aim to provide clear up-to-date information to the public while facilitating ongoing evaluation and adjustments for transitioning to medium term and longer-term responses. Recovery communication must also be sensitive to the diverse realities intersecting with hurricane related concerns. It is imperative that a multidirectional communication approach is used because despite local communities' need for specific aid from various stakeholders, they are also an asset for recovery and hold critical information that can expedite recovery efforts. Therefore, recovery communication not only facilitates the provision of aid to affected people, but also creates an opportunity to build strong relationships across communities and bridge knowledges gaps for all stakeholders thereby enhancing hurricane preparedness for the next hurricane season.

13 Conclusion

The Caribbean has made significant progress in organizing its ability to respond to the fallouts that can be caused by hurricanes during the annual Atlantic Hurricane Season. However, a major impediment is the translation of regional technical capacity to sustainable on the ground capacity and knowledge exchange. Lessons from hurricanes Ivan, Irma, Maria, and Beryl, underscore critical vulnerabilities that persist, including the increasing unpredictability of the development of hurricanes for a season and the risk perception of people on the ground. It is also apparent that a passive top down approach to information sharing about impending hurricanes without adequate resource mobilization and capacity building is not sufficient for catalysing localized hurricane preparedness However, by actively fostering a dialogic exchange between local communities and technical experts, integrating community-based knowledge and enhanced scientific understanding as well as, leveraging capacity building through a multipronged communication strategy, the Caribbean, as a region can enhance hurricane preparedness and bolster the resilience capacities of local communities against the escalating threats of more severe hurricane events due to escalating climate change.

References

ACAPS (2018). ACAPS Dominica lessons learned from hurricane Maria, January 2018.

Alatas, S. F. (2022). Knowledge hegemonies and autonomous knowledge. *Third World Q.* 1–18.https://doi.org/10.1080/01436597.2022.2124155

Baptiste, A. K., & Rhiney, K. (2016) Climate justice and the Caribbean: An introduction. Geoforum, *73*, 17–21. Available at: https://doi.org/10.1016/j.geoforum.2016.04.008

Barker, D., & Miller, D. (1990). Hurricane Gilbert: Anthropomorphising a natural disaster. *Area, 22*, 107–116.

Benz, P., Wei, H., & Chiu-Tat Wang, J. (2022) Alternative knowledges: Communities, creativity, narration. *Cubic Journal*, *5*(5), 4–13. Available at: https://doi.org/10.31182/cubic.2022.5.44

Bishop, M. L., Bouhia, R., Carter, S. G., Corbett, J., Lindsay, C., Scobie, M., & Wilkinson, E. (2025). *Sustaining development in small islands: Climate change, geopolitical security, and the permissive liberal order*. Cambridge University Press.

Bonilla, Y. (2020) The coloniality of disaster: Race, empire, and the temporal logics of emergency in Puerto Rico, USA. *Political Geography*, *78*, 102181. Available at: https://doi.org/10.1016/j.polgeo.2020.102181

Bostrom, A., Morss, R. E., Lazo, J. K., Demuth, J. L., Lazrus, H., & Hudson, R. (2016). A mental models study of hurricane forecast and warning production, communication, and decision-making. *Weather Climate Society, 8*, 1111–1129. https://doi.org/10.1175/WCAS-D-15-0033.1

Caribbean Disaster Emergency Management Agency. (2014). Regional Comprehensive Disaster Management (CDM) Strategy and Results Framework 2014–2024.

Derbyshire, J., & Aven, T. (2025). Out of sight but still in mind: Developing an expectation for surprises by formalizing unknowledge in a contemporary risk-assessment framework. *Risk Analysis, 45*, 1199–1206. https://doi.org/10.1111/risa.17661

Dookie, D. S., & Spence-Hemmings, J. (2022). The timing of storm awareness in the Caribbean: The utility of climate information for improved disaster preparedness. *Disasters, 46*.https://doi.org/10.1111/disa.12540

Fugarino, V. (2015) *Every hurricane leaves you with different priorities: Preparation, cultural response, and narrative entitlement.* Memorial University of Newfoundland. Available at: https://memorial.scholaris.ca/server/api/core/bitstreams/810c5da0-7f2c-4afe-809d-12ba60d4c516/content.

Government of Canada, Global Affairs Canada. (2025). Hurricane season: Be aware of the risks to your safety before you go [WWW Document]. Glob. Aff. Can. https://www.international.gc.ca/world-monde/stories-histoires/2025/hurricane-season-risks-saison-ouragans-risques.aspx?lang=eng (Accessed 28 Aug 25)

Hao, Z., & Lun, Y. (2024). Using traditional knowledge to reduce disaster risk—A case of Tibetans in Deqen County, Yunnan Province. *International Journal Disaster Risk Reduction, 108*, Article 104492. https://doi.org/10.1016/j.ijdrr.2024.104492

Haynes, K., Barclay, J., & Pidgeon, N. (2008). The issue of trust and its influence on risk communication during a volcanic crisis. *Bulletin of Volcanology, 70*, 605–621. https://doi.org/10.1007/s00445-007-0156-z

Hurricane Ivan Documentary: Past. Present. Resilient. (2024).

Jones, E., Doughorty, K., & Brown, P. (2022). 'Building back better' in the context of multi-hazards in the Caribbean. *Disasters, 46*, S151–S165. https://doi.org/10.1111/disa.12545

Kellens, W., Terpstra, T., & De Maeyer, P. (2013). Perception and communication of flood risks: A systematic review of empirical research. *Risk Analysis, 33*, 24–49. https://doi.org/10.1111/j.1539-6924.2012.01844.x

Machen, R., & Nost, E. (2021). Thinking algorithmically: The making of hegemonic knowledge in climate governance. *Transactions of the Institute of British Geographers, 46*, 555–569. https://doi.org/10.1111/tran.12441

Meeks, B., & Lindahl, F. (2001). New Caribbean thought: A reader. University of the West Indies Press.

Meteorological Service Jamaica, Branch of the Ministry of Economic Growth and Job Creation. (2025). Bulletin no:17. Office of the Prime Minister. Available at: https://opm.gov.jm/wp-content/uploads/2025/10/Bulletin-17.pdf

UN Migration. (2024). Situation report 3: Beryl in The Caribbean. United Nations.

Morss, R. E., Cuite, C. L., & Demuth, J. L. (2024). What predicts hurricane evacuation decisions? The importance of efficacy beliefs, risk perceptions, and other factors. *Npj Natural Hazards, 1*, 1–15. https://doi.org/10.1038/s44304-024-00025-8

Nakhid-Chatoor, M. et al. (2018). Exploring liming and ole talk as a culturally relevant methodology for researching with Caribbean people. *International Journal of Qualitative Methods, 17*(1), 160940691881377. Available at: https://doi.org/10.1177/1609406918813772

NOAA predicts above-normal 2025 Atlantic Hurricane Season | National Oceanic and Atmospheric Administration [WWW Document]. (2025). https://www.noaa.gov/news-release/noaa-predicts-above-normal-2025-atlantic-hurricane-season (Accessed 28Aug 25)

Prasad, V., & Nigam, B. (2023). Role of traditional and indigenous knowledge in disaster management. In G.K. Panda, U. Chatterjee, N. Bandyopadhyay, M.D. Setiawati, & D. Banerjee, (Eds.),

Indigenous knowledge and disaster risk reduction: Insight towards perception, response, adaptation and sustainability (pp. 19–34). Springer International Publishing. https://doi.org/10.1007/978-3-031-26143-5_2

Rhiney, K. (2020) Dispossession, disaster capitalism and the post-hurricane context in the Caribbean. *Political Geography, 78*, 102171. Available at: https://doi.org/10.1016/j.polgeo.2020.102171

Rivosecchi, A., & Singh, M. (2023). Small Island city flood risk assessment: The case of Kingston, Jamaica. *Water, 15*.https://doi.org/10.3390/w15223936

Sindoni, M. G. (2010). Creole in the Caribbean: How oral discourse creates cultural identities. *Journal of African*, 217–236.https://doi.org/10.4000/africanistes.2563

Smith, R.-A. (2025). Beyond conventional metrics: Exploring psychosocial impacts and coping strategies in hurricane-affected communities in Jamaica.https://doi.org/10.2139/ssrn.5291489

UN Office for Disaster Risk Reduction (UNDRR). (2025). United Nations office for disaster risk reduction: Annual report 2024. United Nations.

UNDRR and ECLAC (Caribbean Economic Commission for Latin America and the Caribbean) (2021) The Coronavirus Disease (COVID-19) Pandemic—An Opportunity for a Systematic Approach to Disaster Risk for the Caribbean. United Nations.

United Nations Development Programme. (2018). From early recovery to long-term resilience in the Caribbean Hurricanes Irma and Maria: One year on.

United Nations Office for Disaster Risk Reduction (UNDRR)-Regional Office for the Americas and the Caribbean, Economic Commission for Latin America and the Caribbean (ECLAC/CEPAL). (2021).

United Nations. (2025). UN80 Initiative: Workstream 2 report of the mandate implementation review. United Nations.

US Department of Commerce, N. (n.d.) Watch/Warning/Advisory Definitions [WWW Document]. https://www.weather.gov/lwx/warningsdefined (Accessed 14 Sept 25).

Villani, M. (2024). Calgary's Caribbean community pleads for support as Hurricane Beryl causes widespread damage [WWW Document]. CTVNews. https://www.ctvnews.ca/calgary/article/calgarys-caribbean-community-pleads-for-support-as-hurricane-beryl-causes-widespread-damage/ (Accessed 14 Sept 25).

Whitmer, D. E., & Sims, V. K. (2021). An implicit—Not explicit—Understanding of Hurricane storms. *Weather Climate Society, 13*, 1043–1053. https://doi.org/10.1175/WCAS-D-20-0148.1

World Meteorological Organization. (2024). Record-breaking Hurricane Beryl threatens Caribbean [WWW Document]. World Meteorol. Organ. https://wmo.int/media/news/record-breaking-hurricane-beryl-threatens-caribbean (Accessed 13 Sept 25)

Hurricane-Induced Agricultural Disruptions in the Gulf Region: A Framework for Risk Mitigation and Resilience

Mohammad A. Shovon and Rubayet Bin Mostafiz

Abstract The U.S. Gulf Coast is a vital contributor to national agricultural output, producing crops such as corn, rice, soybeans, and cotton. However, this region has become increasingly vulnerable to extreme weather events, particularly hurricanes, due to its geographical location, meteorological conditions, and anthropogenic factors. Recent hurricanes have inflicted widespread damage on Gulf Coast agriculture, affecting crops and farm infrastructure, exposing the fragility of production systems and weaknesses in distribution networks. These disruptions pose significant risks to national food security and rural livelihoods. This work reviews the economic and operational impacts of previous hurricanes on agricultural systems in the Gulf Coast region, focusing especially on Louisiana. It proposes a comprehensive framework for assessing risk and outlining both mitigation and adaptation strategies aimed at reducing vulnerability and enhancing resilience under a changing climate. This framework explores short-term measures, including crop diversification, early harvesting, cover cropping, and improved drainage, while emphasizing long-term strategies such as infrastructure reinforcement, climate-resilient crop development, farmer networking and training, and institutional mechanisms including early warning systems and community-based adaptation efforts. Bridging research gaps and addressing implementation challenges through integrated and inclusive planning is vital to safeguarding Gulf Coast crop production and food systems under escalating hurricane threats.

Keywords Hurricanes · Gulf coast agriculture · Food security · Resilient strategies · Risk mitigation

M. A. Shovon (✉) · R. B. Mostafiz
Department of Biological and Agricultural Engineering, LaHouse Research and Education Center, Louisiana State University Agricultural Center, Baton Rouge, Louisiana, USA
e-mail: achowdhury.shovon@gmail.com

J. Collins et al. (eds.), *Tropical Cyclone Risk and Resilience*,
https://doi.org/10.1007/978-3-032-28745-8_10

1 Introduction

Climate change is expected to increase the intensity of tropical cyclones, the most intense of which in the Atlantic-Caribbean-Gulf basin are commonly known as hurricanes, making their impacts on agricultural systems, rural livelihoods, and food security a critical focus of scientific investigation. Balaguru et al. (2016), for example, project that mean hurricane intensity in the Atlantic will increase by 1.8–4.2% by the late twenty-first century. Mengel et al. (2016), on the other hand, noted that sea-level rise will also continue accelerating due to human-driven climate change, and even if global temperatures stabilize, ocean temperatures will keep rising for centuries because of long-term climate system responses. With tropical cyclones becoming more intense and sea levels continuing to rise, storm surges are projected to grow far more destructive in the Gulf Coast, posing a greater hazard than the hurricanes' severe wind or heavy precipitation impact (Balaguru et al., 2016). Employing GIS and spatial analysis, Shao et al. (2020) suggest that increased hurricane and flood exposure in the Gulf Coast will significantly affect agricultural productivity and rural livelihoods.

Tropical cyclones are considered one of the United States' most severe and destructive types of natural disasters (Biram et al., 2025). They are referred to as hurricanes in the North Atlantic and Eastern Pacific Oceans, typhoons in the Northwest Pacific Ocean, willy-willy in Australia, or simply cyclones in the Indian Ocean and other ocean basins (Ginis, 2020; Raátz, 2006). When winds in the Atlantic and northeastern Pacific exceed 119 km/h, the resulting intense tropical cyclones are classified as hurricanes (Spiridonov et al., 2021). Under climate change, tropical cyclones have become markedly more intense, posing greater risks to vulnerable regions and significantly amplifying disaster-related damages and disruptions (Bandyopadhyay, 2022; Park, 2021). In the U.S., tropical cyclones have become the most economically disruptive and physically damaging severe phenomena (Motha, 2011), in part because they deliver destructive, multi-hazard impacts, including coastal storm surge, inland flooding, strong winds, including tornadoes, power outages that impact marginalized communities disproportionately, and often subsequent drought from storm-defoliated vegetation (Nelson & Miller, 2023). Consequently, numerous high-impact storms in recent decades have made landfall in the Gulf Coast region, a critical ecological and economic hub for the Americas that supports diverse industries, agricultural systems, and ecosystems (McKinney et al., 2021).

Along the U.S. Gulf Coast, climate change is expected to further increase hurricane frequency by altering atmospheric steering patterns and reducing vertical wind shear (Motha, 2011). According to Damoah et al. (2024), the Gulf Coast has been particularly vulnerable to severe hurricanes in recent years due to meteorological, topographical, and anthropogenic factors. Similarly, Hu et al. (2012) noted that the United States, especially the northern Gulf of Mexico and the Louisiana Gulf Coast, is highly vulnerable to frequent tropical storms and hurricanes because of its tropical–subtropical setting, shallow bathymetry, coastal geometry, and distinctive landscape characteristics. A single intense, Category 5 hurricane can wipe out decades of development gains, especially with a storm surge that damages coastal facilities and erodes shorelines (Lewis, 2022).

The U.S. Gulf Coast region, characterized by a broad, flattened U-shaped coastline, extends more than 2000 km (1200 mi) from southwestern Florida across southern Alabama, Mississippi, and Louisiana, and southwestward along southeastern Texas (Damoah et al., 2024). The region is a vital contributor to the nation's economy, with agriculture being one of its important sectors (Tchounwou, 1999; Turner & Rabalais, 2019; Xu et al., 2025). Major crops grown in the Gulf Coast region include corn, cotton, soybeans, wheat, peanuts, rice, and sorghum (Bundy et al., 2023; Webster & Nichols, 2012). Although coastal regions offer fertile soils and favorable climatic conditions suited to agriculture, their exposure to recurring tropical cyclones threatens long-term agricultural sustainability and causes significant damage to essential infrastructure (Selva et al., 2025). Because many tropical cyclones make landfall near the end of the growing season, they often cause substantial losses in agricultural crop yields. However, their impacts on agricultural production show a varying pattern across the Gulf Coast. Bundy et al. (2023), for instance, reported that tropical cyclones can provide some benefits to crops when they occur at the right time, as they may enhance overall crop conditions and yield prospects during early and mid-growing stages by supplying needed soil moisture for development and reproduction, although the advantages are typically short-lived. Using data from the United States Department of Agriculture's National Agricultural Statistics Service, their study assessed the impacts of tropical cyclones on major Gulf Coast crops. The results show that during the growing season, tropical cyclones initially degrade crop conditions, reducing yield expectations by up to 6 percent (Bundy et al., 2023). However, partial recovery over subsequent months often occurs, limiting final yield losses to approximately 3 percent by harvest (Bundy et al., 2023). Earlier empirical studies provide additional evidence of hurricane-induced crop yield losses across the Gulf Coast region. Using regression analysis of 1951–2004 crop-yield data, Chen and McCarl (2009) report hurricane-related yield reductions in Florida (grapefruit, cotton, potatoes, sugarcane), Louisiana (corn, cotton, sorghum, wheat, sugarcane), Mississippi (corn, cotton, soybeans, rice), and Texas (corn, cotton, soybeans, rice, oats, potatoes, tomatoes). While this study provides an important historical benchmark, recent evidence indicates that the agricultural impacts of hurricanes have increased substantially in both magnitude and economic significance. In 2024, hurricanes caused approximately $6.7 billion in U.S. crop losses, with nearly 38% of this damage falling outside federal crop insurance coverage (Munch, 2025). More than $3 billion in direct crop production losses were reported across the Southeast, disproportionately affecting corn, soybeans, cotton, fruits, and sugarcane. Agricultural losses were especially severe across Gulf Coast and Southeastern states, reflecting heightened exposure of crop systems to hurricane-related wind and rainfall damage (Munch, 2025). In 2005, Hurricane Katrina alone caused approximately US$1 billion in production losses across sugarcane, corn, soybeans, and cotton in the Gulf Coast (Schnepf et al., 2005).

Despite the considerable impact of hurricanes on agriculture in the U.S. Gulf Coast, there has been comparatively limited advancement in quantitative or qualitative information regarding the scope and effectiveness of mitigation and adaptation strategies aimed at reducing the effects of hurricanes on agricultural systems in the region. This is particularly concerning because the top five U.S. crop exports, corn, soybeans and their derivative products, wheat products, cotton, and tree nuts, represent a substantial share of total U.S. agricultural export value and are directly linked to the Gulf region through the Mississippi—Missouri River drainage basin (Frittelli, 2005; Padilla et al., 2023). Prior studies have largely focused on individual hurricane events, such as Katrina or Harvey, within specific hurricane-prone regions, including Florida, Texas, Puerto Rico, and the Caribbean. In contrast, this study provides a comprehensive assessment of hurricane impacts on agriculture across multiple hurricane-prone regions and countries, explicitly linking agricultural damage to food security and rural livelihoods, and proposes both short-term and long-term adaptation and mitigation strategies. Focusing on major Gulf Coast hurricanes between 2000 and 2024, with particular emphasis on Louisiana, this study offers an integrated perspective that has not been systematically addressed in prior research. Our framework integrates preparedness, on-farm practices, infrastructure, insurance, and policy measures to bolster resilience and recovery.

2 Hurricane Impacts on the Agricultural System

Climate and agriculture are so obviously deeply intertwined, with extreme events such as hurricanes posing serious challenges to agricultural farming systems (Twumasi et al., 2024). Hurricane effects on agriculture are dependent on a number of variables, including the type of crop, its stage of development, its type, and whether it is directly in the path of the hurricane (Chikodzi & Nhamo, 2021). Agricultural damage by hurricanes is driven by multiple factors, including soil salinization due to storm surges, loss of productive topsoil, outbreak of diseases, and damage to farming equipment, irrigation systems, and associated infrastructure (Zafar et al., 2024). Hurricanes exert long-lasting detrimental effects on agricultural output, primarily through high winds, heavy rainfall, and flooding that directly damage crops (Her et al., 2018; Zhang & Zhu, 2023). Hurricane storm surges that flood agricultural land can affect crops and soil structure in the short and long term (Williams, 2010). Moreover, elevated soil or water salinity restricts water and nutrient absorption in salt-sensitive crops like rice, stunting growth and drastically lowering yields in coastal farming systems (Kaniewski et al., 2016). The long-lasting salinization of coastal lands caused by hurricanes poses one of the most serious agricultural challenges, leading to significant yield loss (Her et al., 2018). Tropical cyclones also intensify soil fertility loss and devastate agricultural fields, exceeding the impacts of typical land use practices; for instance, hurricane-induced heavy rains accelerate nutrient depletion in slash-and-burn systems, leaving soils deficient in nitrogen, phosphorus, potassium, and other cations, which can reduce corn yields by as much as 75% (Gay-des-Combes et al., 2017).

Agricultural infrastructure, such as drying sheds, storage facilities, irrigation systems, and fencing, is often constructed to lower standards than residential structures, making it highly susceptible to hurricane and cyclone damage (Dorfman et al., 2018). In many regions, storm-surge-induced embankment breaches allow seawater to remain trapped on farmland for extended periods, causing prolonged agricultural disruption and elevating long-term food security risks (Barui et al., 2025). These impacts extend beyond farms, as widespread agricultural losses are compounded by severe damage to transportation networks that move goods to market and to the oil and gas infrastructure that supplies energy for mechanized agriculture.

Evidence from diverse regions illustrates that hurricane and cyclone impacts on agriculture are highly context-specific, yet consistently severe across vulnerable farming systems worldwide (Vink & Ahsan, 2018). Using online news sources, (Yuen et al., 2022) compiled a database of cyclone-induced rice damage from 1970 to 2018 across four major Asian rice-producing countries, namely the Philippines, Vietnam, Bangladesh, and Myanmar, documenting affected locations, damaged areas, and losses in production and value. Of the 1046 recorded cyclone events, 138 were associated with rice damage across all four countries, with particularly severe impacts observed in the Philippines and Vietnam, and substantial losses also reported in Bangladesh and Myanmar; the average rice area damaged per event ranged from over 40,000 ha in Vietnam to more than 420,000 ha in Myanmar, while production losses reached hundreds of thousands to several million metric tons per cyclone (Yuen et al., 2022), underscoring the widespread vulnerability of rice systems in cyclone-prone regions. In Myanmar, where agriculture is the primary livelihood, rice occupies nearly half of all cultivated land and is predominantly grown in low-lying coastal areas. As tropical cyclones intensify due to global warming, these regions are increasingly exposed to storm-related risks, raising serious concerns about the future sustainability of rice production (Hirano, 2021). In regions such as the Bay of Bengal, cyclonic storms are increasing in frequency and intensity during pre- and post-monsoon periods, disrupting agricultural crop at critical growing stages (Selva et al., 2025). Comparable patterns have been observed elsewhere, such as in Cuba, where Hurricane Ike caused extensive agricultural devastation through extreme rainfall and flooding, destroying up to 50% of the sugar crop, eliminating as much as 80% of banana and plantain production in some regions, damaging more than 350,000 fruit trees, and wiping out approximately 10,000 hectares of vegetables and root crops (Messina, 2009). Similarly, Central America, particularly nations such as Guatemala and Honduras, faces growing threats from droughts and hurricanes that endanger coffee production and other key agricultural exports (Bunn et al., 2015; Jamil, 2024).

3 Agricultural Damage, Food Security, and Rural Livelihoods under Hurricane Risk

Climate change is expected to reduce the productivity of crops, fisheries, and livestock on a global scale, potentially threatening food security (Islam & Nursey-Bray, 2017; Gopalakrishnan et al., 2019). Extreme weather events associated with climate change, including hurricanes, floods, droughts, and storms, increasingly disrupt agricultural systems, impact agricultural commodity returns, and contribute to global food system instability (Kianersi et al., 2021; Thompson et al., 2023; Shovon, 2024). According to (Marcelin et al., 2016), extreme climate events including hurricanes intensify food insecurity by limiting food production and increasing food prices. Harper and Fine (2024) noted that extreme weather events like hurricanes pose significant challenges to agricultural animal operations, leading to loss of animals and commodities due to disruptions and environmental extremes. Hurricanes also disrupt terrestrial and aquatic ecosystems vital to agriculture and aquaculture by threatening vital ecosystem services (Her et al., 2018; Williams, 2009). Fisheries reliant on coastal wetlands, along with forestry and agronomic crops, face serious risks during extreme events (Shovon et al., 2025). Several other studies noted that natural disasters, particularly hurricanes, are increasingly intensifying the risk of food insecurity, with their frequent occurrence triggering cascading effects across physical, human, and response systems that undermine agricultural productivity, food security, economic stability, and overall human well-being (Álvarez-Berríos et al., 2021; Mark et al., 2025). Moreover, communities are facing ever-increasing risks from droughts, tropical cyclones (including tropical storms, typhoons, and hurricanes), and floods, leading to lower incomes, increased food insecurity, and growing environmental stress (Cottrell et al., 2019; Parajuli et al., 2019; Vogel et al., 2019; Thompson et al., 2023). These extreme weather events interrupt farming, destroy crops, and threaten food security around the world (Jamil, 2024).

Against this backdrop of increasing climate-driven disruptions to agricultural systems, understanding food security and its vulnerability to extreme events is essential for assessing the broader implications of hurricanes and tropical cyclones. Food security depends on consistent physical, social, and economic access to adequate, safe, and nutritious food (MacNabb and Fletcher, 2019). This condition is increasingly threatened by disasters, including hurricanes, which disrupt the four core dimensions of food security, namely availability, access, utilization, and stability, through their impacts on agricultural systems. Tropical cyclones increase risks and uncertainty in agricultural production, thereby threatening food production (Chen & McCarl, 2009). Consequently, the effects of tropical cyclones on agriculture are closely linked to progress toward Sustainable Development Goal 2 (SDG-2), which aims to ensure food security, promote sustainable agriculture, and reduce global hunger (Álvarez-Berríos et al., 2021; Chikodzi & Nhamo, 2021). Several studies suggest that tropical cyclones can reduce crop production across successive seasons by eroding productive soil, introducing saltwater onto farmland, fostering post-storm disease outbreaks, damaging farm machinery, and destroying irrigation systems critical for restoring agricultural productivity (Chikodzi and Nhamo, 2021; Mohan, 2017). Consequently, disasters like hurricanes severely disrupt food systems, impacting production, distribution, access, and consumption patterns (Subaiya et al., 2019; Clay & Ross, 2020). Reliance on food imports worsens the vulnerability, as natural disasters like hurricanes can easily interrupt international supply chains, heightening insecurity in regions already sensitive to climate change (Gould et al., 2015). This is because hurricanes cause severe, cascading disruptions to agricultural infrastructure, with widespread damage to transportation networks, storage facilities, and critical service systems. According to (Cruz & Krausmann 2013), Hurricanes inflict substantial damage on critical infrastructure, including roads, bridges, and railway networks, through storm surges and flooding. Such disruptions weaken delivery and distribution systems, further undermining food access. Hurricanes Katrina and Rita, for instance, destroyed critical rail bridges, halted barge traffic, and disrupted the distribution and export of grain through the Port of New Orleans, significantly impacting national freight movement and agricultural trade (Grenzeback et al., 2008). To address these challenges, promoting sustainable agricultural development is essential for safeguarding global food security and building resilience against future climate-related threats.

Furthermore, hurricanes cause severe agricultural damage that directly undermines rural livelihoods through crop destruction, damage to agricultural infrastructure, reduced access to relief resources, economic disruption, and long-term vulnerability (McGinley et al., 2022). Analyzing the impact of Hurricane Mitch on rural livelihoods, Loebach (2019) revealed that households dependent on agricultural wage employment experienced a lagged recovery compared to other livelihood groups, with agricultural workers and owner-producers suffering more substantial losses due to the disaster than business owners and non-agricultural livelihoods. In addition, as a consequence of Hurricane Matthew, Cook and Beachy (2019) found that hurricane-induced agricultural damage led to significant income losses, with adverse effects on rural livelihoods. (Biram et al., 2025), alternatively, noted that hurricanes increase production risk in U.S. agriculture, prompting greater reliance on crop insurance, which in turn raises insurance costs over time and imposes additional financial burdens on farmers, thereby adversely affecting rural livelihoods.

A systematic review finds that microcredit and micro saving initiatives can improve rural livelihoods, strengthen household food security, and potentially mitigate long-term food insecurity following hurricanes (Stewart et al., 2010). At the policy level, Louisiana's Climate Resilient Agriculture Program promotes adaptive varieties, efficient water management, insurance solutions, and farmer outreach, aiming to protect food security, boost resilience, and maintain sustainable agricultural production (Twumasi et al., 2024).

4 Hurricane Impacts on Agriculture Across Louisiana and Other Gulf Coast Regions

The effects of hurricanes differ widely by country, economic sector, and even across individual U.S. states, shaped by varying geographic conditions, resource availability, and adaptive capacities. Regions in the USA particularly vulnerable to hurricanes include the lower East Coast, Florida, the Gulf Coast, and the coastal areas of Texas and Louisiana (Balaguru et al., 2023). Figure 1 illustrates the number of state-level landfall impacts from tropical cyclones along the Gulf Coast between 2000 and 2024. Because a single storm may affect multiple states, impacts are counted separately for each state. Florida experienced the highest frequency with nearly 30 hurricanes, while Texas had the lowest, and Alabama, Louisiana, and Mississippi experienced intermediate frequencies of hurricane landfalls. Recent tropical cyclones in the Gulf of Mexico include Hurricanes Isidore and Lili (2002), Ivan (2004), Dennis, Emily, Katrina, Rita, and Wilma (2005), Gustav and Ike (2008), Harvey (2017), Laura (2020), and Ida (2021), which collectively highlight the region's recurring exposure to powerful storms (Hu et al., 2012). Among all the Gulf Coast states, Louisiana is particularly vulnerable to tropical cyclones due to its distinctive geography of bayous, marshes, and inlets, with major hurricanes frequently causing severe damage, especially in the low-lying areas surrounding New Orleans (Merem et al., 2012). Moreover, Louisiana's geographic proximity to the Gulf of Mexico has exposed it to frequent extreme weather events over the years (Twumasi et al., 2024), and hurricanes have repeatedly battered Louisiana's coastline, leaving behind widespread economic and environmental impacts, which pose serious challenges for its people, particularly those engaged in agriculture (Mostafiz et al., 2022). These challenges are further exacerbated by increasing flood risks, intensified tropical storms, and ongoing coastal erosion that continues to contribute to land loss (Jones & Brown, 2022; Twumasi et al., 2024), with consequences of hurricanes to agricultural revenues, operational costs, and the health of humans and animals alike (Guidry & Pruitt, 2012).

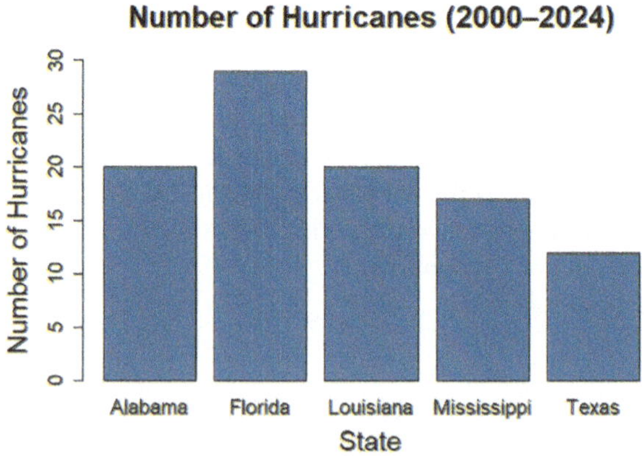

Fig. 1 Number of billion-dollar tropical cycle events affecting Gulf Coast, 2000–2024 (NOAA, 2025a)

As a result, Louisiana's major row crops, including rice, soybeans, cotton, corn, and sugarcane, have all been significantly affected by hurricanes between 2000 and 2024. Among these crops, rice and sugarcane hold particular strategic importance in Louisiana agriculture due to their long production history, extensive acreage, and substantial contributions to farm income, employment, and regional food systems, while soybeans, cotton, and corn play critical complementary roles in crop rotations and diversified farming systems across the state (Westra et al., 2015).

Hurricane Ivan was the first major hurricane in the post-2000 period to cause significant damage to agricultural crops. According to the (NOAA, 2004), Hurricane Ivan impacted Louisiana on 15–16 September 2004, bringing hurricane-force winds, flooding, and structural damage, and again on 22–23 September 2004, when its remnants regenerated into a tropical storm, causing minor coastal flooding. Hurricane Ivan in 2004 caused approximately US$15 million in damage to domestic crops, highlighting the storm's devastating impact on agriculture and the vulnerability of farming communities (Spencer & Polachek, 2015). Table 1 presents all Saffir–Simpson Category 3 or higher hurricanes that affected Louisiana between 2000 and 2024, having a significant impact on the state's agricultural sector.

In 2005, the Gulf region was hit by five hurricanes, and even though only Hurricanes Katrina (Category 5) and Rita (Category 4) caused substantial damage (Kaiser & Kasprzak, 2008), the 2005 hurricane season was the most expensive season ever, with economic damage exceeding $200 billion (NOAA, 2025b). Hurricane Katrina struck the Louisiana coast as a Category 3 storm on 29 August 2005, marking the first of a rapid succession of highly destructive hurricanes affecting the state. Sustained winds of 130 mph breached New Orleans' levees, causing catastrophic flooding (Gordon et al., 2010). Hurricane Katrina seriously damaged several agricultural areas, grain export facilities, and transportation infrastructure (Lara-Chavez & Alexander, 2006). Storm surge from Hurricane Katrina caused severe inland saltwater intrusion, damaging vegetation far from the coast and contaminating soils and waterways essential for rice farming (Williams, 2009). As a result, Louisiana's rice production dropped by 20%, with 80,000–90,000 acres left unplanted due to unsuitable growing conditions. Moreover, the storm's surge and debris rendered the Mississippi River completely impassable (Cruz & Krausmann, 2013). Consequently, Katrina disrupted agricultural trade by temporarily halting shipments through New Orleans, a critical hub for U.S. agricultural exports (Schnepf et al., 2005). In the case of Hurricane Katrina, food insecurity persisted at a high level for two years and started improving in the following year yet still exceeded the average state prevalence (Clay et al., 2018). Less than a month after Hurricane Katrina, on 24 September 2005, Hurricane Rita made landfall as a Category 3 hurricane along Louisiana's extreme southwestern coast near the Texas border (Beuzelin et al., 2009; Knabb et al., 2006). Rita proved equally destructive and deadly, devastating southeast Texas, southwest Louisiana, and leaving major impacts as far as the Florida Keys (Roth, 2010). Hurricane Rita caused severe agricultural losses across multiple crops due to flooding, lodging, wind damage, and prolonged saltwater intrusion that reduced yields and increased production and harvesting costs. Sugarcane, cotton, rice, soybeans, and other crops experienced significant revenue losses and higher replanting and recovery

Table 1 Major Louisiana hurricanes, 2000–2024

Name of hurricane	Start date	End date	Category	Landfall details
Hurricane Ivan	12 Sep 2004	21 Sep 2004	3	Made landfall on gulf coast of Alabama
Hurricane Katrina	25 Aug 2005	30 Aug 2005	3	Initially impacted the U.S. as a category 1 near Miami, FL, then as a strong category 3 along the eastern Louisiana-western Mississippi coastlines
Hurricane Rita	20 Sep 2005	24 Sep 2005	3	Made landfall near the Texas–Louisiana border, the storm produced destructive winds and storm surge across the coastal region
Hurricane Harvey	25 Aug 2017	31 Aug 2017	4	Made landfall near Rockport, Texas, causing severe damage, with historic flooding from extreme rainfall devastating Houston and nearby areas
Hurricane Laura	27 Aug 2021	28 Aug 2021	4	Made landfall in Cameron parish, Louisiana, with 150-mph winds and 15-ft storm surge, heavily damaging coastal areas and Lake Charles
Hurricane Ida	29 Aug 2021	1 Sep 2021	4	Made landfall near port Fourchon, Louisiana, with maximum sustained winds of 150 mph (240 km/hr) and a recorded minimum central pressure of 930 mb

expenses (Guidry, 2005). An estimated 12,000–16,000 hectares of Louisiana sugarcane were flooded with saltwater following the storm surge caused by Rita in September 2005 (Beuzelin et al., 2009).

Three years later, Hurricanes Gustav and Ike struck the Louisiana coast on 1 September and 13 September 2008, respectively (Kaiser & Yu, 2010; Galván et al., 2012). Excessive rainfall from Hurricane Gustav caused severe flooding across Northeast Louisiana, damaging multiple crops, including corn, cotton, soybeans, sorghum, rice, and sweet potatoes (Hutchinson et al., 2008). Sweet potatoes, for instance, suffered an estimated 40% statewide loss due to prolonged soil saturation and oxygen deprivation, representing a $26 million impact (Hutchinson et al., 2008). Hurricane Ike also caused extensive agricultural losses in Louisiana and, together with Hurricane Gustav, contributed to nearly $950 million in combined damages through severe disruptions to crop production, infrastructure damage, and reduced farm revenues (Guidry & Caffee, 2008). In contrast, despite heavy winds and rainfall from Hurricane Isaac that damaged crops in parts of Louisiana, most notably in the southeastern region, the statewide agricultural impact in 2012 was less severe than anticipated. Subsequent favorable weather conditions helped stressed crops recover, contributing to strong and near-record yields (LSU AgCenter, 2012).

Another disastrous hurricane season for the U.S. Gulf Coast was 2017, with losses reaching $330 billion (Misuri et al., 2019). Tropical Cyclone Cindy and the remnants of Hurricane Harvey brought heavy rain and wind to Louisiana, disrupting pollination, delaying input applications, and limiting overall production potential despite improved weather later in the season (LSU AgCenter, 2017). Harvey caused severe flooding and widespread destruction of agricultural areas along the Gulf Coast, with rainfall totals surpassing 125 cm (50 in) in some locations, submerging thousands of acres planted with cotton, corn, and soybeans (Zhang et al., 2018). In Louisiana, Harvey's heavy rainfall and flooding inundated croplands and surrounding communities, particularly in Calcasieu Parish (i.e., the Lake Charles area) in the southwestern part of the state, leading to extensive agricultural losses and prolonged recovery efforts (Loh et al., 2023). Although Hurricane Harvey passed close to Louisiana's sugarcane-producing regions in late August, its overall effect on the state's 2017/18 sugar production was minimal, with no significant disruptions expected (USDA, 2017). However, Harvey caused agricultural losses in Texas exceeding $200 million, of which nearly $50 million was attributed to cotton alone (Hegar, 2018; Davenport, 2020).

Hurricane Michael made landfall in the Florida panhandle in 2018, pushing inland into Georgia and parts of Alabama with Category 3 strength 320 km (200 mi). Striking during early cotton and peanut harvests and critical vegetable growth stages, Michael inflicted severe agricultural damage across the region (Dorfman et al., 2018). Market-ready crops were destroyed, infrastructure related to irrigation and storage was damaged, and access to markets for produce was disrupted. Hurricane Michael caused substantial economic damage to agriculture in Florida, wiping out nearly the entire cotton crop in the core production area, where more than 90% of the crop remained unharvested (University of Florida Institute of Food and Agricultural Sciences, 2018). The storm affected about 405,000 ha (one million acres) of field crops overall, including an estimated 40% loss of the region's peanut crop (University of Florida Institute of Food and Agricultural Sciences, 2018). One estimate indicates that all vegetable crops in Florida incurred $180 million in direct damage from the hurricane (Miller et al., 2021).

Between August and October 2020, Louisiana experienced a highly active hurricane season, with Hurricanes Laura, Delta, and Zeta each making landfall in different parts of the state and producing widespread damage along their paths (LSU AgCenter, 2020). Hurricane Laura was a powerful Category 4 storm that made landfall near Cameron, Louisianaon 27 August 2020, with sustained winds of approximately 150 mph (240 km/h) and a minimum central pressure of 939 mb (National Weather Service, 2020). Following rapid intensification in the Gulf of Mexico, Laura became the strongest hurricane to strike southwest Louisiana since records began in 1851, causing widespread and severe damage to communities, infrastructure, and agricultural systems across the region (National Weather Service, 2020; Schultz, 2020). Category 2 Delta followed Laura by six weeks, making landfall only 20 km (12 mi) from Laura's southwestern Louisiana landfall. Both storms caused infrastructure damage, agricultural losses, and tragic human deaths (Salarieh et al., 2023). While Hurricanes Laura and Delta reduced yields for some commodities, their most significant effect was the extensive infrastructure damage across the food and fiber sector, impairing facilities, supply chains, and essential systems vital to agricultural production and distribution (LSU AgCenter, 2020). Following Laura, another category 4 Hurricane, Ida, made landfall in Louisiana on 29 August 2021. Hurricane Ida caused an estimated $35.4 million loss on Louisiana's cane producers, representing about 7% of the industry's annual value, and contributed to at least $584 million in statewide agricultural damage (LSU AgCenter, 2021). According to Reuters (2021), Hurricane Ida severely impacted the industry, damaging approximately 48,000 ha (118,000 acres), nearly 26% of Louisiana's sugarcane crop.

Gulf ports play a critical role in the U.S. agricultural trade network, functioning as major gateways for U.S. grain, soybeans, and other agricultural products to reach global markets (Guan et al., 2019, Sledge, 2019). Hurricanes and other tropical systems represent a significant interruption to this agricultural trade, which provides livelihoods, revenue, and support for downstream sectors (Mohan, 2017). Barrier islands from Texas to Florida shield the Gulf Coast from tropical cyclones, but repeated storms and human infrastructure have weakened them, disrupting natural rebuilding and leaving the region increasingly exposed to future hurricane threats (Adam, 2021; Williams, 2010). Bulk grain exports depend on the ports of South Louisiana, Baton Rouge, and New Orleans, which transfer Midwest grain from barges to ships on the Mississippi River (Frittelli, 2005). Hurricanes also severely disrupt these critical operations, halting shipments and straining the supply chain (Li et al., 2022). While railroads offer an alternative, they are more expensive and lack the capacity to handle large volumes of agricultural cargo suddenly in a hurricane emergency (Frittelli, 2005).

5 Resilience Strategies for Gulf Coast Agriculture

Hurricanes remain an inevitable reality for many regions of the world (Hay & Mimura, 2010; Helgason, 2020). Although outcomes vary across studies, the interquartile range generally indicates a positive effect of climate change on tropical cyclone intensity, with projections pointing to an increased proportion of Category 4–5 storms over the twenty-first century and potential shifts in large-scale atmospheric steering flows that could bring Atlantic cyclone tracks closer to the U.S. coast (Balaguru et al., 2023; Knutson et al., 2020; Knutson et al., 2022; Seneviratne et al., 2021). The Gulf Coast, in particular, has endured more than fifteen deadly hurricanes between 2005 and 2024, leaving communities increasingly vulnerable to environmental and socioeconomic challenges (Damoah et al., 2024). Therefore, strategies should be implemented to minimize the consequences of hurricanes on agriculture and enhance resilience against them. Hurricane-resilient agricultural strategies are increasingly essential in cyclone-prone regions, where the rising intensity of storms disrupt farming systems through flooding, salinity intrusion, soil degradation, and infrastructure damage (Selva et al., 2025). This study employs a hurricane-specific resilience framework to structure the assessment of risk mitigation and adaptation strategies within Gulf Coast agricultural and food systems.

The framework conceptualizes hurricane risk as the interaction of hazard, exposure, and vulnerability within agricultural and food systems. Hazard refers to the physical processes generated by tropical cyclones, including extreme winds, storm surge, flooding, and salinity intrusion, which frequently occur simultaneously and compound overall damage (Do & Kuleshov, 2023; Lipari et al., 2024; Song et al., 2020). Exposure includes crops, livestock, soil systems, and agricultural infrastructure located in hazard-prone areas, while vulnerability reflects biophysical sensitivity, management practices, coastal proximity, limited adaptive capacity, and socioeconomic constraints that determine the extent to which exposure translates into actual losses during cyclone events (Ming et al., 2014; Romero & León-Cruz, 2024; Velásquez-Espinoza & Alcántara-Ayala, 2024). Within this risk context, our study structures agricultural resilience across three interrelated dimensions: ecological resilience, farm management and preparedness, and financial, technological and institutional capacity systems.

Ecological Resilience Strategies:

Ecological resilience encompasses land-use and production practices that reduce physical damage and accelerate recovery, such as crop diversification, agroecology, agroforestry, windbreaks, soil conservation, drainage management, and salinity-tolerant cropping systems. Research highlights that diversified farms and those adopting agroecological practices, such as shade trees, crop diversity, and soil-conserving techniques, tend to withstand hurricane impacts more effectively (Philpott et al., 2008; Rosset et al., 2011; Mohan & Strobl, 2017). Crop diversification involves cultivating multiple crop types within the same area, a practice that enhances biodiversity, promotes nutrient recycling, and strengthens soil health, making it more resilient to environmental stresses (Williams et al., 2023). By growing diverse crops, farmers can lower the risk of complete crop failure from adverse weather, pests, or diseases, while maintaining a reliable food supply and steady income, thereby contributing to greater agricultural sustainability (Sarma et al., 2024).

Diversification has long been promoted as a vital agricultural adaptation strategy, particularly in regions where economies are heavily dependent on climate-sensitive farming (Birthal & Hazrana, 2019; Nazir & Das Lohano, 2022). Previous studies indicate that increased diversity can strengthen farmers' resilience to climate shocks such as hurricanes (Abson et al., 2013; Aguilar-Støen et al., 2009; Melvani et al., 2020). In Zambia, Wang et al. (2025) report that diversification helps mitigate the effects of climate shocks. Similarly, studies from India and Pakistan demonstrate that diversification improves the stability and adaptability of agricultural systems under climatic stress (Birthal & Hazrana, 2019; Nazir & Das Lohano, 2022). Following

Hurricane Mitch, American hillsides revealed that farmers who employed diversification techniques, including agroforestry, intercropping, and cover crops, saw less damage than their neighbors who practiced traditional monoculture (Altieri et al., 2012). Marrero et al., (2022) focused on diversification as a core climate-resilient farm management strategy, emphasizing its role in reducing agricultural vulnerability and enhancing resilience to hurricane impacts. Therefore, land planning and management for agriculture should focus on maintaining diversity and complexity in the agroecosystem through practices such as perennial crops and windbreaks, which can also enhance resilience against hurricanes (Her et al., 2018). Similarly, Her et al. (2018) reported that land planning and management can reduce future hurricane losses by promoting agricultural diversity and complexity through practices such as perennial crops and strategically placed windbreaks. However, Thompson et al. (2023) found that although diversification, including practices such as intercrops and boundary crops, is promoted and to some extent practiced, it does not play a significant functional role in farmers' livelihoods and therefore does not enhance resilience.

According to Machín-Sosa et al. (2010), agroecologically managed farms recovered more productively than monoculture farms, achieving recovery levels of 80–90% following hurricane impacts. Agroecology refers to sustainable farming practices that adopt a holistic and integrated approach, maximizing the interactions between humans, plants, animals, and the environment to build resilience (Warburton, 2022). Agroecology, beyond its direct contributions to the environmental sustainability of farming systems, has been recognized as an exemplary agricultural approach capable of more equitably addressing health and food insecurity compared to other formally regulated and higher-cost food systems (Marrero et al., 2022). Félix and Holt-Giménez (2017) also emphasized that agroecology played a central role in agricultural recovery following a hurricane. Evidence suggests that compared to conventional farms, agroecological systems experience fewer financial losses during extreme events and are less vulnerable to future occurrences of this kind. One key agroecological practice is agroforestry, which integrates trees, crops, and sometimes livestock into the same farming system. Identified by Sarma et al. (2024) as hurricane-resilient, agroforestry practices such as alley cropping not only provide shade and function as natural windbreaks but also improve soil quality through the organic matter contributed by fallen tree leaves, thereby supporting long-term sustainability.

At the same time, structural and ecological strategies play a vital role. Planting trees as shelterbelts or windbreaks is a widely adopted practice in successful agricultural systems worldwide, contributing to both sustainability and profitability. Windbreaks also serve to protect crops and soils from wind erosion and to reduce damage caused by typhoons (Brandle et al., 2004; Cai et al., 2021). Less intensively managed farms with greater shade tree cover show higher resistance to hurricane impacts than intensively managed farms, as shade trees provide protection through wind breaks, weed suppression, and natural fertilization (Perfecto et al., 2019). Hurricanes pose persistent challenges to agricultural systems by generating storm surges and coastal flooding that accelerate soil salinization, making the identification and mapping of salt-prone areas critical for long-term land management (Gibson et al., 2021). Rice, one of the world's most important staple crops, is cultivated on more than 163 million hectares across over 100 countries, yet its production is particularly vulnerable to saltwater intrusion in coastal farming systems (Laborte et al., 2017). In countries such as Bangladesh and Vietnam, frequent saltwater flooding has compelled farmers to shift from rice cultivation toward shrimp farming, raising concerns about long-term sustainability and food security (Gopalakrishnan et al., 2019). In response to chronic salinity pressures, the adoption of salt-tolerant crops has emerged as an essential adaptation strategy, enabling farmers to maintain productivity and profitability under worsening conditions (Gibson et al., 2021). At the same time, natural recovery processes play an important role, as rainfall gradually removes sodium and chloride through runoff and leaching, allowing most affected agricultural lands to regain their prior productivity within approximately two years.

Additionally, hurricanes bring intense rainfall that heightens soil erosion risks. Soil erosion affects crop productivity by reducing nitrogen, phosphorus, and SOM reserves, varying clay content of the soil, depleting available water capacity, and reducing soil aggregation (Chalise et al., 2019). Soil erosion is also considered as one of the significant threats to the ecosystem (Routschek et al., 2014; Ali et al., 2021). No-till agriculture and cover cropping, can reduce the severity of flooding downstream caused by hurricane and other extreme climatic event (Smith, 2018). Cover crop is an effective solution to soil erosion (Quintarelli et al., 2022). The cover crop protects the soil during the hurricane season (Klassen et al., 2006). Establishing cover crops with substantial biomass helps reduce soil erosion and nutrient leaching, protecting farmland from extreme rain events and fertilizer losses (Kaye & Quemada, 2017).

Farm Management and Preparedness:

Farm management and preparedness strategies include pre and post hurricane actions that limit immediate losses and support rapid recovery, including seed bank/reserves, early harvesting, storage infrastructure, and farmer training.

Accelerated sea-level rise and more intense tropical storms have challenged traditional approaches to coastal ecosystem conservation and restoration, threatening historic species assemblages in vulnerable communities. Seed banks can act as ecological buffers to storm surge disturbances by preserving species capable of adapting or migrating inland, thereby supporting recovery and enhancing resilience in degraded coastal ecosystems (Kalk, 2011). Farmers frequently emphasized seed reserves as one of the most vital resources for rapid recovery when crops were damaged or destroyed by a hurricane (McGinley et al., 2022). Moreover, seed banking serves as a risk-management strategy against future catastrophes, potentially strengthening ecosystem resilience by preserving genetic diversity and supporting species coexistence (Evans & Dennehy, 2005). The National Seed Strategy aims to ensure the availability of native seeds for ecological restoration, a goal that is increasingly important for enhancing resilience to extreme weather events such as hurricanes (Oldfield, 2019).

Additionally, investments in advanced agricultural storage facilities and training on proper storage of perishable products are crucial for reducing post-disaster spoilage (Campbell & Beckford, 2009). Short-term adaptive actions are equally important, including securing generator fuel, potable water, and animal feed, relocating equipment from flood-prone zones, harvesting crops early, and pruning plants to minimize wind damage (Wiener et al., 2020). Short-term storm preparation and recovery plans were scored higher than long-term hurricane preparedness plans in terms of the perceived prevalence and significance of hurricane resilience methods (Wiener et al., 2020). Together, these strategies highlight both proactive and reactive approaches that strengthen resilience within the agricultural sector.

Hovis et al. (2021) identify land drainage systems as a practical hurricane-resilience strategy for agriculture. When designed to slow and store water rather than accelerate surface and subsurface runoff, these systems, both surface and subsurface types, can help manage heavy rainfall from hurricanes, reduce erosion, and protect croplands, offering farmers a low-cost, well-understood method of controlling excess water.

To further enhance preparedness, early warning systems serve as a critical strategy for mitigating the impacts of hurricanes by providing timely, reliable, and actionable information to at-risk communities. These systems integrate a network of meteorological observation stations, advanced radar technologies, and satellite imagery to monitor atmospheric conditions, precipitation trends, and storm trajectories (Boer et al., 2017; Chavula et al., 2025; Zhang et al., 2022). During Hurricane Maria, for example, the U.S. National Weather Service effectively employed its weather stations, radar systems, and satellite data to track the storm's path and issue timely alerts to both the Caribbean Islands and the U.S. mainland, significantly reducing casualties and enhancing community preparedness (Chavula et al., 2025).

Hurricanes remain a persistent threat to agriculture-dependent households, communities, and economies, making it essential to understand how farmers respond to these events in order to design effective adaptation strategies and build resilience (Campbell & Beckford, 2009). Advisors, therefore, emphasized the necessity of enhanced training and expanded educational resources to strengthen the effectiveness and efficiency of hurricane recovery efforts. Prior to hurricanes, farmers often adopt precautionary measures such as spraying protective treatments, securing nurseries, and harvesting or storing crops in advance to minimize losses (Campbell & Beckford, 2009). Strategic land-use adjustments, such as reducing acreage of vulnerable crops like rice, soybeans, and cotton near coastal areas, may also help mitigate risks to U.S. agriculture (Chen & McCarl, 2009).

Financial, Technological and institutional resilience

Financial and institutional resilience reflects mechanisms that buffer income shocks and sustain livelihoods, particularly crop insurance, parametric weather insurance, and collective resource pooling.

Crop insurance serves as an effective adaptation strategy for managing agricultural systems associated with hurricanes and climate change risk. Risk management through crop insurance remains an important buffer against financial losses caused by hurricanes (Motha, 2011). Jørgensen et al. (2020) identify crop insurance as an attractive climate change adaptation strategy for arable farmers, especially those with low soil quality and prior crop damage experience. Empirical evidence further demonstrates that crop insurance improves farm income stability and reduces production risk (Birthal et al., 2022).

Similarly, weather insurance is recognized as a vital strategy to address hurricane-related risks in regions that are highly vulnerable to natural disasters and heavily dependent on the agricultural sector. It operates in three main forms: indemnity-based, parametric-based, and insurance-linked securities. Indemnity insurance requires verification of actual losses, often involving costly on-site assessments, and faces challenges such as asymmetric information, high adjustment costs, and moral hazard when insured parties reduce preventive efforts (Bucheli et al., 2023; Vroege et al., 2021). To address these limitations, Charles (2025) proposed the use of parametric insurance. Parametric insurance provides payouts triggered by predefined events, such as hurricane landfall or rainfall thresholds, enabling rapid financial response without assessing individual damages. This mechanism reduces administrative burdens, ensures timely relief, and supports resilience building. As such, parametric insurance offers a more efficient and scalable approach than indemnity insurance for mobilizing recovery finance and sustaining long-term hurricane preparedness (Charles, 2025).

Additionally, Farmer networks and resource pooling have emerged as critical resilience strategies for agricultural communities facing hurricane risks, providing collaborative mechanisms to mitigate disaster impacts (Marrero et al., 2022). These networks facilitate resource sharing and information exchange (White et al., 2023), enable collective problem-solving and adaptive strategies (Begho & Ogisi, 2025), and support post-disaster recovery through community-level collaboration (McGinley et al., 2022). Evidence indicates that such networks are particularly valuable in small-scale agricultural systems, allowing farmers to share surviving crops and resources, develop innovative coping mechanisms, and overcome individual resource limitations through collective action (Marrero et al., 2022).

Structural and technological adaptation measures also play a critical role in strengthening hurricane resilience and safeguarding long-term food security. Physical interventions such as sea-dyke construction, gabion reinforcement, and mangrove restoration have proven highly effective in protecting agricultural land from cyclone-driven storm surges, with projections suggesting that these strategies could reduce average annual agricultural losses by nearly 99% under future climate scenarios in Viet Nam (Rana et al., 2022). Complementing these measures, advances in disaster monitoring and early warning systems enhance preparedness and response capabilities. The integration of optical and infrared satellite sensors with weather radar enables real-time detection and tracking of hurricanes, while GPS technologies combined with deep learning improve location accuracy and evacuation planning (Neog et al., 2024). In agricultural contexts, these technologies support targeted recovery planning, efficient resource allocation, and the protection of rural livelihoods, thereby strengthening overall resilience to hurricane impacts. Table 2 highlights strategies that strengthen the agricultural system in the face of hurricanes.

While many resilience strategies are discussed globally, their relevance to the Gulf Coast lies in the region's exposure to storm surge, salinity intrusion, and repeated

Table 2 Resilience strategies for agriculture in hurricane-prone regions

Resilient strategy	Short description/role	Reference
Crop diversification	Growing diverse crops to spread risk and stabilize yields/income	Williams et al. (2023); Sarma et al. (2024)
Agroecology (holistic systems)	Integrate ecological practices to reduce losses and speed recovery	Warburton (2022); Rosset et al. (2011); Machín-Sosa et al. (2010)
Agroforestry/alley cropping	Mixing trees with crops provides wind protection, creates shade, and improves soil quality.	Sarma et al. (2024); Altieri et al. (2012)
Windbreaks (shelterbelts)	Reduce wind damage and erosion; enhance farm resistance	Brandle et al. (2004); Cai et al. (2021); Perfecto et al. (2019)
No-till and cover cropping	Cut erosion/flood impacts; protect soil, reduce nutrient loss	Smith (2018); Quintarelli et al. (2022); Kaye and Quemada (2017); Klassen et al. (2006)
Seed banking/reserves	Enable rapid replanting and recovery after crop loss	McGinley et al. (2022)
Drainage systems (surface and subsurface)	Slow/store stormwater, reduce erosion, protect fields	Hovis et al. (2021)
Advanced storage and training	Improve post-disaster handling to limit spoilage	Campbell and Beckford (2009)
Farmer networks and resource pooling	Collective action to share seeds, knowledge, and recovery inputs	Marrero et al. (2022)
Early warning systems	Timely, actionable alerts to reduce losses and casualties	Boer et al. (2017); Zhang et al. (2022); Chavula et al. (2025)
Remote sensing and radar, GPS-AI	Detect/track hazards, optimize evacuations and response	Neog et al. (2024)
Weather insurance (parametric)	Fast payouts on triggers; lowers admin costs, boosts recovery	Vroege et al. (2021); Bucheli et al. (2023); Charles (2025)
Crop insurance (indemnity)	Financial buffer against hurricane-related farm losses	Motha (2011); Vroege et al. 2021; Bucheli et al. 2023
Salinity risk mapping	Identify salt-intrusion hotspots for long-term planning	Gibson et al. (2021)

hurricane landfalls. Together, ecological, managerial, financial, and technological resilience strategies interact to reduce hurricane-induced agricultural losses, accelerate recovery, and stabilize food production systems in the Gulf Coast region. In particular, Louisiana's low-lying coastal agriculture makes integrated ecological and financial resilience strategies essential.

6 Conclusion

Climate change, through warming waters and rising sea levels, is projected to intensify the destructive potential of hurricanes in the Gulf Coast and beyond (Deng, 2022). The recurring impacts of hurricanes extend beyond immediate crop damage, undermining agricultural productivity, food security, economic stability, and human wellbeing (Álvarez-Berríos et al., 2021). It is therefore critically important to deliver precise, timely predictions of hurricane winds, surge, and waves to minimize risks and strengthen coastal resilience. With intensifying climate events, strengthening resilience is crucial for safeguarding livelihoods, food security, and ecosystem services. As demonstrated in this study, hurricanes have repeatedly devastated staple crops such as rice, soybeans, cotton, sugarcanes, and corn, disrupted trade flows through the Mississippi River and Gulf ports, and weakened the resilience of rural livelihoods. Given the strategic role of Gulf Coast agriculture in national and global food supply chains, enhancing resilience in this region is not only a local concern but a matter of global importance. This study highlights agricultural risks while showcasing opportunities to reduce vulnerabilities among farmers and rural communities. The findings highlight pathways to reduce hurricane and disaster risks through improved understanding, adoption, and support of effective agricultural mitigation and adaptation measures.

Enhancing climate-resilient crop calendars, introducing salt-tolerant varieties, and strengthening local adaptive capacities are essential to maintain agricultural productivity and safeguard livelihoods in cyclone-exposed coastal regions. Sustaining agriculture in hurricane-prone environments will require both structural and institutional innovations. Short-term strategies such as early harvesting, improved drainage, and diversified crop management can reduce immediate vulnerabilities, while long-term investments in agro ecology, early warning systems, climate-resilient crop varieties, and infrastructure reinforcement are essential for adaptation. Policy interventions already underway, such as USDA's hurricane-specific crop insurance expanded to cover all tropical cyclones (Bundy et al., 2023), represent important financial safeguards for producers. Similarly, Louisiana's constitutional amendment to allow Wetlands Conservation and Restoration Funds to support hurricane protection demonstrates the value of policy innovation in risk mitigation (Austin, 2006).

Policy implications from this study emphasize the need for integrated disaster management frameworks that link federal, state, and local efforts. Expanding agricultural extension services to include hazard mitigation training will be critical for farmer preparedness. Another important strategy for reducing vulnerability is to apply lessons learned from past experiences to guide planning, preparedness, and mitigation measures at multiple levels. Additionally, policies must prioritize equitable access to insurance, credit, and technical resources, particularly for small and mid-sized farms that face disproportionate risks.

Future research should extend beyond crops to include aquaculture and livestock, and more robust quantification of hurricane impacts is necessary to guide evidence-based policymaking. Strengthening these policies, institutional, and technical measures will be essential for safeguarding agricultural resilience in the face of escalating hurricane risks.

References

Abson, D. J., Fraser, E. D. G., & Benton, T. G. (2013). Landscape diversity and the resilience of agricultural returns: a portfolio analysis of land-use patterns and economic returns from lowland agriculture. *Agriculture & Food Security, 2*, 1–15. https://doi.org/10.1186/2048-7010-2-2

Adam J. (2021). Assessing morphological response and vulnerability of barrier islands to extreme storms in Northwest Florida. Master's thesis. University of South Florida. https://digitalcommons.usf.edu/etd/9055/

AgCenter, L. S. U. (2017). Louisiana summary: agriculture & natural resources. Louisiana State University Agricultural Center publishes, 2382.https://doi.org/10.1007/s10333-021-00842-x

Aguilar-Støen, M., Moe, S. R., & Camargo-Ricalde, S. L. (2009). Home gardens sustain crop diversity and improve farm resilience in Candelaria Loxicha, Oaxaca, Mexico. *Human Ecology, 37*(1), 55–77. https://doi.org/10.1007/s10745-008-9197-y

Ali, M. G., Ali, S., Arshad, R. H., Nazeer, A., Waqas, M. M., Waseem, M., et al. (2021). Estimation of potential soil erosion and sediment yield: a case study of the transboundary Chenab River catchment. *Water, 13*(24), 3647. https://doi.org/10.3390/w13243647

Altieri, M. A., Funes-Monzote, F. R., & Petersen, P. (2012). Agroecologically efficient agricultural systems for smallholder farmers: contributions to food sovereignty. *Agronomy for Sustainable Development, 32*(1), 1–13. https://doi.org/10.1007/s13593-011-0065-6

Álvarez-Berríos, N. L., Wiener, S. L., McGinley, K. A., Lindsey, A. B., & Gould, W. A. (2021). Hurricane effects, mitigation, and preparedness in the Caribbean: Perspectives on high-importance-low-prevalence practices from agricultural advisors. *Journal of Emergency Management, 19*(8). https://doi.org/10.5055/jem.0585

Austin, D. E. (2006). Coastal exploitation, land loss, and hurricanes: a recipe for disaster. *American Anthropologist, 108*(4), 671–691. https://www.jstor.org/stable/4496511

Balaguru, K., Judi, D. R., & Leung, L. R. (2016). Future hurricane storm surge risk for the US gulf and Florida coasts based on projections of thermodynamic potential intensity. *Climatic Change, 138*(1), 99–110. https://doi.org/10.1007/s10584-016-1728-8

Balaguru, K., Xu, W., Chang, C. C., Leung, L. R., Judi, D. R., Hagos, S. M., et al. (2023). Increased US coastal hurricane risk under climate change. *Scientific Advances, 9*(14), eadf0259. https://doi.org/10.1126/sciadv.adf0259

Bandyopadhyay, S. (2022). Impact assessment of cyclone Amphan on agriculture over parts of West Bengal using remote sensing. *Journal of the Indian Society of Coastal Agricultural Research*. https://doi.org/10.54894/JISCAR.40.1.2022.111626

Barui, I., Bhakta, S., & Ghosh, K. (2025). Storm surge-induced soil salinization and its impact on agriculture in the coastal area of the Indian Sundarban. *Environmental Development, 2025*, 101250. https://doi.org/10.1016/j.envdev.2025.101250

Begho, T., & Ogisi, O. D. (2025). A systematic review of the role of networks in information dissemination and resource pooling among farmers in sub-Saharan Africa: impact on farm improvements, empowerment and behaviour change. *Journal of International Development, 37*(5), 1070–1081. https://doi.org/10.1002/jid.4004

Beuzelin, J. M., Reagan, T. E., Akbar, W., Cormier, H. J., Flanagan, J. W., & Blouin, D. C. (2009). Impact of hurricane Rita storm surge on sugarcane borer (Lepidoptera: Crambidae) management in Louisiana. *Journal of Economic Entomology, 102*(3), 1054–1061. https://doi.org/10.1603/029.102.0325

Biram, H. D., Cameron-Harp, M., & Tack, J. (2025). Measuring the impact of hurricane incidence on agricultural production risk using insurance data. *American Journal of Agricultural Economics, 107*(5), 1438–1456. https://doi.org/10.1111/ajae.70008

Birthal, P. S., & Hazrana, J. (2019). Crop diversification and resilience of agriculture to climatic shocks: evidence from India. *Agricultural Systems, 173*, 345–354. https://doi.org/10.1016/j.agsy.2019.03.005

Birthal, P. S., Hazrana, J., Negi, D. S., & Mishra, A. K. (2022). Assessing benefits of crop insurance Vis-a-Vis irrigation in Indian agriculture. *Food Policy, 112*, 102348. https://doi.org/10.1016/j.foodpol.2022.102348

Boer, M. M., Nolan, R. H., Resco De Dios, V., Clarke, H., Price OF, & Bradstock, R. A. (2017). Changing weather extremes call for early warning of potential for catastrophic fire. *Earth's Future, 5*(12), 1196–1202. https://doi.org/10.1002/2017EF000657

Brandle, J. R., Hodges, L., & Zhou, X. H. (2004). Windbreaks in North American agricultural systems. *Agroforestry Systems, 61*(1), 65–78. https://doi.org/10.1023/B:AGFO.0000028990.31801.62

Bucheli, J., Conrad, N., Wimmer, S., Dalhaus, T., & Finger, R. (2023). Weather insurance in European crop and horticulture production. *Climate Risk Management, 41*, 100525. https://doi.org/10.1016/j.crm.2023.100525

Bundy, L. R., Gensini, V. A., & Van Den Broeke, M. S. (2023). Tropical cyclone impacts on crop condition ratings and yield in the coastal southern United States. *Agricultural and Forest Meteorology, 340*, 109599. https://doi.org/10.1016/j.agrformet.2023.109599

Bunn, C., Läderach, P., Rivera, O., & Kirschke, D. (2015). A bitter cup: Climate change profile of global production of Arabica and Robusta coffee. *Climatic Change, 129*(1–2), 89–101. https://doi.org/10.1007/s10584-014-1306-x

Cai, X., Henderson, M., Wang, L., Su, Y., & Liu, B. (2021). Shelterbelt structure and crop protection from increased typhoon activity in Northeast China. *Agriculture, 11*(10), 995. https://doi.org/10.3390/agriculture11100995

Campbell, D., & Beckford, C. (2009). Negotiating uncertainty: Jamaican small farmers' adaptation and coping strategies, before and after hurricanes—a case study of hurricane dean. *Sustainability, 1*(4), 1366–1387. https://doi.org/10.3390/su1041366

Cook, A. C., & Beachy, D. (2019). The impact of hurricane matthew on school attendance: an analysis from rural Haiti. *International Journal of Environmental Research and Public Health, 16*(1), 55. https://doi.org/10.3390/ijerph16010055

Chalise, D., Kumar, L., & Kristiansen, P. (2019). Land degradation by soil erosion in Nepal: a review. *Soil systems, 3*(1), 12. https://doi.org/10.3390/soilsystems3010012

Charles, D. (2025). Parametric insurance: a mechanism to finance disaster resilience and transitioning to a low-carbon economy in the Caribbean. *Green and Low-Carbon Economy, 3*(1), 26–34. https://doi.org/10.47852/bonviewGLCE3202961

Chavula, P., Kayusi, F., Lungu, G., & Uwimbabazi, A. (2025). The current landscape of early warning systems and traditional approaches to disaster detection. *LatIA., 2025*(3), 77. https://doi.org/10.62486/latia202577

Chen, C. C., & McCarl, B. (2009). Hurricanes and possible intensity increases: effects on and reactions from US agriculture. *Journal of Agricultural and Applied Economics, 41*(1), 125–144. https://doi.org/10.1017/S1074070800002595

Chikodzi, D., & Nhamo, G. (2021). Linking the impacts of tropical cyclones to the sustainable development goals. In *Cyclones in southern Africa: Volume 3—implications for the sustainable development goals* (pp. 3–16). Springer. https://doi.org/10.1007/978-3-030-74303-1_1

Clay, L. A., Papas, M. A., Gill, K. B., & Abramson, D. M. (2018). Factors associated with continued food insecurity among households recovering from hurricane Katrina. *International Journal of Environmental Research and Public Health, 15*, 1647. https://doi.org/10.3390/ijerph15081647

Clay, L. A., & Ross, A. D. (2020). Factors associated with food insecurity following hurricane Harvey in Texas. *International Journal of Environmental Research and Public Health, 17*(3), 762. https://doi.org/10.3390/ijerph17030762

Cottrell, R. S., Nash, K. L., Halpern, B. S., Remenyi, T. A., Corney, S. P., Fleming, A., Fulton, E. A., Hornborg, S., Johne, A., Watson, R. A., & Blanchard, J. L. (2019). Food production shocks across land and sea. *Nature Sustainability, 2*, 130–137. https://doi.org/10.1038/s41893-018-0210-1

Cruz, A. M., & Krausmann, E. (2013). Vulnerability of the oil and gas sector to climate change and extreme weather events. *Climatic Change, 121*(1), 41–53. https://doi.org/10.1007/s10584-013-0891-4

Damoah, B., Adu, E., Ofori, E., & Apraku, A. (2024). Aggravation of climate change crisis in gulf regions of United States: a review of anthropogenic factors. *Journal of Ecohumanism, 3*(4), 438–449. https://www.ceeol.com/search/article-detail?id=1272111

Davenport A. (2020). Hurricane Harvey causes $200M in ag losses. AgWeb. Available from https://www.agweb.com/news/crops/cotton/hurricane-harvey-causes-200m-ag-losses. Accessed on December 24, 2025.

Deng, X. (2022). The economic impact of hurricanes and hurricane forecast accuracy on the US Gulf Coast oil and gas industry. *Texas A&M University*. https://www.proquest.com/openview/b1306f18d65d05511723f67808bff30b/1?pq-origsite=gscholar&cbl=18750&diss=y

Do, C., & Kuleshov, Y. (2023). Multi-hazard tropical cyclone risk assessment for Australia. *Remote Sensing, 15*(3), 795. https://doi.org/10.3390/rs15030795

Dorfman, J. H., Liu, Y., & Rabinowitz, A. N. (2018). Estimating agricultural losses from a hurricane. *Journal of Agribusiness, 36*(2), 199–206. https://scholar.google.com/scholar?hl=en&as_sdt=0%2C19&q=Estimating+agricultural+losses+from+a+hurricane.+&btnG=

Evans, M. E., & Dennehy, J. J. (2005). Germ banking: bet-hedging and variable release from egg and seed dormancy. *The Quarterly Review of Biology, 80*(4), 431–451. https://www.journals.uchicago.edu/doi/abs/10.1086/498282

Félix, G. F., & Holt-Giménez, E. (2017). Hurricane María: an agroecological turning point for Puerto Rico?. https://archive.foodfirst.org/publication/hurricane-maria-an-agroecological-turning-point-for-puerto-rico/

Frittelli, J. (2005). Hurricane Katrina: Shipping disruptions. *Congressional Research Service, the Library of Congress*. https://www.policyarchive.org/handle/10207/4208

Galván, K., Bargu, S., White, J. R., Li, C., Sullivan, M., & Weeks, E. (2012). The effects of two consecutive hurricanes on basal food resources in a shallow coastal Lagoon in Louisiana. *Journal of Coastal Research, 28*(2), 407–420. https://doi.org/10.2112/JCOASTRES-D-11-00009.1

Gay-des-Combes, J. M., Robroek, B. J., Hervé, D., Guillaume, T., Pistocchi, C., Mills, R. T., & Buttler, A. (2017). Slash-and-burn agriculture and tropical cyclone activity in Madagascar: Implication for soil fertility dynamics and corn performance. *Agriculture, Ecosystems & Environment, 239*, 207–218. https://doi.org/10.1016/j.agee.2017.01.010

Gibson, N., McNulty, S., Miller, C., Gavazzi, M., Worley, E., Keesee, D., & Hollinger, D. (2021). Identification, mitigation, and adaptation to salinization on working lands in the US Southeast. In *General technical report SRS-259. Asheville, NC: US Department of Agriculture Forest Service, Southern Research Station* (Vol. 259, pp. 1–69). https://doi.org/10.2737/SRS-GTR-259

Ginis, I. (2020). Tropical cyclones. In *From hurricanes to epidemics: The Ocean's evolving impact on human health-perspectives from the US* (pp. 121–128). Springer International Publishing. https://doi.org/10.1007/978-3-030-55012-7_10

Gopalakrishnan, T., Hasan, M. K., Haque, A. T. M. S., Jayasinghe, S. L., & Kumar, L. (2019). Sustainability of coastal agriculture under climate change. *Sustainability, 11*, 7200.

Gordon, P., Park, J., & Richardson, H. W. (2010). Short-run economic impacts of hurricane Katrina (and Rita). In CESifo forum (Vol. 11, no. 2, pp. 73–79). München: Ifo Institut für Wirtschaftsforschung an der Universität München. https://hdl.handle.net/10419/166394

Gould, W. A., Fain, S. J., Pares, I. K., McGinley, K., Steele, R. F., & Perry, A. (2015). Caribbean regional climate sub hub assessment of climate change vulnerability and adaptation and mitigation strategies. https://doi.org/10.22004/ag.econ.338003

Grenzeback, L. R., Lukman, A. T., & Systematics, C. (2008). *Case study of the transportation sector's response to and recovery from hurricane's Katrina and Rita*. Transportation Research Board.

Guan, C., Yahalom, S. S., Germanakos, L., Lapage, S., & McKeever, B. (2019). Global soybean trade, supply chain and tariffs. *WIT Transactions on the Built Environment, 187*, 239–250. https://doi.org/10.2495/MT190221

Guidry, K. M. (2005). *Assessment of damage to Louisiana agricultural, forestry, and fisheries sectors by hurricane Rita*. LSU AgCenter.

Guidry, K., Caffey, R.H. (2008). Preliminary estimates of losses in Louisiana agriculture, forestry and fisheries from Hurricane Gustav and Hurricane Ike. LSU AgCenter Report. Available at: https://www.lsuagcenter.com/topics/family_home/hazards_and_threats/recovery_assistance/agdisaster/preliminary-estimates-of-losses-in-louisiana-agriculture-forestry-and-fisheries-from-hurricane-gusta. Accessed on December 3, 2025

Guidry, K. M., & Pruitt, J. R. (2012). Damages to Louisiana agriculture from natural disasters. *Choices, 27*(3), 1–6. https://www.jstor.org/stable/10.2307/choices.27.3.09

Hay, J., & Mimura, N. (2010). The changing nature of extreme weather and climate events: Risks to sustainable development. *Geomatics, Natural Hazards and Risk, 1*(1), 3–18. https://doi.org/10.1080/19475701003643433

Hegar, G. (2018). A storm to remember. *Texas Comp of Public Accts*. https://scholar.google.com/scholar?hl=en&as_sdt=0%2C19&q=Hegar+G+%282018%29+A+storm+to+remember&btnG=

Helgason, K. S. (2020). The economic and political costs of population displacement and their impact on the SDGs and multilateralism. United Nations,. https://efaidnbmnnnibpcajpcglclefindmkaj/; https://www.un.org/esa/desa/papers/2020/wp167_2020.pdf

Her, Y. G., Smyth, A., Fletcher, P., Bassil, E., Stingl, U., Brym, Z., & Qui, J. (2018). Hurricane impacts on Florida's agriculture and natural Resources: AE528/AW528, 10/2018. *EDIS, 2018*(5). https://doi.org/10.32473/edis-ae528-2018

Hirano, A. (2021). Effects of climate change on spatiotemporal patterns of tropical cyclone tracks and their implications for coastal agriculture in Myanmar. *Paddy and Water Environment, 19*(2), 261–269. https://doi.org/10.1007/s10333-021-00842-x

Hovis, M., Hollinger, J. C., Cubbage, F., Shear, T., Doll, B., Kurki-Fox, J. J., & Potter, T. (2021). Natural infrastructure practices as potential flood storage and reduction for farms and rural communities in the North Carolina coastal plain. *Sustainability, 13*(16), 9309. https://doi.org/10.3390/su13169309

Hu, K., Chen, Q., & Fitzpatrick, P. (2012). Assessment of a parametric hurricane surface wind model for tropical cyclones in the Gulf of Mexico. In *Advances in hurricane research-modelling, meteorology, preparedness and impacts*. IntechOpen. https://doi.org/10.5772/51288

Hutchinson, R. L., Burns, D., Levy, R., Pinnell-Alison, C. L., Van Osdell, M. A., Schultz, B., & Smith, T. (2008). Gustav affects N.E. La. crops. LSU AgCenter.https://www.lsuagcenter.com/portals/communications/news/news_archive/2008/september/headline%20news/gustav-affects-ne-la-crops

Islam, M. T., & Nursey-Bray, M. (2017). Adaptation to climate change in agriculture in Bangladesh: The role of formal institutions. *Journal of Environmental Management, 200*, 347–358. https://doi.org/10.1016/j.jenvman.2017.05.092

Jamil, S. (2024). Climate change and its impact on crop yields: Strategies for adaptation. *Frontiers in Agriculture, 1*(1), 72–102. https://sprcopen.org/index.php/FA/article/view/11

Jones, S., & Brown, T. (2022). Environmental challenges to agricultural production in Louisiana. *Environmental Science and Policy, 12*, 321–335.

Jørgensen, S. L., Termansen, M., & Pascual, U. (2020). Natural insurance as condition for market insurance: Climate change adaptation in agriculture. *Ecological Economics, 169*, 106489. https://doi.org/10.1016/j.ecolecon.2019.106489

Kaiser, M. J., & Kasprzak, R. A. (2008). The impact of the 2005 hurricane season on the Louisiana artificial reef program. *Marine Policy, 32*(6), 956–967. https://doi.org/10.1016/j.marpol.2008.02.006

Kaiser, M. J., & Yu, Y. (2010). The impact of hurricanes Gustav and Ike on offshore oil and gas production in the Gulf of Mexico. *Applied Energy, 87*(1), 284–297. https://doi.org/10.1016/j.apenergy.2009.07.014

Kalk, H.J. (2011). The role of coastal plant community response to climate change: implications for restoring ecosystem resiliency. https://scholar.google.com/scholar?hl=en&as_sdt=0%2C19&q=The+role+of+coastal+plant+community+response+to+climate+change%3A+implications+for+restoring+ecosystem+resiliency&btnG=

Kaniewski, D., Marriner, N., Morhange, C., Faivre, S., Otto, T., & Van Campo, E. (2016). Solar pacing of storm surges, coastal flooding and agricultural losses in the Central Mediterranean. *Scientific Reports, 6*, 25197. https://doi.org/10.1038/srep25197

Kaye, J. P., & Quemada, M. (2017). Using cover crops to mitigate and adapt to climate change. *A review. Agronomy for Sustainable Development, 37*(1), 4. https://doi.org/10.1007/s13593-016-0410-x

Kianersi, S., Jules, R., Zhang, Y., Luetke, M., & Rosenberg, M. (2021). Associations between hurricane exposure, food insecurity, and microfinance: A cross-sectional study in Haiti. *World Development, 145*, 105530. https://doi.org/10.1016/j.worlddev.2021.105530

Klassen, W., Codallo, M., Zasada, I. A., & Abdul-bakil, A. A. (2006). Characterization of velvetbean (Mucuna pruriens) lines for cover crop use. In *Proceedings of the Florida State Horticultural Society* (Vol. 119, pp. 258–262) https://scholar.google.com/scholar?hl=en&as_sdt=0%2C19&q=Characterization+of+velvetbean+%28Mucuna+pruriens%29+lines+for+cover+crop+use&btnG=

Knabb, R. D., Brown, D. P., & Rhome, J. R. (2006). Tropical cyclone report hurricane Rita 18–26 September 2005. *National Hurricane Center, 17*. https://scholar.google.com/scholar?hl=en&as_sdt=0%2C19&q=Knabb+RD%2C+Brown+DP%2C+Rhome+JR+%282006%29+Tropical+cyclone+report+Hurricane+Rita&btnG=

Knutson, T., Camargo, S. J., Chan, J. C., Emanuel, K., Ho, C. H., Kossin, J., Mohapatra, M., Satoh, M., Sugi, M., Walsh, K., & Wu, L. (2020). Tropical cyclones and climate change assessment: Part II: Projected response to anthropogenic warming. *Bulletin of the American Meteorological Society, 101*(3), E303–E322. https://doi.org/10.1175/BAMS-D-18-0194.1

Knutson, T. R., Sirutis, J. J., Bender, M. A., Tuleya, R. E., & Schenkel, B. A. (2022). Dynamical downscaling projections of late twenty-first-century US landfalling hurricane activity. *Climatic Change, 1*(3), 28. https://doi.org/10.1007/s10584-022-03346-7

Laborte, A. G., Gutierrez, M. A., Balanza, J. G., Saito, K., Zwart, S. J., Boschetti, M., Murty, M., Villano, L., Aunario, J. K., & Reinke, R. (2017). RiceAtlas, a spatial database of global rice calendars and production. *Scientific Data, 4*, 170074. https://doi.org/10.1038/sdata.2017.74

Lara-Chavez, A., & Alexander, C. E. (2006). The effects of hurricane Katrina on corn, wheat and soybean futures prices and basis. https://doi.org/10.22004/ag.econ.18994

Lewis, C. T. (2022). Climate change and the caribbean: challenges and vulnerabilities in building resilience to tropical cyclones. *Climate, 10*(11), 178. https://doi.org/10.3390/cli10110178

Li, L., Xiong, X., & Yuan, H. (2022). Ships' response strategies to port disruptions caused by hurricanes. *Ocean & Coastal Management, 227*, 106275. https://doi.org/10.1016/j.ocecoaman.2022.106275

Loebach, P. (2019). Livelihoods, precarity, and disaster vulnerability: Nicaragua and hurricane Mitch. *Disasters, 43*(4), 727–751. https://doi.org/10.1111/disa.12402

Lipari, S., Balaguru, K., Rice, J., Feng, S., Xu, W., & K. Berg L, Judi D. (2024). Amplified threat of tropical cyclones to US offshore wind energy in a changing climate. *Communications Earth & Environment, 5*(1), 755. https://doi.org/10.1038/s43247-024-01887-6

Loh, P. M., Twumasi, Y. A., Ning, Z. H., Anokye, M., Oppong, J., Armah, R., & Namwamba, J. B. (2023). Analyzing the impact of sea level rise on coastal flooding and shoreline changes along the coast of Louisiana using remote sensing imagery. *The International Archives of the*

Photogrammetry, Remote Sensing and Spatial Information Sciences, 48, 139–145. https://doi.org/10.5194/isprs-archives-XLVIII-M-3-2023-139-2023
LSU AgCenter (2012). Louisiana summary: agriculture and natural resources. Available from https://repository.lsu.edu/agnatresources/. Accessed on December,15 2025.
LSU AgCenter (2020). Louisiana summary: agriculture and natural resources. Available from https://repository.lsu.edu/agnatresources/. Accessed December 6, 2025.
LSU AgCenter (2021). AgCenter experts: Louisiana agriculture suffers at least $584M in damage from Hurricane Ida. Available at: https://www.lsuagcenter.com/articles/page1632415649946. Accessed on August 9, 2025.
Machín-Sosa, B., Roque-Jaime, A. M., Avila-Lozano., D. R., & Rosset, P. (2010). *Revolución Agroecológica: el Movimiento de Campesino a Campesino de la ANAP en Cuba*. ANAP. https://www.sidalc.net/search/Record/KOHA-OAI-ECOSUR:53665/Description
Marcelin, L. H., Cela, T., & Shultz, J. M. (2016). Haiti and the politics of governance and community responses to hurricane Matthew. *Disaster Health, 3*, 151–161. https://doi.org/10.1080/21665044.2016.1263539
Mark, J., de Angel, S. D., Rosario-Matos, N., & Wang, L. (2025). Post-disaster food insecurity: Hurricane Maria as a case study. *The Journal of Climate Change and Health, 2*, 100363. https://doi.org/10.1016/j.joclim.2024.100363
Marrero, A., López-Cepero, A., Borges-Méndez, R., & Mattei, J. (2022). Narrating agricultural resilience after hurricane María: how smallholder farmers in Puerto Rico leverage self-sufficiency and collaborative agency in a climate-vulnerable food system. *Agriculture and Human Values, 39*(2), 555–571. https://doi.org/10.1007/s10460-021-10267-1
McKinney, L. D., Shepherd, J. G., Wilson, C. A., Hogarth, W. T., Chanton, J., Murawski, S. A., & Caffey, R. (2021). The Gulf of Mexico. *Oceanography, 34*(1), 30–43. https://www.jstor.org/stable/27020059
McGinley, K. A., Gould, W. A., Álvarez-Berríos, N. L., Holupchinski, E., & Díaz-Camacho, T. (2022). READY OR NOT? Hurricane preparedness, response, and recovery of farms, forests, and rural communities in the US Caribbean. *International Journal of Disaster Risk Reduction, 82*, 103346. https://doi.org/10.1016/j.ijdrr.2022.103346
MacNabb, E., & Fletcher, B. J. (2019). Hurricanes, disasters, and food insecurity: The intersection of two social events. In *Emerging voices in natural hazards research* (pp. 121–144). Butterworth-Heinemann. https://doi.org/10.1016/B978-0-12-815821-0.00012-6
Melvani, K., Bristow, M., Moles, J., Crase, B., & Kaestli, M. (2020). Multiple livelihood strategies and high floristic diversity increase the adaptive capacity and resilience of Sri Lankan farming enterprises. *Science of the Total Environment, 739*, 139120. https://doi.org/10.1016/j.scitotenv.2020.139120
Mengel, M., Levermann, A., Frieler, K., Robinson, A., Marzeion, B., & Winkelmann, R. (2016). Future Sea level rise constrained by observations and long-term commitment. *Proceedings of the National Academy of Sciences, 113*(10), 2597–2602. https://doi.org/10.1073/pnas.1500515113
Merem, E. C., Twumasi, Y., Foster, D., Richardson, C., & Yeramilli, S. (2012). Using GIS and climate risks information to analyze the vulnerability of coastal counties in Louisiana and Mississippi. *Delta, 1*(2), 8–10. https://doi.org/10.5923/j.re.20120201.01
Messina, Jr W. A. (2009). The 2008 hurricane season and its impact on Cuban agriculture and trade. Cuba in Transition: Association for the Study of the Cuban Economy, *18*, 421. https://efaidnbmnnnibpcajpcglclefindmkaj/; https://www.ascecubadatabase.org/wp-content/uploads/2014/09/v19-messina.pdf
Miller, A. J., Arias, M. E., & Alvarez, S. (2021). Built environment and agricultural value at risk from hurricane Irma flooding in Florida (USA). *Natural Hazards, 109*(2), 1327–1348. https://doi.org/10.1007/s11069-021-04880-w
Ming, X., Du, J., Xu, W., Li, Y., Shi, P., & Liu, B. (2014). Quantitative multi-hazard risk assessment with vulnerability surface and hazard joint return period. *Stochastic Environmental Research and Risk Assessment, 29*(1), 35–44. https://doi.org/10.1007/s00477-014-0935-y

Misuri, A., Moreno, V. C., Quddus, N., & Cozzani, V. (2019). Lessons learnt from the impact of hurricane Harvey on the chemical and process industry. *Reliability Engineering & System Safety, 190*, 106521. https://doi.org/10.1016/j.ress.2019.106521

Mohan, P., & Strobl, E. (2017). A hurricane wind risk and loss assessment of Caribbean agriculture. *Environment and Development Economics, 22*, 84–106. https://doi.org/10.1017/S1355770X16000176

Mohan, P. (2017). Impact of hurricanes on agriculture: Evidence from the Caribbean. *Natural Hazards Review, 18*(3), 04016012. https://doi.org/10.1061/(ASCE)NH.1527-6996.0000235

Mostafiz, R. B., Rohli, R. V., Friedland, C. J., Gall, M., & Bushra, N. (2022). Future crop risk estimation due to drought, extreme temperature, hail, lightning, and tornado at the census tract level in Louisiana. *Frontiers in Environmental Science, 10*, 919782. https://doi.org/10.3389/fenvs.2022.919782

Motha, R. P. (2011). The impact of extreme weather events on agriculture in the United States. In *Challenges and Opportunities in Agrometeorology* (pp. 397–407). Springer Berlin Heidelberg. https://doi.org/10.1007/978-3-642-19360-6_30

Munch, D. (2025). Hurricanes, heat and hardship: Counting 2024's crop losses. Market Intel. American Farm Bureau Federation. Available fromhttps://www.fb.org/intel/markets/hurricanes-heat-and-hardship-counting-2024s-crop-losses#:~:text=Overview%20of%20Estimated%20Crop%20Losses,$291%20million%20in%20crop%20losses. Accessed on December 30, 2025

National Weather Service (2020). Hurricane Laura 2020. U.S. Department of Commerce,. https://www.weather.gov/lch/Laura2020. Accessed on January 2, 2026.

Nazir, A., & Das, L. H. (2022). Resilience through crop diversification in Pakistan. In A. K. E. Haque, P. Mukhopadhyay, M. Nepal, & M. R. Shammin (Eds.), *Climate change and community resilience*. Springer. https://doi.org/10.1007/978-981-16-0680-9_28

Nelson, S. A., & Miller, P. W. (2023). Understanding meteorological changes following severe defoliation during a strong hurricane landfall: insights from hurricane Michael (2018). *Earth Interactions, 27*(1), e220012. https://doi.org/10.1175/EI-D-22-0012.1

Neog, D. R., Singha, G., Dev, S., & Prince, E. H. (2024). Artificial intelligence and its application in disaster risk reduction in the agriculture sector. In *Disaster risk reduction and rural resilience: With a focus on agriculture, water, gender and technology* (pp. 279–305). Springer Nature Singapore. https://doi.org/10.1007/978-981-97-6671-0_15

NOAA (2004). Hurricane Ivan post storm report. National Weather Service. Available from https://www.weather.gov/lix/psh_ivan. Accessed on September 8, 2025.

NOAA (2025a). Billion—dollar weather and climate disasters. Available from https://www.ncei.noaa.gov/access/billions/events. Accessed on September 14, 2025.

NOAA (2025b). Costliest U.S. tropical cyclones. Available from https://www.ncei.noaa.gov/access/billions/dcmi.pdf. Accessed on December 25, 2025.

Oldfield, S. (2019). The US national seed strategy for rehabilitation and restoration: progress and prospects. *Plant Biology, 21*(3), 380–382. https://doi.org/10.1111/plb.12851

Padilla, S., Ufer, D. J., Morgan, S., & Link, N. (2023). *US export competitiveness in select crop markets*. Economic Research Service. https://www.ers.usda.gov/publications/pub-details?pubid=106157

Parajuli, R., Thoma, G., & Matlock, M. D. (2019). Environmental sustainability of fruit and vegetable production supply chains in the face of climate change: a review. *Science of the Total Environment, 650*, 2863–2879. https://doi.org/10.1016/j.scitotenv.2018.10.019

Park, G. (2021). A comprehensive analysis of hurricane damage across the US gulf and Atlantic coasts using geospatial big data. *ISPRS International Journal of Geo-Information, 10*(11), 781. https://doi.org/10.3390/ijgi10110781

Perfecto I, Hajian-Forooshani Z, Iverson A, Irizarry AD, Lugo-Perez J, Medina N, …, Vandermeer J. (2019). Response of coffee farms to hurricane Maria: resistance and resilience from an extreme climatic event. Scientific Reports 9(1):15668. doi:https://doi.org/10.1038/s41598-019-51416-1.

Philpott, S. M., Arendt, W. J., Armbrecht, I., Bichier, P., Diestch, T. V., Gordon, C., & Zolotoff, J. M. (2008). Biodiversity loss in Latin American coffee landscapes: review of the evidence on ants, birds, and trees. *Conservation biology, 22*(5), 1093–1105. https://doi.org/10.1111/j.1523-1739.2008.01029.x

Quintarelli, V., Radicetti, E., Allevato, E., Stazi, S. R., Haider, G., Abideen, Z., & Mancinelli, R. (2022). Cover crops for sustainable cropping systems: a review. *Agriculture, 12*(12), 2076. https://doi.org/10.3390/agriculture12122076

Rana, A., Zhu, Q., Detken, A., Whalley, K., & Castet, C. (2022). Strengthening climate-resilient development and transformation in Viet Nam. *Climatic Change, 170*(1), 4. https://doi.org/10.1007/s10584-021-03290-y

Raátz, J. (2006). A ciklonok, hurrikánok, tájfunok neveiről. *Névtani Értesítő, 28*, 155–159. https://doi.org/10.29178/NevtErt.2006.17

Reuters. (2021). Hurricane Ida damages 118,000 acres of Louisiana sugarcane crops. *Reuters*, 14 September 2021. Reporting by M. Teixeira and P.J. Huffstutter; edited by M. Choy. Available from https://www.reuters.com/world/us/ida-damaged-118000-acres-sugarcane-crops-louisiana-report-2021-09-14/. Accessed on December 15, 2025.

Romero, D., & León-Cruz, J. F. (2024). Spatiotemporal changes in hurricane-force wind risk assessment in the Yucatan peninsula, Mexico. *Natural Hazards, 120*(5), 4675–4698. https://doi.org/10.1007/s11069-023-06397-w

Rosset, P. M., Machin Sosa, B., Roque Jaime, A. M., & Ávila Lozano, D. R. (2011). The campesino-to-campesino agroecology movement of ANAP in Cuba: Social process methodology in the construction of sustainable peasant agriculture and food sovereignty. *Journal of Peasant Studies, 38*, 161–191. https://doi.org/10.1080/03066150.2010.538584

Roth, D. (2010). Texas hurricane history (Vol. 17).: National Weather Service. chrome-extension://efaidnbmnnnibpcajpcglclefindmkaj/; https://www.weather.gov/media/crp/txhur.pdf

Routschek, A., Schmidt, J., & Kreienkamp, F. (2014). Impact of climate change on soil erosion—a high-resolution projection on catchment scale until 2100 in Saxony/Germany. *Catena, 121*, 99–109. https://doi.org/10.1016/j.catena.2014.04.019

Salarieh, B., Ugwu, I. A., & Salman, A. M. (2023). Impact of changes in sea surface temperature due to climate change on hurricane wind and storm surge hazards across US Atlantic and gulf coast regions. *SN Applied Sciences, 5*(8), 205. https://doi.org/10.1007/s42452-023-05423-7

Sarma, H. H., Paul, A., Kakoti, M., Talukdar, N., & Hazarika, P. (2024). Climate resilient agricultural strategies for enhanced sustainability and food security: a review. *Plant Archives, 24*(1), 787–792. https://doi.org/10.51470/PLANTARCHIVES.2024.v24.no.1.108

Schnepf, R., & Chite, R. M. (2005). Resources, Science, and Industry Division. In *US agriculture after hurricanes Katrina and Rita: status and issues*. Library of Congress, Congressional Research Service. https://www.policyarchive.org/handle/10207/2556

Schultz B. (2020). AgCenter estimates ag, forestry losses from hurricane Laura exceed $1.6 billion. LSU AgCenter News. Louisiana State University Agricultural Center,. https://www.lsuagcenter.com/articles/page1600179793946. Accessed January 2, 2026

Selva KM, Geethalakshmi V, Pazhanivelan S, Subrahmaniyan K, Dheebakaran G, Saravanakumar V, …, Pugazenthi K. (2025). Review on evolving cyclone patterns in bay of Bengal: Challenges for coastal agriculture. Plant Science Today 12: 11369. DOI: https://doi.org/10.14719/pst.11369.

Seneviratne, S. I., Zhang, X., Adnan, M., Badi, W., Dereczynski, C., Luca, A. D., Ghosh, S., Iskandar, I., Kossin, J., Lewis, S., & Otto, F. (2021). *Weather and Climate Extreme Events in a Changing Climate*. https://doi.org/10.1017/9781009157896.013

Shao, W., Jackson, N. P., Ha, H., & Winemiller, T. (2020). Assessing community vulnerability to floods and hurricanes along the Gulf coast of the United States. *Disasters, 44*(3), 518–547. https://doi.org/10.1111/disa.12383

Shovon, M. A. C., Mostafiz, R. B., Al Assi, A., & Rohli, R. V. (2025). Hydroclimatic extremes and aquaculture: A review of impact and response strategies. *Aquaculture International, 33*(6), 439. https://doi.org/10.1007/s10499-025-02108-3

Shovon, M. (2024). *Extreme climate events and agricultural commodity markets. Master's thesis.* Louisiana State University and Agricultural & Mechanical College. https://doi.org/10.31390/gradschool_theses.5992

Sledge, J. S. (2019). The Gulf of Mexico: a maritime history. *Univ of South Carolina Press.* https://www.google.com/books/edition/The_Gulf_of_Mexico/SWWzDwAAQBAJ?hl=en&gbpv=0

Smith, C. W. (2018). Effects on soil water holding capacity and soil water retention resulting from soil health management practices implementation. USDA Natural Resources Conservation Service. https://efaidnbmnnnibpcajpcglclefindmkaj/; https://www.nrcs.usda.gov/sites/default/files/2022-10/AWC_Effects_on_Soil_Water_Holding_Capacity_and_Retention.pdf

Song, J. Y., Moftakhari, H. R., Alipour, A., & Moradkhani, H. (2020). Toward a more effective hurricane hazard communication. *Environmental Research Letters, 15*(6), 064012. https://doi.org/10.1088/1748-9326/ab875f

Spencer, N., & Polachek, S. (2015). Hurricane watch: battening down the effects of the storm on local crop production. *Ecological Economics, 120*, 234–240. https://doi.org/10.1016/j.ecolecon.2015.10.006

Spiridonov, V., Ćurić, M., Spiridonov, V., & Ćurić, M. (2021). Tropical storms. In *Fundamentals of meteorology* (Vol. 1, pp. 275–288). https://doi.org/10.1007/978-3-030-52655-9_18

Stewart, R., van Rooyen, C., Dickson, K., Majoro, M., & de Wet T. (2010). What is the impact of microfinance on poor people?: A systematic review of evidence from sub-Saharan Africa. chrome-extension://efaidnbmnnnibpcajpcglclefindmkaj/; https://discovery.ucl.ac.uk/id/eprint/1500246/1/Stewart2010What(Report).pdf

Subaiya, S., Stillman, J., & Pumpalova, Y. (2019). A modified community assessment for public health emergency response (CASPER) four months after hurricane sandy. *Disasters, 43*(1), 206–207. https://doi.org/10.1111/disa.12299

Tchounwou, P. B. (1999). Climate change and its potential impacts on the Gulf coast region of the United States. *Reviews on Environmental Health, 14*(2), 91–102. https://doi.org/10.1515/REVEH.1999.14.2.91

Thompson WJ, Varma V, Joerin J, Bonilla-Duarte S, Bebber DP, Blaser-Hart W, …, Krütli P. (2023). Smallholder farmer resilience to extreme weather events in a global food value chain. Climatic Change 176(11): 152. doi:https://doi.org/10.1007/s10584-023-03586-1.

Turner, R. E., & Rabalais, N. N. (2019). The Gulf of Mexico. In *World seas: an environmental evaluation* (pp. 445–464). Academic. https://doi.org/10.1016/B978-0-12-805068-2.00022-X

Twumasi, Y. A., Merem, E. C., Ning, Z. H., Yeboah, H. B., Loh, P. M., Osei, J. D., & Armah, R. N. (2024). The impacts of Louisiana's changing climate on food crop production. *Agricultural Sciences, 15*(11), 1195–1222. https://doi.org/10.4236/as.2024.1511065

University of Florida Institute of Food and Agricultural Sciences (2018). UF/IFAS economists estimate hurricane Michael caused $158 million in Florida agricultural production losses. UF/IFAS News. https://blogs.ifas.ufl.edu/news/2018/10/26/uf-ifas-economists-hurricane-michael-caused-158-million-in-florida-agricultural-production-losses/

USDA (2017). Sugar and sweeteners outlook. Available from https://www.ers.usda.gov/sites/default/files/_laserfiche/outlooks/85182/SSSM-349.pdf?v=41103. Accessed on September 8, 2025

Velásquez-Espinoza, G., & Alcántara-Ayala, I. (2024). Tropical cyclones in Nicaragua: historical impact and contemporary exposure to disaster risk. *International Journal of Disaster Risk Science, 15*(4), 579–593. https://doi.org/10.1007/s13753-024-00581-7

Vink, K., & Ahsan, M. N. (2018). The benefits of cyclones: a valuation approach considering ecosystem services. *Ecological Indicators, 95*, 260–269. https://doi.org/10.1016/j.ecolind.2018.07.035

Vogel, E., Donat, M. G., Alexander, L. V., Meinshausen, M., Ray, D. K., Karoly, D., & Frieler, K. (2019). The effects of climate extremes on global agricultural yields. *Environmental Research Letters, 14*(5), 054010. https://doi.org/10.1088/1748-9326/ab154b

Vroege, W., Vrieling, A., & Finger, R. (2021). Satellite support to insure farmers against extreme droughts. *Nature Food, 2*(4), 215–217. https://doi.org/10.1038/s43016-021-00244-6

Wang, J., Konar, M., Anderson, P. N., Hadunka, P., & Mulenga, B. (2025). Impact of current and historical climate shocks on crop diversification in Zambia: Insights from household- and district-level observations. *Climate Risk Management, 47*, 100683. https://doi.org/10.1016/j.crm.2024.100683

Warburton, E. J. (2022). Evaluating the impact and recovery of hurricane Matthew on the agricultural sector of the Sud and grand'anse departments Haiti. Master of science thesis, Conventry university, Haiti. 166pp. chrome-extension://efaidnbmnnnibpcajpcglclefindmkaj/; https://pure.coventry.ac.uk/ws/portalfiles/portal/82580288/Warburton2023MScR.pdf

Webster, T. M., & Nichols, R. L. (2012). Changes in the prevalence of weed species in the major agronomic crops of the southern United States: 1994/1995 to 2008/2009. *Weed Science, 60*(2), 145–157. https://doi.org/10.1614/WS-D-11-00092.1

Westra, J. V., Niu, H., & Fannin, J. M. (2015). Economic forces affecting Louisiana agriculture production through space and time. https://doi.org/10.22004/ag.econ.205325

White, A., Faulkner, J. W., Niles, M. T., Conner, D., & Ernesto, M. V. (2023). The role of farmer networks in supporting adaptive capacity: opening the door for innovation and transformation in the Northeastern United States. *Elementa, 11*(1), 00039. https://doi.org/10.1525/elementa.2022.00039

Wiener, S. S., Álvarez-Berríos, N. L., & Lindsey, A. B. (2020). Opportunities and challenges for hurricane resilience on agricultural and forest land in the US southeast and Caribbean. *Sustainability, 12*(4), 1364. https://doi.org/10.3390/su12041364

Williams, V. J. (2009). The ecological effects of salt water intrusion on the agriculture industry after hurricane Katrina. In *Proceedings of the 2007 national conference on environmental science and technology* (pp. 97–102). Springer New York.

Williams, V. (2010). Identifying the economic effects of salt water intrusion after hurricane Katrina. *Journal of Sustainable Development, 3*(1), 29–37. https://scholar.google.com/scholar?hl=en&as_sdt=0%2C19&q=Identifying+the+economic+effects+of+salt+water+intrusion+after+Hurricane+Katrina&btnG=

Williams, A. G., Birt, H. W., Raghavendra, A., & Dennis, P. G. (2023). Cropping system diversification influences soil microbial diversity in subtropical dryland farming systems. *Microbial Ecology, 85*(4), 1473–1484. https://doi.org/10.1007/s00248-022-02074-w

Xu, Y., Elbakidze, L., Gassman, P. W., Yen, H., Ma, Y., Hubbart, J. A., & Arnold, J. G. (2025). US–China agricultural trade and environmental outcomes: The case of nutrient runoff to the Gulf of Mexico. *Applied Economic Perspectives and Policy, 47*(3), 983–1002. https://doi.org/10.1002/aepp.13501

Yuen, K. W., Switzer, A. D., Teng, P. P., & Lee, J. S. H. (2022). Assessing the impacts of tropical cyclones on rice production in Bangladesh, Myanmar, Philippines, and Vietnam. *Natural Hazards and Earth System Sciences Discussions, 2022*, 1–28. https://doi.org/10.5194/nhess-2022-4

Zafar, S., Shah, A. A., Ashraf, M. A., Rasheed, R., Muddasar, M., Khan, I. M., & Iqbal, R. (2024). Plant growth under extreme climatic conditions. In S. Fahad, S. Saud, T. Nawaz, L. Gu, M. Ahmad, & R. Zhou (Eds.), *Environment, climate, plant and vegetation growth* (pp. 133–178). Springer Nature Switzerland. https://doi.org/10.1007/978-3-031-69417-2_5

Zhang, Y., et al. (2018). The impact of hurricane Harvey on Texas agriculture: a case study. *Journal of Agricultural and Applied Economics, 50*(2), 226–241.

Zhang, K., Chen, G., Xia, Y., & Wang, S. (2022). An ensemble-based, remote-sensing-driven, flood-landslide early warning system. In K. Zhang, Y. Hong, & A. AghaKouchak (Eds.) *Remote sensing of water-related hazards* (pp. 123–134). https://doi.org/10.1002/9781119159131.ch7.

Zhang, G., & Zhu, W. (2023). Characteristics and predictive modeling of short-term impacts of hurricanes on the US employment. arXiv preprint arXiv 2307: 13686.

Advancing Cyclone Resilience Through Community-Centered and Glocally-Informed AI Systems

Poulomi Chakravarty, Sai Gattupalli, Urjani Chakravarty, Gulab Chand, and Betty Bitengo Atandi

Abstract Cyclones are intensifying under climate change, creating compound hazards where flooding, heatwaves, and infrastructure failures converge to overwhelm communities. As cyclones intensify under climate change, resilience demands systems that navigate compound hazards, close the forecast-to-action gap, and embed trust in communication. In this chapter, we discuss a glocally informed, community-centered framework that integrates Artificial Intelligence (AI) with Indigenous Knowledge Systems, ecological signals, visual cues, and local communication networks. Anchored in the Climate Communication Channels (3C) knowledge integration model, we explore how hybrid intelligence combining AI with human interpretation, cultural metaphors, and ecological wisdom can transform early warning into actionable preparedness. The framework emphasizes multilingual, culturally attuned, and co-designed AI tools that embed and embody local metaphors, trusted communication networks, and ecological signals into risk communication. Key design principles include participatory co-creation, human-in-the-loop adaptability, and user interfaces that align with local semiotics of trust. Ethical foundations, including Indigenous data sovereignty, free prior informed consent, and benefit-sharing, ensure that digital tools amplify rather than appropriate traditional knowledges. Case insights from cyclone-prone regions illustrate how plural epistemologies enhance accuracy, legitimacy, and equity in disaster risk management. The chapter reframes AI as civic infrastructure, modular, place-based, and culturally

P. Chakravarty
Global Climate Association, Hyderabad, India

S. Gattupalli
University of Massachusetts, Amherst, MA, USA

U. Chakravarty
Department of Humanities and Liberal Arts, Indian Institute of Management Bodh Gaya, Bodh Gaya, India

G. Chand
Department of Liberal Arts and Social Sciences, Manipal University Jaipur, Jaipur, India

P. Chakravarty (✉) · B. B. Atandi
Biodiversity for a Livable Climate, Boston, MA, USA
e-mail: cpoulomi25@gmail.com

J. Collins et al. (eds.), *Tropical Cyclone Risk and Resilience*,
https://doi.org/10.1007/978-3-032-28745-8_11

attuned, uniting scientific precision, ecological stewardship, and social meaning to advance pluriversal resilience in a warming world.

Keywords Cyclone resilience · Community-centered AI · Climate communications channels · Disaster informatics · Crisis communication · Hybrid intelligence; disaster informatics · Pluriversal resilience

1 Introduction

1.1 An Overview of Cyclones in a Changing Climate

In a world rife with vicious natural disasters, cyclones are among the most destructive climate-related hazards. Cyclones, in the broadest meteorological sense, are large-scale air masses that rotate around a strong center of low atmospheric pressure. They are generally classified based on their origin, structure, and location, including extratropical cyclones, subtropical cyclones, and tropical cyclones (Emanuel, 2003; Holland, 1993). Extratropical cyclones, driven primarily by horizontal temperature gradients, typically form in mid-latitudes and can produce powerful winds, heavy precipitation, and significant storm surges. Subtropical cyclones exhibit characteristics of both tropical and extratropical systems, and often serve as transitional stages in cyclone development. Tropical cyclones (TCs), in contrast, are warm-core systems that derive their energy from the latent heat of condensation over warm ocean waters and include regional variants such as hurricanes in the Atlantic and Northeast Pacific, typhoons in the Northwest Pacific, and simply "cyclones" in the South Pacific and Indian Ocean basins (Emanuel, 1987). Scientific evidence indicates that anthropogenically forced climate change is influencing the intensity, precipitation, and associated impacts of tropical cyclones. Observations since the 1980's reveal a global increase in the proportion of intense TCs (Category 3 or higher), with the most pronounced changes in the North Atlantic, and projections suggest further increases in the proportion of Category 4–5 storms under continued warming scenarios (Knutson et al., 2020; Wang et al., 2024). Rising sea levels very likely driven in large part by anthropogenic forcing are compounding TC risks by elevating storm surge and inundation levels, while enhanced atmospheric moisture is projected to amplify TC rainfall rates (Bindoff et al., 2014). Additional emerging trends, such as poleward migration of peak intensity, slowing forward motion, and increasing rates of rapid intensification in some basins, further indicate a shifting risk profile potentially linked to climate change (Bhatia et al., 2019). Figure 1 depicts the classification of cyclones and impacts of climate change on cyclogenesis. These observed and projected changes demand the urgent need for adaptive, science-informed strategies to manage cyclone risk in a warming world. Observed changes in cyclone behavior show clear basin-specific patterns that motivate regionally calibrated risk modeling approaches. In the North Atlantic, the probability that a tropical cyclone reaches major hurricane intensity (Category 3–5) has increased by approximately 25% between 1979 and 2017, indicating a higher likelihood of the most damaging storms under recent climate conditions (Kossin et al., 2020). The Western North Pacific, the most active tropical cyclone basin globally, produces an average of about 26 tropical storms per year and has experienced a statistically significant increase in the rate of rapid intensification events in recent decades, increasing forecasting and preparedness challenges (Kossin et al., 2020). In the Arabian Sea, while the total number of cyclones has not increased substantially, the frequency of very severe cyclonic storms ($\geq$ 64 knots) has risen by more than 50% during the period 2001–2019 compared to 1979–2000, representing a marked shift toward higher-impact events in a region with historically limited exposure (Murakami et al., 2017). In the Mediterranean, tropical-like cyclones (medicanes) remain rare, occurring on average 1–2 times per year, but research indicates increasing rainfall intensity associated with these systems, amplifying flood risk in densely populated coastal areas even as overall frequency remains stable or slightly decreases (Cavicchia et al., 2014). In the North Atlantic, hurricanes increasingly generate blackout–heat exposure, where prolonged power outages following landfall coincide with extreme post-storm heat, substantially elevating health risks for vulnerable populations (Kossin et al., 2020). In the Western North Pacific, rapidly intensifying typhoons frequently trigger inland flooding and transport-access disruption, isolating communities and delaying emergency response far beyond coastal impact zones (Kossin et al., 2020). In the Arabian Sea, the recent rise in very severe cyclonic storms has amplified coastal flooding–livelihood disruption, particularly affecting port infrastructure, fisheries, and coastal settlements with limited historical preparedness (Murakami et al., 2017). In the Mediterranean, rare but intense medicanes (Mediterranean Tropical like cyclones) increasingly lead to extreme rainfall–urban flash flooding, overwhelming drainage systems and disrupting mobility and emergency access in densely populated coastal cities (Cavicchia et al., 2014). Together, these statistics highlight the heterogeneity of cyclone risk and the importance of multi-scale, context-aware modeling frameworks.

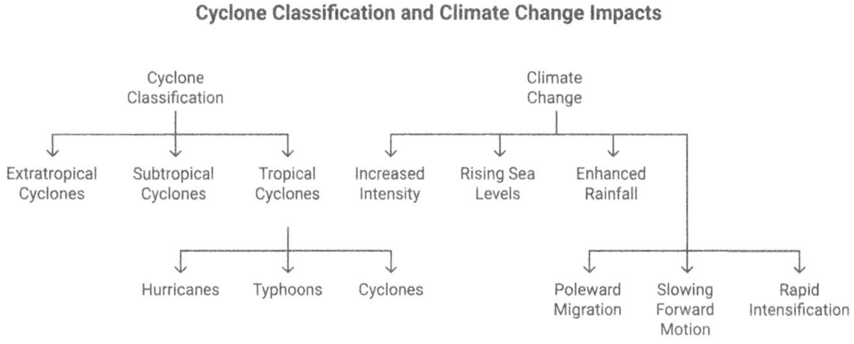

Fig. 1 Classification of cyclones and associated climate change impacts

1.2 Need for Resilience Strategies Beyond Early Warning

While advances in meteorological science have significantly improved the accuracy of cyclone forecasts, recent research highlights that forecasting alone cannot sufficiently address the escalating risks posed by climate change–intensified storms. Rising sea levels, amplified rainfall rates, poleward migration of storm tracks, and increased rapid intensification collectively elevate the scale of destruction in ways that exceed the capacity of warning systems alone (Bhatia et al., 2019) For example, Hurricane Harvey revealed how slow-moving cyclones can overwhelm urban infrastructure despite accurate forecasts, leading to unprecedented flooding in Houston (van Oldenborgh et al., 2017). Cyclones are multi-hazard events that produce cascading risks across social and ecological systems: violent winds compromise roofs and power lines, storm surge and extreme rainfall inundate settlements and cropland, and saturated soils trigger landslides together causing immediate fatalities and prolonged disruption to livelihoods, health services, and food security (IPCC, 2021, Ch. 11). The compound impacts of Cyclone Idai in March 2019 exemplify these hazards; major flooding and infrastructure losses across Mozambique, Zimbabwe, and Malawi were followed by elevated communicable-disease risks and documented cholera outbreaks (ECDC, 2019; WHO Regional Office for Africa, 2019). There is robust assessment evidence that climate change is altering these hazards. The IPCC finds heavy-precipitation extremes have increased in many regions and that the proportion of intense tropical cyclones has increased (global-scale confidence: *medium*), with further increases in TC rainfall rates projected under continued warming (IPCC, 2021; Knutson et al., 2020). Risk reduction extends beyond forecasting. Ecosystem-based adaptation can attenuate impacts: global analyses show mangrove forests substantially reduce coastal flood damages by damping waves and storm surge and stabilizing shorelines (Menéndez et al., 2020). At the building scale, engineering research and empirical post-storm analyses indicate that strong roof-to-wall connections and roof system upgrades markedly reduce wind damage to housing (Feng et al., 2020). Together, these findings support a multidimensional resilience strategy that integrates early warnings with actionable capacity, ecosystem protection, and structural retrofits a portfolio shown to reduce cyclone losses today while preparing communities for a warmer, wetter future (IPCC, 2021; Knutson et al., 2019, 2020; Menéndez et al., 2020). These realities prove that resilience requires a broader framework that not only anticipates hazards but also equips communities with adaptive strategies, infrastructure readiness, and culturally relevant preparedness tools that extend far beyond meteorological prediction (Chakravarty & Gattupalli, 2024; Knutson et al., 2020).

1.3 Introduction to the Glocally-Informed, Community-Centered AI Framework

Emerging approaches emphasize the need for resilience strategies that integrate global scientific insights with localized knowledge systems. A glocally-informed, community-centered Artificial Intelligence (AI) framework provides one such pathway, combining meteorological data with ecological signals, Indigenous weather indicators, and trusted communication networks (Chakravarty et al., 2025). Unlike top-down early warning systems, this approach situates AI as a participatory and adaptable tool designed with local stakeholders, embedding cultural metaphors, multilingual accessibility, and ecological wisdom into risk communication (Chakravarty & Gattupalli, 2024). By anchoring resilience in the Climate Communication Channels framework, these AI-driven tools foster place-based adaptation, ensuring cyclone preparedness strategies resonate with local semiotics of trust while maintaining technical precision. This pluriversal approach reframes resilience as a shared endeavor that unites scientific rigor, cultural meaning, and ecological stewardship to manage cyclone risks in a warming world.

2 Climate Change and Cyclone Risk

2.1 Climate Change Impacts on Storm Intensity, Frequency, and Patterns

A growing body of scientific evidence shows that anthropogenic climate change is altering the characteristics of tropical and extra-tropical cyclones, with profound implications for coastal and inland communities. Rising sea surface temperatures, higher atmospheric moisture content, and accelerating sea-level rise together create the conditions for storms of greater destructive potential. Observations since the late twentieth century indicate an increase in the intensity and rapid intensification of tropical cyclones, especially in the North Atlantic and parts of the Indian Ocean. The proportion of Category 4–5 storms has risen, and modeling studies project a further ~ 5% increase in mean storm intensity under a + 2 °C warming scenario (Knutson et al., 2020). The destructive power of storms is being amplified by higher rainfall rates, fueled by greater atmospheric moisture, and by stronger storm surges driven by rising seas (World Meteorological Organization, 2025). While global totals of cyclone occurrences have not shown consistent upward trends, their distribution and intensity are shifting. Some basins, like the Arabian Sea, have seen marked increases in the probability of extremely severe cyclones in the post-monsoon season, partly attributed to anthropogenic forcing. Conversely, models often project a slight decrease or no significant change in overall cyclone numbers, but a clear rise in the share of high-intensity storms. Regional studies highlight how these changes manifest differently. Along U.S. Atlantic and Gulf coasts, tropical cyclones dominate extreme compound flooding events, combining intense rainfall with storm surges. Future projections suggest joint extreme flood events could become *7–195 times more frequent* by 2100, with storm climatology changes contributing more than sea-level rise to these risks (Gori et al., 2022). In the Mediterranean basin, altered precipitation regimes and rising seas are exacerbating hydrological extremes, compounding droughts with flash floods and storm surges (Granata et al., 2025). Climate change is also intensifying *compound hazards* situations where multiple climate drivers interact to amplify risks. For instance, Hurricane Ida (2021) in Louisiana demonstrated a compound cyclone–blackout–heatwave event, where infrastructure failures overlapped with extreme heat, magnifying human health risks. Under high-emissions scenarios, such compound events could become up to 17 times more frequent by the end of the century (Feng et al., 2025). The underlying drivers of these changes stem from both global and regional processes. Warming oceans provide greater thermal energy to cyclones, while shifts in atmospheric circulation contribute to poleward migration of storm tracks and slower storm translation speeds, leading to prolonged rainfall and flooding over affected areas. Data science and climate informatics approaches further confirm that the climate crisis is reshaping storm dynamics, with computational models consistently linking anthropogenic greenhouse gas emissions to heightened storm hazards (Ali & Thakkar, 2023).

2.1.1 Regional Vulnerability Trends and Projected Risk Scenarios

Regional vulnerabilities to cyclones are highly uneven, with exposure shaped by geographic, climatic, and socio-economic factors. In South Asia, warming seas in the Arabian Sea and Bay of Bengal have increased the probability of extremely severe cyclones, posing escalating risks for densely populated, low-lying coastal regions. In the Mediterranean basin, climate change has intensified both prolonged droughts and sudden extreme rainfall, compounding risks of flash floods and storm surges for coastal populations (Granata et al., 2025). Along the U.S. Atlantic and Gulf coasts, projections suggest compound flood hazards from tropical cyclones could become 7–195 times more frequent by 2100, with storm climatology change contributing more to risk escalation than sea-level rise (Gori et al., 2022). Furthermore, compound events such as blackout–heatwave crises triggered by hurricanes are expected to become markedly more common in the Gulf Coast region, magnifying health and infrastructure vulnerabilities (Feng et al., 2025). These findings underscore the necessity of region-specific adaptation strategies informed by localized hazard interactions and socio-economic conditions.

2.1.2 Integrating Climate Models with Local Hazard Data

Integrating global climate models with local hazard datasets enhances the accuracy and usability of risk projections for communities facing cyclone threats. High-resolution downscaling of CMIP6 projections has demonstrated how local rainfall and storm surge intensities can significantly amplify compound flooding risks across the U.S. eastern seaboard, with local hazard dependence emerging as a critical driver of risk (Gori et al., 2022). In Europe, model-based simulations highlight how climate change is expected to increase flood risk and hydrodynamic loading on infrastructure, emphasizing the need for site-specific adaptation strategies. In the Mediterranean, combining atmospheric circulation modeling with observational precipitation data reveals escalating extremes in both drought and rainfall, pointing to the need for improved early warning systems and adaptive water governance (Granata et al., 2025). Complementarily, advances in climate informatics show that integrating machine learning with regional hazard observations can uncover hidden patterns of vulnerability and optimize resilience planning (Ali & Thakkar, 2023). Together, these approaches highlight the importance of bridging large-scale projections with fine-grained, place-based data to ensure that adaptation measures are scientifically rigorous and locally actionable.

2.2 Examples from Cyclone-Prone Regions to Contextualize the Framework

Cyclone-prone regions across the globe illustrate why resilience must be both globally informed and locally grounded. In the North Indian Ocean, the Arabian Sea has experienced a marked increase in extremely severe cyclones, such as Cyclone Amphan (2020), attributed in part to anthropogenic warming and reduced vertical wind shear (Mohapatra 2016; Murakami et al., 2017). These storms exposed the vulnerability of coastal populations in India and Bangladesh, where traditional knowledge of mangrove ecosystems and community-led evacuation networks played a critical role in reducing loss of life. Similarly, along the U.S. Gulf Coast, Hurricane Ida (2021) not only caused catastrophic flooding and power outages but also produced compound hazards when extreme heatwaves overlapped with widespread blackouts, exacerbating health risks for vulnerable populations (Feng et al., 2025). In the western North Pacific, storms making landfall in Southeast Asia have intensified significantly since the late 1970's, with rainfall-driven flooding compounding urban vulnerability (Mei & Xie, 2016). Meanwhile, Mediterranean regions, though historically less cyclone-prone, are experiencing compound hydrological extremes where intense rainfall converges with sea-level rise to magnify flood hazards (Granata et al., 2025). Together, these cases underscore the urgency of a framework that merges global scientific modeling with Indigenous ecological knowledge and culturally rooted communication strategies, enabling AI-driven early warning systems that resonate with local realities and foster community trust (Chakravarty & Gattupalli, 2024; Chakravarty et al., 2025). Here, the Climate Communications Channels (3C) knowledge integration framework provides a systematic approach for synthesizing information from diverse disciplinary sources, including environmental sciences, meteorology, communications, Indigenous Knowledge Systems, linguistics, disaster management, and education. Through AI-mediated knowledge integration, this framework enables the convergence of multidisciplinary expertise to develop comprehensive cyclone preparedness and resilience strategies that bridge scientific precision with community-centered approaches.

Operationally, the 3C framework implements this integration as a continuous loop 1. Nature's signals alert to risk 2. Community wisdom strengthens interpretation 3. AI integrates all knowledge to produce precise, multilingual forecasts 4. Trusted voices and visual cues disseminate warnings and 5. Response plus adaptive learning feedback to improve future sensing and decisions aligning technical, social, and ecological resilience in practice (Fig. 2).

Currently we observe a persistent *forecast-to-action gap* as warnings often arrive, but siloed systems, language/access barriers, and limited integration of Indigenous knowledge and trust networks mean communities still lack clear, culturally legible steps and resources to act. The 3C framework proposes a system that bridges the forecast-to-action gap by integrating nature's signals and Indigenous knowledge with

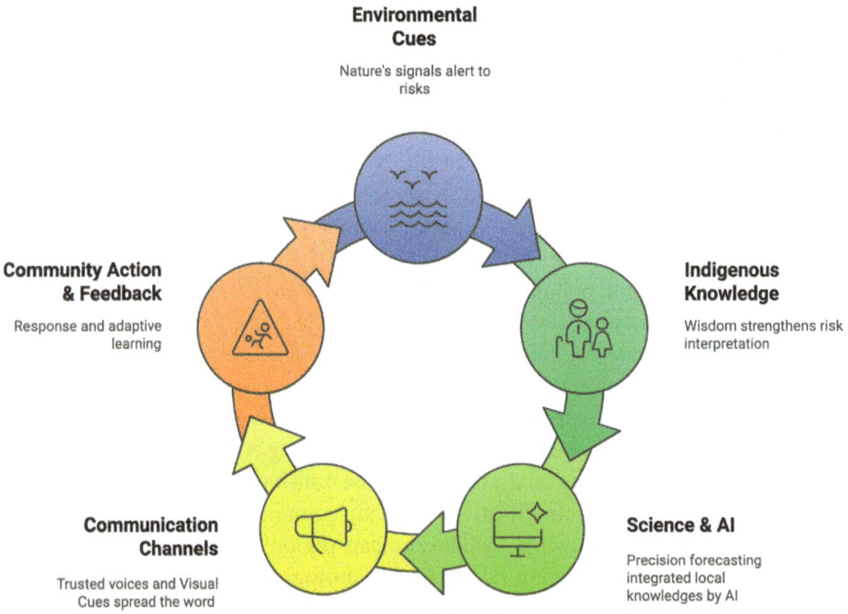

Fig. 2 Cyclical knowledge integration in the climate communication channels (3C) framework: bridging ecological signals, indigenous wisdom, AI-supported forecasting, and community feedback

AI-refined, multilingual forecasts, then routing them through trusted messengers and clear visual cues. This is carried out by turning signals into action through a continuous loop that is—nature's cues → community interpretation → AI-supported, multilingual forecasts → trusted messengers and visual cues → creating practiced response with feedback so information, capacity, and financing are linked before, during, and after a storm. In doing so, 3C proposes a resilience strategy that is a coordinated technical-social-ecological cycle rather than a one-way forecast.

3 AI-Integrated Disaster Resilience

3.1 Foundations of AI-Integrated Disaster Resilience

The integration of artificial intelligence into disaster resilience represents a paradigmatic shift from reactive emergency response to proactive, adaptive systems that anticipate, prepare for, and respond to complex hazard scenarios (Pak & Mostafavi, 2025; Hamid & Abedlmajid, 2025). At its core, AI-integrated disaster resilience encompasses the systematic application of machine learning algorithms, data analytics, and computational intelligence to enhance community preparedness, response coordination, and recovery processes while maintaining human agency in critical decision-making (Ali Maher et al., 2025; Miranda et al., 2022; Park & Kang, 2024).

Disaster informatics, as defined in contemporary literature, refers to the interdisciplinary field that applies information and communication technologies to support disaster management activities across the disaster cycle (Palen & Anderson, 2016). This field has evolved from simple data collection systems to sophisticated hybrid intelligence frameworks that combine computational power with human expertise, local knowledge, and community wisdom (Moore & Purohit, 2019; Fischer-Preßler et al., 2024).

The concept of intelligent disaster management (IDM) emerges as a central organizing principle, representing systems that leverage AI technologies while addressing ethical considerations such as privacy protection, bias mitigation, and ensuring equitable access to resources. Unlike traditional disaster management approaches that rely primarily on top-down command structures, IDM emphasizes distributed intelligence and multi-stakeholder collaboration (Belcastro et al., 2025; Vollmer et al., 2025).

Hybrid intelligence systems, as defined by Kamar (2016), represent collaborative frameworks where human and machine intelligence are integrated to overcome the shortcomings of existing AI systems, with human intelligence complementing machine capabilities throughout the system's lifecycle. Hybrid intelligence systems in disaster management context refer to collaborative frameworks where human cognition and artificial intelligence work synergistically, with each component contributing complementary strengths. These systems recognize that disaster scenarios involve complex social, cultural, and environmental factors that require both computational analysis and human interpretation (Gao et al., 2025; Hamid & Abedlmajid, 2025; Maceika et al., 2024).

Transformational Potential and Current Applications

Recent advances demonstrate AI's capacity to transform disaster management through multi-modal data integration, real-time situational awareness, and predictive analytics (Chen et al., 2025; Krichen et al., 2025; Pak & Mostafavi, 2025). Contemporary applications include remote sensing analysis using convolutional neural networks for rapid damage assessment (Krichen et al., 2025), natural language processing for social media monitoring (Pal et al., 2024), and machine learning algorithms for resource optimization (Reddy et al., 2024; Upadhyay et al., 2024).

Geospatial intelligence enhanced by AI enables more accurate hazard mapping and vulnerability assessments (Elshoukry, 2024). Systems now integrate satellite imagery, sensor networks, and crowdsourced data to create dynamic risk models that update in real-time. The BRIGHT dataset (Chen et al., 2025), representing one of the most comprehensive multimodal building damage assessment resources, exemplifies how AI can support all-weather disaster response across diverse geographic contexts.

Predictive modeling capabilities have advanced significantly, with AI systems now capable of extended forecast horizons and reduced false alarm rates (Rolnick et al., 2022). These improvements are particularly crucial for cyclone prediction, where lead time directly correlates with life-saving potential (Bhatia et al., 2019; Knutson et al., 2020). However, the effectiveness of these predictions depends heavily on their integration with local knowledge systems and community preparedness infrastructure (Chakravarty & Gattupalli, 2024; Hiwasaki et al., 2014).

Critical Limitations and Systemic Challenges

Despite technological advances, several fundamental limitations constrain the effectiveness of AI in disaster management (Cao, 2023). According to a report from the Nature Tech Collective (2025), data scarcity remains a persistent challenge, particularly for rare extreme events and in resource-constrained environments. Many AI models require extensive training data, yet disaster events by definition represent low-probability, high-impact scenarios with limited historical records (Camps-Valls et al., 2024).

Algorithmic bias presents significant ethical concerns, particularly when AI systems make decisions affecting vulnerable populations (Jonker & Rogers, 2024). Research indicates that AI models often perform poorly for marginalized communities due to training data limitations and algorithmic assumptions that fail to capture diverse social realities (Farnadi et al., 2024; Faust et al., 2025; Marko et al., 2025). This bias can exacerbate existing inequalities during disaster response when resources are allocated based on biased algorithmic recommendations (Chen & Li, 2025; Choong et al., 2025).

Explainability and trust represent crucial barriers to adoption (Gevaert et al., 2021). Emergency responders and community leaders require interpretable decision support rather than "black box" recommendations (Zentner, 2025). The high-stakes nature of disaster response demands transparent algorithmic processes that can be audited and understood by diverse stakeholders (Visave, 2025). At the same time, infrastructure dependencies create vulnerabilities in disaster scenarios when communication networks, power systems, and computing resources may be compromised (EIS Council, 2025). AI systems must be designed for operational resilience under degraded conditions, requiring edge computing capabilities and offline functionality.

Theoretical Foundations for Human-AI Collaboration

Human-centered AI (HCAI) provides foundational principles for designing AI systems that augment rather than replace human decision-making (Auernhammer, 2020; Larouche, 2024; Shneiderman, 2022). In disaster contexts, this framework emphasizes preserving human agency, supporting domain expertise, and enabling community participation in AI-driven processes (Senarath et al., 2023).

Human-in-the-loop (HITL) (Stryker, 2025) architectures represent essential design patterns for disaster management applications. These systems maintain human oversight while leveraging AI capabilities for data processing, pattern recognition, and scenario modeling. Effective HITL implementations require intuitive interfaces, clear uncertainty communication, and override mechanisms that preserve human authority in critical decisions.

Participatory AI design frameworks emphasize the importance of stakeholder engagement throughout system development (OECD, 2023; Park, 2024). This approach recognizes that disaster management involves diverse communities with varying needs, capabilities, and cultural contexts (Marciniak-Nuqui, 2024). Community involvement in AI system design helps ensure cultural appropriateness, local relevance, and sustainable adoption (Khan & Park, 2024; OECD, 2023).

Value alignment in AI systems requires explicit consideration of ethical principles, community priorities, and cultural values. For disaster management, this includes balancing efficiency with equity, individual privacy with collective safety, and technological capability with human dignity (Comes, 2024).

3.2 Knowledge Integration Methodologies

3.2.1 Integrating Environmental Cues, Traditional Weather Indicators, and Meteorological Data

The integration of environmental sensing, traditional ecological knowledge, and meteorological data represents a cornerstone of glocally-informed AI systems. This multi-source data fusion approach recognizes that disaster prediction and preparedness benefit from diverse information streams that capture different temporal and spatial scales of environmental change (Leong, 2025; Zeng et al., 2023).

Environmental sensor networks provide continuous monitoring of atmospheric, hydrological, and ecological parameters that precede disaster events (Zeng et al.,

2023). Modern Internet of Things (IoT) deployments can detect subtle environmental changes—such as atmospheric pressure fluctuations, soil moisture variations, or seismic activity—that serve as early warning indicators (Salam, 2024; Zeng et al., 2023). However, sensor data alone provides insufficient context for disaster preparedness without integration with human knowledge systems. Traditional weather indicators encompass centuries of accumulated observations about phenological cycles, animal behavior patterns, atmospheric conditions, and ecological relationships that communities have used for environmental prediction (Motsumi & Nemakonde, 2025; Radeny et al., 2019). These indicators often detect environmental changes at timescales and spatial resolutions that complement technical monitoring systems (Radeny et al., 2019).

Data fusion methodologies must address the challenge of integrating quantitative sensor data with qualitative traditional knowledge (Xiang et al., 2022). Hybrid ontologies that represent both technical measurements and cultural categories provide frameworks for meaningful integration. Machine learning approaches, particularly ensemble methods and multi-modal architectures, can learn to correlate technical indicators with traditional observations to improve prediction accuracy. Similarly, temporal integration requires recognizing different timescales inherent in various knowledge systems. Meteorological models operate on hourly to weekly timescales, while traditional ecological knowledge often incorporates seasonal and multi-year patterns that reflect longer-term environmental cycles (Price et al., 2024; Waqas et al., 2025). AI systems must integrate these different temporal perspectives to provide comprehensive risk assessments.

3.2.2 Hybrid Intelligence Systems: Bridging Scientific and Indigenous Knowledge

Hybrid intelligence refers to the synergistic integration of human cognition and artificial intelligence to achieve outcomes that neither could accomplish alone (Liu & Fu, 2024). This paradigm emphasizes collaboration rather than replacement, leveraging human strengths such as creativity, empathy, ethical reasoning, and contextual understanding, alongside AI's capabilities in data processing, pattern recognition, and computational efficiency. Hybrid intelligence frameworks in disaster management represent sophisticated knowledge integration architectures that combine computational analysis with human expertise and cultural knowledge systems. These systems recognize that effective disaster resilience requires both technical precision and social meaning-making.

Indigenous knowledge systems (IKS) contribute essential insights about local hazard patterns, environmental relationships, and community response strategies developed through generations of experience (Kugara et al., 2022). Research demonstrates that Indigenous communities often possess predictive capabilities that complement scientific forecasting, particularly for localized hazard conditions and ecological indicators (Dube et al., 2024; Motsumi & Nemakonde, 2025).

Knowledge co-production methodologies provide frameworks for integrating IKS with scientific approaches while respecting intellectual property rights and cultural protocols (Robinson et al., 2022). These methodologies emphasize collaborative research designs, community consent processes, and benefit-sharing arrangements that ensure Indigenous communities retain control over their knowledge.

Ontological integration requires developing conceptual frameworks that can represent diverse knowledge systems without forcing artificial translations between incompatible worldviews (Lewis et al., 2024). Multi-ontology architectures allow AI systems to maintain separate but interlinked representations of different knowledge domains. Validation and verification processes must accommodate different epistemological standards for knowledge reliability. Scientific knowledge emphasizes statistical significance and reproducibility, while Indigenous knowledge prioritizes cultural validation and community consensus (Whaanga & Mato, 2020). Hybrid systems require multi-criteria evaluation frameworks that respect diverse standards for knowledge legitimacy.

3.3 Human-Centered Design and Implementation

3.3.1 Human-in-the-Loop Adaptability Frameworks

HITL systems represent essential architectures for disaster management where human judgment, domain expertise, and ethical oversight remain critical for effective decision-making. These frameworks recognize that AI systems should augment human capabilities rather than replace human decision-makers in high-stakes scenarios. For instance, adaptive interfaces enable AI systems to adjust their interaction patterns based on user expertise, situational context, and task requirements. For disaster management, this includes providing detailed technical information to expert responders while offering simplified risk communication to community members (Hanashima et al., 2024). Interface adaptation must occur dynamically as situations evolve and different stakeholders engage with the system.

Feedback mechanisms allow human operators to correct AI predictions, update model assumptions, and incorporate new information that may not be captured in training data. Effective HITL systems implement active learning approaches that use human feedback to continuously improve AI performance, such as the HERMES system developed by Avvenuti et al. (2020).

Override capabilities ensure that humans retain ultimate authority over critical decisions while benefiting from AI analysis and recommendations. These mechanisms must be designed to prevent automation bias while maintaining the efficiency benefits of AI assistance. Workload management considers how AI systems can help distribute cognitive load appropriately between human and machine components. During high-stress disaster scenarios, AI can handle routine data processing while humans focus on strategic decisions and interpersonal coordination. Trust calibration represents a crucial challenge where users must develop appropriate reliance on AI systems—neither over-trusting automated recommendations nor dismissing useful insights. Trust calibration requires transparency mechanisms, explainable outputs, and consistent system performance across diverse scenarios.

3.3.2 Participatory Design Processes and Community Co-creation

Participatory design (PD) methodologies (Schuler & Namioka, 1993) provide established frameworks for community-engaged AI development that prioritize stakeholder involvement, democratic participation, and user empowerment. In disaster management contexts, PD approaches help ensure that AI systems reflect community needs, cultural values, and local knowledge. Community-centered design emphasizes power redistribution, where affected communities become co-designers rather than merely end-users of AI systems. This approach requires long-term engagement, capacity building, and shared decision-making authority throughout the development process. For example, co-creation methodologies include design workshops, participatory mapping, storytelling sessions, and prototype evaluation activities that enable communities to shape AI system functionality. These methods help surface tacit knowledge, cultural preferences, and contextual requirements that may not be apparent to external developers.

Stakeholder mapping identifies the diverse actors involved in disaster management, including emergency responders, community leaders, vulnerable populations, and traditional knowledge holders (Mishra & Dahal, 2024). Each stakeholder group brings different perspectives, needs, and capabilities that must be considered in system design. Democratic deliberation processes enable communities to discuss tradeoffs, negotiate priorities, and establish governance mechanisms for AI systems that will affect their lives. These deliberations address questions about data use, algorithmic fairness, resource allocation, and accountability mechanisms.

Capacity building ensures that communities develop technical literacy, critical evaluation skills, and ongoing governance capabilities, to maintain oversight of AI systems. This includes training in data interpretation, bias recognition, and system evaluation.

3.3.3 Cultural Fidelity and Ethical Considerations

Cultural fidelity in AI systems requires deep understanding of local worldviews, communication patterns, and decision-making processes (Komene et al., 2023; Silano, 2024). For disaster management, this includes respecting traditional authority structures, cultural protocols for information sharing, and community-specific definitions of risk and resilience.

Indigenous data sovereignty principles ensure that communities retain control over data related to their territories, knowledge systems, and cultural practices. The CARE Principles (Collective benefit, Authority to control, Responsibility, Ethics) provide frameworks for ethical data governance that complement technical FAIR (Findable,

Accessible, Interoperable, Reusable) principles (Lindgren et al., 2024). In Brazil, benefits sharing mechanisms ensure that communities receive tangible advantages from AI systems developed with their participation and knowledge (de Lima et al., 2024). This includes direct benefits such as improved disaster preparedness, as well as indirect benefits such as capacity building and economic opportunities.

Cultural protocols for AI system design include considerations about sacred knowledge, appropriate disclosure levels, and community-specific communication norms. Some traditional knowledge may require restricted access or specific contextual framing to prevent misuse or misinterpretation. Similarly, bias mitigation strategies must address how AI systems might perpetuate stereotypes, marginalize minority perspectives, or privilege dominant cultural viewpoints (Rezaei et al., 2024). This requires diverse development teams, community validation processes, and ongoing bias monitoring.

3.4 Synthesis and Conclusion

Taken together, AI-integrated disaster resilience is most effective when engineered as hybrid intelligence: computational models that are continually calibrated by human judgment, local ecological knowledge, and community governance. Rather than scaling "black box" solutions, this chapter shows how human-centered, participatory, and culturally faithful design—paired with edge-capable, offline-robust infrastructure and clear data-sovereignty commitments—turns prediction into preparedness and information into equitable action. In this framing, algorithms are instruments, not arbiters; communities retain agency over risk definitions, thresholds, and trade-offs; and validation draws on both statistical evidence and lived consensus.

Bending modern science to these aims means re-tooling our methods and incentives. It entails glocally informed data fusion (sensors + satellite + traditional indicators), explainable and auditable models embedded in human-in-the-loop workflows, open standards for interoperability, and "living-lab" deployments that iteratively learn from drills, near-misses, and events without amplifying bias. By centering equity, accountability, and co-creation, AI becomes a civic technology that strengthens cyclone resilience: extending lead times, improving allocation under constraint, and—crucially—building trust. This synthesis reframes resilience as a socio-technical capacity we can design: one that integrates scientific rigor with community wisdom to anticipate hazards, reduce harm, and recover with dignity.

4 Climate Communication and Trust Networks

4.1 The Climate Communication Channels Framework

Over the years, communication models, based on theoretical frameworks explicating how information is transmitted and received, have evolved significantly. They became robust, incorporating advancements in technologies and research methodologies. These developments in communication models helped enhance the clarity and efficiency of critical message delivery. However, despite these advancements, the actual test of any communication model often occurs during times of crisis, when the stakes are high and conditions are challenging. Traditional communication modules deal with the established ways of exchanging information, such as verbal communication or written messages, and fail to capture the human aspect of healing in pre- and post-disaster situations.

In one particularly demanding context, the unpredictable nature of extreme weather events presents unique problems that can overwhelm even the most sophisticated communication models. In moments of crisis, when communities are thrown into turmoil, the process of conveying timely and accurate information becomes paramount. Disaster management does not simply conclude once a crisis has been addressed. To effectively plan for the future, it is crucial for disaster management teams to learn from their experiences and prepare for upcoming challenges. This ability can significantly impact the difference between saving lives and experiencing loss in future emergencies.

During a cyclone, the urgency of decision-making escalates, and the need for reliable communication intensifies. This is a crucial aspect of disaster management, as factors such as power outages, which can disrupt electronic communication, infrastructure damage, which can render traditional communication channels unusable, and the sheer volume of information to be processed, which can lead to information overload, complicate the situation further. Moreover, the emotional stress experienced by those affected can hinder their ability to receive and comprehend critical information. Consequently, the struggle to disseminate clear, actionable messages is exacerbated by these real-world limitations.

In this section of the chapter, we explore the intricacies of communication during crises, focusing on the specific challenges posed by cyclones. We examine how the 3C model of communication not only meets the demands of emergency situations, but also identifies common pitfalls, and analyzes case studies such as the communication strategies used during a few selected cyclones, like Cyclone Nargis in Myanmar, 2008, but also, with the help of visual metaphors from the crisis situation create chances of healing. By understanding the dynamics at play, we can better prepare for future crises and improve the resilience of our communication models when these are needed most, instilling a sense of responsibility and motivation for all stakeholders.

4.2 Embedding Local Metaphors and Semiotics in Alerts and Advisories

Crisis visuals play a critical role in not just conveying information, but also in shaping public perception during emergencies. These images, whether photographs, infographics, or videos, serve as powerful tools that can evoke strong emotional responses and drive action. For example, striking visuals of natural disasters like hurricanes or wildfires can encapsulate the urgency of the situation, prompting individuals to seek safety or assist those affected. This underscores the weight of responsibility in crisis communication. The meaning behind these visuals often extends beyond their immediate content. They can reflect cultural values, highlight vulnerabilities, and most importantly, illustrate the interconnectedness of communities. This shared responsibility and unity are key aspects of the crisis communication model. Additionally, the use of iconography such as warning signs or symbolic colors can convey messages quickly and effectively, allowing people to grasp the severity of a crisis at a glance. In a digital age, where information spreads rapidly, crisis visuals must be carefully crafted and studied to avoid misinformation and panic. Clear, accurate, and contextually relevant images not only guide the public in understanding the situation and the necessary responses but also empower them to take informed action, ultimately fostering resilience in the face of adversity.

4.3 Leveraging Trusted Local Media, Social Networks, and on-the-Ground Communicators

4.3.1 Communication Channels-Trusted Voices and Visual Metaphors

The 3C model, with its communication channel as a crucial stage, amplifies trusted voices that can guide people during a disaster. Can any voice be more impactful and accurate than a picture? How does this unique feature aid in the message or mitigation of the issues during or after a disaster? This section of the chapter revisits the 3C model, introducing a visual metaphoric level under the communication channel stage in the model. This level makes the 3C model particularly relevant for cyclone rehabilitation in the affected areas, highlighting its practical application.

According to Roland Barthes, visual imagery can be analyzed through a semiotic lens, where each image holds layers of meaning that extend beyond the literal interpretation (Barthes, 1977). He emphasized the distinction between denotation and connotation to illustrate how images convey ideas. Denotation refers to the explicit meaning of an image, while connotation encompasses the emotional and cultural associations it invokes. Another foundational research by Lakoff and Johnson (1980) on metaphor shows how these are effective instruments in communication and comprehension, conveying intricate ideas and emotions through imagery. Forceville (1996, 2024) examined the use of pictorial metaphors in advertisements and their impact on consumer perception. Several other researchers have also offered new perspectives on how visual metaphors create meaning and their implications for visual communication.

Borgo et al. (2012), in the context of media and communication, modelled visual metaphors that play a crucial role in shaping viewers' perceptions, emotions, and overall visualization. The 3C model, with its potential to improve disaster communication, can use the given keywords to map disaster-related visual images to trusted voices, facilitating better cyclone response and resilience. The keywords of compactness, vividness, and inexpressibility should map onto or lead to memorization, visual search, and concept grasping.

As Chouliaraki (2006, 2012) notes, the way the media present images can influence the audience's emotional responses and understanding of events. By translating narratives into visual formats, creators can enhance engagement and foster a deeper emotional connection with the audience. This process not only shapes their interpretation of reality but also enlightens them, making them more aware of the power of visual metaphors.

Moreover, the use of visual metaphors in media and communication raises ethical considerations. For instance, the deliberate use of particular visual metaphors to manipulate or misrepresent information can be seen as unethical. Therefore, creators need to use visual metaphors responsibly and with the audience's best interests in mind.

By applying the visual metaphor analyses discussed by several researchers, we can appreciate how visual metaphors not only represent concepts but also shape perceptions and influence interpretations in nuanced ways even in case of disaster communication. Visual metaphors, with their power to evoke emotions, play a crucial role in effective communication. The analysis below illustrates how visual metaphors play a crucial role in enhancing understanding. By drawing parallels between different concepts, viewers can understand and interpret unfamiliar or abstract ideas, such as devastation and possible rehabilitation, through familiar visual representations. For instance, a broken tree might symbolize destruction and loss, while a blooming flower could represent hope and renewal. This cognitive connection helps evoke feelings and sentiments more effectively than literal representations, leading to better comprehension and resilience. The analysis examined in Fig. 3 various visuals captured from different cyclones to collectively identify underlying features and metaphors that can be used to develop the 3C model. By understanding the visual cues of resilience in these cyclones, disaster response professionals can better prepare for and respond to future disasters.

Visual metaphors are crucial in crisis communication, particularly during events such as cyclones, where immediate understanding is vital (Fitrianie & Rothkrantz, 2007). For example, describing a cyclone as a "fierce beast" can evoke a strong emotional reaction, emphasizing the urgency and danger of the situation. This imagery helps audiences comprehend the seriousness of the threat and encourages them to take proactive measures.

Additionally, visual tools such as maps illustrating storm paths or infographics detailing safety protocols can simplify complex information, making it easier for the

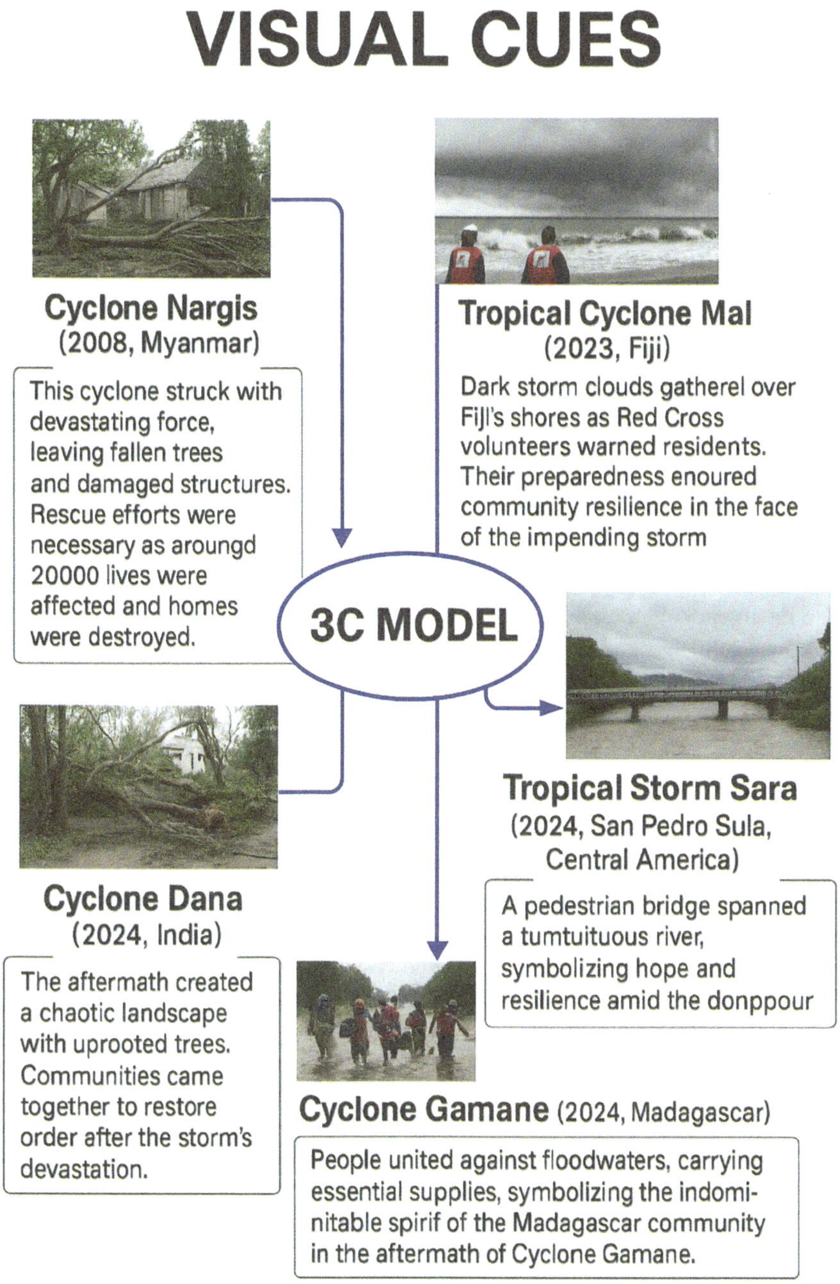

Fig. 3 Enhancing communication channels through visual cues

public to understand. During a crisis, clear and engaging visuals not only inform but also build confidence and motivate communities to respond effectively (Lindell, 2017). By using impactful visual metaphors and effective communication strategies, organizations can better manage the challenges posed by natural disasters, ultimately saving lives and enhancing community resilience (Patel et al., 2017).

4.4 Operationalization of the 3C Framework

4.4.1 Stakeholder Engagement and Co-design Framework

The proposed model is developed through a structured co-design process that integrates expertise from multiple stakeholder groups to ensure scientific validity, contextual relevance, and operational applicability. Local stakeholders may include coastal community organizations, fisher cooperatives, resident associations, and indigenous or traditional knowledge holders, who contribute place-based knowledge, historical cyclone experience, and locally relevant vulnerability indicators. Institutional stakeholders may include municipal disaster management authorities, regional or state emergency operations centers, national meteorological services, and coastal zone management agencies responsible for preparedness, response, and recovery planning. Academic and research stakeholders may include universities, climate research centers, and artificial intelligence or data science laboratories that support climate modeling, algorithm development, and model validation. In addition, non-governmental and humanitarian organizations such as disaster risk reduction NGOs and international relief agencies, support inclusive engagement and ethical deployment, while technology and data partners, including satellite data providers and early warning system developers, enable system infrastructure, deployment, and long-term maintenance. Figure 4 depicts the community-centered and glocally-informed 3C framework for cyclone resilience, illustrating the integration of multi-scale data inputs, AI-enabled analytical processes, stakeholder co-design, and feedback-driven decision support across global, regional, and local scales.

4.4.2 Spatiotemporal Domain of the Modeling Framework

The framework is explicitly designed to operate across multiple spatial and temporal scales to support comprehensive cyclone risk management. Spatially, the model integrates global-scale climate drivers (e.g., large-scale atmospheric circulation patterns and sea surface temperature anomalies), regional-scale hazard dynamics (e.g., cyclone genesis, track projections, rainfall intensity, and storm surge modeling at basin or coastline scales), and local-scale exposure and vulnerability data (e.g., settlement-level population distribution, critical infrastructure, evacuation routes, and social vulnerability indicators). Temporally, the model spans short-term horizons

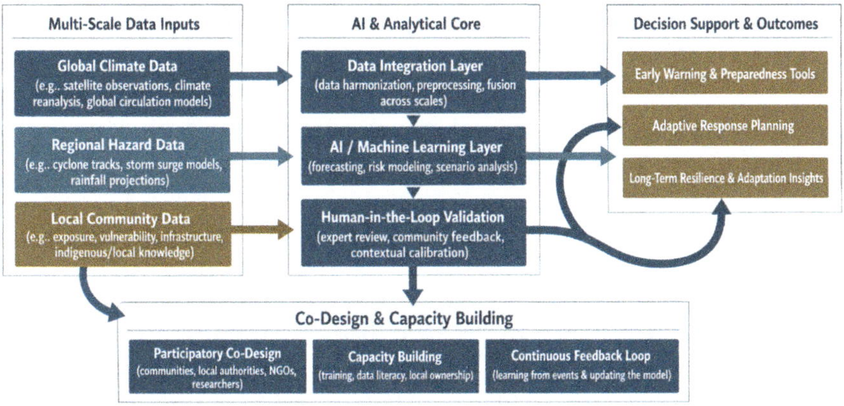

Fig. 4 Schematic of the climate communication channels (3C) framework

(hours to days) to support early warning and emergency response, seasonal to interannual horizons (weeks to months) for preparedness planning and resource allocation, and long-term horizons (years to decades) to inform adaptation strategies, land-use planning, and resilience-building under projected climate change scenarios. This multi-scale spatiotemporal design enables alignment between immediate operational decision-making and long-term resilience planning.

4.4.3 Local Capacity Development and Sustained Model Use

Local capacity development is embedded as a core component of the framework to promote ownership, interpretability, and long-term sustainability. Capacity-building activities include targeted training workshops for local officials and community representatives on data literacy, risk interpretation, and the use of AI-enabled decision-support tools; participatory development of locally meaningful indicators, thresholds, and response triggers; and scenario-based planning exercises and drills that integrate model outputs into existing preparedness and response protocols. Additional efforts include the development of accessible user interfaces and documentation in local languages, mentorship for local institutions to manage and update data inputs, and structured feedback mechanisms through which community observations and post-event experiences inform iterative model refinement. Together, these activities strengthen institutional capacity, enhance trust in AI-assisted decision-making, and support the sustained, adaptive use of the framework in diverse local contexts.

5 Integration of Indigenous and Traditional Knowledges for Cyclone Resilience

5.1 Role of Indigenous Knowledge Systems in Cyclone Prediction and Preparedness

Indigenous Knowledge Systems (IKS) have long played a critical role in environmental management and disaster preparedness, particularly in regions prone to climate hazards such as cyclones. Unlike scientific forecasting, which relies on advanced technology and meteorological data, IKS is grounded in centuries of observation, cultural practices, and ecological knowledge passed down through generations. In the context of cyclone prediction and preparedness, indigenous communities across Africa and Asia have developed reliable systems for interpreting environmental signals, mobilizing collective responses, and safeguarding their livelihoods (Dada, 2021).

5.2 Indigenous Knowledge in Cyclone Prediction

Cyclone prediction within indigenous communities often relies on ecological and atmospheric signals. In South Asia, for example, traditional fishers have developed acute observational skills for reading ocean conditions. They interpret unusual swells, changes in the color of seawater, and shifts in wind direction as precursors to cyclones (Nasreen et al., 2023). These methods, though not technologically advanced, provide real-time and locally relevant information that helps communities prepare well before official forecasts are available. Similar traditions exist in other coastal regions of the world. Communities observe animal behavior, such as birds flying inland, insects clustering abnormally, or fish changing migration patterns, as signs of incoming storms (Dada, 2021). While these practices may not rival the precision of satellite models, they are rooted in generations of lived experience and remain vital in contexts where scientific meteorology is inaccessible. In Africa, IKS plays an equally important role. Along the coast of Tanzania, for example, fishers on Mafia Island use monsoon cycles, sea state, the lunar calendar, and careful readings of winds and cloud formations to decide whether it is safe to go to sea (Chengula, 2023). In Kenya, pastoralists and farmers rely on animal and insect behavior, celestial signs, and flowering patterns to forecast rainfall and potential floods. Among the Borana, such observations often trigger early movement to higher ground or reinforcement of structures before hazards strike. In Malawi, households watch for bird migrations, ant behavior, river color changes, and unusual cloud formations as indicators of imminent flooding, a knowledge system that was particularly useful during Cyclone Idai (Kalanda-Joshua et al., 2011). Likewise, in Mozambique, families read the speed and level of rivers like the Búzi and Limpopo to anticipate incoming flood pulses during cyclones, combining current observations with memory of past flood events (Prevention Web UNDRR, 2023). Across East and Southern Africa, such systems are well-established, even if under-documented. They demonstrate that IKS is not a relic of the past but a practical tool for cyclone forecasting, offering locally grounded knowledge that complements scientific predictions (Hiwasaki et al., 2014).

5.3 Preparedness Strategies Rooted in Indigenous Knowledge Systems

Indigenous Knowledge also guides preparedness measures that reduce harm once cyclones occur. In South and Southeast Asia, communities have long designed houses with aerodynamic thatched roofs, stilted bases in flood-prone regions, and reinforced bamboo structures that flex under strong winds (Mercer et al., 2009). Social preparedness is equally important, with oral traditions, drumming, horn blowing, and communal songs functioning as grassroots communication systems that alert entire villages and mobilize collective action (Dada, 2021). In Africa, preparedness measures reflect similar ingenuity. In Madagascar, communities traditionally construct light, flexible houses with steep roof pitches that can be quickly lashed down before storms. Recent collaborations have improved these techniques with simple reinforcements like stronger joints and bracing, which remain affordable and rooted in local materials (Edinburgh Innovations, 2024). Communities also maintain mangrove forests as natural barriers against storm surges, reducing coastal vulnerability while supporting local fisheries. After Cyclone Enawo displaced nearly half a million people in Madagascar, the government began blending local and scientific insights into preparedness. Indigenous knowledge was used in mapping flood-prone zones and designing early-warning communication systems (radio, mobile alerts, and community liaisons). This integration allowed communities to prepare more effectively for later storms by knowing which areas to avoid and which shelters to prioritize (UNDRR, 2024). These practices blend traditional coping strategies with modern recovery guidance, creating a stronger resilience framework. In Chimanimani, Zimbabwe, communities prepared and responded to Cyclone Idai using spiritual and cultural frameworks. Beliefs in ancestral displeasure or mermaid curses shaped evacuation behaviors, collective rituals, and community solidarity. Village disaster committees combined ecological signals with simple, effective communication methods such as flags, whistles, and door-to-door messaging to ensure timely evacuations.

5.4 *Examples of Ecological Signals and Cultural Practices for Weather Forecasting*

Communities around the world, especially rural, agropastoral and Indigenous groups have long read the landscape for clues about coming weather. Oral traditions are a significant form of climate education. Elders narrate stories about past famines, devastating floods, or miraculous rains, teaching children survival strategies. Women focus on household-centered lessons, such as food preservation and saving crop seeds for future use, while men recount tales of migration routes, coping with drought, and surviving cyclones. Storytelling ensures that resilience is passed down, embedding ecological wisdom into the social fabric. These systems of local and Indigenous knowledge combine repeated observation of plants, animals, insects, sky patterns and other environmental cues into rules of thumb that inform planting, herding and water use decisions. Such knowledge is usually place-based, learned over generations, and used alongside (or to triangulate against) formal meteorological forecasts where these exist (Radeny et al., 2019). Many communities interpret changes in plant life, timing of flowering, early leafing, unusual fruiting or sudden shedding of leaves as proxies for humidity, impending rains or drought. For example, flowering or early fruiting of particular trees can indicate increased soil moisture or shifts in seasonal humidity that precede rains; conversely, premature leaf drop may signal an approaching dry spell. These phenological cues work because plants often respond to microclimatic changes (soil moisture, ground humidity) before those changes are obvious to human senses, so farmers treat the timing or intensity of particular plant events as a localized, empirically tested "forecast" (Balehegn et al., 2019; Radeny et al., 2019). Shifts in animal movement, vocalization or body condition are widely used as weather indicators. Farmers and pastoralists watch bird migration patterns, roosting behaviour, or unusual calls (e.g., prolonged calls before storms) as well as livestock behaviour (restlessness, seeking shelter) to anticipate storms or cold snaps. For example, unusual congregation or early departure of particular bird species has been associated locally with upcoming rains or wind changes. Because animals often respond rapidly to atmospheric pressure and humidity shifts, communities have found them valuable short-term predictors for timing field operations or moving herds (Radeny et al., 2019; case studies across East Africa).

In rural India, men, especially herders and elders, rely on keen observation of animals to predict weather changes. For example, when camels sit restlessly or when goats refuse to graze, it is interpreted as a sign of an approaching storm. Birds flying unusually low or ants shifting their nests are considered reliable indicators of rainfall. Women, too, contribute by observing domestic animals, such as cows or hens, within the household. These subtle interpretations form an indigenous "forecast system" passed down through generations, guiding decisions on preparing homes, storing food, or protecting cattle before cyclones or floods (Enock, 2013).

Insects and other small fauna frequently provide sharp, localised signals. Termite mound activity, emergence patterns of ants, cicada calls, swarms of particular species, or mass appearances of butterflies or locusts are interpreted as precursors to rainfall, humidity change, or drought. Because many insects have life cycles tightly coupled to soil moisture and temperature, their sudden abundance (or absence) can be a reliable short-range signal to experienced observers. Field studies and surveys have documented dozens of insect species used as biotic indicators in places as diverse as Tamil Nadu, Zimbabwe, and East Africa more broadly (local studies and systematic reviews). Not all indicators are living. Changes in wind direction and strength, the smell or "weight" of the air (local senses of humidity), cloud shapes and movement, and unusual lunar or stellar appearances (e.g., haloed moon, bright/diffuse stars) form a core set of abiotic and astronomical signals used in many cultures. Traditional astrologists or seasonal calendars based on constellations remain common in some areas and are used to time planting seasons or seasonal migrations. These cues are often combined, for example, a particular wind plus a plant sign to increase confidence in a forecast (Radeny et al., 2019; Huntington 2020). Indigenous forecasting is rarely a single person's hobby; it's embedded in social institutions. Communities often rely on a network of scouts, elders, ritual specialists or "astrologists" who collect, compare and interpret indicators. Decisions are usually reached by triangulating multiple signals, biological, astronomical, meteorological and by applying customary verification routines (e.g., elders' councils, range-scout missions). These institutions also calibrate, validate and transmit knowledge: when a predicted outcome succeeds or fails, the interpretations are discussed, and the rules adjusted, an iterative, community-level knowledge system (Balehegn et al., 2019).

5.5 Language as a Medium for Embedding Local Worldviews in AI Systems

Language shapes what a community notices and how it organizes knowledge; therefore, language is a powerful medium for embedding local worldviews into AI systems. Vocabulary, metaphors, and grammar encode relationships, not just labels between people, animals, places, and spiritual entities. When AI systems are trained only on datasets derived from dominant languages and Western ontologies, they inherit a flattened, decontextualized model of the world that can miss essential relational meanings used in local decision-making. Embedding local worldviews therefore begins with recognizing that language is not neutral data: it carries epistemologies that determine what counts as relevant information and how it should be acted on (Lewis et al., 2024). Practical work to embed worldview through language includes creating community-rooted lexicons, ontologies and narrative corpora that capture local categories and priorities. For example, Indigenous language projects show that preserving oral genres, ritual speech, place-names, and ecological terms supplies AI with richer signals than scraping generic web text; those resources enable models to reflect relational descriptions (land as kin, seasons as agents) rather than treating entities as isolated tokens. Building such resources requires linguists, cultural experts and technologists to co-curate datasets so that the training data preserves nuance (Lewis et al., 2020). Embedding worldview is more than lexical coverage, it requires reshaping modelling assumptions. Standard ML approaches emphasize decontextualized, de-referenced tokens and universalizing loss functions. To respect local meaning, researchers propose hybrid architectures that couple symbolic ontologies (crafted with communities) with statistical models, and training objectives that preserve relational structure (e.g., role relationships, non-linear temporality). Recent framing papers argue for research programs that reconceptualize "intelligence" from plural epistemologies, showing how Indigenous knowledge systems can supply alternative primitives for AI reasoning and evaluation (Lewis et al., 2024; Ofosu-Asare, 2024). Language efforts must be governed by ethical, legal and cultural protocols. Data collection is never neutral. Socially responsible multilingual data practices stress consent, the right of refusal, data sovereignty and community benefit (Smart et al., 2024). That means any language corpus, dictionary, or speech dataset should be authorized by communities, stored and shared under agreed terms, and paired with capacity-building, so communities maintain control. Without these safeguards, language datasets risk reproducing extractive dynamics where cultural meaning is appropriated, misinterpreted, or monetized without benefit to knowledge holders. There are technical and social challenges: many local languages are primarily oral, have multiple overlapping varieties, or encode culturally specific phenomena that lack one-to-one translations. Large models trained naively will misrepresent or erase those phenomena. Empirical work assessing AI performance in Indigenous languages confirms poor out-of-the-box performance and culturally biased outputs, underscoring the need for tailored evaluation metrics that include cultural coherence and pragmatic appropriateness (Lucas et al., 2025). These findings mean technical fixes (better tokenizers, multilingual corpora) must be paired with ethnographic methods to evaluate whether outputs respect local meaning. Embedding local worldviews via language benefits from participatory, long-term partnerships rather than one-off data pulls. Co-design, local governance arrangements, Indigenous-led research agendas, and investments in community capacity produce systems that communities can use, adapt, and critique. Scholars recommend programmatic approaches, institutionally supported collaborations, local language engineering training, and co-owned digital infrastructure so language becomes infrastructure for culturally-grounded AI rather than a resource to be mined. When done well, language-centered design can help transform AI into tools that amplify, rather than override, local worldviews (Lewis et al., 2024; Ofosu-Asare, 2024; Smart et al., 2024).

5.6 Co-design Approaches with Indigenous Communities to Ensure Cultural Fidelity

Co-design with Indigenous communities must start from *relationships and rights*, not from tools. Practitioners repeatedly stress that ethical co-design begins with building long-term, trustful partnerships and recognizing Indigenous governance over decisions that affect them, including data use, benefit-sharing and intellectual property (Kukutai & Walter, 2017). Community consent and the option for refusal are core: projects that treat co-design as a one-off consultation risk reproducing extractive research patterns, whereas projects that invest in sustained dialogue and shared decision-making better support Indigenous self-determination and culturally faithful outcomes (Collins et al., 2018). Practical co-design processes combine participatory research methods (e.g., CBPR), iterative prototyping, and culturally adapted facilitation so that Indigenous priorities drive choices about what gets built and why. Community-based participatory research frameworks emphasize co-learning, equitable labour, and capacity building so community members are co-researchers, not merely data sources (Collins et al., 2018). Recent co-design case studies further show that using storywork, visual activities, and on-Country workshops rather than abstract technical meetings helps surface relational values, seasonal knowledge, and customary protocols that must shape any AI system (Ofosu-Asare, 2024). Protecting Indigenous data sovereignty (IDS) is essential for cultural fidelity: co-design must explicitly embed IDS principles (e.g., the CARE principles: Collective benefit, Authority to control, Responsibility, and Ethics) into project governance so communities retain control over how knowledge is collected, stored, and shared (Kukutai & Walter, 2017). This shifts power from external technologists to Indigenous decision-makers and creates enforceable agreements (MOUs, data-governance protocols) that specify when data can be used, by whom, and with what safeguards a practical foundation for culturally respectful AI (Carroll et al., 2020). Methodologically, co-design must translate Indigenous epistemologies into technical requirements rather than force Indigenous knowledge to fit existing algorithms. Scholars recommend hybrid approaches pairing symbolic ontologies and domain rules derived with knowledge holders with statistical models so that relational categories, seasonal logics, or kin-based land relationships are represented in system architecture and evaluation metrics (Ofosu-Asare, 2024; "counteracting bias" literature). In practice, this means co-creating glossaries, developing localized ontologies, and designing evaluation tests that measure cultural coherence and impact (Ofosu-Asare, 2024). Capacity building and community ownership are essential to co-designing projects including training local practitioners in data stewardship, model interpretation, and system maintenance to generate durable, autonomous capability rather than temporary interventions (Collins et al., 2018). Co-training also helps surface hidden harms early (mislabelling, misclassification) and enables community members to adapt models to changing cultural or environmental contexts; this iterative stewardship is what preserves cultural fidelity over time (Kukutai & Walter, 2017). Transparent evaluation and reflexive governance make cultural fidelity measurable and improvable. The co-design literature calls for mixed-method evaluation, combining technical performance metrics with community-defined success indicators (e.g., respect for protocols, usefulness to elders, cultural safety) and for governance mechanisms (tribal review boards, community steering committees) that can pause or terminate projects if harms appear. In short, culturally faithful AI emerges when co-design is relational, rights-respecting, technically plural, capacity-building, and accountable not when it's merely a checkbox on a project plan (Collins et al., 2018; Ofosu-Asare, 2024).

5.7 *Addressing Ethical Considerations in Integrating Traditional Knowledges into Digital Tools*

5.7.1 Free, Prior and Informed Consent (FPIC) and Community Agency

Integrating TK into digital systems must begin with meaningful, informed consent that is community-level, iterative, and revocable, not a one-time checkbox. FPIC requires that communities understand how knowledge will be stored, processed, shared, and potentially commercialised, and that they have the authority to accept, negotiate conditions, or refuse participation (Carroll et al., 2020; Whaanga et al., 2015). Practical projects show consent works best when it is negotiated through local governance institutions and recorded in accessible agreements (e.g., MOUs or biocultural protocols) that specify scope, duration, and benefit terms, and which explicitly allow communities to withdraw permission later if harms or misuses arise (Carroll et al., 2020; Whaanga et al., 2015).

5.7.2 Indigenous Data Sovereignty (IDS) and Governance Frameworks

Ethical TK digitization must be governed by IDS principles so that data practices support collective self-determination rather than reproduce colonial extraction. The CARE Principles (Collective benefit, Authority to control, Responsibility, Ethics) establish a people-and-purpose orientation for data stewardship and complement FAIR technical goals; operationalizing FAIR with CARE requires organisational changes, local governance mechanisms, and legally meaningful control points (Carroll et al., 2020). Embedding IDS means communities set rules about who can access data, for what purposes, and under what conditions, not merely relying on institutional ethics boards alone (Carroll et al., 2020).

5.7.3 Provenance, Provenance Metadata, and Benefit-Sharing

Technical provenance (clear metadata about origin, custodianship and permitted uses) and legal mechanisms for benefit-sharing are central ethical safeguards. Initiatives like Traditional Knowledge and Biocultural Labels and provenance metadata systems let communities append machine-readable labels and licences that travel with digital objects, signalling permitted uses and provenance to downstream users (e.g., researchers, repositories) and supporting claims under access-and-benefit-sharing regimes such as the Nagoya Protocol. These measures make misappropriation harder and help ensure any commercial or scientific benefits are tracked, negotiated and shared with rightful knowledge holders.

5.7.4 Cultural Protocols, Contextualization, and Avoiding Decontextualisation

Digitizing TK risks stripping practices, ceremonial uses, and context that make those knowledge forms meaningful. Case studies from museum and library projects demonstrate that ethical digitization follows cultural protocols (e.g., access restrictions for sacred materials, community review of public displays), uses culturally appropriate metadata schemas, and preserves contextual narratives, so materials are not misread or misused by outsiders (Whaanga et al., 2015; issues in Indigenous digital heritage literature). In practice, this means co-curation (elders and knowledge holders co-design metadata and display rules), tiered access controls (private, community-only, public), and culturally informed user interfaces that reflect local ontologies rather than imposing external categorisations (Whaanga et al., 2015).

5.7.5 Sustainability, Capacity Building, and Governance Instruments

Ethical integration must plan for long-term stewardship, not a one-off export of content. Many digitization projects have failed because of outdated software, lack of funding for maintenance, or absent local capacity to manage collections (digital heritage sustainability reviews). Ethical practice therefore includes training community members in data stewardship and system administration, creating locally hosted or community-controlled repositories, and establishing governance instruments (tribal review boards, community steering committees, binding MOUs) that can enforce terms and manage disputes so communities retain ongoing control and benefit (Whaanga et al., 2015).

5.7.6 Technical Safeguards and Policy Alignment

A layered ethical approach combines technical tools such as provenance metadata, biocultural labels, tiered access controls, encryption, and audit logs with policy and legal instruments like biocultural protocols, benefit-sharing agreements, and adherence to Indigenous Data Sovereignty (IDS) and CARE principles. Recent reviews emphasize that system designs should "bake in" these protections by default, for instance, through metadata that travels with files, smart contracts, or licence enforcement mechanisms (Ruckstuhl et al., 2023). These technical layers must align with broader governance instruments such as the Nagoya Protocol on Access and Benefit-Sharing and the United Nations Declaration on the Rights of Indigenous Peoples (UNDRIP) to ensure rights are enforceable beyond project partners (Carroll et al.,

2020; Carroll et al., 2022). By operationalizing the CARE principles in tandem with international policy frameworks, this hybrid technical-plus-policy model reduces risks of appropriation, improves transparency, and helps ensure that Indigenous and local communities actually gain the social, cultural, and economic benefits they expect (Carroll et al., 2022; Roberts & Montoya, 2023).

6 Conclusion Final Reflections on the Path Forward

Cyclone resilience advances fastest when climate science, AI, communication strategies, and Indigenous wisdom are synthesized into a single practice, operationalized through knowledge integration frameworks like the Climate Communication Channels (3C) loop. We propose a modular, place-based AI framework built for multilingual communities, low-bandwidth settings, governed by communities, and linked to clear early-action triggers for evacuation, shelters, and relief. Trusted local messengers and culture-aligned visual cues translate probabilistic forecasts into steps households can actually take, closing the forecast-to-action gap. Within this approach, trusted local messengers and Indigenous and locally grounded knowledge systems such as culturally familiar symbols, color codes, flags, icons, and narrative metaphors embedded in local languages serve as no-regrets measures that translate probabilistic forecasts into intuitive visual cues and actionable guidance households can immediately understand and act upon. Another measure lies in low-tech, AI-integrated early warning and decision-support systems aligned with the 3C framework, combining lightweight, on-device or offline-capable models with radio, SMS, community notice boards, and siren networks to deliver timely, interpretable alerts even under connectivity constraints.

Future work should focus on higher-resolution climate downscaling, compound-risk analytics, and systematic auditing and explainability of multilingual AI models, alongside the development of mixed success metrics that jointly evaluate hazard-loss reduction with community trust, understanding, and uptake of early-action guidance. Our vision of resilience is of accurate models, empowered communities, and protected ecosystems reinforcing each other. We therefore call for sustained, interdisciplinary collaboration and co-governance scaling community-led pilots into durable public systems so that the path forward is one of pluriversal, just, and practically actionable resilience.

References

Ali, A. H., & Thakkar, R. (2023). Climate changes through data science: Understanding and mitigating environmental crisis. *Mesopotamian Journal of Big Data, 2023*(1), 125–137. https://doi.org/10.58496/mjbd/2023/017

Ali Maher, O., Cegolon, L., & Bellizzi, S. (2025). Artificial intelligence as a tool for enhancing the performance of public health emergency operation centres (EOC). *BMJ Global Health, 10*(3), e016097. https://doi.org/10.1136/bmjgh-2024-016097

Auernhammer, J. (2020, August 11). Human-centered AI: The role of human-centered design research in the development of AI. In *Proceedings of DRS*. https://doi.org/10.21606/drs.2020.282

Avvenuti, M., Bellomo, S., Cresci, S., Nizzoli, L., & Tesconi, M. (2020). Towards better social crisis data with HERMES: Hybrid sensing for Emergency Management system. *Pervasive and Mobile Computing, 67*, 101225. https://doi.org/10.1016/j.pmcj.2020.101225

Balehegn, M., Balehey, S., Fu, C., & Liang, W. (2019). Indigenous weather and climate forecasting knowledge among Afar pastoralists of north eastern Ethiopia: Role in adaptation to weather and climate variability. *Pastoralism, 9*(1). https://doi.org/10.1186/s13570-019-0143-y

Barthes, R. (1977). *Image, music, text*. HarperCollins UK.

Belcastro, L., Cosentino, C., Marozzo, F., Gündüz-Cüre, M., & Öztürk-Birim, S. (2025, March 30). *Multi-stakeholder disaster insights from social media using large language models*. arXiv.Org. https://arxiv.org/abs/2504.00046

Bhatia, K. T., Vecchi, G. A., Knutson, T. R., Murakami, H., Kossin, J., Dixon, K. W., & Whitlock, C. E. (2019). Recent increases in tropical cyclone intensification rates. *Nature Communications, 10*(1). https://doi.org/10.1038/s41467-019-08471-z

Bindoff, N., Stott, P. A., AchutaRao, K. M., Allen, M. R., Gillett, N., Gutzler, D., Hansingo, K., Hegerl, G., Hu, Y., Jain, S., & Sebbari, R. (2014). Detection and attribution of climate change: From global to regional. In *Climate change 2013—The physical science basis* (pp. 867–952). Cambridge University Press. https://doi.org/10.1017/cbo9781107415324.022

Borgo, R., Abdul-Rahman, A., Mohamed, F., Grant, P. W., Reppa, I., Floridi, L., & Chen, M. (2012). An empirical study on using visual metaphors in visualization. *SSRN Electronic Journal*. https://doi.org/10.2139/ssrn.3849224

Camps-Valls, G., Fernández-Torres, M.-Á., Cohrs, K.-H., Höhl, A., Castelletti, A., Pacal, A., Robin, C., Martinuzzi, F., Papoutsis, I., Prapas, I., Pérez-Aracil, J., Weigel, K., Gonzalez-Calabuig, M., Reichstein, M., Rabel, M., Giuliani, M., Mahecha, M., Popescu, O.-I., Pellicer-Valero, O. J., & Williams, T. (2024, June 28). *AI for extreme event modeling and understanding: Methodologies and challenges*. arXiv.Org. https://arxiv.org/abs/2406.20080

Cao, L. (2023). AI and data science for smart emergency, crisis and disaster resilience. *International Journal of Data Science and Analytics, 15*(3), 231–246. https://doi.org/10.1007/s41060-023-00393-w

Carroll, S. R., Garba, I., Figueroa-Rodríguez, O. L., Holbrook, J., Lovett, R., Materechera, S., Parsons, M., Raseroka, K., Rodriguez-Lonebear, D., Rowe, R., Sara, R., Walker, J. D., Anderson, J., & Hudson, M. (2020). The CARE principles for indigenous data governance. *Data Science Journal, 19*. https://doi.org/10.5334/dsj-2020-043

Carroll, S. R., Garba, I., Plevel, R., Small-Rodriguez, D., Hiratsuka, V. Y., Hudson, M., & Garrison, N. A. (2022). Using indigenous standards to implement the CARE principles: Setting expectations through tribal research codes. *Frontiers in Genetics, 13*. https://doi.org/10.3389/fgene.2022.823309

Cavicchia, L., von Storch, H., & Gualdi, S. (2014). Mediterranean tropical-like cyclones in present and future climate. *Journal of Climate, 27*(19), 7493–7501. https://doi.org/10.1175/JCLI-D-14-00339.1

Chakravarty, P., & Gattupalli, S. (2024). Integration of indigenous traditional knowledge and AI in hurricane resilience and adaptation. In *Hurricane risk* (pp. 125–158). Springer Nature Switzerland. https://doi.org/10.1007/978-3-031-63186-3_6

Chakravarty, P., Gattupalli, S., Chakravarty, U., Chand, G., & Lee, W. (2025). Advancing earth resource management through AI enhanced early warning systems and crisis communication. In *Data analytics and artificial intelligence for earth resource management* (pp. 77–95). Elsevier. https://doi.org/10.1016/b978-0-443-23595-5.00005-x

Chen, H., Song, J., Dietrich, O., Broni-Bediako, C., Xuan, W., Wang, J., Shao, X., Wei, Y., Xia, J., Lan, C., Schindler, K., & Yokoya, N. (2025). *BRIGHT: A globally distributed multimodal building damage assessment dataset with very-high-resolution for all-weather disaster response*. Copernicus GmbH. https://doi.org/10.5194/essd-2025-269

Chen, Y., & Li, Q. (2025). Scale-dependent exposure bias: Assessing disaster risk in less economically developed regions. *International Journal of Disaster Risk Reduction, 121*, 105406. https://doi.org/10.1016/j.ijdrr.2025.105406

Chengula, F. I. (2023). *Knowing weather, strengthening livelihoods the role of Indigenous knowledge weather forecasting in fishing communities in coastal Tanzania*. University of Roehampton Repository. https://doi.org/https://pure.roehampton.ac.uk/portal/files/23118102/Chengula_Fasco_-_Final_Thesis.pdf

Choong, J. J., Wagenaar, D., Rabonza, M. L., Hamel, P., Switzer, A. D., & Lallemant, D. (2025). Shared hazards, unequal outcomes: Income-driven inequities in disaster risk. *Npj Natural Hazards, 2*(1). https://doi.org/10.1038/s44304-025-00087-2

Chouliaraki, L. (2006). *The spectatorship of suffering*. Sage Publications Ltd.

Chouliaraki, L. (2012). The theatricality of humanitarianism: A critique of celebrity advocacy. *Communication and Critical/Cultural Studies, 9*(1), 1–21. https://doi.org/10.1080/14791420.2011.637055

Collins, S. E., Clifasefi, S. L., Stanton, J., The LEAP Advisory Board, Straits, K. J. E., Gil-Kashiwabara, E., Rodriguez Espinosa, P., Nicasio, A. V., Andrasik, M. P., Hawes, S. M., Miller, K. A., Nelson, L. A., Orfaly, V. E., Duran, B. M., & Wallerstein, N. (2018). Community-based participatory research (CBPR): Towards equitable involvement of community in psychology research. *American Psychologist, 73*(7), 884–898. https://doi.org/10.1037/amp0000167

Comes, T. (2024). AI for crisis decisions. *Ethics and Information Technology, 26*(1), 1–14. https://doi.org/10.1007/s10676-024-09750-0

Dada, U. (2021). *Exploring the integration of indigenous knowledge systems in disaster risk reduction: A case study from South Africa*. University of the Witwatersrand. https://www.proquest.com/openview/94c61de4ad27526a20bc58c9abca0212/

de Souza de Lima, A., Julião, C. G., da Silva, E. M., Dias, A. F., de Jesus, J. M., da Silva, M. T., Almada, E. D., de Andrade, J. P., Bensusan, N., Zank, S., & Soldati, G. T. (2024). Proposals of indigenous peoples and local communities from Brazil for multilateral benefit-sharing from digital sequence information. *Nature Communications, 15*(1), 8933. https://doi.org/10.1038/s41467-024-53421-z

Dube, T., Huhn, A. L., Guimarães Nobre, G., Moyo, E. N., & Enenkel, M. (2024). Incorporating indigenous knowledge systems-based climate services in anticipatory action in Zimbabwe: An ex-ante assessment. *Frontiers in Climate, 6*. https://doi.org/10.3389/fclim.2024.1301908

Edinburgh Innovations. (2024, October 17). *Building cyclone-resistant homes together*. Edinburgh Innovations. https://edinburgh-innovations.ed.ac.uk/case-studies/cyclone-resistant-homes-madagascar

EIS Council. (2025, August 10). *AI predictive resilience could cut disaster losses by 15%*. EIS. https://eiscouncil.org/new-study-shows-ai-predictive-resilience-could-cut-disaster-infrastructure-losses-by-15/

Elshoukry, H. (2024). Enhancing disaster management by leveraging Smart cities technologies. In *2024 Mediterranean Smart cities conference (MSCC)* (pp. 1–8). https://doi.org/10.1109/mscc62288.2024.10697072

Emanuel, K. (2003). Tropical cyclones. *Annual Review of Earth and Planetary Sciences, 31*(1), 75–104. https://doi.org/10.1146/annurev.earth.31.100901.141259

Emanuel, K. A. (1987). The dependence of hurricane intensity on climate. *Nature, 326*(6112), 483–485. https://doi.org/10.1038/326483a0

Enock, C. M. (2013). Indigenous knowledge systems and modern weather forecasting: Exploring the linkages. *Journal of Agriculture and Sustainability, 2*(2) https://doi.org/https://www.infinitypress.info/index.php/jas/article/view/72

European Centre for Disease Prevention and Control. (2019). Cyclone Idai: Risk of communicable diseases in southern Africa. https://doi.org/https://www.ecdc.europa.eu/sites/default/files/documents/RRA-cyclone-idai-10-Apr-2019.pdf
Farnadi, G., Havaei, M., & Rostamzadeh, N. (2024, June 3). *Position: Cracking the code of cascading disparity towards marginalized communities*. arXiv; arXiv. https://arxiv.org/html/2406.01757v1
Faust, O., Salvi, M., Barua, P. D., Chakraborty, S., Molinari, F., & Acharya, U. R. (2025). Issues and limitations on the road to fair and inclusive AI solutions for biomedical challenges. *Sensors, 25*(1). https://doi.org/10.3390/s25010205
Feng, C., Gan Chowdhury, A., Elawady, A., Chen, D., Azzi, Z., & Vutukuru, K. S. (2020). Experimental assessment of wind loads on roof-to-wall connections for residential buildings. *Frontiers in Built Environment, 6*. https://doi.org/10.3389/fbuil.2020.00010
Feng, K., Lin, N., Gori, A., Xi, D., Ouyang, M., & Oppenheimer, M. (2025). Hurricane Ida's blackout-heatwave compound risk in a changing climate. *Nature Communications, 16*(1), 4553. https://doi.org/10.1038/s41467-025-59737-8
Fischer-Preßler, D., Bonaretti, D., & Bunker, D. (2024). Digital transformation in disaster management: A literature review. *The Journal of Strategic Information Systems, 33*(4), 101865. https://doi.org/10.1016/j.jsis.2024.101865
Fitrianie, S., & Rothkrantz, L. J. (2007). A visual communication language for crisis management. *International Journal of Intelligent Control and Systems, 12*(2), 208–216.
Forceville, C. (1996). *Pictorial metaphor in advertising*. Psychology Press.
Forceville, C. (2024). Identifying and interpreting visual and multimodal metaphor in commercials and feature films. *Metaphor and Symbol, 39*(1), 40–54. https://doi.org/10.1080/10926488.2023.2271544
Gao, Q., Xu, W., Pan, H., Shen, M., & Gao, Z. (2025, May 28). *Human-Centered human-ai collaboration (HCHAC)*. arXiv.Org. https://arxiv.org/abs/2505.22477
Gevaert, C. M., Carman, M., Rosman, B., Georgiadou, Y., & Soden, R. (2021). Fairness and accountability of AI in disaster risk management: Opportunities and challenges. *Patterns, 2*(11), 100363. https://doi.org/10.1016/j.patter.2021.100363
Gori, A., Lin, N., Xi, D., & Emanuel, K. (2022). Tropical cyclone climatology change greatly exacerbates US extreme rainfall–surge hazard. *Nature Climate Change, 12*(2), 171–178. https://doi.org/10.1038/s41558-021-01272-7
Granata, F., Zhu, S., & Di Nunno, F. (2025). Hydrological extremes in the Mediterranean basin: Interactions, impacts, and adaptation in the face of climate change. *Regional Environmental Change, 25*(3). https://doi.org/10.1007/s10113-025-02432-7
Hanashima, M., Sano, H., & Usuda, Y. (2024). Designing and developing dynamic decision support information for disaster response. *Proceedings of IAHS, 386*, 189–195. https://doi.org/10.5194/piahs-386-189-2024
Hamid, H., & Abedlmajid, E. (2025). Artificial intelligence in disaster risk management: A scientometric mapping of evolution, collaboration, and emerging trends (2003–2025). *International Journal of Advanced Computer Science and Applications, 16*(6). https://doi.org/10.14569/ijacsa.2025.01606101
Hiwasaki, L., Luna, E., Syamsidik, & Shaw, R. (2014). Process for integrating local and indigenous knowledge with science for hydro-meteorological disaster risk reduction and climate change adaptation in coastal and small island communities. *International Journal of Disaster Risk Reduction, 10*, 15–27. https://doi.org/10.1016/j.ijdrr.2014.07.007
Holland, G. J. (1993). *The Global Guide to Tropical Cyclone Forecasting*. World Meteorological Organization. https://n2t.org/ark:/85065/d76m3830
Intergovernmental Panel on Climate Change. (2021). *Chapter 11: Weather and climate extreme events in a changing climate*. IPCC. https://www.ipcc.ch/report/ar6/wg1/chapter/chapter-11/
Jonker, A., & Rogers, J. (2024, September 20). Algorithmic bias. *IBM*. https://www.ibm.com/think/topics/algorithmic-bias
Kalanda-Joshua, M., Ngongondo, C., Chipeta, L., & Mpembeka, F. (2011). Integrating indigenous knowledge with conventional science: Enhancing localised climate and weather forecasts in

Nessa, Mulanje, Malawi. *Physics and Chemistry of the Earth, Parts A/B/C, 36*(14–15), 996–1003. https://doi.org/10.1016/j.pce.2011.08.001

Kamar, E. (2016). Directions in hybrid intelligence: Complementing AI systems with human intelligence. *Microsoft*. https://doi.org/https://www.microsoft.com/en-us/research/wp-content/uploads/2016/11/hi.pdf

Khan, T., & Park, T. (2024, July 25). *AI needs inclusive stakeholder engagement now more than ever*. Partnership on AI. https://partnershiponai.org/ai-needs-inclusive-stakeholder-engagement-now-more-than-ever/

Knutson, T., Camargo, S. J., Chan, J. C. L., Emanuel, K., Ho, C.-H., Kossin, J., Mohapatra, M., Satoh, M., Sugi, M., Walsh, K., & Wu, L. (2019). Tropical cyclones and climate change assessment: Part I: Detection and attribution. *Bulletin of the American Meteorological Society, 100*(10), 1987–2007. https://doi.org/10.1175/bams-d-18-0189.1

Knutson, T., Camargo, S. J., Chan, J. C. L., Emanuel, K., Ho, C.-H., Kossin, J., Mohapatra, M., Satoh, M., Sugi, M., Walsh, K., & Wu, L. (2020). Tropical cyclones and climate change assessment: Part II: Projected response to anthropogenic warming. *Bulletin of the American Meteorological Society, 101*(3), E303–E322. https://doi.org/10.1175/bams-d-18-0194.1

Komene, E., Pene, B., Gerard, D., Parr, J., Aspinall, C., & Wilson, D. (2023). Whakawhanaungatanga—Building trust and connections: A qualitative study indigenous Māori patients and whānau (extended family network) hospital experiences. *Journal of Advanced Nursing, 80*(4), 1545–1558. https://doi.org/10.1111/jan.15912

Kossin, J. P., Knapp, K. R., Olander, T. L., & Velden, C. S. (2020). Global increase in major tropical cyclone exceedance probability over the past four decades. *Proceedings of the National Academy of Sciences, 117*(22), 11975–11980. https://doi.org/10.1073/pnas.1920849117

Krichen, M., Abdalzaher, M. S., Shaaban, M., & Aburukba, R. (2025). Using aerial videos for natural disaster damage assessment. In *2025 7th international youth conference on radio electronics, electrical and power engineering (REEPE)* (pp. 1–6). https://doi.org/10.1109/reepe63962.2025.10970856

Kugara, S. L., Kugedera, A. T. T., Sakadzo, N., Chivhenge, E., & Museva, T. (2022). The role of indigenous knowledge systems (IKS) in climate change. In *Advances in religious and cultural studies* (pp. 1–23). IGI Global. https://doi.org/10.4018/978-1-7998-7492-8.ch001

Kukutai, T., & Walter, M. (2017). Indigenous statistics. In *Handbook of research methods in health social sciences* (pp. 1–16). Springer Singapore. https://doi.org/10.1007/978-981-10-2779-6_40-1

Lakoff, G., & Johnson, M. (1980). Metaphors we live by. Chicago and London: The University of Chicago Press. *Canadian Journal of Linguistics/Revue Canadienne de Linguistique, 27*(1), 81–82. https://doi.org/10.1017/s0008413100023744

Larouche, P. (2024). Human-Centered artificial intelligence (HCAI). In *Human-Centered AI* (pp. 58–69). Chapman and Hall/CRC. https://doi.org/10.1201/9781003320791-7

Leong, W. (2025). Internet of things for enhancing public safety, disaster response, and emergency management. *Engineering Proceedings, 92*(1). https://doi.org/10.3390/engproc2025092061

Lewis, J. E., Abdilla, A., Arista, N., Baker, K., Benesiinaabandan, S., Brown, M., Cheung, M., Coleman, M., Cordes, A., Davison, J., Duncan, K., Garzon, S., Harrell, D. F., Jones, P.-L., Kealiikanakaoleohaililani, K., Kelleher, M., Kite, S., Lagon, O., Leigh, J., & Whaanga, H. (2020, January 1). *Indigenous protocol and artificial intelligence position paper*. Indigenous Protocol and Artificial Intelligence Working Group and the Canadian Institute for Advanced Research. https://spectrum.library.concordia.ca/id/eprint/986506/

Lewis, J. E., Whaanga, H., & Yolgörmez, C. (2024). Abundant intelligences: Placing AI within indigenous knowledge frameworks. *AI & SOCIETY, 40*(4), 2141–2157. https://doi.org/10.1007/s00146-024-02099-4

Lindell, M. K. (2017). Communicating imminent risk. In *Handbooks of sociology and social research* (pp. 449–477). Springer International Publishing. https://doi.org/10.1007/978-3-319-63254-4_22

Lindgren, C. A., Yunes, E., & Itchuaqiyaq, C. U. (2024). Technical communication's fight against extractive large language modeling by applying FAIR and CARE principles of data. *Journal of Business and Technical Communication, 39*(1), 11–25. https://doi.org/10.1177/10506519241280587

Liu, Y., & Fu, Z. (2024). Hybrid intelligence: Design for sustainable multiverse via integrative cognitive creation model through human–computer collaboration. *Applied Sciences, 14*(11), 4662. https://doi.org/10.3390/app14114662

Lucas, M., Burgueño, A., Buenadicha Sánchez, C., Carazas, M., Ramirez Rufino, S., Rosales Torres, C. S., Korn, D., Tesfaye, H., & Deo, G. (2025). *El desempeño de la inteligencia artificial en el uso de lenguas indígenas americanas.* Inter-American Development Bank. https://doi.org/10.18235/0013542

Maceika, A., Bugajev, A., & Šostak, O. R. (2024). Enhancing organizational resilience: Sustainable development scenarios incorporating disaster impacts and AI tools. *Sustainability, 16*(24), 11147. https://doi.org/10.3390/su162411147

Marciniak-Nuqui, Z. (2024, May 24). Culturally appropriate care in the age of AI. RAND Corporation. https://www.rand.org/pubs/commentary/2024/05/culturally-appropriate-care-in-the-age-of-ai.html

Marko, J. G. O., Neagu, C. D., & Anand, P. B. (2025). Examining inclusivity: The use of AI and diverse populations in health and social care: A systematic review. *BMC Medical Informatics and Decision Making, 25*(1). https://doi.org/10.1186/s12911-025-02884-1

Mei, W., & Xie, S.-P. (2016). Intensification of landfalling typhoons over the Northwest Pacific since the late 1970s. *Nature Geoscience, 9*(10), 753–757. https://doi.org/10.1038/ngeo2792

Menéndez, P., Losada, I. J., Torres-Ortega, S., Narayan, S., & Beck, M. W. (2020). The global flood protection benefits of mangroves. *Scientific Reports, 10*(1). https://doi.org/10.1038/s41598-020-61136-6

Mercer, J., Kelman, I., Taranis, L., & Suchet-Pearson, S. (2009). Framework for integrating indigenous and scientific knowledge for disaster risk reduction. *Disasters, 34*(1), 214–239. https://doi.org/10.1111/j.1467-7717.2009.01126.x

Miranda, E. J., Kumarji, K. N., Ramesan, S., Varghese, T., Panicker, V. V., & Yadav, D. K. (2022). Application of machine learning algorithms in the mitigation phase of disaster management. *International Journal of Social Ecology and Sustainable Development, 13*(1), 1–13. https://doi.org/10.4018/ijsesd.292079

Mishra, B. K., & Dahal, K. (2024, June 15). *Collaborative framework with shared responsibility for relief management in disaster scenarios.* arXiv.Org. https://arxiv.org/abs/2406.10572

Mohapatra, M. (2016). Monitoring and forecasting of tropical cyclones over north indian ocean. In *Advanced numerical Modeling and data assimilation techniques for tropical cyclone prediction* (pp. 409–447). Springer. https://doi.org/10.5822/978-94-024-0896-6_16

Moore, K., & Purohit, H. (2019). Discovering requirements for the technology design to support disaster resilience analytics. *International Journal of Information Systems for Crisis Response and Management, 11*(2), 20–37. https://doi.org/10.4018/ijiscram.2019070102

Motsumi, M. M., & Nemakonde, L. D. (2025). Indigenous early warning indicators for improving natural hazard predictions. *Jamba (Potchefstroom, South Africa), 17*(1), 1754. https://doi.org/10.4102/jamba.v17i1.1754

Murakami, H., Vecchi, G. A., & Underwood, S. (2017). Increasing frequency of extremely severe cyclonic storms over the Arabian Sea. *Nature Clim Change, 7*, 885–889. https://doi.org/10.1038/s41558-017-0008-6

Nasreen, M., Hossain, K. M., & Khan, M. M. (2023). *Coastal disaster risk management in Bangladesh.* Routledge. https://doi.org/10.4324/9781003253495

Nature Tech Collective. (2025, February 26). AI for nature data, planetary monitoring & earth observation. Nature Tech Collective. https://www.naturetechcollective.org/stories/artificial-intelligence-earth-observation-nature-data

OECD. (2023, March 27). *Participatory AI framework.* OECD.AI. https://oecd.ai/en/catalogue/tools/participatory-ai-framework

Ofosu-Asare, Y. (2024). Cognitive imperialism in artificial intelligence: Counteracting bias with indigenous epistemologies. *AI & SOCIETY, 40*(4), 3045–3061. https://doi.org/10.1007/s00146-024-02065-0

Pak, H., & Mostafavi, A. (2025, August 21). *Situational awareness as the imperative capability for disaster resilience in the era of complex hazards and artificial intelligence.* arXiv.Org. https://arxiv.org/abs/2508.16669

Pal, S., Tanushka, Jha, K., Sharma, D., Sahani, P., & Awathi, A. (2024). Crowdsourced and social media data for environmental monitoring and disaster response: A dual case study of real-time flood tracking and air quality surveillance. In *2024 international conference on emerging technologies and innovation for sustainability (EmergIN)* (pp. 264–269). https://doi.org/10.1109/emergin63207.2024.10961460

Palen, L., & Anderson, K. M. (2016). Crisis informatics—New data for extraordinary times. *Science, 353*(6296), 224–225. https://doi.org/10.1126/science.aag2579

Park, J., & Kang, D. (2024). Artificial intelligence and smart technologies in safety management: A comprehensive analysis across multiple industries. *Applied Sciences, 14*(24). https://doi.org/10.3390/app142411934

Park, T. (2024, September 17). *Stakeholder engagement for responsible AI: Introducing pai's guidelines for participatory and inclusive AI.* Partnership on AI. https://partnershiponai.org/stakeholder-engagement-for-responsible-ai-introducing-pais-guidelines-for-participatory-and-inclusive-ai/

Patel, S. S., Rogers, M. B., Amlôt, R., & Rubin, G. J. (2017). What do we mean by "community resilience"? A systematic literature review of how it is defined in the literature. *PLoS Currents, 9*, ecurrents.dis.db775aff25efc5ac4f0660ad9c9f7db2. https://doi.org/10.1371/currents.dis.db775aff25efc5ac4f0660ad9c9f7db2

Price, I., Sanchez-Gonzalez, A., Alet, F., Andersson, T. R., El-Kadi, A., Masters, D., Ewalds, T., Stott, J., Mohamed, S., Battaglia, P., Lam, R., & Willson, M. (2024). Probabilistic weather forecasting with machine learning. *Nature, 637*(8044), 84–90. https://doi.org/10.1038/s41586-024-08252-9

Radeny, M., Desalegn, A., Mubiru, D., Kyazze, F., Mahoo, H., Recha, J., Kimeli, P., & Solomon, D. (2019). Indigenous knowledge for seasonal weather and climate forecasting across East Africa. *Climatic Change, 156*(4), 509–526. https://doi.org/10.1007/s10584-019-02476-9

Reddy, M. S., Vamsi, C., & Kathambari, P. (2024). Rescue me: AI emergency response and disaster management system. In *2024 2nd international conference on artificial intelligence and machine learning applications theme: Healthcare and internet of things (AIMLA)* (pp. 1–5). https://doi.org/10.1109/aimla59606.2024.10531386

Rezaei, M., Pironti, M., & Quaglia, R. (2024). AI in knowledge sharing, which ethical challenges are raised in decision-making processes for organisations? *Management Decision, 63*(10), 3369–3388. https://doi.org/10.1108/md-10-2023-2023

Roberts, J. S., & Montoya, L. N. (2023, September 19). *In consideration of indigenous data sovereignty: Data mining as a colonial practice.* arXiv.Org. https://arxiv.org/abs/2309.10215

Robinson, C., Macdonald, J., Perry, J., Nadji, J., Nadji, S., McCartney, S., Taylor, A., Hunter, F., Cooper, D., Moyle, F., Drummond, A., Borovac, C., van Bodegraven, S., Gilfedder, M., Setterfield, S., & Douglas, M. (2022). Coproduction mechanisms to weave indigenous knowledge, artificial intelligence, and technical data to enable indigenous-led adaptive decision making: Lessons from Australia's joint managed Kakadu National Park. *Ecology and Society, 27*(4). https://doi.org/10.5751/es-13747-270436

Rolnick, D., Donti, P. L., Kaack, L. H., Kochanski, K., Lacoste, A., Sankaran, K., Ross, A. S., Milojevic-Dupont, N., Jaques, N., Waldman-Brown, A., Luccioni, A. S., Maharaj, T., Sherwin, E. D., Mukkavilli, S. K., Kording, K. P., Gomes, C. P., Ng, A. Y., Hassabis, D., Platt, J. C., et al. (2022). Tackling climate change with machine learning. *ACM Computing Surveys, 55*(2), 1–96. https://doi.org/10.1145/3485128

Ruckstuhl (Ngāi Tahu, Rangitāne ki Wairau), K., Rout, M., Macpherson, E., & Reid (Ngāti Pikiao, Tainui), J. (2023). Safeguarding indigenous biocultural resources in a global context: A case study of taramea. *AlterNative: An International Journal of Indigenous Peoples, 19*(4), 771–783. https://doi.org/10.1177/11771801231198562

Salam, A. (2024). Internet of things for environmental sustainability and climate change. In *Internet of things* (pp. 33–69). Springer International Publishing. https://doi.org/10.1007/978-3-031-62162-8_2

Schuler, D., & Namioka, A. (1993). Participatory design: Principles and practices.

Senarath, Y., Pandey, R., Peterson, S., & Purohit, H. (2023). Citizen-helper system for human-centered AI use in disaster management. In *International handbook of disaster research* (pp. 477–497). Springer Nature Singapore. https://doi.org/10.1007/978-981-19-8388-7_34

Shneiderman, B. (2022). *Human-centered AI*. https://doi.org/10.1093/oso/9780192845290.001.0001

Silano, J. A. (2024). Towards abundant intelligences: Considerations for indigenous perspectives in adopting artificial intelligence technology. *Healthcare Management Forum, 37*(5), 329–333. https://doi.org/10.1177/08404704241257144

Smart, A., Hutchinson, B., Amugongo, L. M., Dikker, S., Zito, A., Ebinama, A., Wudiri, Z., Wang, D., van Liemt, E., Sedoc, J., Olojo, S., Uwakwe, S., Wornyo, E., Schmer-Galunder, S., & Smith-Loud, J. (2024, September 8). *Socially responsible data for large multilingual language models*. arXiv.Org. https://arxiv.org/abs/2409.05247

Stryker, C. (2025, July 8). Human in the loop. *IBM*. https://www.ibm.com/think/topics/human-in-the-loop

United Nations Office for Disaster Risk Reduction. (2023, June 13). *Cyclone Freddy puts Mozambique's early warning system to the test*. PreventionWeb. https://www.preventionweb.net/feature/cyclone-freddy-puts-mozambique-s-early-warning-system-to-the-test

United Nations Office for Disaster Risk Reduction. (2024, January 30). *Regional office for Africa*. UNDRR. https://www.undrr.org/about-undrr/where-we-work/regional-office-africa

Upadhyay, G. M., Joshi, M., Rathi, P., Vats, P., Narula, M., & Gupta, S. K. (2024). A comprehensive framework for unmanned aerial vehicle (uav)-enabled real-time human detection system for disaster management. In *2024 international conference on electrical electronics and computing technologies (ICEECT)* (pp. 1–6). https://doi.org/10.1109/iceect61758.2024.10739154

van Oldenborgh, G. J., van der Wiel, K., Sebastian, A., Singh, R., Arrighi, J., Otto, F., Haustein, K., Li, S., Vecchi, G., & Cullen, H. (2017). Attribution of extreme rainfall from hurricane Harvey, august 2017. *Environmental Research Letters, 12*(12), 124009. https://doi.org/10.1088/1748-9326/aa9ef2

Visave, J. (2025). Transparency in AI for emergency management: Building trust and accountability. *AI and Ethics, 5*(4), 3967–3980. https://doi.org/10.1007/s43681-025-00692-x

Vollmer, M., Berchtold, C., & Anniés, J. (2025). Co-creating solutions for disaster risk reduction in multi-country research projects—Opportunities and challenges. *International Journal of Disaster Risk Reduction, 117*, 105187. https://doi.org/10.1016/j.ijdrr.2025.105187

Wang, J., Chang, Y., & Tim Tse, K. (2024). Synthesis of tropical cyclones: Understanding, modeling, and adapting to climate change impacts. In *Sustainable development*. IntechOpen. https://doi.org/10.5772/intechopen.114390

Waqas, M., Humphries, U. W., Chueasa, B., & Wangwongchai, A. (2025). Artificial intelligence and numerical weather prediction models: A technical survey. *Natural Hazards Research, 5*(2), 306–320. https://doi.org/10.1016/j.nhres.2024.11.004

Whaanga, H., Bainbridge, D., Anderson, M., Scrivener, K., Cader, P., Roa, T., & Keegan, T. T. (2015). He matapihi mā Mua, mō Muri: The ethics, processes, and procedures associated with the digitization of indigenous knowledge—The pei jones collection. *Cataloging & Classification Quarterly, 53*(5–6), 520–547. https://doi.org/10.1080/01639374.2015.1009670

World Health Organization Regional Office for Africa. (2019). *Tropical cyclones idai and kenneth mozambique—National situation report 5*. WHO | Regional Office for Africa. https://www.afro.who.int/publications/tropical-cyclones-idai-and-kenneth-mozambique-national-situation-report-5

World Meteorological Organization (WMO). (2025). *State of the climate in Asia.* World Meteorological Organization. https://wmo.int/publication-series/state-of-climate-asia

Xiang, X., Li, K., Huang, B., & Cao, Y. (2022). A multi-sensor data-fusion method based on cloud model and improved evidence theory. *Sensors (Basel, Switzerland), 22*(15), 5902. https://doi.org/10.3390/s22155902

Zeng, F., Pang, C., & Tang, H. (2023). Sensors on the internet of things systems for urban disaster management: A systematic literature review. *Sensors (Basel, Switzerland), 23*(17), 7475. https://doi.org/10.3390/s23177475

Zentner, A. (2025). *From opacity to transparency: Advancing explainable AI through black-box, white-box, and glass-box models.* Elsevier BV. https://doi.org/10.2139/ssrn.5266967

Cyclone Risk and Livelihood Recovery in Coastal South India: Narratives of Small-Scale Fishers of Andhra Pradesh

Bala Raju Nikku, Dalibandhu Pukkalla, Srinivasu Kodi, and Navya Vikraman Nair

Abstract This chapter examines the intersecting challenges of climate change, cyclone disasters, and livelihood vulnerabilities and resilience among the small-scale fishing (SSF) communities along the coast of Andhra Pradesh, India. Andhra Pradesh's 974 km coastline, vital for coastal and marine fisheries, is among South Asia's most cyclone-prone areas, making it crucial for studying climate vulnerability and resilience. As cyclones in the Bay of Bengal become more frequent and intense due to changing climatic conditions, Small-scale fishing is already grappling with socio-economic marginalization, coastal erosion, tenure issues, and declining fish stocks. Nevertheless, they actively develop resilience strategies that foster hope and empowerment, despite increased risks. Small-scale fishing includes artisanal, traditional, subsistence, and family-based activities. The chapter further explores how SSF communities in Andhra Pradesh experienced, interpreted, and responded to past cyclone events through a case-based, collaborative ethnographic approach grounded in community conversations, oral narratives, and policy analysis. This chapter documents narratives of loss, survival, and recovery from major cyclones such as Hudhud (2014) and Titli (2018), highlighting how local and traditional ecological knowledge (TEK), social capital, and adaptation strategies intertwine to strengthen community resilience and foster respect for local expertise and resourcefulness. Taking the context of state-led disaster management and post-disaster relief mechanisms, this chapter explores, particularly with respect to inclusivity, equity, and

---

B. R. Nikku (✉)
Faculty of Education and Social Work, Thompson Rivers University, Kamloops, BC, Canada
e-mail: bnikku@tru.ca

D. Pukkalla
Department of Studies in Anthropology, University of Mysore, Manasagangotri, Mysuru, Karnataka, India
e-mail: anthrodbp@gmail.com

S. Kodi
Department of Anthropology, Andhra University, Visakhapatnam, Andhra Pradesh, India
e-mail: anthrosrinu.kodi@gmail.com

N. V. Nair
University of Victoria, Victoria, Canada
e-mail: navya.nair@uwaterloo.ca

J. Collins et al. (eds.), *Tropical Cyclone Risk and Resilience*,
https://doi.org/10.1007/978-3-032-28745-8_12

long-term rehabilitation and resilience. It argues that, while Andhra Pradesh State policies often overlook SSF-specific needs and knowledge systems, these communities actively develop resilience through mobility, diversification, gendered labor, and localized governance practices. Ultimately, the chapter calls for a reimagining of coastal disaster risk governance in India that foregrounds community-led, co-produced adaptation, recognizes the agency and local expertise of SSF communities, and integrates their lived experiences into climate, fisheries, and blue justice frameworks.

Keywords Climate change · Economy of fishing · Marine livelihoods · Small scale fisheries (SSF) · Vulnerability · Well-being · Resilience · Blue justice · Andhra Pradesh · India

1 Introduction

Globally, coastal and marine fishing communities are struggling to sustain their livelihoods due to the impacts of climate change and variability, which are affecting global food security (Agnishwaran et al., 2024). Several research and policy contributions have been made to build community relationships, examine the impacts of local ecological changes on these communities, and understand how these changes shape their livelihoods. While managing their sustainable livelihoods, SSF communities have often struggled to cope with and adapt to the ecological and economic challenges in the discourse of their occupational livelihood development.

Climatic variations along the coast of Andhra Pradesh (AP) are resulting in increased coastal erosion and shoreline changes, driven by sea-level rise and storm surges, thereby degrading beaches and estuaries where fishers live and work. Furthermore, the loss of protective mangroves and natural buffers increases the susceptibility of fisher villages to flooding and habitat loss. Rao et al. (2020) noted an approximately 5% increase in overall climate variations, highlighting more intense and frequent precipitation events across peninsular South India during the northeast monsoons. For example, seasonal rainfall (October to December) routinely exceeds 800 mm along the north Tamil Nadu and south coastal Andhra Pradesh (AP), making the eastern Bay of Bengal coast especially sensitive to monsoon variability and flood risk (IMD, 2020).

Coastal South India encompasses the diverse, scenic, and culturally rich shorelines of Karnataka, Kerala, Tamil Nadu, Andhra Pradesh, and Pondicherry, stretching from the Arabian Sea to the Bay of Bengal. The Bay of Bengal is a highly energetic basin characterized by warm sea-surface temperatures, cyclonic circulation, and complex sediment dynamics. Its relatively shallow, funnel-shaped coast amplifies storm surges and wave heights, exacerbating impacts on coastal communities' livelihoods. Rising sea levels, saline intrusion, and coastal erosion, exacerbated by land reclamation, port development, and sand mining, led to vulnerabilities in livelihoods. In addition, the intense and frequent nature of tropical cyclones is a major driver of coastal damage and fishers' wellbeing (Jeevamani et al., 2021; Lacheheb et al., 2025).

Small-scale fishing is the backbone of Andhra Pradesh's marine economy. The sector provides local employment and food security, supporting thousands of fishing households along the 974 km coastline (Devaraj, 1997). SSF communities are concentrated in coastal villages and rely on traditional fishing methods with shore seines, gill nets, hook-and-line, and small plank or catamaran boats. Unlike distant industrial trawlers, artisanal fishers plan their trips based on deep local knowledge of sea currents, fish behavior, and seasons and lunar cycles. These embodied knowledge and cultural practices (rituals, sea deities, and gods) guide their risk management in fish harvesting along the coast of the Bay of Bengal. Nonetheless, there is a need to explore community disaster governance and resilience practices further to understand how these vulnerable fishers take risks. Further, shifts in ocean temperatures (in this case, in the Bay of Bengal) and salinity, combined with acidification and ecosystem change, alter fish species composition and seasonal availability, forcing fishers to adjust their fishing patterns (e.g., deeper or more extended trips), raising costs and increasing economic debts and risks for SSF. This chapter documents and analyzes cyclone vulnerabilities through memory-based stories and narratives to understand the consequences of frequent and intense cyclones and related calamities, as well as the Coastal fishers' interpretations and narratives collected during fieldwork.

Few studies have documented fishing communities in the context of climate change and cyclones (Adger et al., 2009; Hoeppe, 2007; Kapapa et al., 2024; Madhanagopal & Pattanaik, 2020). This chapter addresses this gap by presenting narratives of cyclones and their impacts on coastal Andhra fishers. Despite its social and ecological significance, the SSF sector faces increasing threats. There are significant threats to the livelihoods of fishing communities, including overfishing, habitat degradation, conflicts with mechanized fleets, unregulated coastal development, pollution, and climate variability (Basheer Ahammed et al., 2016; Malakar et al., 2018). The sector also suffers institutional neglect: formal credit, insurance, cold storage, and early warning systems are often lacking (Bosher et al., 2007; Vipinkumar et al., 2013). These structural vulnerabilities and more intense storms make fishing communities susceptible to disaster-induced displacement and livelihood loss. There are also significant contributions to both the *Jalari* and *Vadabalija* communities in north coastal Andhra Pradesh (Pukkalla & Gangadhar, 2025; Kodi, 2021; Pukkalla & Sharma, 2018 and 2020; Nikku et al., 2021, 2025; Nuckolls, 1996; Kodanda Rao, 1990; Sridevi, 1989, and others). Furthermore, the spatial organization of coastal settlements, seasonal migration patterns, and divisions of labor (often gendered) are intricately tied to the rhythms of fishing in social life. Women, though not always involved in fishing at sea, play vital roles in fish drying, processing, vending, and managing household economies (Vennela, 2017).

George and Subbaraju (1979) provide a historical overview of the fishing settlement and catch-landing patterns of fishers in Lawson's Bay, which include fishing villages such as Jalaripeta, Vasavanipalem, and Moolapalem. Their study demonstrates the importance of small-scale fisheries for local food systems and livelihoods. Immanuel and Syda Rao (2012) describe the socio-economic profile of fishermen in

Peda Jalaripeta, particularly those using hook-and-line gear. Their study details age groups, income levels, boat ownership (wooden and fiber catamarans), caste (*Jarali and Vada Balija*), and daily earnings.

The study by Kodi and Mohan (2015) explores the traditional and local ecological knowledge among fishing communities along the Visakhapatnam coast. Peda Jalaripeta is among the communities studied, and a substantial portion of the community's demographic, economic, and technological characteristics is documented. Additionally, various news reports from local and national newspapers, including The Hindu, Times of India, and Deccan Chronicle, covered conflicts between traditional fishermen from Peda Jalaripeta and neighboring fishers using ring nets within restricted areas, highlighting local responses to overfishing, gear regulation, and internal conflicts.

The chapter is divided into six sections. After the introductory section, coastal landscape and cyclone trajectories in Andhra Pradesh are presented, and a brief critique of Andhra Pradesh's disaster management is given. Section 3 discusses the research methods used, and Sect. 4 discusses changing coastal ecosystems and livelihoods in India. Section 5 analyses climatic risks, adaptation, and community Resilience using fishers' stories from the community perspective. The chapter concludes with Sect. 6.

2 Coastal Landscape and Cyclone Trajectories in Andhra Pradesh

Andhra Pradesh's coastline, extending approximately 974 km along the Bay of Bengal, includes sandy beaches, lagoons, and fertile river deltas (Godavari, Krishna, Penna) that support dense populations and agriculture. Rising sea levels and local pressures (land reclamation, port development, sand mining) increase coastal vulnerability, underscoring the importance of sustainable management.

Historically, Andhra Pradesh has been a cyclone hotspot. The IMD (2022) reports that more than 60 major tropical storms have made landfall in the past century, resulting in extensive loss of life and property. The following notable cyclones illustrate the evolving risk:

1. **1977 Diviseema Cyclone**: One of India's deadliest storms, it struck the Krishna and Guntur deltas with winds of about 200 km/h. It killed more than 10,000 people and affected about 2.4 million across eight districts, devastating road, rail, and communication networks. Inadequate evacuation at the time worsened social and economic losses.
2. **1990 Machilipatnam Cyclone**: A powerful cyclone over the Krishna-Godavari region destroyed roads, bridges, power lines, and telecom networks. Proactive evacuations limited the death toll, but property damages were estimated at INR 2300 crore.

3. **2010 Cyclone Laila**: Landfall near Bapatla (wind about 120 km/h) on May 21, 2010. It affected 1.78 million people across 1,492 villages, resulting in 22 deaths and damaging 102 cell towers. The Andhra Pradesh government sought INR 1,357 crores in relief from the Central government.
4. **2013 Cyclone Phailin** made landfall on October 12 near Odisha, but also impacted Srikakulam, Vizianagaram, and Visakhapatnam districts in the North Coast of Andhra Pradesh. Gusts up to 220 km/h caused prolonged power and communication outages (336 villages) despite the evacuation of 150,000 people, keeping fatalities very low.
5. **2014 Cyclone Hudhud**: With about 180 km/hour winds, Hudhud made landfall on October 12 near Visakhapatnam and caused massive destruction. About 10 million people were affected in Andhra Pradesh (AP), with 61 deaths. Roughly 50% of all cell towers (3447 towers) were damaged. The World Bank estimated total damage at INR 13,263 crores.
6. **2018 Cyclone Titli** made landfall on October 11 in north Andhra and southern Odisha. It affected about 1.7 million people across eight districts, killed 16, and damaged about 62% of telecom towers in the region. Reported losses were about INR3,673 crores.
7. **2019 Cyclone Fani**: Although Fani made landfall in Odisha (May 3), its effects reached Andhra Pradesh's northern districts. Approximately 33 villages in Srikakulam, Vizianagaram, and Visakhapatnam experienced winds of up to 140 km/h (Srikakulam district reported the highest wind speed). There were no reported fatalities locally, but approximately 218 cell towers were disrupted.

Although improved early warnings and evacuations have reduced fatalities, significant infrastructure damage and long-term livelihood disruptions persist. Climate projections indicate stronger, more erratic cyclones (often rapidly intensifying) in the Bay of Bengal (Emanuel, 2017).

2.1 Andhra Pradesh Disaster Management

Andhra Pradesh has made notable advances in disaster preparedness over the past two decades, particularly through the Andhra Pradesh State Disaster Management Authority (APSDMA). Investments in cyclone shelters, early warning dissemination, and evacuation planning have significantly reduced cyclone-related fatalities, as demonstrated during recent events such as Cyclones Phailin, Hudhud, and Titli. These improvements represent an apparent success of state-led disaster governance in protecting human life. However, these gains have not translated into comparable outcomes for livelihood recovery among artisanal and small-scale fishing communities. Post-disaster assistance mechanisms continue to prioritize housing and general infrastructure. In contrast, fishing-specific assets, such as boats, nets, engines, and shore-based processing facilities, often receive delayed, partial, or no compensation. Moreover, informal tenure arrangements, caste-based exclusions, and the absence of insurance coverage for small-scale fishers limit access to formal recovery schemes. As a result, many fishers rely on informal credit, kinship networks, and women's market labor to rebuild livelihoods after cyclones, revealing a persistent gap between disaster preparedness and livelihood resilience.

Despite improvements in preparedness, disaster governance in Andhra Pradesh continues to engage unevenly with the socio-economic realities of small-scale fishing communities. Scholars note that post-disaster interventions often emphasize physical infrastructure rehabilitation while neglecting deeper social vulnerabilities, including caste-based exclusions, gendered labor burdens, and limited access to formal credit or insurance for damaged boats and nets (Balaram & Katta, 2024; Bosher et al., 2007; Vipinkumar et al., 2013). Relief distribution is frequently delayed and poorly targeted, resulting in inequities in which marginalized fishing communities benefit the least (Srinivasa Rao, 2022). Scholarly work documenting community responses and recovery strategies can foster appreciation for local resilience and inform more effective disaster governance.

Moreover, disaster policies rarely integrate local knowledge or participatory governance mechanisms, reducing community trust in official institutions. As a result, fishers rely on informal finance, kinship support, and traditional weather observation to cope with cyclones, highlighting the need for more inclusive, community-centered disaster governance in the State. These structural gaps in post-disaster recovery are reflected in the lived experiences of fishing households, where recovery from cyclones such as Hudhud relied less on formal state assistance and more on informal credit, kinship networks, and women's market labor.

Disjointed disaster policies and a focus on the blue economy (not justice) have led to the systematic exclusion of SSF communities in Andhra Pradesh from governance of coastal commons, resulting from the privatization of coastlines, industrial encroachment, tourism development, and top-down bureaucratic fisheries management. The following sections draw on community narratives to illustrate how these policy gaps are (re)negotiated and contested in practice by local fisher communities. We further argue that disaster risk governance cannot be separated from broader struggles for coastal justice, including secure tenure, access to resources, participation in decision-making, and recognition of customary rights and practices.

3 Research Methods

This chapter builds on ongoing collaborative research with the *Jalari* and *Vadabalija* fishing communities of Visakhapatnam District, North Andhra Pradesh. We employ a narrative ethnographic approach, conducting focus group discussions, case studies, and interviews with active fishermen, fisherwomen, and senior fishermen to capture community experiences of cyclones and weather variability, and to examine livelihood strategies to address risks.

Jalaripeta, Visakhapatnam
N
RK Beach
Bay of Bengal
Beach Road
Jalaripeta
VUDA Park
Visakha Museum
Visakhapatnam Fishing Harbour
Harbour Quay
0 200 400 m
ANDHRA PRADESH
Visakhapatnam Location Map
VISAKHAPATNAM
ODISHA
CHHATTISGARH
TELANGANA
MAHARASHTRA
KARNATAKA
TAMIL NADU
Amaravati
Bay of Bengal
LEGEND

Using Narrative analysis, a critical qualitative research method that helped us systematically interpret stories (or narratives) to understand how fisher communities in North Coastal Andhra Pradesh make sense of their experiences, construct meaning, and relate to others. Our analysis is grounded in their first-person stories and local observations, complemented by a review of historical cyclone data and policy reports. Narrative experiences, heavily relying on participants' descriptions, led us to employ an unstructured interview method to allow the conversations to proceed at their own pace. We were able to make diverse yet equally substantial and meaningful interpretations and conclusions by focusing on different elements of the stories we collected and curated during 2024–2025.

3.1 Research Site

Jalaripeta is a historic coastal neighbourhood in the city of Visakhapatnam, along the shores of the Bay of Bengal. The settlement developed as a traditional fishing community adjacent to the city's historic port and shoreline, where sandy beaches and erodible coastal sediments shape the local littoral landscape. The settlement's proximity to the sea has historically supported marine-based livelihoods, particularly artisanal fishing and fish processing linked to nearby harbor and landing sites.

Jalaripeta is one of the oldest and most densely populated fishing settlements in the city, with a population of about 26,262 residents and over 6700 households recorded in the 2011 Census. Currently, the neighbourhood is predominantly inhabited by the Jalari community, a traditional Telugu-speaking fisher caste engaged in marine fishing and related occupations. However, smaller groups from other local communities also reside. Social organization within the settlement is typically structured around extended and nuclear family households associated with fishing livelihoods, in which men are commonly involved in offshore fishing. At the same time, women play key roles in fish processing, drying, marketing, and household financial management. Over time, the rapid urban expansion of Visakhapatnam city has integrated Jalaripeta into the broader metropolitan fabric. Nevertheless, its coastal geography and occupational structure continue to reflect its origins as a traditional fisher settlement closely tied to the Bay of Bengal.

Environmentally, the coastal zone of Jalari Peta forms part of the broader wetland system of Visakhapatnam, which includes estuaries,tidal flats, lagoons, creeks, and backwaters covering creeks, and backwaters covering about 59 fishing villages with 132kms coastline length (GOAP, 2018). Intensive port and urban development along the city's shoreline have resulted in the near absence of mangrove ecosystems within the immediate harbor zone. The low-lying coastal terrain heightens exposure to cyclones and storm surges originating in the Bay of Bengal. Geomorphologically, the Visakhapatnam coast is characterized by sandy beaches, wave-cut platforms, and dune systems shaped by monsoon driven wave regimes and longshore sediment transport (Raju & Vaidyanadhan, 1978).

The average erosion and accretion rates were − 1.80 m/year and 1.47 m/year respectively. Erosion is observed along transects of Sagar Nagar Beach, Jalari Peta, VUDA Park, R K Beach, Vizag, Relli Veedhi coast, whereas Yendada and Pedda Rushikonda show accretion (Baig et al., 2020). These changes reflect the combined influence of natural coastal processes, such as seasonal wave energy and littoral drift, and anthropogenic modifications, including harbor breakwaters and coastal infrastructure that alter sediment dynamics. Continued urban expansion along the shoreline has intensified coastal vulnerability, prompting management responses such as beach nourishment and other shoreline stabilization measures to mitigate erosion and storm-surge impacts on adjacent communities (Baig et al., 2020).

Seasonal migration is a critical feature of coastal fishing villages in Andhra Pradesh. There is an increased pattern of internal displacement and forced migration trends evident in other parts of the country. Various studies of the north-coastal Andhra fishing communities in Visakhapatnam, Vizianagaram, and Srikakulam districts reveal significant seasonal migration patterns. An estimated 12,000 to 30,000 fishermen from the three districts migrate in search of work to western Indian ports such as Veraval (Gujarat), Ratnagiri (Maharashtra), and Mangalore (Karnataka). The main reasons for this distress migration are reported as dwindling local catches, increased sedimentation from infrastructure projects, and the unviability of traditional fishing in their home villages. The demand for labor in mechanized, large-scale fishing operations in other Indian states lures a higher earning potential than the local, small-scale fishery. The International Collective in Support of Fishworkers (ICSF) has highlighted the importance of addressing the vulnerabilities of these migrant fishers through improved policies, documentation, and safety measures (Khadar, 2021; Palli, 2025; Roshan, 2016).

4 Changing Coastal Ecosystems and Livelihoods

Coastal ecosystems are complex and fragile interfaces between land and sea that support biodiversity, regulate climate, and sustain human well-being, making their conservation central to climate adaptation and disaster resilience strategies. These ecosystems are facing growing pressures from climate change (sea-level rise, stronger cyclones, ocean acidification), habitat loss due to ports and coastal development projects (referring to Blue Economy or Blue Ocean strategies), destructive fishing practices (due to corporate power and uncertain legal provisions) and increasing pollution such as plastic waste and untreated sewage (Nagelkerken et al., 2008; Unnikrishnan et al., 2006). Coastal ecosystems provide vital ecological services, including carbon sequestration (often referred to as blue carbon), shoreline stabilization, and water filtration (McLeod et al., 2011). They buffer coastlines against cyclones, storm surges, and erosion (Das & Vincent, 2009), while supporting livelihoods such as small-scale fisheries, aquaculture, salt production, and eco-tourism. They also hold cultural, spiritual, and recreational significance for local communities (MEA, 2005).

4.1 Local and Traditional Knowledge Systems

Coastal Fishermen from Andhra Pradesh link climate and ecosystem changes directly to their daily experiences, increased risks, and vulnerabilities. Many young fishers shared that unsustainable practices (pollution, mechanized overfishing) have altered fish behaviour and weather patterns over the past decades.

Middle-aged (40 years and older) fishermen reported more "sudden cloud formations," unexpected rains, and untimely cyclones that their traditional indicators (winds, tides, etc.) sometimes fail to predict. In response, Coastal fishers closely monitor environmental cues. For example, Jalaripeta fishers track the direction of local winds (*galulu*), the shape of cloud formations (*mammulu*), sea currents (*odishi*), water temperature, and tide patterns before each fishing expedition. A day before sailing, they make careful weather observations—a traditional early-warning system. If these indicators suggest danger, fishermen avoid going to sea, safeguarding themselves and their gear. In the past, thick clouds or rain at sea did not deter them, but fishers now err on the side of caution.

The elderly fishers (aged 60 and older) link technological changes to ecological shifts they observed over the past two decades. They noted that diesel pollution and extensive nearshore fishing have driven fish populations deeper. Coupled with warming seas, these changes have disrupted longstanding seasonal phenomena. For instance, fishermen shared a decline in the historic "*sallirlu*" upwelling (when cold, deep water comes near shore, drawing fish and triggering crab migrations). They also noted fewer shore crabs and seagulls today than in previous decades, attributing this to polluted, warmer waters. Such observations of vanishing fish species and changing fish patterns reinforce fishers' sense of increased risk and inform the exploration and application of new strategies, e.g., deeper-water fishing and the use of zero net, which may have negative consequences, disrupt migration again, and lead to negative coping as an alternate livelihood.

Coastal elders in Jalaripeta shared their observations on environmental degradation in their habitats and along the coast where they fish for their livelihoods. They remember a time when beaches were clean and plastic was absent. Now, plastic and garbage wash ashore. They recall abundant red and white crabs along the sand—sightings that have sharply declined—and fewer sea birds nearshore. Many attribute these changes to pollution, rising sea temperatures, and more erratic weather, altering marine ecosystems and fish behaviour.

Fishing, as an occupational identity, remains closely coupled with knowledge of the weather and bravery. Fishermen say that the four wind types in the Bay, locally called *sita gali, deddi gali, pai gali, and kaccim gali*, determine boat movement. During cyclones or heavy rain, wind direction dictates drift, so they constantly monitor sudden changes in wind direction. Through community discussion and shared observations, Jalaripeta fishers co-produce local knowledge on wind shifts, tides, and fish cues. These collective insights into altered wind patterns, sea currents, or fish migrations are actively used to determine whether to sail or shelter, serving as a resilient practice for mitigating occupational hazards.

4.2 *Fisherwomen Workers' Economic Resilience Through Entrepreneurship Skills*

Studies across India emphasize that fisherwomen workers' embedded position in the supply chain stabilizes community economies under disaster risk. At the same time, their agency in market negotiations challenges gender hierarchies within fishing societies. Furthermore, social well-being is integral to understanding the gendered influences within the fisheries sector and to driving transitions in livelihood perspectives (Biswal & Johnson, 2023; Kleiber et al., 2015; Sridevi, 1989).

Fisherwomen workers in Andhra Pradesh play a critical yet often overlooked role in sustaining coastal livelihoods by driving the fish supply chain from shore to market. Beyond traditional roles in fish drying, vending, and household provisioning, women have increasingly assumed entrepreneurial functions, including investing in boats, organizing auctions, managing credit through chit funds, and forging direct links between fishing harbours and urban consumers (Vennela, 2017). Fisherwomen learned about market preferences and fish prices at the nearby Visakhapatnam Harbor and became traders (*beragatte/berakatte*). Initially, the fisherwomen act as informal financial intermediaries, extending credit to fishermen for fuel, repairs, and daily expenses, thereby securing access to fish catches (Vipinkumar et al., 2013), symbolizing gendered relationships tied to economic strategies.

The story of Potti Ramulamma from the Peda Jalaripeta community (see below) exemplifies how fisherwomen combine entrepreneurship, social capital, and gendered labour to shape resilient livelihoods in volatile marine fisheries.

4.2.1 Story of Ramulamma

Ramulamma (age 47) of Peda Jalaripeta became a prominent fisherwoman-financier (known as a marakatte) after her husband's death. She now invests in 11 fishing boats and earns approximately INR 3000–5000 (USD 40–65) per-vessel auction and approximately INR 10,000 per day in total from commission on catches. During lean seasons, she diversifies into freshwater fishing and markets prawns, earning about INR 2500 to 3500 per day.

Ramulamma reported that she obtained an INR 2 lakh (200 thousand) loan and repaid it at INR 2000 per day over 110 days, including interest, leveraging her social network and community goodwill to expand her business. Through these strategies, she financed significant family expenses over time. For example, she built a three-story house (INR 1.5 million) and purchased homes for her three daughters and son. Ramulamma's story illustrates how fisherwomen leverage gendered labor roles, informal finance, and entrepreneurial skills to support household resilience in the face of disaster risk. Understanding such gendered dimensions is crucial for disaster governance, as the integration of women's experiences and knowledge of climate impacts into policy remains a gap. We argue that Andhra Pradesh's SSF integrates traditional ecological knowledge and local institutions to address rising risks. As the climate changes, it is imperative to recognize and integrate the voices of women fisher workers into disaster planning and policy, which currently do not reflect their perspectives. Using her earnings, Ramulamma funded major family expenses, including building a three-story house (INR 1.5 million) and purchasing homes for her three daughters and son. She repaid construction loans via *a rotating savings scheme (locally known as a chit fund).*

Ramulamma's story illustrates how fisherwomen leverage gendered market roles and informal finance to sustain their livelihoods. As a fisherwoman-entrepreneur, she leveraged her agency and hard work to accumulate wealth and provide for her family, even after personal loss. Her success helped buffer the community against shocks: by giving credit and stable fish sales, she strengthened local economic strategies under disaster risk.

4.3 Intergenerational Fishing Knowledge and Post-Cyclone Recovery

Fishers cite various reasons for changes in climate and their impact on marine ecosystems and living organisms. They have been closely eyeing regular weather observations, noting the present and documenting the past in their minds to predict risks and develop resilience strategies for successful fishing expeditions. These predictions enable them to avoid occupational risks during fishing expeditions by anticipating sudden changes in local weather conditions. Based on these changes, they can conceive strategies for the subsequent fishing expeditions. These strategies are based on wind directions (*galulu*), cloud formations (*mammulu*), the movement of sea currents (*odishi*), sea tides, directions, temperature, and humidity. These are significant indicators for successful fishing expeditions. Suppose fishermen observe any consequential changes or variations in local weather conditions and mostly avoid going fishing, enabling them to safeguard themselves from occupational accidents during fishing. A day before a fishing expedition, the fishermen unfailingly conduct weather calculations and observations, drawing on embodied local knowledge to help them avoid occupational risks. Failure to understand local weather conditions can lead to risks, difficulties, and fatalities in decision-making related to fishing expeditions. Furthermore, local weather conditions and effective decision-making by fishermen intersect to enable successful fishing expeditions (Kodi, 2021; Pukkalla & Sharma, 2018, 2020; Pukkalla & Rama Mohan, 2021).

The close connection between local knowledge and understanding of ecological systems and fishers' daily fishing activities is also associated with fishers' well-being. Furthermore, fishermen have observed that their local knowledge has not enabled them to correctly identify weather conditions in response to sudden changes. Sometimes, they experience sudden cloud formations and rainfall, winds blowing in different directions, temperature changes, and premature cyclones. The elderly fishermen and fisherwomen also noted that these changes were not limited to the past and were observed primarily during seasonal periods. Unsustainable practices in marine and terrestrial environments drive resource exploitation and directly affect the livelihoods of vulnerable communities, including those engaged in marine fishing worldwide.

The connection between the technological intervention and the contending local knowledge practices among the fishers. There have been significant changes in fishing practices since the introduction of mechanized motorboats into the sector. These can alter local marine ecosystems in response to local climatic variations. Traditional fishing boats were used to catch fish near the shore within 10 fathoms (*baralu*) of the shore. Pollutants released from mechanized boats, particularly diesel particles, enter seawater and alter its composition, affecting fish behavior and reproductive systems. These changes are restricted to fish behavior and other marine organisms within marine ecosystems. There is a need for a balanced approach to healthy, sustainable living patterns that enhance the well-being of organisms in marine ecosystems. After the introduction of mechanized fishing boats, these living creatures cannot sustain themselves in the sea and therefore move to the deep sea. They are affecting the marine food web and driving shifts in marine ecosystems. Identifying fish behavior and movements depends on wind direction, sea currents, and the nature of seawater in the given marine ecosystem. The greater influence of external agents, particularly water pollutants, leads migrating fish to move to different locations in the sea and to alter their behavior. Thus, the fishermen ventured into deeper seas to catch fish on mechanized boats. They were also aware that the use of mechanized boats for fishing expeditions significantly threatened marine ecosystems and that there were no viable alternative livelihood options. In addition, fishers reported that changes in marine ecosystems are due to the gradual rise in seawater pollution from the dumping of domestic waste and other anthropogenic activities, which are driving structural changes in marine ecosystems.

Increasing water pollution consistently negatively affects fish behavior. It has led to changes in seawater conditions, particularly the gradual decline of *sallirlu/saliviri* in the sea over the past few years. During *sallirlu*, the fish come from the deep sea towards the seashore. During this time, the seawater cools, and fish move towards the shore to warm up. The declining presence of *sallirlu* is due to changes in seawater composition. A few reasons are cited, such as the rise of water pollutants, mechanized boats, and other unsustainable fishing practices, which are responsible for altering seawater characteristics. The interconnections among seawater, wind direction, tidal direction, and current direction depend on local marine environments and their ecological systems. If any variations occur in marine ecology, changes in fish behavior can be observed simultaneously. One of the senior fishermen noted that one can assess marine ecological conditions from fish behavior. Furthermore, he noted that there is always a connection between sea behavior and fish behavior that informs ecosystem evaluation. Sea temperatures affect the behavior of fish and other marine life. Due to moderate temperatures, fishermen can catch more fish in the early morning and late evening than at other times of day. Rising temperatures these days may also be one reason to move fish from one location to another, along with other factors.

4.3.1 Story of Jaggayya

Jaggayya (82) is a veteran fisher of Peda Jalaripeta. He learned to fish with his father and brothers from age 11 and has four sons who fish with him. Together, the family traditionally shared boats and catch according to caste customs. Jaggayya shared his memories of Cyclone Hudhud (Oct 2014), which destroyed three of the family-owned fiber boats. Without insurance, Jaggayya and his sons incurred INR 45,000 in repair costs. The family borrowed INR 80,000 from private moneylenders to cover costs. Fishing resumed about 26 days after the storm, illustrating how sudden cyclones can disrupt livelihoods and push people into debt. During the 2019 and 2020 COVID-19 pandemic lockdowns, the family again could not fish for nearly two years and borrowed another INR 60,000. These informal loans (often repaid over many years from catch revenues) exemplify the debt cycles that fishers face. Jaggayya's narrative shows that, in the absence of savings or insurance, community members rely on private credit and collective support to cover disaster-related losses.

From 2022 to 2024, Jaggayya's family gradually rebuilt the fleet. They bought new fiber boats (costing roughly INR 80,000–160,000 each) and outboard engines, so that by 2024, the combined family will operate six boats. These investments were financed through pooled savings and further informal credit, reflecting a community-level recovery strategy. Even today, at 82, Jaggayya continues to fish with his sons and grandsons and teaches them traditional net-making, boat navigation, and catch-sharing rules. Kinship arrangements play an important role in fishing together at sea. When they return, the members divide the day's catch according to long-held cultural norms. At 82, he continues to pass on decades of local ecological knowledge (weather cues, fish behavior) to the younger generation, ensuring that family fishing expertise endures. Jaggayya's story highlights the power of family-based resilience. His inherited knowledge and leadership guided recovery after Hudhud, while informal financial arrangements provided immediate relief. The combination of intergenerational knowledge transfers and community lending allowed his family to rebound. His experience underscores that for fishers, social capital (kin networks and know-how) and informal finance are as crucial as any formal relief for returning from disasters.

Jaggayya continues his reflections, noting that identifying rocks beneath the sea once helped fishers find their way. Numerous stones are found under the sea, providing habitat for small plants and fish. The small fish live near these rocks and plants, where they grow in response to local weather conditions. Small fish eat plants, and big fish eat small fish. A rise in weather temperatures and seawater pollution, or any imbalances in the sea, damages marine conservation systems. Previously, fishermen fished from traditional boats; there were no mechanized boats. With the introduction of mechanized and modern fishing practices, the rise of external factors has accelerated, threatening marine ecosystems and livelihoods.

From the story, we learned that the Jalaripeta fishermen reported experiencing more thunderstorms (*pidugulu*) at sea during their fishing expeditions, a significant weather change they have observed in recent years, as reported to the local authorities. Earlier, even with thick cloud cover and rain during fishing expeditions, they did not panic and continued catching fish. However, the situation has been changing, and when they observe changes in the weather, they mostly avoid fishing expeditions. Furthermore, there is no need to assume additional risks if anything goes wrong, as families would become vulnerable.

The elderly fishermen noted that a few years back, the shorelines were clean and they found no external materials, particularly plastic items and other garbage, along the shorelines. With the rise of urbanization and industrial activity, waste accumulation has increased, contributing to changes in the Bay of Bengal's marine ecosystem. Elder fishers also noted that, earlier, there were many red and white crabs along shorelines, but in the past few years their numbers have decreased. They also reported seeing fewer seagulls at the sea, indicating that there is not enough fish. They argue that changes in weather conditions have likely affected the behavior of crabs and seabirds.

Despite all these efforts, Jaggayya mentioned in his narration rather sadly that young fishers from their communities are forced to migrate, leaving wives, children, and older parents in search of jobs, sometimes even non-fishing jobs like quarry work, painting, and seasonal agricultural labor, which we are not related to fishing skills (See story of Ravi below). Young fishers are forced to work in risky conditions, facing mental health issues, becoming alcoholics, and becoming severely ill. Many scholars have well documented this precarious situation. Due to a combination of dwindling local fish stocks, a lack of local infrastructure (harbors and jetties), and the promise of better, more stable wages in the large fishing industries on India's west coast. Being a fisherman in Andhra Pradesh is often described as a "gamble," with uncertain daily catches and irregular income. The lack of government loans to purchase better equipment exacerbates this financial instability (see Roshan, 2016; ICSF, 2020; Palli, 2025).

Jaggayya's story reveals that coastal fishing in Andhra Pradesh (South India) is closely tied to local weather conditions, and fishing expeditions are planned accordingly. In the event of any changes in these weather conditions, strategies are developed to venture out to fish while avoiding occupational risks. Furthermore, weather conditions, particularly wind and sea-current directions, are the primary determinants of successful fishing expeditions.

4.3.2 The Story of Ravi

Ravi, aged 30, is an active fisherman with two children. He completed tenth grade at the local government school and discontinued his studies because his family could not afford to continue his education. He began fishing with other family members at age 12, but at 18, he began fishing at sea with his kin groups regularly to earn a livelihood and support his family. Ravi reported that he and his father often went fishing together on the same boat, learning fishing techniques and applying the skills he had acquired during the expeditions. He also revealed that fishermen have been experiencing gradual changes, particularly in weather conditions, that are becoming detrimental to their fishing expeditions. Furthermore, he noted that fishing practices differ between his father's generation and his own, particularly in the transition from traditional to modern boat fishing. In the past, the father and grandfather generations were entirely dependent on traditional boats for fishing, which were navigated according to local weather conditions and wind directions. After modern fishing boats entered their lives, fishing became dependent on the direction of sea currents, which play a significant role in navigating vessels during fishing expeditions.

His father purchased a motorboat for deep-sea fishing and borrowed money from the local financier at an interest rate (vaddi in Telugu) that varied with the amount borrowed. The boat was purchased ten years ago for one lakh twenty thousand rupees, which was considered low cost at the time. Later, they repaid the amount they had borrowed from the financier and, in 2017, bought a boat for INR 240,000. In addition, his father bought suitable fishing nets, hooks, and lines, and recruited a crew. The total cost of the fishing boat, nets, and other items was INR 360, 000 and later stages put them into a financial crisis due to poor fish catch. Later, they could clear only one lakh rupees of the total debt. In addition, the COVID-19 pandemic led to a year of reduced fishing activities due to irregular fishing engagements, resulting in economic distress among their family members.

After his father died due to an illness, the economic issues were becoming more severe. While he was there with them, he also went on fishing expeditions, and their income from the fish catches was used to cover household expenses. Now, he said that the economic burden had doubled and was impacting his mental well-being. After his father's death, the boat was sold at half price to other fishermen, which cleared some debt, and the rest was spent on his father's funeral. After a few months of settling all his father-related funerals, he went to Gujarat, Mumbai, Mangalore, Kerala, Chennai, for fishing on big vessels and stayed for less than three years, experienced different kinds of healthcare issues, and was unable to cope with the local weather conditions, and returned to his hometown and resumed fishing there. However, he felt that continuing to fish in his village and earn lower economic returns would not constitute a clear debt, so he moved to the Visakhapatnam fishing harbour, where his wife's father (uncle) already goes on fishing expeditions in Sona boats. They used to make fishing expeditions for more than two weeks at a time. It is also reported that, while fishing in his village, he earned lower returns than from Sona boat fishing, which enabled him to meet his daily household needs and gradually reduce his debts.

He further shares that "Sona" boats (sona is gold in the Hindi language), a type of mechanized fishing vessel, larger than existing small, mechanized boats, became more common along the Andhra Pradesh coast about twenty-five years ago, particularly in Visakhapatnam, where hundreds (about 600) operate often for prawn and other types of fishing. These boats are known for their high fish-hold capacity (up to 5 tons), although they often lack onboard refrigeration and therefore rely on ice blocks carried at the start of the voyage. These boats are often owned by non-fishing-community investors with ties to party politics and other businesses. A small fraction of Sona boats is also owned by fishers from the Vadabalija community, owing to their access to political and banking resources. These boats spend about 8 to 15 days at sea and often utilize sonar technology to detect fish shoals. While older boats relied on simple navigation, wealthy boat owners increasingly equipped them with sonar and GPS to locate fish and enhance safety. Ravi further reports that, sometimes, if the crew finds a large catch (10-70 m in depth), a portion of the fish bycatch is discarded during the first four days of the voyage. In the remaining period of the voyage, the fish bycatch is dried or salted.

Despite the risks, Ravi laments that economic distress has not only reduced returns from fishing but has also been exacerbated by frequent cyclones and wind formations, which have disrupted fishing expeditions and affected their livelihoods. In addition, without any safety net or welfare schemes, fishers are perpetually in a debt cycle and are forced to sell their fish at lower prices to moneylenders.

Ravi hopes that, by pursuing regular fishing expeditions without interruptions, they can earn a suitable income, manage their household finances, and avoid being forced to migrate to other Indian states to work under harsh labour conditions. Ravi states that the number of cyclones in the Bay of Bengal has increased, significantly affecting their livelihoods and forcing many young fishers to leave their homes for other ports and fishing harbours, such as Veraval in Gujarat.

Ravi's story illustrates that fishing activities are interconnected across multiple political, economic, technological, and marine dimensions, influencing their decisions and livelihood practices, particularly in the context of disasters such as cyclones, heavy winds, and prolonged rains, which disrupt fish growth and movements and lead to economic distress among fishers who depend on marine fishing.

5 Interlinking Climatic Risks, Adaptation, and Community Resilience

Andhra Pradesh's extensive coast has improved warning systems and shelters, but significant gaps remain. Cyclone alerts often do not reach non-literate or isolated communities, and many fishers lack resilient housing or reliable evacuation routes. Post-disaster relief has been criticized for delays and failing to address fishing-specific losses (e.g., damaged nets or boats). Consequently, many fishers report falling into cycles of debt, selling assets (boats and nets), or migrating temporarily for wage labor, a pattern termed negative resilience. These patterns indicate that although disaster management in the State has reduced loss of life, it has not sufficiently mitigated the long-term economic vulnerabilities of small-scale fishers, particularly regarding debt accumulation, asset loss, and labor exploitation.

The stories shared in this chapter are not simply data or illustrations of abstract concepts; they are forms of knowledge, embodied, relational, and deeply contextualized. The stories of Ramulamma, Jaggayya, and Ravi provide critical insights and evidence of community resilience attributes. Ramulamma's story illustrates how fisherwomen leverage gendered labor roles, informal finance, and entrepreneurial skills to build household resilience in the face of disaster risk. Understanding such gendered dimensions is crucial for disaster governance, as the integration of women's experiences and knowledge of climate impacts into policy remains a gap.

Jaggayya experienced cyclones and remained an active mentor and knowledge keeper of the community throughout these ordeals. Even at the age of 82, he teaches his sons and grandchildren traditional fishing techniques and oversees their catch-sharing customs. His experience shows Fisher families mobilize kinship and credit to recover after cyclones.

Ravi's narration underscored the struggles he experienced during the fishing expeditions and in managing the household economy. Ravi's case also highlights the dynamics of the economic burden, which stem not only from lower returns from fishing expeditions but also from the impacts of frequent cyclones, heavy rains, and other unprecedented weather conditions. Ravi also shared experiences related to healthcare and well-being during his migration to different places and the management of distress, which leads to poor mental well-being. Furthermore, his case illustrates resilience and reliance in economic management, as he adapts to local weather conditions to sustain livelihoods and support family well-being through marine fishing.

These stories revealed patterns of economic exclusion and vulnerability, as SSF communities lacked access to formal insurance and timely government aid. These fisher workers relied mainly on informal savings (*chit funds*), rotating credit, or moneylenders to finance boat repairs and living expenses after a storm. After disasters, men come together in small groups, often based on family clans, to repair nets or seek help from relatives. Women's enterprises (like Ramulamma's fish marketing business) and off-season wage labor further diversify incomes and provide buffers against shocks. These grassroots adaptations, for example, blending traditional ecological knowledge, social networks, and informal finance, constitute the community's de facto resilience.

These narratives provide us with a more nuanced understanding of vulnerability as a product of systemic marginalization and structural inequality. SSF communities in Andhra Pradesh are vulnerable not simply because they live on the coast, but because they are excluded from secure land tenure, formal credit systems, social protection schemes, and meaningful participation in disaster policy making. Their vulnerability is compounded by caste-based discrimination, gender inequality, and the neglect of small-scale fisheries in national development agendas. SSF integrates local and traditional ecological knowledge and rituals and mobilizes local resources from traditional fishermen's cooperatives to address rising risks. By foregrounding the structural dimensions of vulnerability, the chapter aligns with critical disaster studies scholarship (Oliver-Smith, 2004; Wisner et al., 2004) that emphasizes the social production of disaster risk. It challenges technocratic approaches that focus narrowly on hazard mitigation and infrastructure development, arguing instead for interventions that address the root causes of marginalization: land rights, social protection, equitable access to resources, and political voice.

6 Conclusion and Way Forward: From Narratives to Governance Praxis

This chapter has journeyed through the lived experiences of small-scale fishing (SSF) communities along the cyclone prone coastline of Andhra Pradesh, documenting how these communities navigate the intersecting crises of climate change, disaster vulnerability, and livelihood precarity. Through collaborative ethnographic engagement with communities and fish worker-centered narratives, this chapter traced the impacts of devastating cyclones, particularly Hudhud (2014) and Titli (2018), on the social, economic, political, and ecological fabric of fishing villages. These are not merely stories of loss and suffering; they are testimonies to adaptive ingenuity, collective solidarity, and the persistent assertion of small-scale fishers' dignity in the face of systemic marginalization.

The narratives shared by fishers, fisherwomen, and community elders reveal a complex tapestry of vulnerability and resilience. These stories unravel how cyclone disasters compound pre-existing challenges: socio-economic marginalization, declining fish stocks, coastal erosion, insecure land tenure, and inadequate access to formal credit and social protection. Despite these vulnerabilities, SSF communities have demonstrated remarkable coping skills and adaptive capacity. They mobilize traditional ecological knowledge (TEK) to read weather patterns, adjust fishing practices, and protect their boats and nets. In addition, SSF communities leveraged social capital, kinship networks, caste-based solidarity, and community cooperation to share resources and provide mutual aid, thereby attempting to rebuild their livelihoods after disasters. Fisher Women, often invisible in formal disaster management frameworks, emerge as critical actors in household survival, income diversification, and post-disaster recovery, even as they bear disproportionate burdens of care work and economic vulnerability.

By documenting and synthesizing the risk and recovery narratives, the chapter has also critically examined the state of Andhra Pradesh's response to cyclone disasters. While early warning systems, evacuation protocols, and relief distribution have improved over the years, significant gaps remain. Post-disaster relief mechanisms often focus on disjointed efforts that fail to recognize the specific recovery needs of SSF communities, including the loss of fishing gear, boats, and access to the sea; the disruption of credit cycles and market linkages; the gendered dimensions of disaster impacts; and the erosion of cultural and spiritual landmarks. Rehabilitation efforts are frequently short-term, top-down, and disconnected from the lived realities and knowledge systems of fishing communities.

In this chapter, we argue that this disconnect reflects a broader failure of disaster governance, one that continues to privilege technocratic, state-centric models of disaster management over community-led, participatory, and justice-oriented approaches. The chapter thus contributes to a growing body of scholarship that calls for 'resilience with justice' (Matin et al., 2018), an approach that recognizes the agency and knowledge of marginalized SSF communities while addressing the structural inequalities that constrain their adaptive capacity.

The chapter further highlights the role of mobility and migration of young fishers as both family coping strategies and markers of deepening economic vulnerability. Fishers from Andhra Pradesh increasingly migrate to Gujarat, Kerala, and other states in search of better catches and income, often facing exploitation, debt bondage, stigma, social isolation, and dislocation. This pattern of distress migration underscores the limits of local adaptation and the need for structural interventions that address the root causes of livelihood insecurity, including overfishing, pollution, industrial encroachment on coastal commons, and the neglect of small-scale fisheries in national and state policy frameworks.

This chapter also makes important theoretical contributions to the fields of disaster studies, climate adaptation, livelihood resilience, and blue justice. These contributions are both empirical and conceptual, grounded in the lived experiences of SSF communities while engaging with broader scholarly work about vulnerability, agency, and justice in the context of climate change and disaster risk.

Finally, this chapter calls for a fundamental reimagining of coastal disaster risk governance in India. It advocates a shift from reactive, relief-oriented models to proactive, rights-based, and community-centered approaches that recognize SSF communities as knowledge holders, decision-makers, and agents of change. It emphasizes the importance of integrating traditional ecological knowledge (TEK) with scientific expertise, ensuring a two-eyed community science approach and gender-responsive disaster management, and embedding principles of equity, inclusion, and blue justice into climate adaptation and fisheries policy.

Theoretically, the chapter contributes to a growing body of work that seeks to decolonize disaster studies, climate adaptation research, and development practice. It challenges the universalizing assumptions of mainstream frameworks, the idea that vulnerability, resilience, and adaptation can be understood in the same way across contexts, or that technical solutions can be applied without attention to power, culture, and history. Instead, it argues for situated, context-specific, and justice-oriented approaches that recognize the diversity of experiences, knowledge systems, and adaptive strategies. In practical terms, the chapter offers a roadmap for more equitable and effective coastal disaster governance in India and beyond. It calls for the 5R approach:

- *Recognition* of SSF communities as rights-holders and knowledge-holders, not merely beneficiaries, victims, or vote banks
- *Redistribution* of resources, power, and opportunities to address structural inequalities and ensure equitable access to coastal commons
- *Representation* of SSF communities in decision-making processes at all levels, from village-level disaster management committees to national climate adaptation policies
- *Reconciliation* and *Recognition* of traditional ecological knowledge and scientific expertise through processes of co-production and collaborative learning
- *Reparation* for historical injustices and ongoing harm, including land dispossession, environmental degradation, and neglect by the state and non-state actors

Acknowledgements Bala Raju Nikku, the first author, gratefully acknowledges financial support from Canada's New Frontiers Research Fund (NFRFR-2021-00194), which enabled the field research underpinning this chapter.

References

Adger, W. N., Dessai, S., Goulden, M., Hulme, M., Lorenzoni, I., Nelson, D. R., & Wreford, A. (2009). Are there social limits to adaptation to climate change? *Climate Change, 93*(3), 335–354.

Agnishwaran, R., Hariharan, M., Quintoil, M. N., Varun, A., Rekha, B. V. V., & Kumar, A. V. J. (2024). Impact of climate change on fisheries and food security: A comprehensive review. *Chronicle of Aquatic Science, 1*(10), 199–211.

Baig, M. R. I., Ahmad, I. A., Shahfahad, T. M., & Rahman, A. (2020). Analysis of shoreline changes in Vishakhapatnam coastal tract of Andhra Pradesh, India: An application of digital shoreline analysis system (DSAS). *Annals of GIS, 26*(4), 361–376. https://doi.org/10.1080/19475683.2020.1815839

Balaram, P., & Katta, R. M. R. (2024). The role of disaster-affected people on natural disaster management in Godavari Districts of Andhra Pradesh—An overview. *Library of Progress International, 44*(3), 13806–13813.

Basheer Ahammed, K. K., Mahendra, R. S., & Pandey, A. C. (2016). Coastal vulnerability assessment for the Eastern Coast of India, Andhra Pradesh, using geospatial techniques. *Geoinformatics and Geostatistics: An Overview, 4*(3).

Biswal, R., & Johnson, D. S. (2023). A social wellbeing approach to the gendered impacts of fisheries transition in Gujarat, India. *Maritime Studies, 22*(2), 13.

Bosher, L., Penning-Rowsell, E., & Tapsell, S. (2007). Resource accessibility and vulnerability in Andhra Pradesh: Caste and non-caste influences. *Development and Change, 38*(4), 615–640.

Das, S., & Vincent, J. R. (2009). Mangroves protected villages and reduced the death toll during the Indian super cyclone. *Proceedings of the National Academy of Sciences, 106*(18), 7357–7360.

Devaraj, M. (1997). *Vision 2020: CMFRI perspective plan.* CMFRI.

Emanuel, K. (2017). Will global warming make hurricane forecasting more difficult? *Bulletin of the American Meteorological Society, 98*(3), 495–501.

George, M. J., & Subbaraju, G. (1979). Small-scale fisheries at Lawson's Bay, Waltair. *Marine Fisheries Information Service, Technical and Extension Series, 6*, 10–13.

Government of Andhra Pradesh. (2018). District survey report: Visakhapatnam district. Department of Mines and Geology, Government of Andhra Pradesh (GOAP).

Hoeppe, G. (2007). *Conversations on the beach: Fishermen's knowledge, metaphor and environmental change in South India* (1st ed.). Berghahn Books.

IMD (2020, 2022). Ministry of Earth Sciences (MoES). Observed monsoon rainfall variability and changes during the recent 30 years (1989–2018). [Press releases]. Accessed 26 December 2025.

Immanuel, S., & Syda Rao, G. (2012). Social status of hook and line fishermen in Pedajalaripeta Village, Visakhapatnam District, Andhra Pradesh. *Fishery Technology, 49*(2), 204–209.

International Collective in Support of Fishworkers (ICSF). (2020, December). At sea, out of sight: Migrant fishers from Andhra Pradesh. https://www.icsf.net/wp-content/uploads/2022/07/42.pdf

Jeevamani, J. J. J., Priya, P., Infantina, J. A., Abhilash, K. R., Behera, D. P., Samuel, V. D., & Ramesh, R. (2021). An integrated approach to assess coastal vulnerability versus fisheries livelihood sustainability: Strategies for climate change adaptation in Sindhudurg, west coast of India. *Environment, Development and Sustainability, 23*(3), 4011–4042.

Kapapa, L., Onyango, P., Bwanthondi, P., & Mfilinge, P. (2024). The vulnerability of fisheries-based livelihoods to climate variability and change in Tanzania's coastal small pelagic fishing communities. *Marine Policy, 169*, Article 106344.

Khadar, K. V. A. (2021, August 16). How climate change is taking a toll on the livelihoods of small-scale fishers in Kerala. Conserving Agro-biodiversity. Mongabay India. https://india.mongabay.com/2021/08/how-climate-change-is-taking-a-toll-on-livelihoods-of-small-scale-fishers-in-kerala

Kleiber, D., Harris, L. M., & Vincent, A. C. J. (2015). Gender and small-scale fisheries: A case for counting women and beyond. *Fish and Fisheries, 16*(4), 547–562.

Kodanda Rao, M. (1990). *Cultural and structural dimensions of family: A study of Jalari fishermen.* Concept Publishing.

Kodi, S. (2021). Indigenous knowledge and engaging marine fishing practices in North Coastal Andhra Pradesh. *Anthropologist, 44*(1–3), 1–9.

Kodi, S., & Murali Mohan, M. (2015). Traditional technological aspects of fishing crafts in North Coastal Andhra Pradesh, East Coast of India. *International Journal of Advanced Research, 3*(10), 698–703.

Lacheheb, M., Noy, I., & Pundit, M. (2025). Impact of tropical cyclones on fishing activity in the Philippines. *Fisheries Research, 281*, Article 107221.

Madhanagopal, D., & Pattanaik, S. (2020). Exploring fishermen's local knowledge and perceptions in the face of climate change: The case of coastal Tamil Nadu, India. *Environment, Development and Sustainability, 22*, 3461–3489.

Malakar, K., Mishra, T., & Patwardhan, A. (2018). Perceptions of multiple stressors affecting the livelihoods of marine fishermen. *Marine Policy, 97*, 18–26.

Matin, N., Forrester, J., & Ensor, J. (2018). What is equitable resilience? *World Development, 109*, 197–205.

McLeod, E., Chmura, G. L., Bouillon, S., Salm, R., Björk, M., Duarte, C. M., & Silliman, B. R. (2011). A blueprint for blue carbon: Toward an improved understanding of the role of vegetated coastal habitats in sequestering CO_2. *Frontiers in Ecology and the Environment, 9*(10), 552–560.

Millennium Ecosystem Assessment (MEA). (2005). *Ecosystems and human well-being: Synthesis.* Island Press.

Nagelkerken, I., Blaber, S. J., Bouillon, S., Green, P., Haywood, M., Kirton, L. G., & Somerfield, P. J. (2008). The habitat function of mangroves for terrestrial and marine fauna: A review. *Aquatic Botany, 89*(2), 155–185.

Nair, N. V., Teerthala, M., Ghosh, P., Attipoe, E., Norberg, E. M., Djatayu, M. A., & Emon, M. (2024). *Reflections from Chilika-V2V field school: Exploring local ecologies and social-ecological dynamics.* Working paper series, University of Waterloo.

Nikku, B. R., Nepali, B., & Khatiwada, H. (2021). Earthquakes of Nepal: Making the case for Indigenous resilience. In *The routledge international handbook of indigenous resilience* (pp. 285–299). Routledge.

Nikku, B. R., Galimberti, S., Gurung, K., & Kodi, S. (2025). International social work, community science, and disaster justice. In *Pushing boundaries in social work around the World, Vol. 2: Policy and global perspectives* (pp. 29–42). Springer Nature.

Nikku, B. R. (2020). Nepal: Protracted displacement from Haku Village, Rasuwa District, after the 2015 earthquakes. In *Climate change, disasters, and internal displacement in Asia and the Pacific* (pp. 181–201). Routledge.

Nuckolls, C. (1996). *The cultural dialectics of knowledge and desire.* The University of Wisconsin Press.

Oliver-Smith, A. (2004). Theorizing vulnerability in a globalized world: A political ecological perspective. In G. Bankoff, G. Frerks, & D. Hilhorst (Eds.), *Mapping vulnerability: Disasters, development and people* (pp. 10–24). Routledge.

Palli, I. C. (2025, March 14). *Better catch pulls east coast fishers westwards.* Mongabay India. https://india.mongabay.com/2025/03/better-catch-pulls-east-coast-fishers-westwards/

Pukkalla, D., & Gangadhar, M. (2025). Marine ecological adaptation and fishing livelihoods in a South Indian fishing community. *The Eastern Anthropologist, 78*(1–2), 151–160.

Pukkalla, D., & Rama Mohan, K. R. (2021). Local knowledge and marine livelihoods among the South Indian fishing community. *Journal of Asian and African Studies, 56*(3), 549–557.

Pukkalla, D., & Sharma, B. V. (2018). Occupational health risks and etiologies among the Jalari community of the Northern District of Andhra Pradesh, India. *Journal of Human Ecology, 61*(1–3), 64–71.

Pukkalla, D., & Sharma, B. V. (2020). Transmission and erosion of local knowledge practices in a fishing village in South India. *Journal of Ecological Anthropology, 22*(1), 34–42.

Raju, D. S. N., & Vaidyanadhan, R. (1978). Coastal geomorphology of the Visakhapatnam region, east coast of India. *Journal of the Geological Society of India, 19*, 97–106.

Roshan, M. (2016). *A study of migrant fishers from Andhra Pradesh in the Gujarat marine fishing industry,* ICSF Occasional Paper, International Collective in Support of Fishworkers. https://www.icsf.net/wp-content/uploads/2016/08/930.ICSF071.pdf

Sridevi, C. (1989). The fisherwoman financier: A study of status-role nexus in a peasant community. *Economic and Political Weekly, 24*(17), 6–9.

Srinivasa Rao, D. (2022). Disaster resilient infrastructure: A study of the impact of cyclones on the telecom sector in the Andhra Pradesh coastal zone. *Disaster Advances, 15*(5).

Unnikrishnan, A. S., Kumar, K. R., Fernandes, S. E., Michael, G. S., & Patwardhan, S. K. (2006). Sea level changes along the Indian coast: Observations and projections. *Current Science, 90*(3), 362–368.

Vennela, V. V. (2017). *Dynamics of gender roles: A study on the maritime fisherwomen of Visakhapatnam District.* Department of Anthropology, Andhra University, Visakhapatnam, Doctoral Dissertation. Retrieved: http://hdl.handle.net/10603/387095

Vipinkumar, V. P., Swathi Lekshmi, P. S., Johnson, B., & Narayanakumar, R. (2013). Coastal indebtedness and impact of microfinance in the marine fisheries sector of Andhra Pradesh. *Indian Journal of Fisheries, 60*(4), 81–86.

Wisner, B., Blaikie, P., Cannon, T., & Davis, I. (2004). *At risk: Natural hazards, people's vulnerability, and disasters* (2nd ed.). Routledge.

Changing Climate and Gendered Dynamics in Tropical Cyclone Evacuation Behaviors in Bangladesh: Who Will Leave Before the Storm?

Musabber Ali Chisty

Abstract This study addresses two principal gaps in the existing literature on tropical cyclone evacuation under a unified research question. First, it notes that the vast majority of prior research has been retrospective, focusing on the analysis of documented behavior during past events. Second, it observes that within these analyses, gender has almost universally been treated as a demographic control variable rather than as a central determinant of the decision-making process. To advance beyond this descriptive tradition, this study explicitly investigates gender as a primary predictor of evacuation intention, seeking to identify whether male or female-headed households are more likely to state an intention to comply with a future evacuation order. Using data from the Bangladesh Environment and Migration Survey (BEMS) 2019, this study conducts a predictive analysis of evacuation intention by employing multiple regression models alongside marginal effect estimations. The findings indicate that female-headed households exhibit a strong and significant association with evacuation intention, demonstrating significantly higher odds of evacuation (OR = 1.51–1.70, $p < 0.001$) and a 9–12 percentage point increase in predicted probability ($p < 0.001$). This study finds that, even after controlling for variables like direct cyclone experience, remittance access, and social capital, gender—specifically female-headed households—remains a robust, independent predictor. This study concludes with a list of policy and practice recommendations aimed at ensuring that strategies, resources, and networks utilize females' greater evacuation propensity during tropical cyclones in Bangladesh.

Keywords Gender · Evacuation · Cyclone · Social Vulnerability · Human Behavior · Response

---

M. A. Chisty (✉)
Department of Sociology, University of Colorado Boulder, Boulder, USA
e-mail: Musabber.chisty@colorado.edu

Natural Hazards Center, University of Colorado Boulder, Boulder, USA

Institute of Disaster Management and Vulnerability Studies, University of Dhaka, Dhaka, Bangladesh

J. Collins et al. (eds.), *Tropical Cyclone Risk and Resilience*,
https://doi.org/10.1007/978-3-032-28745-8_13

1 Background

Frequency and intensity of extreme weather events are changing drastically due to anthropogenic climate change (IPCC, 2021). Whether it is flooding in central Texas (Plail et al., 2025) or a tropical cyclone affecting the Philippines (UN Ocha, 2024), catastrophic weather events will become more frequent around the world. Scientists from different parts of the world are working vigorously to present how natural climate change is different from anthropogenic climate change and how attribution science can be used to model, predict, and refer to present and future extreme weather events like tropics cyclones, floods, droughts, heatwaves, among other hazard types (Thompson & Otto, 2015). Researchers and scientists are using extreme event attribution (EEA) science to model that anthropogenic climate change is the main agent for consistent catastrophes (Janković & Schultz, 2017; Osaka & Bellamy, 2020).

However, an exclusive focus on anthropogenic climate change when analyzing disaster risks promotes a reductionist adaptation framework and obscures the multi-dimensional nature of vulnerability that transforms natural hazards into catastrophes (Albright & Crow, 2015; Brody et al., 2008). Anthropogenic climate change is a key driver of intensified extreme weather events, including tropical cyclones and floods (Albright & Crow, 2019); however, sociodemographic and economic vulnerabilities critically undermine mitigation, preparedness, response, and recovery efforts following such disasters (Lahsen & Ribot, 2022). Researchers have been presenting the arguments related to the "no natural disasters" idea for a long time. We, who identify ourselves as disaster and environmental sociologists, assert that there is no such thing as a "natural disaster" and the consequences of extreme weather events are fundamentally shaped by the unequal and unjust organization of societies (Kelman, 2020; Mileti, 1999; Tierney, 2019; Wisner et al., 2004). The "no natural disaster" paradigm emphasizes the agency of communities, institutions, and government bodies to take proactive measures to mitigate and reduce disaster impacts. By recognizing that disasters are socially constructed, households can understand that actions such as mitigation, preparedness, evacuation, and emergency planning can enhance resilience and minimize losses.

The empirical recognition of anthropogenic climate change, coupled with an epistemological shift toward human vulnerability frameworks (Adger, 2006), establishes the necessary foundation for developing socio-culturally grounded mitigation and adaptation paradigms. This people-centered ontological positioning reveals structural pathways for addressing climate crises through justice-oriented interventions (Malm & Hornborg, 2014). In Bangladesh, communities perceive climate change as amplifying natural hazard intensity; however, empirical evidence suggests pre-existing socioeconomic vulnerabilities, including entrenched poverty, gendered marginalization, and uneven development, constitute the primary barrier to effective hazard response (Haq & Ahmed, 2020; Hasan & Kumar, 2020; Karim, 2024). As a climate-vulnerable deltaic nation, Bangladesh faces increasing exposure to intensified tropical cyclones and other hydro-meteorological hazards under anthropogenic climate change (Intergovernmental Panel on Climate Change (IPCC), 2023). This climatic forcing interacts with geophysical precarity and systemic socioeconomic vulnerabilities to create compound risks that will likely escalate throughout the twenty-first century (Brammer, 2014; Islam & Amstel, 2021).

The World Meteorological Organization reports that Bangladesh experiences approximately 1% of global cyclonic activity, with tropical cyclones causing between 750,000 and 1.23 million fatalities since 1900, representing one of the highest mortality rates from such events globally (Alam et al., 2003; Islam, 2025). While Bangladesh has significantly reduced cyclone-related deaths through improved early warning systems and evacuation infrastructure, coastal communities must strengthen adaptive measures as climate change increases the variability and intensity of cyclones (Ahsan et al., 2016b; Paul, 2009; Paul & Dutt, 2010; Paul et al., 2010; Rai et al., 2014). Scholars contend that, under climate change conditions, communities must emphasize adaptation strategies, including maintaining preparedness plans, understanding warning systems, and executing prompt evacuations to minimize tropical cyclone consequences (Das, 2019).

Extensive research confirms that timely evacuation during tropical cyclones remains a critical life-saving intervention and loss-reduction strategy across empirical studies. The Government of Bangladesh (GoB) implements the Cyclone Preparedness Program (CPP), which mobilizes volunteers to disseminate early warnings and facilitate community evacuations to designated cyclone shelters prior to tropical cyclone events (Akhand, 2003). However, the rate of non-compliance with tropical cyclone evacuation warnings and directions is still high (Saha & James, 2017). Researchers have found that multiple factors shape evacuation behavior in Bangladesh, including: limited cyclone shelter centers; lack of transportation; weak social capital; low risk perception; protection of livelihood and livestock; lack of awareness about warnings; cultural and gender-based obstacles; age- and health-related issues; and financial barriers (Paul, 2012; Ahsan et al., 2016a; Parvin et al., 2019; Chakma et al., 2020; Shoji & Murata, 2021; Alam & Chakraborty, 2021; Alam et al., 2023; Alam, 2024).

The majority of existing research has predominantly examined either retrospective analyses of non-evacuation decisions or evaluated evacuation patterns through risk perception frameworks. Furthermore, gender has typically been treated as a control variable rather than examined as a determinant of differential evacuation behaviors. However, empirical evidence demonstrates that gender constitutes a critical analytical dimension for understanding differential evacuation behaviors during cyclones (Bateman & Edwards, 2002; Shen et al., 2014; Cahyanto & Pennington-Gray, 2015; Akter & Tasnim, 2023). Previous studies have shown that gender significantly influences evacuation behavior. Women are often more likely to evacuate before a disaster, driven by heightened risk perception and a strong sense of responsibility to protect family members (Bateman & Edwards, 2002). However, other studies have found that women may be less willing to evacuate due to concerns about shelter conditions, safety, and family obligations (Gaillard et al., 2017). These mixed findings highlight the need for ongoing, country-specific research to better understand and address these gendered differences. This study addresses the critical gap by analyzing evacuation behavior through gendered dimensions while adopting a forward-looking perspective. This study primarily examines the association between household head gender and prospective evacuation decisions under tropical cyclone scenarios in coastal Bangladesh.

2 Data and Method

Before diving into the differences of evacuation behavior among different genders, the author wants to operationalize the definition of gender to use it throughout the chapter instead of using the definition of sex. The author adapted the definition of gender stated in the National Social Security Strategy (NSSS) by the Government of Bangladesh. According to this policy, *gender refers to socially constructed roles and relationships, personality traits, attitudes, behaviors, values, relative power, and influence that society ascribes to women and men on a differential basis. Whereas biological sex is determined by genetic and anatomical characteristics, gender is an acquired identity that is learned, changes over time, and varies widely within and across cultures* (Government of Bangladesh, 2018). Although this definition challenges binary gender identities traditionally tied to biological sex, Bangladesh's policies, constrained by Islamic laws and constitutional provisions, do not recognize the LGBTQI + community (Human Dignity Trust, 2025). The Government of Bangladesh officially recognized the *Hijra* community in 2013; however, this recognition is based on biological attributes rather than social or behavioral identity, distinguishing it from the broader concept of LGBTQI + identities (Akter & Saha, 2024). The dataset used in this study adhered to the same definition, aligning with the established laws and policies of Bangladesh by categorizing gender as either female or male throughout the research process. Accordingly, this book chapter employs the term gender in the same manner across its analysis and discussion.

The study used Bangladesh Environment and Migration Survey (BEMS) 2019 dataset (Carrico & Donato, 2023), which is publicly available through the ICPSR data repository. The main objective of the BEMS was to review the internal and external migration scenario in Bangladesh. The survey also included environmental hazards-related questions, which were mainly used for the current study. Data were collected from 4382 households across 13 districts (Table1). After rigorous data cleaning to address errors and missing responses, the final sample comprised 3,995 valid household head responses, including 3431 male and 564 female respondents. Although this study covered a wide range of districts in Bangladesh, some cyclone-prone areas were not included. The sample size is relatively representative given the dataset's objective of examining environmental migration; however, certain limitations remain. Greater representation of female-headed households would have strengthened the data's rigor. Additionally, since the dataset was collected in 2019, shifts in current social conditions may not be reflected. As the data are entirely quantitative, the study cannot capture the underlying reasons for specific evacuation behaviors without qualitative insights. Future research should aim to address and minimize these limitations.

Although the Bangladesh Environmental Migration Survey (BEMS) was conducted in 2019, the dataset remains highly relevant for examining gender-based evacuation behaviors. Theoretical frameworks linking gender, social norms, and household decision-making in evacuation contexts are grounded in structural and cultural dynamics that evolve slowly over time. Therefore, the patterns observed in 2019 continue to provide valuable insight into the underlying mechanisms shaping evacuation decisions. Moreover, the 2019 BEMS data offer an important pre-pandemic baseline for understanding gendered evacuation behavior prior to the disruptions brought by COVID-19 and subsequent major cyclones. This baseline not only strengthens the interpretation of current findings but also serves as a useful reference point for future studies seeking to assess whether and how gendered evacuation patterns have shifted in the post-pandemic and post-cyclone context.

The study acknowledges a minor limitation related to the geographic coverage of the study areas. Evacuation-related information was drawn from the Bangladesh Emergency Management Survey (BEMS) 2019 for all districts listed in Table 1. This coverage reflects the structure of the dataset rather than a selective choice by the study. While several included districts, such as Meherpur, Jhenaidah, and Gopalganj, are not formally classified as coastal, they have nonetheless experienced significant cyclone-related impacts due to Bangladesh's deltaic setting, interconnected river systems, and their relative proximity to the coast. Conversely, the BEMS 2019 dataset does not include certain coastal districts, including Chattogram, Noakhali, and Cox's Bazar, which are recognized as highly cyclone-prone. The inclusion of some non-coastal districts alongside the exclusion of these coastal districts may introduce variability in observed patterns of tropical cyclone exposure, evacuation

Table 1 Study area (District level)

Name of the district	Number of sample
Bagerhat	226
Barguna	427
Barisal	436
Bhola	206
Faridpur	456
Gopalganj	224
Jhalokati	393
Jhenaidah	243
Khulna	208
Madaripur	260
Meherpur	211
Patuakhali	434
Shariatpur	658

experience, and preparedness. As this constraint originates from the data source rather than the study methodology, the clarification is provided to assist readers in interpreting the findings with appropriate caution and to highlight the importance of comprehensive geographic coverage in future research on cyclone-related evacuation and preparedness in Bangladesh.

2.1 Dependent and Independent Variables

For this study, a set of variables was extracted that supported assessment of the primary objective. The dependent variable here is future evacuation behavior for a probable tropical cyclone. Respondents were asked if they would evacuate in the future if they received an evacuation notice. Respondents replied in binomial responses (1 = Will evacuate, 0 = Will stay home). The independent variable was gender. As our main goal is to see if gender has any association with future evacuation behavior, we coded female as 1 and male as 0. Historically, research has shown that the rate of non-compliance with evacuation warnings is higher among women (Akter & Tasnim, 2023; Khan et al., 2025). The primary factors contributing to this non-compliance included limited access to evacuation resources and information, lack of personal transportation, and inadequate social support networks (West & Orr, 2007; K.C. & Hilhorst, 2022). This study predicts that *Hypothesis 1: Female-headed households are more likely to intend to evacuate than male-headed households,* and *Hypothesis 2: The association between female-headed households and evacuation intention persists after controlling for disaster experience, socioeconomic factors, and institutional trust.*

Other than the dependent and independent variables, the study also included several covariates to assess the reliability of the association. These covariates included, previous cyclone experience (1 = Experienced cyclone impacts, 0 = Never experienced cyclone impacts); international remittance (1 = Don't receive remittance, 0 = Receive remittance); Lost property due to cyclone (1 = Loss property, 0 = Didn't loss property); Trust in government (1 = I trust the government for help, 0 = I don't trust the government for help); Trust in neighbors (1 = I trust neighbors for help, 0 = I don't trust neighbors for help); Food insecurity (1 = Faced food insecurity in the last 12 months, 0 = Didn't face food insecurity in the last 12 months); Water insecurity (1 = Relied on external water source, 0 = Own the water source). Including these covariates, the study also provided an analysis of sociodemographic indicators to present a brief overview of the respondents.

2.2 Analysis Model

The author used STATA package 18.5 to complete all statistical analyses. Socio-demographic characteristics of the respondents were summarized using frequencies and percentages. To examine bivariate associations between categorical variables, Pearson's chi-square tests were conducted. This analysis identified significant differences in response patterns across demographic subgroups, providing preliminary evidence for further modeling. Additionally, chi-square tests were applied to assess the unadjusted associations between the outcome variable and key covariates. These tests determined whether subsequent regression modeling was necessary to control for potential confounders. To evaluate the adjusted associations between the dependent variable and independent variable, a logistic regression model (LRM) was fitted (Eq. 1). The model included gender as the primary predictor while adjusting for covariates to assess whether observed associations with the independent variable remained significant after accounting for confounding factors (Fekete, 2009; Crowley, 2021). The results were reported as odds ratios (ORs) with corresponding confidence intervals for easy interpretation and understanding.

Equation 1: Logistic Regression Model

$$\log\left(\frac{p(Y=1)}{1-p(Y=1)}\right) = \beta_0 + \beta_1 X_{gender} + \beta_2 X_2 + \beta_k X_k + \varepsilon \tag{1}$$

Here,

- p(Y = 1): Probability that Y = 1 (Will evacuate = 1)
- β_0: Intercept (baseline log-odds when all predictors = 0)
- X_{gender}: Primary independent variable
- β_1: Coefficient for gender identity
- $\beta_2 \ldots \beta_k$: Coefficients for covariates
- ϵ: Error term.

To further interpret the substantive impact of the independent variables on the dependent variable, average marginal effects (AMEs) were calculated from the logistic regression model (Eq. 2). Unlike odds ratios, which provide multiplicative changes, marginal effects quantify the change in predicted probability of the outcome associated with a categorical shift in the predictor, holding other covariates at their observed values (Norton et al., 2019; Howell-Moroney, 2024). This approach offers a more intuitive measure of effect size, particularly for policy or practical implications.

Equation 2: Marginal Effect Model

$$AME = \frac{1}{N}\sum_{i=1}^{N}[\hat{P}\left(Y_i = 1|X_{indepdentvariable} = 1\right) - P(Y_i = 1|X_{independentvariable} = 0\} \tag{2}$$

Here,

- $P\ (Y = 1|X_{\text{independent variable}} = 1)$ = Raw probability of evacuation for the assessment group, here females.
- $P\ (Y = 1|X_{\text{independent variable}} = 0)$ = Raw probability of evacuation for the reference group, here males.

3 Results

The following section outlines the results of the statistical analyses used to examine the determinants of evacuation behavior in tropical cyclone-prone areas of Bangladesh. First, descriptive results provide an overview of the sample characteristics and key variables. Next, logistic regression analyses and average marginal effects presents the effect of demographic, socioeconomic, and vulnerability indicators on the likelihood of evacuation.

3.1 Descriptive Statistics of Sociodemographic Variables

Table 2 provides a comprehensive sociodemographic profile of households disaggregated by gender of the household head, revealing significant structural disparities. The data highlight how gender intersects with multiple dimensions of socioeconomic status in this population. Female-headed households face pronounced disadvantages across several indicators, particularly in asset ownership and economic participation. The gendered division of labor is strikingly apparent in livelihood patterns, with the vast majority of female heads engaged in homemaking roles compared to their male counterparts. Educational attainment shows a concerning gap, suggesting potential barriers to schooling or early dropout rates among women.

Health disparities between genders may reflect differential access to healthcare services or the compounded stresses of caregiving responsibilities among female heads. The substantial differences in house ownership rates likely stem from inheritance laws or traditional property rights favoring male ownership. Access to basic amenities like electricity and technology shows gendered patterns that may contribute to the digital divide. While family size remains comparable across genders, the differential access to productive assets like livestock suggests varying economic capacities. These findings underscore how gender operates as a structural determinant of socioeconomic positioning, with female-headed households systematically disadvantaged across multiple domains.

Table 2 Sociodemographic description based on household head gender (n = 3995)

Items	Household head's gender		
	Male (%) (N = 3431)	Female (%) (N = 564)	Pearson Chi-square
Education level			
Secondary school certificate (SSC) complete or less	78.67	87.23	22.13***
More than the SSC level	21.33	12.77	
Livelihood			
Agricultural or environmental work	28.42	4.79	2.7
Non-agricultural work	71.41	23.40	
Homemaker	0.17	71.81	
Family size			
Small family (1 to 5 members)	55.17	55.85	0.09
Large family (6 or more)	44.83	44.15	
House ownership			
Own the house	81.88	56.76	176.54***
Don't own the house	18.12	43.24	
Household head's health condition			
Healthy	80.06	70.16	27.90***
Unhealthy/sick	19.94	29.84	
Livestock			
Don't have livestock	41.50	52.84	25.37***
Have livestock	58.50	47.16	
Electricity			
Have access to electricity	83.27	76.95	13.33***
Don't have access to electricity	16.73	23.05	
Computer			
Have access to a computer	9.09	6.56	3.90*
Don't have access to a computer	90.91	93.44	

* $p < 0.05$, ** $p < 0.01$, *** $p < 0.001$

3.2 Descriptive Statistics of Outcome Variables and Covariates

This section presents descriptive comparisons of key variables by household head gender using percentage distributions and Pearson's chi-square tests of independence. As shown in Table 3, several statistically significant but modest differences emerged between male- and female-headed households. Female-headed households reported slightly higher intentions to evacuate before a disaster (55.14% vs 44.83%) and were more likely to receive international remittances (30.67% vs 11.09%), suggesting some variation in preparedness attitudes and economic support sources.

Both groups had comparable levels of cyclone experience, trust in government, and trust in neighbors. Female-headed households showed marginally lower reports of property loss (38.12% vs 42.90%). Small differences were also observed in food insecurity (12.41% vs 8.04%) and water source ownership (42.91% vs 50.39%), indicating slight disparities in access to basic resources. These findings highlight the intersection of gender and disaster vulnerability, where female-headed households exhibit greater preparedness behaviors while facing persistent economic and resource challenges.

Table 3 Descriptives of outcome and independent variables (n = 3995)

Items	Household head's gender		
	Male (%) (N = 3431)	Female (%) (N = 564)	Pearson Chi-square
Future evacuation behavior			
Will stay home	55.17	44.86	20.73***
Will evacuate	44.83	55.14	
Cyclone experiences			
Never experienced cyclone impacts	37.63	33.87	2.94
Experienced cyclone impacts	62.37	66.13	
International remittance			
Receive remittance	11.09	30.67	145.67***
Don't receive remittance	88.91	69.33	
Lost property due to the cyclone			
Didn't lose property	57.10	61.88	4.54*
Loss property	42.90	38.12	
Trust in government			
I don't trust the government for help	22.18	19.33	2.32
I trust the government for help	77.82	80.67	
Trust in neighbors			
I don't trust neighbors for help	13.49	12.43	0.46
I trust my neighbors for help	86.51	87.57	
Food insecurity			
Didn't face food insecurity in the last 12 months	91.96	87.59	11.68**
Faced food insecurity in the last 12 months	8.04	12.41	
Water insecurity			
Own the water source	50.39	42.91	10.86**
Relied on external water source	49.61	57.09	

* $p < 0.05$, ** $p < 0.01$, *** $p < 0.001$

3.3 *Gender Differences in Evacuation Intentions: Logistic Regression Results*

Results from logistic regression analysis reveal that female-headed households consistently show stronger evacuation intentions than male-headed households, even after accounting for disaster experience and socioeconomic factors. The models further identify cyclone exposure, resource insecurity, and institutional trust as significant predictors of preparedness behavior.

Table 4 presents the results of a sequential logistic regression analysis examining factors associated with future evacuation intentions, with particular focus on gender differences. The analysis reveals robust and consistent findings: female-headed households demonstrate significantly higher odds of evacuation intention across all model specifications (OR = 1.51–1.70, $p < 0.001$). This gender effect persists even after controlling for multiple vulnerability factors, suggesting an independent association between female household leadership and protective action-taking.

The models identify several other significant predictors of evacuation intention. Previous cyclone experience shows particularly strong effects (OR = 2.14–2.66, p

Table 4 Future evacuation behavior and household head's gender (Logistic Regression Model)

	Odds ratio							
	Unadjusted model	Adjusted model						
Household head's gender (Female)	1.51*** (0.14)	1.49*** (0.14)	1.67*** (0.17)	1.70*** (0.17)	1.70*** (0.17)	1.67*** (0.17)	1.62*** (0.17)	1.54*** (0.20)
Cyclone experience	2.50*** (0.17)	2.50*** (0.17)	2.66*** (0.19)	2.34*** (0.19)	2.34*** (0.19)	2.32*** (0.20)	2.31*** (0.20)	2.14*** (0.19)
Remittance	1.53*** (0.15)		1.86*** (0.19)	1.88*** (0.19)	1.85*** (0.19)	1.67*** (0.18)	1.60*** (0.18)	1.48*** (0.17)
Lost property	1.81*** (0.12)			1.27** (0.10)	1.26** (0.10)	1.24** (0.10)	1.24** (0.10)	1.29** (0.11)
Trust in govt	1.25** (0.10)				1.20* (0.10)	1.21* (0.11)	1.21* (0.11)	1.21* (0.12)
Trust in neighbors	1.50*** (0.15)					1.29* (0.14)	1.32* (0.14)	1.33* (0.15)
Faced food insecurity	1.84*** (0.21)						1.67*** (0.21)	1.58*** (0.20)
Faced with water insecurity	2.11*** (0.14)							1.97*** (0.14)
Intercept		0.45*** (0.03)	0.24*** (0.03)	0.23*** (0.03)	0.20*** (0.03)	0.20*** (0.03)	0.20*** (0.03)	0.14*** (0.02)

* $p < 0.05$, ** $p < 0.01$, *** $p < 0.001$

< 0.001), indicating that direct disaster exposure substantially increases preparedness behaviors. Economic factors, including international remittance receipt (OR = 1.48–1.88, $p < 0.001$) and property loss history (OR = 1.24–1.81, $p < 0.001$), demonstrate significant associations. Institutional trust measures show more modest but still significant effects, with both government trust (OR = 1.20–1.25, $p < 0.05$) and neighbor trust (OR = 1.29–1.50, $p < 0.05$) positively associated with evacuation intentions. Resource insecurity measures show particularly strong effects, with water insecurity (OR = 1.97–2.11, $p < 0.001$) and food insecurity (OR = 1.58–1.84, $p < 0.001$) both substantially increasing evacuation likelihood.

The progressive model specifications demonstrate the stability of the gender effect, which remains significant even as additional controls are introduced. The consistent pattern across models suggests that female household heads exhibit greater evacuation intentions regardless of other vulnerability factors, while simultaneously facing greater resource insecurity that further motivates protective actions.

3.4 Gendered Patterns in Disaster Preparedness: Marginal Effects Analysis

The marginal effects analysis (Table 5) provides critical nuance to the logistic regression results, quantifying how each factor changes evacuation probability. Female-headed households demonstrate a persistent 9–12 percentage point higher evacuation likelihood than male-headed households across all model specifications ($p < 0.001$), an effect that remains robust when controlling for other vulnerability factors. While cyclone experience shows the largest absolute effect, increasing evacuation probability by 17–23 percentage points, the gender disparity proves more consistent, maintaining significance even after accounting for disaster exposure, resource insecurities, and socioeconomic factors.

Resource insecurities emerge as particularly influential predictors, with water insecurity increasing evacuation probability by 15–20 percentage points and food insecurity by 10–15 percentage points. Economic factors and institutional trust show smaller but still significant effects, with remittance receipt (9–15 percentage points) and property loss (5–14 percentage points) demonstrating how financial circumstances shape preparedness decisions. Trust in government and neighbors shows more modest effects (4–10 percentage points), suggesting social capital plays a secondary but still meaningful role.

These results extend the findings from Table 4 by demonstrating that gender differences in evacuation intentions are not just statistically significant but substantively important. The stability of the gender effect across progressively controlled models suggests female household heads have fundamentally different risk calculus that transcends situational factors.

Table 5 Marginal effect model

	Marginal effect (dy/dx)							
	Unadjusted Model	Adjusted Model						
Household head's gender (Female)	0.10*** (0.02)	0.09*** (0.02)	0.12*** (0.02)	0.12*** (0.02)	0.12*** (0.02)	0.12*** (0.02)	0.11*** (0.02)	0.10*** (0.02)
Cyclone experience	0.22*** (0.01)	0.21*** (0.01)	0.23*** (0.01)	0.20*** (0.02)	0.20*** (0.02)	0.20*** (0.02)	0.19*** (0.02)	0.17*** (0.02)
Remittance	0.10*** (0.02)		0.14*** (0.02)	0.15*** (0.02)	0.14*** (0.02)	0.12*** (0.03)	0.11*** (0.03)	0.09*** (0.03)
Lost property	0.14*** (0.02)			0.05** (0.02)	0.05** (0.02)	0.05** (0.02)	0.05** (0.02)	0.06** (0.02)
Trust in govt	0.05** (0.02)				0.04* (0.02)	0.04* (0.02)	0.04* (0.02)	0.04* (0.02)
Trust in neighbors	0.10*** (0.02)					0.06* (0.03)	0.06* (0.03)	0.06* (0.02)
Faced food insecurity	0.15*** (0.03)						0.12*** (0.03)	0.10*** (0.03)
Faced with water insecurity	0.20*** (0.01)							0.15*** (0.02)

* $p < 0.05$, ** $p < 0.01$, *** $p < 0.001$

3.5 *Validation of Hypotheses*

The findings from Tables 4 and 5 provide strong empirical support for both hypotheses. Hypothesis 1 is confirmed, as female-headed households exhibit a strong and significant association with evacuation intention, showing significantly higher odds of evacuation (OR = 1.51–1.70, $p < 0.001$) and a 9–12 percentage point increase in predicted probability ($p < 0.001$) across all models. This result has direct practical implications for disaster preparedness. A 10-percentage-point increase in evacuation intention is substantial; translating this to a community of 10,000 households, it could mean approximately 1,000 more households headed by women would evacuate compared to a scenario where no gender difference existed. This underscores the critical need for gender-sensitive evacuation campaigns that explicitly address the drivers of this propensity, potentially leading to more effective overall community compliance.

Crucially, Hypothesis 2 is also supported. The positive and significant coefficient for female-headed households remains stable even after sequentially controlling for key potential confounders, including direct cyclone experience, socioeconomic status (remittance, lost property), and social capital (trust in government and neighbors). The persistence of this effect from the unadjusted model (OR = 1.51) through to the fully adjusted model (OR = 1.54) and the marginal effect model (AME = 0.10) demonstrates that the gender difference is not an artifact of these other factors. Instead, it establishes gender, specifically, a female-headed household, as an independent and robust predictor of evacuation preparedness.

4 Discussion

Gender has been a central focus of disaster studies for a long time. The number of articles, book chapters, books, blogs, and news articles that covered the gender aspect of disaster is massive. Learnings from these resources show that most of the women are more likely to evacuate before a disaster and comply with warnings than men (Bolin et al., 1996; Fothergill, 1996; Southeast Louisiana Hurricane Taskforce et al., 2005; Morss et al., 2016; Ye, 2024). Analyses of global disaster responses confirm this gendered pattern, with women's risk perception and protective behaviors significantly outpacing men's (Flynn et al., 1994; Brown et al., 2021). Women, most of the time, try to avoid any risk and take necessary actions to keep their family members safe, including evacuating before a disaster (Leik et al., 1981; Hitchcock, 2001; Harris & Jenkins, 2006). However, in Bangladesh, due to cultural and social barriers, vulnerable groups, especially women and children, do not evacuate and are sometimes left behind by their family members (Sultana, 2010; Azad et al., 2013; Ahsan et al., 2016b; Ayeb-Karlsson, 2020).

This study supports existing evidence that women in coastal Bangladesh exhibit strong evacuation intentions due to heightened risk perception and protective tendencies. However, as shown in disaster research, these positive attitudes may not materialize into actual evacuation behavior during cyclones. Prevailing sociocultural norms and safety concerns frequently prevent women from acting on their preparedness intentions, creating a significant disconnect between willingness and ability to evacuate when disasters strike (Huang et al., 2016). Therefore, the higher intention to evacuate we observe is a rational response to a heightened *positional vulnerability,* women are not inherently more vulnerable but are placed in more vulnerable positions by socio-economic systems. The finding that the gender effect persists after controlling for other socioeconomic indicators suggests that this vulnerability is not solely economic but is also embedded in these deeper political and social structures. The contribution of this study, therefore, is not simply identifying a gender gap in intention, but in revealing how that intention is a trapped form of resilience, constrained by the very political-economic and patriarchal systems that create differential vulnerability in the first place.

Studies have found both positive and negative associations between direct disaster exposure experience and evacuation behavior. By "disaster experience," studies refer to individuals' direct encounters with hazardous events and their subsequent consequences, including economic and environmental losses and damages, adverse physical and psychological health impacts, and the disruption or loss of social ties, networks, and relationships (Sattler et al., 2000; Hoffmann & Muttarak, 2017;

Rahman et al., 2022; Karim, 2024). Studies showed that individuals who have previously experienced tropical cyclones have a higher risk perception, which ultimately results in complying with evacuation orders (Paul, 2012; Cahyanto & Pennington-Gray, 2015; Ahsan et al., 2016a; Demuth et al., 2016; Rahman et al., 2021). Some studies also didn't find any significant association between previous experience of a hazard and evacuation decisions (Dow & Cutter, 1998, 2001). However, the number of studies finding a positive association between previous disaster exposure and evacuation decision is wide (Brommer & Senkbeil, 2010; Huang et al., 2016). In terms of future evacuation intentions, researchers also found that. People who have faced a tropical cyclone previously were positive about evacuating if an order is received (Sharma & Patt, 2012; Meyer et al., 2018). In Bangladesh, people who have experienced tropical cyclones are enthusiastic to evacuate as soon as they receive warnings from the government (Alam & Collins, 2010). Results from this study also showed positive intention among people to evacuate if they have previous cyclone experience. Most of the respondents of this study were from districts which got affected by Cyclone Sidr and Cyclone Aila. Thus, cyclone experience and future evacuation intentions showed a strong correlation in this study.

Other than gender and cyclone experience, resource insecurity was one of the major drivers in pushing household heads to show positive intentions for future evacuation. Researchers have already claimed that climate change is worsening food and water security conditions (Maharatna, 2014; Tinazzi, 2023; Pakravan-Charvadeh et al., 2025). In different literature, it has been shown that food and water insecurity force people to leave their homes and migrate (McGregor, 1994; Warziniack, 2013; Basu & Shaw, 2013; Tuholske et al., 2024; Richa, 2025). Resource scarcity, food insecurity, and water shortages are known as push factors that influence people to leave any area. During disasters, such conditions become more consistent (Llull, 2018; Urbański, 2022). People who have already experienced food and water scarcity in normal periods will choose to evacuate during disasters before the conditions worsen. Studies related to Non-Government Organizations (NGOs) and humanitarian agencies presented that a lot of people evacuate and use shelter centers to get access to scarce and lifesaving resources like food, water, medicine, etc. (Islam & Walkerden, 2015; Rodríguez-Espíndola et al., 2018). In this study, respondents who have faced food scarcity and water insecurity showed positive intentions to evacuate if they receive a cyclone warning. In previous research, it has been observed that people who have cyclone experience and a higher level of risk perception know how challenging it can be to ensure the safety of family members and availability of necessary goods during cyclones (Hitchcock, 2001; Demuth et al., 2016). In this regard, the result of this study is highly supported by previous literature.

Remittances, more specifically international remittances, have been seen as a critical lifeline during crises and emergencies. In developing and underdeveloped countries, remittances help households to meet crucial items like food, water, medicine, hygiene products, clothes, etc. (United Nations Development Programme, 2022). Organizations like the International Monetary Fund (IMF) and the World Bank have discussed that remittances contribute to the adaptation and resilience of households in economically disadvantaged countries. Remittances can support families to prepare for disasters and recover from the impacts rapidly (Malpass, 2022; Ratha, 2023). In the context of climate change, vulnerable households require increased financial resources to mitigate the impacts of extreme weather events and enhance preparedness for recurrent hazards such as tropical cyclones and flooding (Reid, 2023). Remittances represent a resilient financial mechanism, providing a steady inflow of capital to support climate risk adaptation and disaster preparedness (Harvey, 2007; Le De et al., 2013; Bettin et al., 2025). This study corroborates existing findings, demonstrating that households receiving international remittances exhibit stronger intentions to evacuate in future disasters compared to those without such financial support. The analysis underscores remittances as a critical economic determinant, enhancing adaptive capacity by facilitating proactive evacuation decisions.

On the other hand, loss of property due to prior cyclones significantly influences future evacuation intentions among affected households. Although closely associated with previous cyclone experience, the extent of damage varies across households. Losses may include casualties, property destruction, livestock depletion, environmental degradation, or agricultural losses. Researchers have presented that there is a strong association between previous losses and damages due to disasters and future evacuation intentions (Smith & McCarty, 1996; Morss et al., 2016). Hasan and colleagues found that as part of tropical cyclone experiences, property damage and loss faced by families motivate them to evacuate (Hasan et al., 2011). Meyer and other researchers showed that cascading losses and damages of tropical cyclones, such as storm surge-related damages, flooding-related damages, and wind-related damages, are highly associated with future evacuation intentions of households (Baker, 1991; Wei et al., 2014; Huang et al., 2016; Meyer et al., 2018). In the current study results, previous losses and damages have a strong association with intentions to evacuate in the future, which aligns with the literature.

Researchers have shown that households evacuate more readily when they trust institutions for help and believe neighbors will assist during disasters (Quarantelli, 1980; Greene et al., 1981; Burton & Silver, 2006; Jacob et al., 2008; Kim & Oh, 2015). The authors conceptualize trust in institutions as the propensity of households to depend on governmental bodies and agencies to implement actions that reduce the consequences of hazardous events. This definition aligns with the performance thesis of trust, which frames citizens' confidence as being predicated on the expectation that relevant authorities will intervene with necessary assistance during a crisis (Huurne & Gutteling, 2008; Malesic, 2019; Kim et al., 2025). People are more likely to evacuate, prepare, and recover efficiently when they trust government evacuation directives, relief assistance, and resilience programs (Alexander Priest, 2023). On the other hand, trust in neighbors acts as vital social capital, aiding evacuation decisions and securing shared resources like transport and supplies during disasters (LaLone, 2012; Fleming et al., 2014; Khankeh et al., 2025). This study operationalizes trust in neighbors through the lens of social capital. It is defined as an individual's belief that neighbors, particularly those with capacity, will provide assistance during a hazardous event (Uslaner, 2016). This construct is framed by the concept of in-group trust, which suggests that such confidence arises from a perceived shared socio-cultural identity and a norm of reciprocity within the community (LaLone, 2012). The findings of this study revealed that respondents with higher institutional trust and stronger social capital exhibited significantly greater evacuation intentions. Given that government agencies in Bangladesh bear primary responsibility for disaster response, including shelter provisions, medical aid, and livestock protection, households with greater institutional trust demonstrate higher evacuation motivation compared to those with lower trust. Moreover, households leverage supportive neighbor relations as social capital, which positively influences evacuation decision-making during and after disaster events.

5 Conclusion

The findings of this study reaffirm the critical role of gender in shaping evacuation intentions during cyclones in coastal Bangladesh. This study demonstrate that female-headed households exhibit a significantly higher intention to evacuate, must be interpreted through a feminist political ecology lens. This framework posits that disaster vulnerability is not a natural condition but a social product, deeply rooted in gendered power relations and political-economic structures that shape access to resources and define responsibilities. While women exhibit heightened risk perception and a strong willingness to evacuate to protect their families, deeply entrenched sociocultural norms often hinder their ability to act on these intentions. This disconnects between intention and action underscores the persistent barriers women face in disaster contexts, where mobility restrictions, safety concerns, and gendered caregiving responsibilities limit their agency in evacuation decisions. Addressing these structural constraints is essential to ensuring that women's proactive risk awareness translates into actual evacuation behavior, ultimately reducing disaster-related vulnerabilities. The study also acknowledges that women in rural areas often engage in home-based or informal income-generating activities, which can serve as important forms of economic and social capital influencing evacuation decisions. However, such information was not captured in the current dataset. Future research should incorporate data on informal livelihoods for both men and women to better understand their role in shaping disaster risk reduction behaviors.

Beyond gender, this study underscores the influence of prior cyclone experience, resource insecurity, remittance income, and institutional trust in shaping evacuation intentions. Households with direct disaster exposure are more likely to perceive risk realistically and take preemptive action, consistent with studies linking experience to heightened preparedness. Food and water insecurity also motivate evacuation, as resource-constrained households recognize their limited capacity to cope with disruptions during a cyclone. Conversely, access to remittance income provides financial flexibility, enabling timely evacuation and recovery. Institutional trust and social capital play equally vital roles. When households trust government agencies, they are more likely to comply with evacuation directives, reflecting confidence in institutional effectiveness. Strong neighborhood ties further enhance cooperation, information sharing, and collective preparedness, reducing uncertainty and hesitation during crises. Taken together, these factors reveal that evacuation decisions are shaped by the intersection of experiential, economic, and social dimensions rather than by individual characteristics alone.

Policy and Practice Recommendations

1. **Implement Structurally Sensitive, Gender-Responsive Evacuation Systems**

Policies must move beyond generic plans to actively dismantle the barriers that trap women's preparedness. This requires mandating and funding the establishment of safe, women-only shelters staffed by female personnel, providing secure, gender-segregated transportation to evacuation centers, and ensuring facilities meet the specific needs of women, children, and the elderly. These measures are non-negotiable for converting high evacuation intention into safe action.

2. **Institutionalize Women's Leadership in Disaster Governance**

To ensure policies address gendered realities, women must be empowered as decision-makers. This involves mandating a minimum 50% representation of women in all local disaster management committees and creating female-led community watch teams for early warning dissemination. This transforms women from vulnerable subjects into active agents of resilience, ensuring evacuation protocols are shaped by their knowledge and experience.

3. **Develop Gender-Targeted Financial Instruments to Overcome Economic Barriers**

Economic vulnerability is a primary constraint. Financial support must be directly accessible to women. This includes designing pre-disaster conditional cash transfers targeted at female-headed households to fund evacuation costs and promoting gender-inclusive microinsurance schemes for lost livelihoods. Furthermore, remittance channels should be formally leveraged to deliver pre-emptive disaster funds, recognizing women's central role in managing household remittances.

4. **Foster Gendered Social Capital Through Women-Centric Community Networks**

While social capital is vital, it must be deliberately cultivated to protect and empower women. Policies should support the formation of women-led disaster preparedness groups and mutual aid networks. These groups provide a trusted, safe space for sharing risk information, coordinating collective evacuation, and reducing the social isolation that often prevents women from seeking help, thereby strengthening community-wide resilience from the bottom up.

By addressing gendered vulnerabilities and fostering an enabling environment for evacuation, policymakers and practitioners can bridge the gap between intention and action, ensuring more equitable and effective disaster resilience in coastal Bangladesh.

References

Adger, W. N. (2006). Vulnerability. *Global Environmental Change, 16*, 268–281. https://doi.org/10.1016/j.gloenvcha.2006.02.006

Ahsan, M. N., Takeuchi, K., Vink, K., et al. (2016a). A systematic review of the factors affecting the cyclone evacuation decision process in Bangladesh. *Journal Disaster Research, 11*, 742–753. https://doi.org/10.20965/jdr.2016.p0742

Ahsan, Md. N., Takeuchi, K., Vink, K., & Warner, J. (2016b). Factors affecting the evacuation decisions of coastal households during cyclone Aila in Bangladesh. *Environmental Hazards, 15*, 16–42. https://doi.org/10.1080/17477891.2015.1114912

Akhand, M. H. (2003). Disaster management and cyclone warning system in Bangladesh. In J. Zschau & A. Küppers (Eds.), *Early warning systems for natural disaster reduction* (pp. 49–64). Springer.

Akter, A., & Tasnim, M. (2023). Mapping for women's evacuation plans during climate-induced disasters. In P. Solís & M. Zeballos (Eds.), *Open mapping towards sustainable development goals* (pp. 251–260). Springer International Publishing.

Akter, S., & Saha, S. (2024). Impact of legal recognition on the lives of the third gender: A study in Khulna district of Bangladesh. *Heliyon, 10*, Article e28671. https://doi.org/10.1016/j.heliyon.2024.e28671

Alam, E. (2024). Reasons for non-evacuation and shelter-seeking behaviour of local population following cyclone warnings along the Bangladesh coast. *Progress Disaster Science, 21*, Article 100307. https://doi.org/10.1016/j.pdisas.2023.100307

Alam, E., & Collins, A. E. (2010). Cyclone disaster vulnerability and response experiences in coastal Bangladesh. *Disasters, 34*, 931–954. https://doi.org/10.1111/j.1467-7717.2010.01176.x

Alam, Md. M., Hossain, Md. A., & Shafee, S. (2003). Frequency of Bay of Bengal cyclonic storms and depressions crossing different coastal zones. *International Journal of Climatology, 23*, 1119–1125. https://doi.org/10.1002/joc.927

Alam, Md. S., & Chakraborty, T. (2021). Understanding the nexus between public risk perception of COVID-19 and evacuation behavior during cyclone Amphan in Bangladesh. *Heliyon, 7*, Article e07655. https://doi.org/10.1016/j.heliyon.2021.e07655

Alam, Md. S., Chakraborty, T., Hossain, Md. Z., & Rahaman, K. R. (2023). Evacuation dilemmas of coastal households during cyclone Amphan and amidst the COVID-19 pandemic: A study of the Southwestern region of Bangladesh. *Natural Hazards, 115*, 507–537. https://doi.org/10.1007/s11069-022-05564-9

Albright, E. A., & Crow, D. A. (2015). Learning processes, public and stakeholder engagement: Analyzing responses to Colorado's extreme flood events of 2013. *Urban Climate, 14*, 79–93. https://doi.org/10.1016/j.uclim.2015.06.008

Albright, E. A., & Crow, D. (2019). Beliefs about climate change in the aftermath of extreme flooding. *Climate Change, 155*, 1–17. https://doi.org/10.1007/s10584-019-02461-2

Alexander Priest, A. (2023). Under pressure: Social capital and trust in government after natural disasters. *Social Current, 10*, 381–400. https://doi.org/10.1177/23294965231153066

Ayeb-Karlsson, S. (2020). When the disaster strikes: Gendered (im)mobility in Bangladesh. *Climate Risk Management, 29*, Article 100237. https://doi.org/10.1016/j.crm.2020.100237

Azad, A. K., Hossain, K. M., & Nasreen, M. (2013). Flood-induced vulnerabilities and problems encountered by women in northern Bangladesh. *International Journal Disaster Risk Science, 4*, 190–199. https://doi.org/10.1007/s13753-013-0020-z

Baker, E. J. (1991). Hurricane evacuation behavior. *International Journal Mass Emergencies Disasters, 9*, 287–310. https://doi.org/10.1177/028072709100900210

Basu, M., & Shaw, R. (2013). Water scarcity and migration: An Indian perspective. In M. A. Abedin, U. Habiba, & R. Shaw (Eds.), *Community, environment and disaster risk management* (pp. 187–211). Emerald Group Publishing Limited.

Bateman, J. M., & Edwards, B. (2002). Gender and evacuation: A closer look at why women are more likely to evacuate for hurricanes. *Natural Hazards Review, 3*, 107–117. https://doi.org/10.1061/(ASCE)1527-6988(2002)3:3(107)

Bettin, G., Jallow, A., & Zazzaro, A. (2025). Responding to natural disasters: What do monthly remittance data tell us? *Journal of Development Economics, 174*, Article 103413. https://doi.org/10.1016/j.jdeveco.2024.103413

Bolin, R., Jackson, M., Crist, A., et al. (1996). Gender inequality, vulnerability, and disaster: Issues in theory and research. In *The gendered terrain of disasters*.

Brammer, H. (2014). Bangladesh's dynamic coastal regions and sea-level rise. *Climate Risk Management, 1*, 51–62. https://doi.org/10.1016/j.crm.2013.10.001

Brody, S. D., Zahran, S., Vedlitz, A., & Grover, H. (2008). Examining the relationship between physical vulnerability and public perceptions of global climate change in the United States. *Environment and Behavior, 40*, 72–95. https://doi.org/10.1177/0013916506298800

Brommer, D. M., & Senkbeil, J. C. (2010). Pre-landfall evacuee perception of the meteorological hazards associated with Hurricane Gustav. *Natural Hazards, 55*, 353–369. https://doi.org/10.1007/s11069-010-9532-7

Brown, G. D., Largey, A., & McMullan, C. (2021). The impact of gender on risk perception: Implications for EU member states' national risk assessment processes. *International Journal Disaster Risk Reduction, 63*, Article 102452. https://doi.org/10.1016/j.ijdrr.2021.102452

Burton, P. W., & Silver, B. D. (2006). The credibility of disaster evacuation warnings: The effects of issue framing and trust in government. https://doi.org/10.13140/2.1.4219.1046

Cahyanto, I., & Pennington-Gray, L. (2015). Communicating hurricane evacuation to tourists: Gender, past experience with hurricanes, and place of residence. *Journal of Travel Research, 54*, 329–343. https://doi.org/10.1177/0047287513517418

Carrico, A., & Donato, K. M. (2023). Bangladesh Environment and Migration Survey (BEMS), 2019: Version 1.

Chakma, S., & Hokugo, A. (2020). Evacuation behavior: Why do some people never evacuate to a cyclone shelter during an emergency? A case study of coastal Bangladesh. *Journal Disaster Research, 15*, 481–489. https://doi.org/10.20965/jdr.2020.p0481

Crowley, J. (2021). Social vulnerability factors and reported post-disaster needs in the aftermath of hurricane florence. *International Journal Disaster Risk Science, 12*, 13–23. https://doi.org/10.1007/s13753-020-00315-5

Das, S. (2019). Evaluating climate change adaptation through evacuation decisions: A case study of cyclone management in India. *Climate Change, 152*, 291–305. https://doi.org/10.1007/s10584-018-2292-1

Demuth, J. L., Morss, R. E., Lazo, J. K., & Trumbo, C. (2016). The effects of past hurricane experiences on evacuation intentions through risk perception and efficacy beliefs: A mediation analysis. *Weather Climate Society, 8*, 327–344. https://doi.org/10.1175/WCAS-D-15-0074.1

Dow, K., & Cutter, S. L. (1998). Crying wolf: Repeat responses to hurricane evacuation orders. *Coastal Management, 26*, 237–252. https://doi.org/10.1080/08920759809362356

Dow, K., & Cutter, S. L. (2001). Public orders and personal opinions: Household strategies for hurricane risk assessment. *Environmental Hazards, 2*, 143–155. https://doi.org/10.3763/ehaz.2000.0220

Fekete, A. (2009). Validation of a social vulnerability index in context to river-floods in Germany. *Natural Hazards and Earth Systems Sciences, 9*, 393–403. https://doi.org/10.5194/nhess-9-393-2009

Fleming, D. A., Chong, A., & Bejarano, H. D. (2014). Trust and reciprocity in the aftermath of natural disasters. *Journal of Development Studies, 50*, 1482–1493. https://doi.org/10.1080/00220388.2014.936395

Flynn, J., Slovic, P., & Mertz, C. K. (1994). Gender, race, and perception of environmental health risks. *Risk Analysis, 14*, 1101–1108. https://doi.org/10.1111/j.1539-6924.1994.tb00082.x

Fothergill, A. (1996). Gender, risk, and disaster. *International Journal Mass Emergencies Disasters, 14*, 33–56. https://doi.org/10.1177/028072709601400103

Gaillard, J. C., Sanz, K., Balgos, B. C., et al. (2017). Beyond men and women: A critical perspective on gender and disaster. *Disasters, 41*, 429–447. https://doi.org/10.1111/disa.12209

Government of Bangladesh. (2018). NSSS Gender Policy. General Economics Division.

Greene, M., Perry, R., & Lindell, M. (1981). The march 1980 eruptions of MT. ST. Helens: Citizen perceptions of volcano threat. *Disasters, 5*, 49–66. https://doi.org/10.1111/j.1467-7717.1981.tb01129.x

Haq, S. M. A., & Ahmed, K. J. (2020). Perceptions about climate change among university students in Bangladesh. *Natural Hazards, 103*, 3683–3713. https://doi.org/10.1007/s11069-020-04151-0

Harris, C. R., & Jenkins, M. (2006). Gender differences in risk assessment: Why do women take fewer risks than men? *Judgment and Decision Making, 1*, 48–63. https://doi.org/10.1017/S1930297500000346

Harvey, P. (2007). *Remittances during crises: Implication for humanitarian response.* ODI Global. https://odi.org/en/about/our-work/remittances-during-crises-implication-for-humanitarian-response/

Hasan, M. K., & Kumar, L. (2020). Meteorological data and farmers' perception of coastal climate in Bangladesh. *Science of the Total Environment, 704*, Article 135384. https://doi.org/10.1016/j.scitotenv.2019.135384

Hasan, S., Ukkusuri, S., Gladwin, H., & Murray-Tuite, P. (2011). Behavioral model to understand household-level hurricane evacuation decision making. *Journal of Transportation Engineering, 137*, 341–348. https://doi.org/10.1061/(ASCE)TE.1943-5436.0000223

Hitchcock, J. L. (2001). Gender differences in risk perception: Broadening the contexts. *Risk Health Safety Environment, 12*, 179–204.

Hoffmann, R., & Muttarak, R. (2017). Learn from the past, prepare for the future: Impacts of education and experience on disaster preparedness in the Philippines and Thailand. *World Development, 96*, 32–51. https://doi.org/10.1016/j.worlddev.2017.02.016

Howell-Moroney, M. (2024). Inconvenient truths about logistic regression and the remedy of marginal effects. *Public Administration Review, 84*, 1218–1236. https://doi.org/10.1111/puar.13786

Huang, S.-K., Lindell, M. K., & Prater, C. S. (2016). Who leaves and who stays? A review and statistical meta-analysis of hurricane evacuation studies. *Environment and Behavior, 48*, 991–1029. https://doi.org/10.1177/0013916515578485

Human Dignity Trust. (2025). Bangladesh. https://www.humandignitytrust.org/country-profile/bangladesh/. Accessed 4 Nov 2025

Huurne, E. T., & Gutteling, J. (2008). Information needs and risk perception as predictors of risk information seeking. *Journal of Risk Research, 11*, 847–862. https://doi.org/10.1080/13669870701875750

Intergovernmental Panel on Climate Change (IPCC) (Ed.). (2023). *Climate change 2022—Mitigation of climate change: Working Group III contribution to the sixth assessment report of the intergovernmental panel on climate change* (1st edn). Cambridge University Press.

IPCC. (2021). Climate change widespread, rapid, and intensifying – IPCC — IPCC.

Islam, M. N., & Amstel, A. V. (2021). *Bangladesh II: Climate change impacts, mitigation and adaptation in developing countries.* Springer International Publishing.

Islam, M. R. (2025). Tropical cyclones in Bangladesh: Retrospective analysis of storm information, disaster statistics, and preparedness. *Environment Research Communications, 7*, Article 015003. https://doi.org/10.1088/2515-7620/ada239

Islam, R., & Walkerden, G. (2015). How do links between households and NGOs promote disaster resilience and recovery?: A case study of linking social networks on the Bangladeshi coast. *Natural Hazards, 78*, 1707–1727. https://doi.org/10.1007/s11069-015-1797-4

Jacob, B., Mawson, A. R., Marinelle, P., & Guignard, J. C. (2008). Disaster mythology and fact: Hurricane Katrina and social attachment. *Public Health Reports®, 123*, 555–566. https://doi.org/10.1177/003335490812300505

Janković, V., & Schultz, D. M. (2017). Atmosfear: Communicating the effects of climate change on extreme weather. *Weather Climate Society, 9*, 27–37. https://doi.org/10.1175/WCAS-D-16-0030.1

Karim, A. (2024). Knowledge, perception or disaster experience? The new determinants of household disaster preparedness behaviour in Bangladesh. *Journal of International Development, 36*, 2557–2580. https://doi.org/10.1002/jid.3922

K. C. L, & Hilhorst, D. (2022). Gendered experience of disaster: Women's account of evacuation, relief and recovery in Nepal. *International Journal Disaster Risk Reduction, 72*, 102840. https://doi.org/10.1016/j.ijdrr.2022.102840

Kelman, I. (2020). *Disaster by choice: How our actions turn natural hazards into catastrophes*. Oxford University Press.

Khan, M. A. H., Mozumder, P., Halim, N., & Meng, S. (2025). Understanding evacuation behavior during cyclones: Evidence from Bangladesh. *Economics Disasters Climate Change, 9*, 107–133. https://doi.org/10.1007/s41885-024-00157-1

Khankeh, H., Farrokhi, M., Saatchi, M., et al. (2025). Influencing factors on building and improving social trust in emergency and disaster relief efforts: A systematic review. *International Journal Disaster Resilient Built Environments, 16*, 129–142. https://doi.org/10.1108/IJDRBE-02-2023-0012

Kim, J., & Oh, S. S. (2015). Confidence, knowledge, and compliance with emergency evacuation. *Journal of Risk Research, 18*, 111–126. https://doi.org/10.1080/13669877.2014.880728

Kim, K., Ciszek, E., Ki, W., & Faust, K. M. (2025). Building institutional trust during disasters: An organization-public relationship examination of public service organizations (PSOs). *International Journal Disaster Risk Reduction, 118*, Article 105253. https://doi.org/10.1016/j.ijdrr.2025.105253

Lahsen, M., & Ribot, J. (2022). Politics of attributing extreme events and disasters to climate change. *Wires Climate Change, 13*, Article e750. https://doi.org/10.1002/wcc.750

LaLone, M. B. (2012). Neighbors helping neighbors: An examination of the social capital mobilization process for community resilience to environmental disasters. *Journal of Applied Social Science, 6*, 209–237. https://doi.org/10.1177/1936724412458483

Le De, L., Gaillard, J. C., & Friesen, W. (2013). Remittances and disaster: A review. *International Journal Disaster Risk Reduction, 4*, 34–43. https://doi.org/10.1016/j.ijdrr.2013.03.007

Leik, R. K., Carter, T. M., & Clark, J. P. (1981). *Community response to natural hazard warnings*. University of Minnesota.

Llull, J. (2018). The effect of immigration on wages: Exploiting exogenous variation at the national level. *The Journal of Human Resources, 53*, 608–662. https://doi.org/10.3368/jhr.53.3.0315-7032R2

Maharatna, A. (2014). Food scarcity and migration: An overview. *Social Research International Quarterly, 81*, 277–298. https://doi.org/10.1353/sor.2014.0026

Malesic, M. (2019). The concept of trust in disasters: The Slovenian experience. *Disaster Prevention Management International Journal, 28*, 603–615. https://doi.org/10.1108/DPM-11-2018-0375

Malm, A., & Hornborg, A. (2014). The geology of mankind? A critique of the Anthropocene narrative. *Anthropology Review, 1*, 62–69. https://doi.org/10.1177/2053019613516291

Malpass, D. (2022). *Remittances are a critical economic stabilizer*. World Bank Blogs. https://blogs.worldbank.org/en/voices/remittances-are-critical-economic-stabilizer

McGregor, J. (1994). Climate change and involuntary migration: Implications for food security. *Food Policy, 19*, 120–132. https://doi.org/10.1016/0306-9192(94)90065-5

Meyer, M. A., Mitchell, B., Purdum, J. C., et al. (2018). Previous hurricane evacuation decisions and future evacuation intentions among residents of southeast Louisiana. *International Journal Disaster Risk Reduction, 31*, 1231–1244. https://doi.org/10.1016/j.ijdrr.2018.01.003

Mileti, D. (Ed.). (1999). *Disasters by design: A reassessment of natural hazards in the United States*. The National Academies Press.

Morss, R. E., Demuth, J. L., Lazo, J. K., et al. (2016). Understanding public hurricane evacuation decisions and responses to forecast and warning messages*. *Weather and Forecasting, 31*, 395–417. https://doi.org/10.1175/WAF-D-15-0066.1

Norton, E. C., Dowd, B. E., & Maciejewski, M. L. (2019). Marginal effects—Quantifying the effect of changes in risk factors in logistic regression models. *JAMA, 321*, 1304. https://doi.org/10.1001/jama.2019.1954

Osaka, S., & Bellamy, R. (2020). Natural variability or climate change? Stakeholder and citizen perceptions of extreme event attribution. *Global Environmental Change, 62*, Article 102070. https://doi.org/10.1016/j.gloenvcha.2020.102070

Pakravan-Charvadeh, M. R., Chamcham, J., & Maleknia, R. (2025). How climate change adaptation strategies and climate migration interact to control food insecurity? *Regional Sustainability, 6*, Article 100229. https://doi.org/10.1016/j.regsus.2025.100229

Parvin, G. A., Sakamoto, M., Shaw, R., et al. (2019). Evacuation scenarios of cyclone Aila in Bangladesh: Investigating the factors influencing evacuation decision and destination. *Progress Disaster Science, 2*, Article 100032. https://doi.org/10.1016/j.pdisas.2019.100032

Paul, B. K. (2009). Why relatively fewer people died? The case of Bangladesh's cyclone Sidr. *Natural Hazards, 50*, 289–304. https://doi.org/10.1007/s11069-008-9340-5

Paul, B. K. (2012). Factors affecting evacuation behavior: The case of 2007 cyclone Sidr, Bangladesh. *The Professional Geographer, 64*, 401–414. https://doi.org/10.1080/00330124.2011.609780

Paul, B. K., & Dutt, S. (2010). Hazard warnings and responses to evacuation orders: The case of Bangladesh's cyclone Sidr*. *Geographical Review, 100*, 336–355. https://doi.org/10.1111/j.1931-0846.2010.00040.x

Paul, B. K., Rashid, H., Islam, M. S., & Hunt, L. M. (2010). Cyclone evacuation in Bangladesh: Tropical cyclones Gorky (1991) versus Sidr (2007). *Environmental Hazards, 9*, 89–101. https://doi.org/10.3763/ehaz.2010.SI04

Plail, M., Liu, L., & Thompson, R. (2025). Flash flooding. *Cell Reports Sustainability, 2*. https://doi.org/10.1016/j.crsus.2025.100462

Quarantelli, E. L. (1980). *Evacuation behavior and problems: Findings and implications from the research literature*. The Ohio State University.

Rahman, M. A., Hokugo, A., Ohtsu, N., & Chakma, S. (2021). Evacuation preparation scenarios of households during early and emergency evacuation: A case study of cyclone bulbul in southwestern coastal Bangladesh. *IDRiM Journal, 11*.https://doi.org/10.5595/001c.29128

Rahman, M. M., Islam Arif, Md. S., Mahdi, I., et al. (2022). Cyclone vulnerability in ethnic minority: A case study among rakhain community of Bangladesh. *International Journal Disaster Risk Reduction, 83*, Article 103449. https://doi.org/10.1016/j.ijdrr.2022.103449

Rai, N., Huq, S., & Huq, M. J. (2014). Climate resilient planning in Bangladesh: A review of progress and early experiences of moving from planning to implementation. *Development in Practice, 24*, 527–543. https://doi.org/10.1080/09614524.2014.908822

Ratha, D. (2023). *Resilient remittances*. International Monetary Fund. https://www.imf.org/en/Publications/fandd/issues/2023/09/B2B-resilient-remittances-dilip-ratha

Reid, E. (2023). How remittances and migrant networks can help mitigate climate disasters. Humanitarian Practice Network. https://odihpn.org/en/publication/how-remittances-and-migrant-networks-can-help-mitigate-climatedisasters/

Richa, S. S. (2025). Migration induced by water scarcity: A brief review. In S. Chakraborty, A. Chatterjee, & P. Kumar (Eds.), *Urban water ecosystems in Africa and Asia: Challenges and opportunities for conservation and restoration* (1st edn). Routledge.

Rodríguez-Espíndola, O., Albores, P., & Brewster, C. (2018). Disaster preparedness in humanitarian logistics: A collaborative approach for resource management in floods. *European Journal of Operational Research, 264*, 978–993. https://doi.org/10.1016/j.ejor.2017.01.021

Saha, S. K., & James, H. (2017). Reasons for non-compliance with cyclone evacuation orders in Bangladesh. *International Journal Disaster Risk Reduction, 21*, 196–204. https://doi.org/10.1016/j.ijdrr.2016.12.009

Sattler, D. N., Kaiser, C. F., & Hittner, J. B. (2000). Disaster preparedness: Relationships among prior experience, personal characteristics, and distress1. *Journal of Applied Social Psychology, 30*, 1396–1420. https://doi.org/10.1111/j.1559-1816.2000.tb02527.x

Sharma, U., & Patt, A. (2012). Disaster warning response: The effects of different types of personal experience. *Natural Hazards, 60*, 409–423. https://doi.org/10.1007/s11069-011-0023-2

Shen, Y., Wang, Q., Yan, W., et al. (2014). Evacuation processes of different genders in different visibility conditions—An experimental study. *Procedia Engineering, 71*, 65–74. https://doi.org/10.1016/j.proeng.2014.04.009

Shoji, M., & Murata, A. (2021). Social capital encourages disaster evacuation: Evidence from a cyclone in Bangladesh. *Journal of Development Studies, 57*, 790–806. https://doi.org/10.1080/00220388.2020.1806245

Smith, S. K., & McCarty, C. (1996). Demographic effects of natural disasters: A case study of hurricane Andrew. *Demography, 33*, 265–275. https://doi.org/10.2307/2061876

Southeast Louisiana Hurricane Taskforce, Survey Research Center, UNO, Center for Hazards Assessment, Response and Technology (CHART), UNO, Department of Geography, UNO. (2005). *Citizen hurricane evacuation behavior in Southeastern Louisiana: A twelve parish survey*. Survey Research Center Publications.

Sultana, F. (2010). Living in hazardous waterscapes: Gendered vulnerabilities and experiences of floods and disasters. *Environmental Hazards, 9*, 43–53. https://doi.org/10.3763/ehaz.2010.SI02

Thompson, A., & Otto, F. E. L. (2015). Ethical and normative implications of weather event attribution for policy discussions concerning loss and damage. *Climate Change, 133*, 439–451. https://doi.org/10.1007/s10584-015-1433-z

Tierney, K. J. (2019). *Disasters: A sociological approach*. Polity Press.

Tinazzi, I. (2023). *Water scarcity, climate change and migration: An assessment of the nexus*. Master's Thesis, Ca'Foscari University of Venice.

Tuholske, C., Di Landro, M. A., Anderson, W., et al. (2024). A framework to link climate change, food security, and migration: Unpacking the agricultural pathway. *Population and Environment, 46*, 8. https://doi.org/10.1007/s11111-024-00446-7

UN OCHA. (2024). *Philippines: Tropical cyclones and floods*. UN OCHA.

United Nations Development Programme. (2022). *Remittances are a lifeline in a crisis*. https://stories.undp.org/remittances-are-a-lifeline-in-a-crisis

Urbański, M. (2022). Comparing push and pull factors affecting migration. *Economies, 10*, 21. https://doi.org/10.3390/economies10010021

Uslaner, E. M. (2016). Disasters, trust, and social cohesion. *Ritsumeikan Language Culture Study, 28*, 183–192.

Warziniack, T. (2013). The effects of water scarcity and natural resources on refugee migration. *Society and Natural Resources, 26*, 1037–1049. https://doi.org/10.1080/08941920.2013.779339

Wei, H.-L., Lindell, M. K., & Prater, C. S. (2014). "Certain Death" from storm surge: A comparative study of household responses to warnings about hurricanes Rita and Ike. *Weather Climate Society, 6*, 425–433. https://doi.org/10.1175/WCAS-D-13-00074.1

West, D. M., & Orr, M. (2007). Race, gender, and communications in natural disasters. *Policy Studies Journal, 35*, 569–586. https://doi.org/10.1111/j.1541-0072.2007.00237.x

Wisner, B., Blaikie, P., Cannon, T., & Davis, I. (2004). In *At risk: Natural hazards, people's vulnerability and disasters*. Routledge.

Ye, Y. (2024). *'Evacuating is a privilege.' Why some stay behind when hurricanes strike*. CU Boulder Today.